W9-ABP-143

Write Like a Chemist

Write Like a Chemist

A Guide and Resource

Marin S. Robinson
Northern Arizona University

Fredricka L. Stoller
Northern Arizona University

Molly S. Costanza-Robinson
Middlebury College

James K. Jones

New York Oxford
Oxford University Press
2010

OXFORD
UNIVERSITY PRESS

Oxford University Press, Inc., publishes works that further
Oxford University's objective of excellence
in research, scholarship, and education.

Oxford New York
Auckland Cape Town Dar es Salaam Hong Kong Karachi
Kuala Lumpur Madrid Melbourne Mexico City Nairobi
New Delhi Shanghai Taipei Toronto

With offices in
Argentina Austria Brazil Chile Czech Republic France Greece
Guatemala Hungary Italy Japan Poland Portugal Singapore
South Korea Switzerland Thailand Turkey Ukraine Vietnam

Published by Oxford University Press, Inc.
198 Madison Avenue, New York, New York 10016

www.oup.com

Library of Congress Cataloging-in-Publication Data
Write like a chemist : a guide and resource / Marin S. Robinson . . . [et al.].
p. cm.
Includes bibliographical references and index.
ISBN 978-0-19-530507-4 (pbk.)
ISBN 978-0-19-536742-3
1. Chemistry—Authorship—Textbooks. 2. Communication in chemistry—Textbooks.
3. Technical writing—Textbooks. I. Robinson, Marin S.
QD9.15.W75 2008
808'.06654—dc22 2007038271

Printed in the United States of America
on acid-free paper

Dedicated to
Jeff and Bill,
Chuck and Ronnie,
and Kara

Preface

Write Like a Chemist is designed to be used as a *textbook* in upper division and graduate-level university chemistry classes and as a *resource book* by chemistry students, postdocs, faculty, and other professionals who want to perfect their chemistry-specific writing skills. To this end, *Write Like a Chemist* focuses on four types of writing:

- The journal article
- The conference abstract
- The scientific poster
- The research proposal

Each type of writing, or genre, is directed toward a distinct audience and written for a different purpose. For example, writing a journal article requires a style and organizational format that are quite different from that of a research proposal. Thus, to write like a chemist requires learning to write for multiple audiences and purposes.

One of the best ways to learn to *write* like a chemist is to *read* like a chemist. Indeed, many of today's chemists taught themselves to write by reading others' works and using those works as templates for their own writing. Corrections (often in red pen) from peers, mentors, reviewers, and editors along the way also played an integral role in the learning-to-write process. Although ultimately successful, this approach was often painful and inefficient for all involved. The goal of *Write Like a Chemist* is to teach writing in a more systematic way. Because reading is integral to writing, we use the chemical literature (and other examples of writing) to model conventional writing practices; indeed, more than 250 excerpts from ACS journal articles and NSF CAREER proposals are included in this book. But we do not stop there. *Write Like a Chemist* endorses a *read-analyze-write* approach that combines the *reading* of authentic passages with the *analysis* of those passages to gain insights into the writing conventions of the targeted genres. Reading and analysis activities are followed by structured *writing* tasks, culminating in authentic writing assignments, using the excerpts as models.[1]

Before going to press, *Write Like a Chemist* was piloted at 16 U.S. colleges and universities. The book was used successfully in a variety of instructional settings (including writing-dedicated courses and lecture, lab, and seminar courses) with a broad cross section of students, including non-native English speakers and students from multicultural backgrounds. Although intended primarily for chemists, the book will also benefit chemistry majors who ultimately decide to pursue other careers. Indeed, learning to write in a well-organized and concise manner requires writing skills that are highly coveted across many fields. Moreover, the read-analyze-write approach featured in *Write Like a Chemist* is readily transferable to other disciplines.

Because it will take more than a single encounter with *Write Like a Chemist* to become a skilled writer, the book can serve as a rich resource over the years when the goal is to communicate effectively in writing with chemists, other scientists, and funding agencies.

Contents

Following the introductory chapter 1, section 1 of *Write Like a Chemist* is divided into three modules:

Module 1: The Journal Article (chapters 2–7)

Module 2: The Scientific Poster (chapters 8–10)

Module 3: The Research Proposal (chapters 11–15)

Each module combines authentic readings with exercises to introduce and reinforce discipline-specific writing skills. At the core of each module is a multistep writing assignment, guided by "Writing on Your Own" tasks, that assists writers in completing the type of writing emphasized in the module.

Section 2 of *Write Like a Chemist* includes three chapters that focus on skills that run across different types of chemistry writing. These chapters guide writers in formatting and finalizing their written work:

Chapter 16: Formatting Figures, Tables, and Schemes

Chapter 17: Formatting Citations and References

Chapter 18: Finalizing Your Written Work

Write Like a Chemist concludes with two appendices. Appendix A provides helpful tips about language areas that often prove troublesome for writers (e.g., easily confused words, scientific plurals, punctuation, and grammar). Each language tip includes exercises and an answer key, facilitating self-study. (For a full listing of tips, see the first page of appendix A.) For ease of consultation, appendix B replicates flow charts (called "move structures" in the book) that illustrate the typical organization of sections of the genres that are emphasized in the book.

The *Write Like a Chemist* Web Site

Accompanying the textbook is the *Write Like a Chemist* Web site (http://www.oup.com/us/writelikeachemist). Notable features of the Web site include the following:

- Web exercises: To prevent users from needlessly retyping full exercises that only need editing, we have duplicated these exercises on the Web site. In this way, students can copy and paste the exercises into a text document and edit them accordingly.
- "Canned" research projects: We recognize that not all users of this textbook will have a sufficiently robust research project to write about. To date, we have developed four "canned" research projects to address this need. These projects, all based on authentic research, provide sufficient data and background information for a mock journal article or poster.
- Peer-Review Memo forms: Writing benefits from peer review. To facilitate this process, we include Peer Review Memo forms, with a list of guiding questions and prompts, for each major section of the journal article.
- Full-color posters: Module 2 of the book ("The Scientific Poster") includes examples of posters in only black-and-white, but full-color versions of these posters are available at the Web site.
- Faculty resources: An answer key and examples of analytical and holistic grading rubrics for major writing assignments (journal-quality paper, poster, and research proposal) are available at the Web site for faculty adopting the book.

Unique Features of *Write Like a Chemist*

Write Like a Chemist is unique in many ways, not the least of which is the fact that it was conceived by a chemist (M.S.R.) and an applied linguist (F.L.S.) and developed with additional help from faculty and students in both disciplines. As part of our interdisciplinary effort, we analyzed chemistry-specific writing practices using tools from corpus linguistics, a discipline that investigates language empirically through computer-based analyses of large collections of texts known as corpora (or *corpus, singular*).[2] A 1.5-million-word corpus of chemistry texts was created, comprising 200 full-length refereed journal articles and 240 sections of refereed journal articles (i.e., 60 abstracts, Introduction, Methods, Results and Discussion sections) from *Anal. Chem.*, *Biochemistry*, *J. Am. Chem. Soc.*, *J. Org. Chem.*, *J. Phys. Chem. A & B*. This database was used to identify common, generalizable patterns[3] in the language of chemistry, a task that would have been virtually impossible without the help of the computer. Later on in the project, the ACS Journals Search (http://pubs.acs.org) was used extensively to accomplish many of the same aims. Findings from both corpora are included here.

Icons Used in *Write Like a Chemist*

Icons, each one with a special meaning, are used throughout the book:

 Exercises

 Exercises that are also posted on the *Write Like a Chemist* Web site

 Definition of a key term or concept

 Reminders and/or elaboration of important points

 Writing on Your Own tasks (chapters 2–18) and proofreading tips (appendix A)

 Reference to a later part of the book for additional information and/or practice

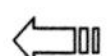 Reference to an earlier part of the book for additional information

 Explanation of a scientific term or concept

 Findings from computer-based analyses of the language of chemistry

 A useful principle (i.e., rule of thumb) with broad applications

 "Road map" to research proposal (module 3) with enumeration of typical headings

Writing Conventions Used in *Write Like a Chemist*

The original text, tables, and figures in this book generally follow Oxford University Press writing conventions, for example,

- Initial paragraphs of sections are not indented but subsequent paragraphs are indented.
- The "F" in figure and "T" in table are in lowercase when calling out a table or figure in the text (e.g., see table 1).
- In tables, column headings are written in title case (e.g., Verb Tense); labels are bolded with no period (e.g., **Table 1** Rates.).

The works cited in this book (i.e., excerpts from ACS journal articles, conference abstracts, and research proposals) are reproduced as written with only slight modifications as needed to adhere to journal-specific (according to the journal's *Information for Authors*) or ACS guidelines (according to *The ACS Style Guide* 3rd ed.), for example,

- Initial paragraphs are indented or not, as they were in the original source.
- The "F" in figure and "T" in table are capitalized when calling out a table or figure in the text (e.g., see Table 1).
- Tables and figures are reproduced as written; hence, different formatting conventions are observed (e.g., some table column headings are in lowercase, others are in title case).
- Table titles and figure captions are reproduced as written; hence, different formatting conventions are observed (e.g., some table titles are centered, bolded, and in title case; others are left-justified, unbolded, and in sentence case).

Because writing conventions vary and are likely to change with time, we urge readers to consult the journal of their choice as they prepare manuscripts for publication.

We wish you good luck. With this book and hard work, you too will be able to write like a chemist!

Notes

1. Although ours is the first chemistry-specific writing textbook and resource that we know of that approaches the reading and analysis of chemistry writing in this way, we have been inspired by a wide body of research into the genres of various fields (e.g., Bhatia, 1993, 2004; Connor and Mauranen, 1999; Hill, Soppelsa, and West, 1982;

Huckin, 1987; Hyland, 1994, 1996, 1998, 2002, 2004a, 2004b, 2006; Johns, 2002; Paltridge, 1997; Swales, 1990, 2004).

2. Of considerable influence were the following publications: Biber, Conrad, and Reppen (1994, 1998) and Bowker and Pearson (2002).
3. Note that we do not always use conventional linguistic terminology when discussing and presenting language-related issues in this book. Rather, we use terminology that best reaches our intended audience (i.e., chemists).

Acknowledgments

We gratefully acknowledge the National Science Foundation for financial support of the *Write Like Chemist* project (DUE 0087570 and DUE 0230913). We are also indebted to Jeremy Lewis, Acquisitions Editor at Oxford University Press, and Eric Slater, Copyright Manager of the American Chemical Society, for their ongoing support of this project. We also thank Paul Hobson, Production Editor, and Edward (Ned) Sears, Editorial Assistant, at Oxford University Press, as well as Patricia Watson, copyeditor.

We are grateful to many individuals at Northern Arizona University who contributed to this project. In particular, we thank William Grabe and Bradley Horn for developing and coordinating project assessment efforts, Sharon Baker for making thoughtful contributions to the answer key, John Rothfork for developing the project Web site, and Liz Grobsmith for providing institutional support. We are also indebted to CHM 300W and CHM 610 students who endured early drafts of the book, in particular, Jennifer Broyles, Lana Chavez, Kevin Pond, and Catherina Salanga.

This project would not have succeeded without additional support and inspiration from many other individuals at Northern Arizona University and elsewhere, including Geoffrey Chase, Beverly Cleland, Ann Eagan, Julie Gillette, Kris Harris, Victor O. Leshyk, Alan Paul, Martha Portree, Kurt Ristinen, Paul Torrence, and Kierstin Van Camp-Horn, as well as the Departments of Chemistry & Biochemistry and English and the Colleges of Engineering & Natural Sciences and Arts & Letters at Northern Arizona University.

We are particularly indebted to chemistry faculty (and their students) who piloted drafts of *Write Like a Chemist* at their home institutions during 2004–2006 (if faculty affiliations have changed, we note the pilot institution in parentheses):

Frances Blanco-Yu	Seton Hill University
David Collins	Colorado State University–Pueblo
Ellen R. Fisher	Colorado State University
Brian Gilbert	Linfield College
Alex Grushow	Rider University

Angela Hoffman	University of Portland
Timm Knoerzer	Nazareth College
Daphne Norton	Emory University
Donald Paulson	California State University–Los Angeles
Dan Philen	Emory University
Jennifer N. Shepherd	Gonzaga University
Betty H. Stewart	Midwestern State University (Austin College)
Joe Vitt	University of South Dakota
Carl Wamser	Portland State University
Barry L. Westcott	Central Connecticut State University

We also thank faculty who served as external evaluators for the *Write Like a Chemist* project:

Jeanne Arquette	Maricopa Community College
Troy Cahou	Coconino Community College
Larry Eddy	Yavapai Community College
Don Gilbert	Northern Arizona University
Sibylle Gruber	Northern Arizona University
Hans Gunderson	Northern Arizona University
Cynthia Hartzell	Northern Arizona University
Pierre Herckes	Arizona State University
Jani Ingram	Northern Arizona University
David F. Nachman	Maricopa Community College
John Pollard	University of Arizona
Scott Savage	Northern Arizona University
Michael Scott	Maricopa Community College
Paul Smolenyak	Yavapai Community College
Diane Stearns	Northern Arizona University
Timothy Vail	Northern Arizona University

We also acknowledge other colleagues who reviewed the book and offered feedback or contributed quotes to the book:

Joseph H. Aldstadt	University of Wisconsin–Milwaukee
Kevin Cantrell	University of Portland
Bert D. Chandler	Trinity University
Joan Curry	University of Arizona

Robert Damrauer	University of Colorado–Denver
Charles H. DePuy	University of Colorado–Boulder
Mari Eggers	Little Big Horn College
Dave Goodney	Willamette University
Nora S. Green	Randolph-Macon College
Suzanne Harris	University of Wyoming
Ann M. Johns	San Diego State University
David B. Knaff	Texas Tech University
Carol Libby	Moravian College
Richard Malkin	University of California–Berkeley
Charlotte Otto	University of Michigan
Pete Palmer	San Francisco State University
Bradley F. Schwartz	Southern Illinois University School of Medicine
Grigoriy Sereda	University of South Dakota
Steve Singleton	Coe College
Gerald Van Hecke	Harvey Mudd College
Gabriela Weaver	Purdue University
James B. Weissman	Pfizer Pharmaceutical Marketing
Thomas J. Wenzel	Bates College
Barry L. Westcott	Central Connecticut State University

We gratefully acknowledge the American Chemical Society Publications Division for granting us permission to use numerous excerpts from journal articles published by the American Chemical Society (including words, phrases, sentences, one or more paragraphs, titles, figures, tables, and, in one instance, a full article). All selections were reprinted with permission from the American Chemical Society, granted by Eric S. Slater, Esq., copyright manager. A citation accompanies each selection (e.g., "from Boesten et al., 2001" or "adapted from Boesten et al., 2001"), and the corresponding bibliographic information is included in the reference list of cited works at the end of the book.

We also thank the following individuals for granting us permission to use excerpts from their research proposals or ACS conference abstracts:

Primary Investigator of an ACS Division of Analytical Chemistry Graduate Fellowship

Amanda J. Haes	University of Iowa

Primary Investigators of NSF CAREER Awards

Diana Aga	State University of New York–Buffalo
Daniel J. Dyer	Southern Illinois University–Carbondale
Howard Fairbrother	Johns Hopkins University
Nathaniel Finney	University of California–San Diego
Anna D. Gudmundsdottir	University of Cincinnati
Karen S. Harpp	Colgate University
Paul Hergenrother	University of Illinois–Urbana-Champaign
Robert P. Houser	University of Oklahoma
Jeffrey S. Johnson	University of North Carolina–Chapel Hill
Gary R. Kinsel	University of Texas–Arlington
Amnon Kohen	University of Iowa
Jeehiun Katherine Lee	Rutgers University
Gary A. Lorigan	Miami University
L. Andrew Lyon	Georgia Tech Research Corporation
David L. Patrick	Western Washington University
Christoph G. Rose-Petruck	Brown University
Andrei Sanov	University of Arizona
Eileen M. Spain	Occidental College
Mark E. Tuckerman	New York University
James R. Vyvyan	Western Washington University
Robert A. Walker	University of Maryland–College Park
Timothy H. Warren	Georgetown University

Corresponding Authors of American Chemical Society Conference Abstracts

Joseph T. Bushey	Syracuse University
Gerald B. Hammond	University of Louisville
Arthur Lee	Wyeth Research
Athanasios Nenes	Georgia Institute of Technology
Catherine C. Neto	University of Massachusetts–Dartmouth
Peter S. Nico	Lawrence Berkeley National Laboratory
Denis J. Phares	University of Southern California
Cynthia Rohrer	University of Wisconsin-Stout

Lynn Russell	Scripps Institution of Oceanography, University of California–San Diego
Kevin M. Smith	Louisiana State University
Yuegang Zuo	University of Massachusetts–Dartmouth

Finally, we thank Dr. Roald Hoffmann for permission to reproduce his poem "Next Slide Please" from *The Metamict State* (1987; Orlando: University of Central Florida Press, pp 51–52).

The opinions, findings, conclusions, and recommendations expressed in this book are those of the authors and do not necessarily reflect the views of the National Science Foundation, the American Chemical Society, or authors whose works are included in *Write Like a Chemist*.

Contents

Section 1 *Writing Modules*

Chapter 1 *Learning to Write Like a Chemist* 5

Module 1 *The Journal Article* 31

Chapter 2 *Overview of the Journal Article* 33
Chapter 3 *Writing the Methods Section* 57
Chapter 4 *Writing the Results Section* 111
Chapter 5 *Writing the Discussion Section* 163
Chapter 6 *Writing the Introduction Section* 199
Chapter 7 *Writing the Abstract and Title* 241

Module 2 *The Scientific Poster* 271

Chapter 8 *Writing the Conference Abstract and Title* 273
Chapter 9 *Writing the Poster Text* 293
Chapter 10 *Designing the Poster* 335

Module 3 *The Research Proposal* 357

Chapter 11 *Overview of the Research Proposal* 359
Chapter 12 *Writing the Goals and Importance Section* 387
Chapter 13 *Writing the Experimental Approach Section* 433
Chapter 14 *Writing the Outcomes and Impacts Section* 479
Chapter 15 *Writing the Project Summary and Title* 501

Section 2 *Graphics, References, and Final Stages of Writing*

Chapter 16 *Formatting Figures, Tables, and Schemes* 523

Chapter 17 *Formatting Citations and References* 543

Chapter 18 *Finalizing Your Written Work* 569

Appendix A *Language Tips* 583

Audience and Purpose 584

Writing Conventions 601

Grammar and Mechanics 612

Word Usage 634

Appendix B *Move Structures* 659

Sources of Excerpts 667

References 685

Index 687

Write Like a Chemist

Section 1

Writing Modules

1 *Learning to Write Like a Chemist*

Writing, more than any other skill developed as a chemistry student, has enabled me to advance my career.

—James B. Weissman, Pfizer Pharmaceutical Marketing

Chapter 1 introduces the basic approach to reading and writing in chemistry used in this textbook. It also provides a brief orientation to the textbook. By the end of this chapter, you should be able to do the following:

- Identify common writing genres in chemistry and in this textbook
- Describe the five essential components of genre analysis and explain why genre analysis is so useful for developing writers
- Explain what is meant by audience, and identify the audiences addressed in this textbook
- Differentiate between broad and fine organizational structure
- Explain the meaning and significance of a move and a move structure
- Understand how the textbook is organized and the approach it takes to help you improve your chemistry writing skills

Many effective writers develop their discipline-specific writing skills by reading and analyzing the works of others in their fields. Learning to write in chemistry is no exception; chemistry-specific writing skills are developed by reading and analyzing the writing of chemists. We coined the phrase "read-analyze-write" to describe this approach and promote this process throughout the textbook. In this chapter, we lay the foundation for the read-analyze-write approach by analyzing a few common, nonscientific examples of writing. We use these everyday examples (e.g., letters, recipes, jokes, used-car ads, poems) to introduce you to the process of analyzing writing and to share with you the tools that you will need to analyze chemistry writing in subsequent chapters.

Genre

I had no idea how much time I would spend writing in my career as a chemistry professor at an undergraduate institution. With course materials, grant proposals, and research papers, I am always writing something.

—Thomas J. Wenzel, Bates College

Unless you are reading this chapter very early in the morning, you have likely already encountered several different types of writing today. Newspaper articles, e-mail messages, novels, letters, and billboards are just a few examples of writing that people view on a daily basis. You may also have glanced at some chemistry-specific writing in textbooks, lab manuals, course notes, reference books, or chemical catalogs. Each of these types of writing is unique and distinguishable from the others. This is true even if they share overlapping content. For example, information about the chemical properties of ethanol is presented differently in an organic chemistry textbook, a chemical catalog, and a chemical dictionary.

The word used to describe these different types of writing is **genre**. For example, there are different genres in literature (e.g., poems, short stories, or romance novels) and in film (e.g., comedy, horror, or mystery films). There are also different genres in chemistry. Although the word may sound a little funny at first, you will soon see that recognizing a chemistry genre is the first step toward writing successfully in that genre.

Genre

A type of writing that is distinguished from other types of writing because of differences in content, form, style, audience, purpose, and context.

This textbook focuses on four distinct genres commonly read and written by chemists; the four genres are addressed in these three textbook modules:

- Journal Article (module 1)
- Scientific Poster and conference abstract (module 2)
- Research Proposal (module 3)

Exercise 1.1

Make a list of five genres that a college student majoring in chemistry might read or write. Make a second list with three to five genres that a professional chemist in academia, industry, or a government lab might read and write. How do these genres differ from one another?

At the core of the read-analyze-write approach is **genre analysis**, a systematic way to read and analyze writing. Through genre analysis, you will identify and examine essential components of a genre, thereby facilitating your ability to write effectively in that genre. This textbook focuses on five such components: audience and purpose, organization, writing conventions, grammar and mechanics, and science content. As shown in table 1.1, each component can be further divided into two or more subcomponents. Our goal is to teach you to analyze chemistry-specific writing according to these components and subcomponents. To get you started, and to illustrate how genre analysis works, we begin by identifying each component in some familiar (nonchemistry) types of writing.

Genre Analysis

A systematic way of analyzing a genre to identify its distinguishing features.

Audience and Purpose

Before you begin to write, you must decide the **audience** that is most likely to read your work and the reason or **purpose** for writing it in the first place. In turn, the audience and purpose will influence the levels of detail, formality, and conciseness that you use in your writing and the words that you choose. To illustrate this, consider two everyday genres: a recipe in a cookbook and a shopping list. The recipe is written to instruct a hopeful chef (audience) how to prepare a meal (purpose); the shopping list is written to remind a shopper (audience) what foods to buy

Table 1.1 Components of genre analysis addressed in this textbook.

Audience and Purpose	Organization	Writing Conventions	Grammar and Mechanics	Science Content
Conciseness	Broad structure	Abbreviations and acronyms	Parallelism	Graphics
Level of detail	Fine structure ("moves")	Formatting	Punctuation	Text
Level of formality		Verb tense	Subject–verb agreement	
Word choice		Voice	Word usage	

(purpose). Because of their different purposes, detailed instructions are needed in the recipe but not in the shopping list. (Imagine how useless a recipe would be if it included only a list of ingredients or how unwieldy a shopping list would be if it included instructions for locating each item in the store!) Moreover, because recipes are often published, the writing is more formal, with titles, headings, lists of ingredients presented in a parallel fashion, and unambiguous, fully punctuated sentences (e.g., Melt 2 tsp. butter in a small saucepan.). Shopping lists, on the other hand, are scrawled out in personal shorthand (e.g., choc, OJ, mlk) with no titles, headings, or punctuation. Thus, we can see how audience and purpose influence the levels of detail, formality, and conciseness of a particular genre.

Audience

The people who will most likely read a specific piece of writing.

Purpose

The aims, goals, or intentions of the writer.

As a second example, consider two genres of letters: a job application cover letter and a sympathy letter to a friend. These two types of letters are sent to different individuals (audience) with whom the writer has different relationships and for entirely different reasons (purpose). These differences are reflected not only in tone (i.e., the job application letter is formal and professional, while the sympathy letter is personal and compassionate) but also in characteristic phrases. A potential employer would be quite surprised to read an application letter signed "Love, Mario" as would a friend reading a sympathy letter beginning with "To Whom It May Concern." **Word choices** such as these are anticipated by readers. Choosing the right word is not easy (see figure 1.1). If expected words are missing, or a wrong phrase is used, readers will have a difficult time following, or even recognizing, the genre. Hence, by learning words and phrases that are characteristic of a genre, you can make your own writing sound more like a typical example of that genre.

Figure 1.1 Even Snoopy struggles for just the right words. PEANUTS: ©United Feature Syndicate, Inc.

Word Choice

Readers expect characteristic words and phrases to be used in a genre (e.g., the word "Discussion" is used to demark the start of a journal article Discussion section). Effective writers must learn to incorporate these words into their writing.

Keep in mind that both audience and purpose define a genre. Two pieces of writing with the same intended audience may be written very differently if they have different purposes. For example, a university catalog and a university student newspaper are both written for a student audience, but the two publications are distinct from one another in many ways (e.g., organization and content).

With these everyday examples in mind, let's consider audience and purpose for chemistry-specific genres. We begin with audience. Chemists write for many different audiences, including students, teachers, and Ph.D. chemists, to name only a few. Thus, it is instructive to divide audience into different categories. For our purposes, we consider four categories: the expert audience, the scientific audience, the student audience, and the general audience. The **expert audience** includes professional chemists with advanced knowledge in a subdiscipline of chemistry, such as biochemistry, analytical chemistry, or organic chemistry. The subdiscipline is often reflected in the name of the journal written for experts in that field (e.g., *Biochemistry*, *Analytical Chemistry*, or *The Journal of Organic Chemistry*). The **scientific audience** comprises readers with scientific backgrounds but not necessarily in the authors' field or subdiscipline. For example, a biologist or geologist asked to review a chemist's research proposal would be considered a scientific audience. The **student audience** consists of individuals who are reading to learn chemistry at any level, such as a high school student reading an introductory chemistry book or a graduate student studying a book on quantum mechanics. The **general audience** includes readers who are interested in a chemistry topic but with little to no formal training in chemistry, such as an English or history teacher reading *Science News* or *Popular Science*.

Expert Audience

Readers with expert-level knowledge in a specific area of chemistry.

Scientific Audience

Readers with significant scientific knowledge, but not in the specific area targeted in the written work.

Student Audience

Readers learning chemistry.

General Audience

Readers with little or no chemistry knowledge.

Together, these four audiences form a continuum that spans a wide range of expertise in chemistry (figure 1.2). In general, journal articles are written for an expert audience, research proposals and scientific posters for a scientific audience, textbooks for a student audience, and popular science articles for a general audience. Of course, these pairings are only guidelines. A genre can change position on the continuum if an audience is expected to have more (or less) chemistry-specific knowledge. For example, a poster presented at a highly technical conference should address an expert audience, but a poster presented at an undergraduate research conference should target a student or general audience. Moreover, a single genre often addresses more than one audience. Although a journal article is written primarily for an expert audience, parts of its Introduction section are often written for a scientific audience. You can see that determining your audience is an integral part of the writing process.

In this textbook, we focus on two audiences. In module 1 ("The Journal Article") we focus on the expert audience, and in modules 2 and 3 ("The Scientific Poster" and "The Research Proposal"), we focus on the scientific audience. There are other genres that target these same audiences, such as technical memos and reports, but they are not covered in this textbook. An important goal of this

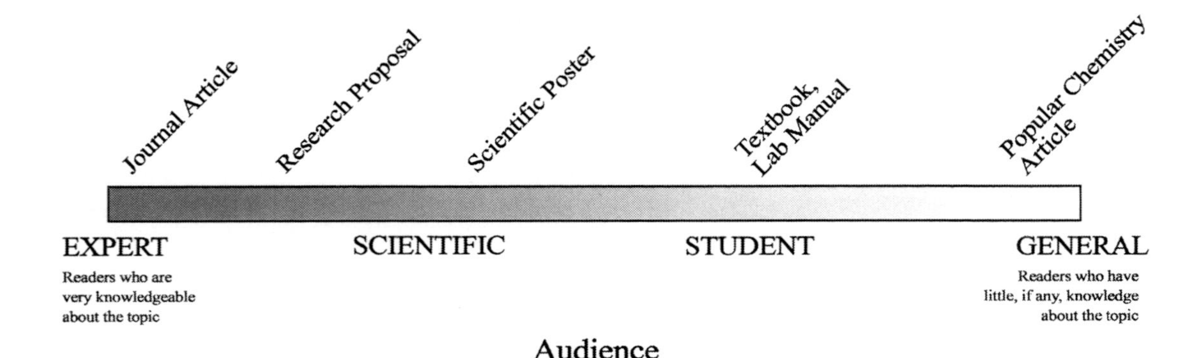

Figure 1.2 A spectrum of genres for audiences with varying degrees of expertise.

textbook is to help you move beyond writing for a student audience (the targeted audience in many undergraduate lab reports) and begin to write for expert and scientific audiences.

Closely linked to audience is purpose; a genre is also influenced by the purpose for the writing. Several different purposes for chemistry-specific writing are listed below. Representative genres are shown in parentheses. This textbook addresses only the first two of these purposes:

- To present research results or convey new scientific insights (journal articles and posters)
- To request funding (research proposals)
- To teach or instruct (textbooks)
- To convey instructions (lab or operating manuals)
- To provide chemical information (safety data sheets)
- To communicate with colleagues (memos or e-mails)

Exercise 1.2

Look back at the lists of chemistry genres that you created in exercise 1.1. Who is the primary audience for each genre: general, student, scientific, and/or expert audience? Some genres will target only a single audience; others will target a range of audiences.

Exercise 1.3

What are some characteristics of writing intended for a student audience? Look at a chemistry textbook. How has the author attempted to address a student audience? Consider features such as examples, illustrations, and definitions, as well as the type of vocabulary used.

Exercise 1.4

What are some characteristics of writing intended for a general audience? Find an article in your local newspaper about a science topic. How has the author attempted to make the article interesting and accessible to nonscientists? Consider features such as illustrations, the lengths of sentences and paragraphs, descriptive language, and the use of direct quotes.

Exercise 1.5

Write a sentence or short statement related to a topic that you are knowledgeable about (e.g., a hobby, favorite sport, type of music) as if you were writing to a friend with a similar interest. Then "translate" that sentence (or statement) for a person who has limited knowledge of the topic.

For example, if you were playing correspondence chess with another person, the two of you might write the following to depict the first three rounds of moves:

(1)	e4	e5
(2)	Nf3	Nc6
(3)	Bc4	Be2

For a newcomer to correspondence chess, you might translate the chess "shorthand" for the first three (of six) moves as follows:

> The first player (white) moves his/her King Pawn (the small white chess piece immediately in front of the white King) forward two spaces. In reply, the second player (black) moves his/her King Pawn two spaces forward. In the second round of moves, white moves his/her King Knight (a larger piece that traditionally looks like a horse) to the open space immediately in front of the King Bishop Pawn (the pawn immediately in front of the Bishop, the piece that stands to the right of the King).

Exercise 1.6

Most readers of this book are already expert enough to interpret the following notation, which summarizes the ^{1}H NMR spectrum of CH_3Br:

$$^1\text{H NMR (TMS) } \delta \text{ 2.68 (s, 3H)}$$

(If necessary, consult an organic textbook to remind yourself what this notation means.) Make a list of the concepts that are required to understand the notation. Which of these concepts would you need to explain to a student starting organic chemistry that a more advanced student (e.g., a junior-level chemistry major) would already know?

Organization

The second essential component of genre analysis is organization. If you decide to write in a particular genre, you implicitly agree to follow the organizational structure of that genre. Such is the case for romance novelists, Disney scriptwriters, and "whodunit" mystery writers; all must adhere to a time-honored formula (or risk having their work remain unpublished). Indeed, one of the best ways to learn to write in a new genre is to analyze the organizational structure of that genre.

Organizational structure can be divided into broad and fine structural features. Broad structural features are indicated, for example, by readily identifiable sections or headings (e.g., Introduction, Results, and Discussion); fine structural features are identified by patterns of organization within paragraphs and within sections (e.g., from more general to more specific) and by transitions between paragraphs. Both sets of features contribute to the readability and flow of the written piece. A joke book, for example, can be broadly organized by type (e.g., knock-knock jokes, light bulb jokes, genie jokes), but a single joke can also be organized into finer segments (e.g., opening line, punch line). We can think of these finer structural features as the many steps (or **moves**) that writers take to progress from the beginning to the end of each section of their writing, always with the purpose of communicating clearly with their intended audience. Writers who make use of conventional moves in their written work meet the organizational expectations of their intended audience. (Although the move concept, like the term "genre," may sound odd, its utility will become clearer as you progress through the textbook.)

A Move

A step taken by writers to achieve part of their overall purpose. Writers who use conventional moves in their written work meet the organizational expectations of their readers.

The move concept is a bit easier to illustrate with examples; hence, we do this with two jokes. The overall purpose of a joke is to make the reader (or listener) laugh. The smaller parts of the joke—the moves—serve as building blocks to help the teller (writer) achieve the overall purpose of the joke.

The first joke is a "three-men" joke. In table 1.2, the joke itself is in the left-hand column, the moves are in the center column, and the sentences that accomplish the moves are in the right-hand column. The joke is told in six moves (or steps): the joke setup; actions 1, 2, and 3; the punch-line setup; and the punch-line delivery. The second joke is a variation of a "guy-walks-into-a-bar" joke (table 1.3). (We could not resist this joke because it pokes fun at incorrect punctuation.) The joke is told in seven moves: the joke setup, a four-step action/reaction sequence between the guy (panda) and the bartender, the punch-line setup, and the punch-line delivery. In both examples, the sequencing of moves plays an important role in achieving the purpose of the jokes; if the moves were sequenced differently (e.g., if the punch line were given first), the jokes would no longer be successful. Thus, the appropriate moves not only must be present but also must be presented in the correct order.

Table 1.2 A three-men joke and its moves.

Joke	Moves	Illustration of Moves
Three men on death row are about to be executed by firing squad. The first man goes before the firing squad, but just as he is about to be shot, he distracts the squad by shouting, "Earthquake!" During the confusion, he makes his escape. On the next day, the second man goes before the firing squad. Just as he is about to be shot, he distracts them by shouting, "Tornado!" In the confusion, he makes his escape. On the third day, the third man goes before the firing squad. Just as he is about to be shot, he yells "Fire!"	Set up the joke	Three men on death row are about to be executed by firing squad.
	Describe action #1	The first man goes before the firing squad, but just as he is about to be shot, he distracts the squad by shouting, "Earthquake!" During the confusion, he makes his escape.
	Describe action #2	On the next day, the second man goes before the firing squad. Just as he is about to be shot, he distracts them by shouting, "Tornado!" In the confusion, he makes his escape.
	Describe action #3	On the third day, the third man goes before the firing squad.
	Set up the punch line	Just as he is about to be shot, he yells
	Deliver the punch line	"Fire!"

Table 1.3 A variation on a guy-walks-into-a-bar joke and its moves.

Joke[a]	Moves	Illustration of Moves
A panda walks into a bar and orders a drink. When he's done, he draws a gun and fires two shots into the air. "What was that for?" asks the confused bartender. The panda produces a badly punctuated wildlife manual and tosses it over his shoulder. "I'm a panda," he says, at the door. "Look it up." The bartender turns to the relevant entry and, sure enough, finds an explanation. "Panda. Large, black-and-white bearlike mammal, native to China. Eats, shoots, and leaves."	Set up the joke	A panda walks into a bar and orders a drink.
	Describe guy action #1	When he's done, he draws a gun and fires two shots into the air.
	Describe bartender reaction #1	"What was that for?" asks the confused bartender.
	Describe guy action #2	The panda produces a badly punctuated wildlife manual and tosses it over his shoulder. "I'm a panda," he says, at the door. "Look it up."
	Describe bartender reaction #2	The bartender turns to the relevant entry and, sure enough, finds an explanation.
	Set up the punch line	"Panda. Large, black-and-white bearlike mammal, native to China.
	Deliver the punch line	Eats, shoots, and leaves."

a. Joke adapted from Truss (2003).

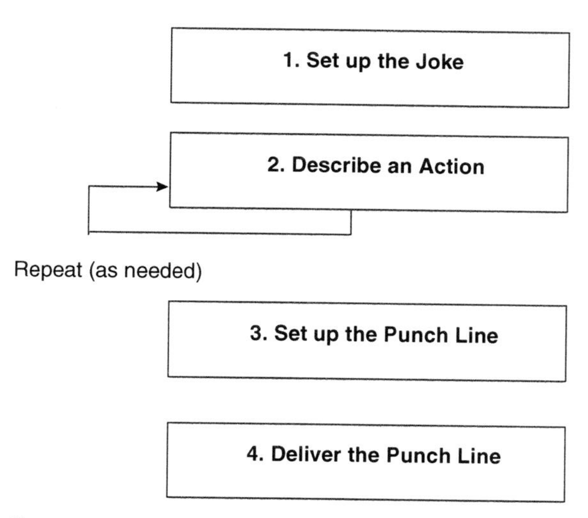

Figure 1.3 A visual representation of the move structure for a typical three-men joke.

Note that the moves and their sequencing are quite similar in both jokes, but because the jokes come from two different genres, they are not identical. (Even jokes within the same genre can have slight variations in moves.) The major difference is in the action steps: the three-men joke reiterates the action three times (once for each man); the guy-walks-into-a-bar joke reiterates the guy action/bartender response sequence twice.

In addition to listing the moves, as we did in tables 1.2 and 1.3, moves can also be represented graphically using a diagram similar to a flow chart. Such a diagram is called a **move structure**. A move structure illustrates required moves, optional moves (when appropriate), and the sequence of moves, including any repeated move patterns. A move structure for the three-men joke is illustrated in figure 1.3. Note that, in the move structure, the three action steps (the second move) comprise a single move that is reiterated as needed (in this case, three times).

Move Structure

A flowchart-like representation of the moves within a genre. The diagram visually depicts required and optional moves, illustrates repeated moves, and shows the sequencing of moves.

Exercise 1.7

Using figure 1.3 for guidance, propose a move structure for the panda version of the "guy-walks-into-a-bar" joke shown in table 1.3.

Table 1.4 Analyzing the moves present in used-car ads.

Car Ad	Information Presented
1995 Ford Aspire, great mpg, good reliable car, $2000. Call 774–3972	Year, make, model, subjective description, price, phone number
94 Ford Mustang conv GT, 5.0L, 5sp, new tires $7200/obo. David 526–0240	Year, make, model, special features, price, name, phone number
88 Ford Taurus wagon, good cond, $2200/obo. 213–1327 evenings	Year, make, model, subjective description, price, phone number, when to call
94 Chrysler LaBaron, red, 142K, sporty, fun, looks/runs grt, $2300 Josh 226–1260	Year, make, model, special features, subjective description, price, name, phone number
1995 Ford Taurus GL, 117K, air bags, 3.0L V6, great running cond, $1600. 522–8272	Year, make, model, special features, subjective description, price, phone number
92 Pontiac Bonneville SSEi, good cond, runs well, 130k mi, asking $2700/firm. 600–1721	Year, make, model, subjective description, special features, price, phone number

We next analyze the moves and move structure for another common genre: the used-car ad. Several examples of newspaper used-car ads are shown in table 1.4. As you examine these ads, you will likely notice that certain information—the year, model of the car, price, and a phone number for contacting the seller—is contained in every ad. Other types of information—such as car features and seller's name—appear in only some of the ads. To keep track of the information, and how often it appears, we list the contents of each ad in the second column in table 1.4. Some combination of this information is needed for the seller to achieve his or her purpose (i.e., to sell the car).

Using the information in table 1.4, the used-car ad can be divided into five moves. In the first move (included in all six ads), the seller states the essential facts about the car (year, make, and model). In the second move (included in all but two ads), the seller highlights select features of the car (e.g., new tires, air bags, five speeds). In the third move (interchangeable with the second), the writer offers a subjective description of the car (e.g., "good cond" or "fun"). In the fourth move, the writer states the price. Finally, in the fifth move, the seller provides contact information: (1) a phone number, (2) a contact name (optional), and (3) when to call (optional).

A move structure that depicts these moves is shown in figure 1.4. Important features about this move structure (and other move structures in this textbook) include the following:

- A box is placed around each move.
- Some moves are divided into submoves (e.g., moves 1 and 5).
- Moves and submoves are numbered to convey their conventional order in the genre. Occasionally, moves are placed side by side (e.g., moves 2 and 3).

1. State Objective Facts

1.1 State year
1.2 Report make
1.3 Indicate model

2. Identify Select Features
(if applicable)
(e.g., new tires, all bags, mileage)

↔

3. Offer Subjective Description
(optional)
(e.g., car condition)

4. State Price

5. Provide Contact Information

5.1 Give seller's name (optional)
5.2 Include phone number
5.3 State when to call (optional)

Figure 1.4 A visual representation of the move structure for a typical used-car ad. Moves that are side by side can occur in either order.

This indicates that the moves can be addressed in either order. (Submoves that can occur in any order are also placed side by side in boxes but are not numbered.)

- Each move and submove begins with an action verb (e.g., state, identify, provide).
- Unless stated otherwise, a move or submove is required; without the move or submove, the genre would be incomplete and ineffective (e.g., imagine a used-car ad that omits the make of the car).
- Some moves or submoves are not required in all instances. Such moves and submoves are followed by the words *if applicable* or *optional* (in parentheses). *If applicable* indicates that the move is required only when appropriate. For example, if a car has special features, the seller should mention them; otherwise, the move should be skipped. *Optional* indicates that the move is left to the discretion of the writer. For example, in the used-car ad, sellers can decide whether to state their cars' condition, their names, or when to call.

Required Moves

Most moves are required; that is, the genre would be incomplete or unrecognizable without them.

Occasionally, moves or submoves are required only in some instances or are left to the discretion of the writer. We label such moves *if applicable* and *optional*, respectively.

Analogous move structures are used throughout this textbook to illustrate major sections of the journal article, poster, and research proposal. The move structures are meant to guide you in reading and writing these genres. Like the used-car ad, most moves are required, but a few are not. For example, in the Methods section of a journal article, the move "Describe Numerical Methods" is labeled *if applicable* (see figure 3.1) because not all authors use numerical methods in their work. Similarly, the submove "preview key findings" in the Introduction section is labeled *optional* (see figure 6.1) because it is the author's prerogative to include that move or not. Of course, we cannot possibly know what is applicable for all individuals reading this textbook, so, as a writer, you will need to decide for yourself which moves and submoves are most relevant for your own purposes.

Exercise 1.8

Moves highlight the fine organizational structure of a genre and help to achieve the purpose of the genre. With this in mind, what is the purpose of the used-car ad genre? Could you achieve this purpose if your ad was missing one of the required moves? On the other hand, if your ad contains all of the required moves, are you guaranteed to achieve your purpose?

The Right Answer?

Exercise 1.8, like many exercises in this textbook, is designed to get you thinking about writing; hence, it will have several "right" answers (although some answers may be better than others).

Writing Conventions

Every genre has its own writing conventions (the third essential component of genre analysis), and chemistry-specific genres are no exception. **Writing conventions**, as the name implies, are generally accepted (and expected) practices; they are not "right" in the absolute sense (unlike most rules of grammar and punctuation). Writing conventions are governed by rules of writing that should be followed within a particular genre but often vary across genres. (Thus, if you write in more than one genre, you will need to learn the writing conventions for each genre.)

Writing Conventions

Rules of writing that are followed within a particular genre, but often vary across genres. Examples include how to format graphics, how to cite references, when to capitalize, and whether to use past or present tense.

One writing convention that varies across genres is **formatting**. Consider, for example, the formatting of business letters; typically either indented paragraphs or blocked paragraphs separated by spaces are used. These variations in formatting lead to noticeable differences in appearance. Or consider the formatting used in a telephone book. Lasts names are bolded and in uppercase, followed by a lowercase first name (e.g., **MILLER** Albert); for an extra fee, you can request additional bolding and/or a larger font size (e.g., **CHICAGO TITLE INSURANCE**). If telephone book entries were formatted differently (e.g., first name followed by last name), the genre would hardly be recognizable, and the information provided would be more difficult to access.

Formatting

Writing conventions specific to a genre that dictate the appearance and physical placement of written elements in, for example, tables, figures, references, headings, and number/unit combinations.

Chemistry-specific genres also have formatting rules. There are formatting rules for tables, figures, in-line citations, references, and number/unit combinations, to name only a few. The rules reflect reader expectations with regard to font size (e.g., in poster titles), bolding (e.g., in labeling, where **1** can be used to represent a chemical compound), italics (e.g., in references, where volume numbers are italicized), and placement (e.g., in citations, where numbers are superscripted). The rules also dictate whether or not to include a space between a number and its unit. For example, which is correct: 10mm or 10 mm, 100° C or 100 °C? Formatting conventions will help you answer this question.

The use of **abbreviations** and **acronyms** is another writing convention that varies across genres. In every genre, we see abbreviations and acronyms for words and phrases that are used repeatedly in that genre (and often, these abbreviations appear to be a foreign language to newcomers to that genre). For example, the shopping list uses "choc," OJ," and "pb" for chocolate, orange juice, and peanut butter, respectively; the used-car ad uses "mpg," "sp," and "obo" for "miles per gallon," "speed," and "or best offer," respectively. To write effectively in chemistry, you need to learn the standard abbreviations. Although you may already be familiar with "m" for meters and "g" for grams, you may be less familiar with "μg" for micrograms and "h" for hours. With abbreviations such as these, chemists omit the "s" for plural units ("g" not "gs" for grams) and seldom use periods ("min" not "min." for minutes).

Abbreviations and Acronyms

Abbreviations and acronyms are agreed-upon short forms for commonly used words and units.

When spoken aloud, abbreviations are often pronounced letter by letter (e.g., A-C-S), whereas acronyms form a pronounceable word (e.g., NASA).

Genres also vary by their conventional uses of verb tense (past, present, or future) and voice (active or passive voice). For example, most jokes (including the three-men joke) are told (or written) in present tense ("Three men *are*... about to be executed" as opposed to "Three people *were*... about to executed"). Present tense is used to make the joke more vivid for the listener or reader. Jokes also tend to be told in active rather than passive voice:

Active A panda walks into a bar and orders a drink.
Passive A bar is entered by a panda, and a drink is ordered.

In this textbook, we examine how tense and voice are used in journal articles, posters, and proposals. As we will see, all tenses and both voices are used, depending on which genre, or section of a genre, is being written.

Writing conventions may seem a bit picky to you at this point; however, by adhering to the writing conventions of chemists, you take an important step toward sounding like an expert chemist. If you submit a journal article, for example, with improperly formatted units and figures, incorrect abbreviations, and inappropriate verb tenses (e.g., present tense in sentences that are conventionally written in past tense), readers may judge you as a careless scientist and dismiss your work.

Exercise 1.9

Look back at the sentence(s) that you wrote in exercise 1.5. List any special writing conventions that you used. Would others interested in this topic know and use the same conventions?

Grammar and Mechanics

The fourth component of genre analysis addressed in this textbook relates to **grammar** and **mechanics**. Unlike writing conventions, which vary across genres, grammar and mechanics are governed by rules that apply to many formal written genres, though variations exist, for example, in spelling and punctuation. (The rules may be altered intentionally in creative writing genres such as poetry or fiction writing.) Although grammar and mechanics are not the focus of this textbook, we do point out common pitfalls experienced by novice writers. As shown in table 1.1, these include errors in parallelism, punctuation, subject–verb agreement, and correct **word usage** (e.g., affect vs. effect).

Grammar and Mechanics

Grammar: Rules for combining words into meaningful sentences (e.g., subject–verb agreement).

Mechanics: Rules for spelling and standard punctuation (including the use of apostrophes, hyphens, and capitalization).

In general, rules of grammar and mechanics are followed across formal written genres, though variations exist (e.g., British and American spelling).

Word Usage

The term *word usage* refers to correct and incorrect uses of words and phrases. For example, there is a right way and a wrong way to use such words as *affect* and *effect* and *spectra* and *spectrum*.

In contrast, the term *word choice* refers to choosing among several conventionally accepted words and phrases for a particular audience.

The panda joke (table 1.3), adapted from a bestselling book entitled *Eats, Shoots & Leaves: The Zero Tolerance Approach to Punctuation* (Truss, 2003), illustrates the importance of correct punctuation (specifically, the troublesome comma). How should the punch line be punctuated in the panda joke to provide a proper definition of a panda? Here is what we recommend:

Punch line	**Panda**. Large, black-and-white bearlike mammal, native to China. Eats, shoots, and leaves.
Corrected	**Panda**. Large, black-and-white bearlike mammal, native to China. Eats shoots and leaves.

Like punctuation, the misuse of commonly confused words (e.g., its/it's, affect/effect, comprise/compose, fewer/less) can result in miscommunication and undermine the message conveyed in your writing.

Exercise 1.10

Consider the following sentences. Choose the correct word in each. What rule guided your choice? (See appendix A for assistance, if needed.)

a. The human wrist is *comprised/composed* of eight bones, but the ankle has only seven.
b. *Fewer/Less* samples were used in the original series.
c. The new procedure *affected/effected* the yield.
d. The instrument was chosen for *it's/its* detection limits.

Science Content

It goes without saying that having a crystal clear understanding of a subject is a prerequisite to effectively writing about it.

—Joseph H. Aldstadt, University of Wisconsin–Milwaukee

The fifth and last essential component of genre analysis addressed in this textbook is science **content**. It is impossible to write a clear and effective paper if you lack a clear understanding of the chemistry involved; understanding the chemistry and writing about that chemistry go hand in hand. Writing is also an effective tool for learning chemistry. Chemists often "think through their hands" (i.e., through writing). You will find that as you write about your science, you will gain deeper insights and knowledge of that science.

Content

The topic(s) covered in a given genre; content is expressed through both text and graphics.

Every genre has rules (often unspoken) restricting appropriate content for that genre. A joke restricts content by appropriateness; depending on the audience, some content may be viewed as offensive rather than funny. A used-car ad restricts content by topic (you must advertise a used car, not a used refrigerator) and by space (you pay by the word, so you describe the car in a precious few lines, using standard abbreviations). Journal articles also restrict content; a chemistry-related journal article must be written about novel research in a subfield of chemistry. Although a chemist could write a paper that describes how a cake is baked ("After mixing, the ingredients were heated in an oven for 60 min at 176 °C."), we all know that such a paper would never be published in a chemistry journal, even if it adhered to all other defining characteristics of that genre. However clear the organization and writing are, if the content differs from the expectations of the genre, it will not be recognizable as an instance of the genre. Thus, a genre requires not only appropriate organization and language, but also appropriate content.

Content is typically expressed in one of two ways: text (prose, written language) and graphics (photographs, drawings, figures, etc.). Used-car ads, for example, may include a photograph of the vehicle, while recipe cards may include pictures of the prepared dish. Chemistry genres are no exception. Chemists express their content with graphics (tables, figures, and schemes) in addition to text. One key to clear chemistry writing is the appropriate and effective use of both forms of expression. In this textbook, we illustrate how chemists use text and graphics to communicate content effectively, and how the authors weave back and forth between the two to tell a story of scientific discovery.

As you analyze the ways in which chemists communicate content, don't be surprised if you learn some new chemistry, too. Although the primary focus of this textbook is writing, we believe that your chemistry knowledge will also expand as you read, analyze, and write.

Exercise 1.11

Read and analyze the following excerpt from a Material Safety Data Sheet (MSDS) for barbecue lighting fluid. Comment on as many of the five essential writing components as you can: audience and purpose, organization, writing conventions, grammar and mechanics, and science content.

SECTION 2: COMPOSITION/INFORMATION ON INGREDIENTS[a]

HAZARDOUS INGREDIENT NAME	CAS No	CONTENT	RISK	CLASS
Petroleum Distillate (Kerosene)	64742-47-8	99%	R65	Xn

Benzene (CAS No 71-43-2) will not normally be present, but always be less than the 0.1% w/w marker level in the 21st ATP to the Dangerous Substance Directive. Barbeque Lighting Fluid is not classified as a carcinogen under 67/548/EEC and the UK CHIP Regulations.

SECTION 3: HAZARDS IDENTIFICATION

INGESTION	Harmful if swallowed in large amounts.
SKIN CONTACT	Unlikely to cause irritation to skin on single exposure. Prolonged exposure may defat the skin leading to dermatitis.
EYE CONTACT	May cause irritation and reddening of the eyes.
INHALATION	Vapour at high concentrations may cause dizziness, headaches, nausea.

PHYSICAL AND CHEMICAL HAZARDS	Toxic to aquatic organisms, may cause long term effects in the aquatic environment

a. Adapted from Bird Brand Material Safety Data Sheet. Product: Barbecue Lighting Fluid. http://www.birdbrand.co.uk/msds/Barbecue%20Lighter%20Fluid.doc (accessed June 2004).

What to Expect

My graduate advisor told me something many times that I now tell my students: Writing is thinking. In order to truly communicate a scientific idea in a precise written form, one really needs to think long and hard about the best way to accomplish that goal.

—Alexander Grushow, Rider University

Each module in Section 1 of this textbook begins with an introduction to the targeted genre, including an overview of the sections that commonly compose the genre. The remainder of the module examines each of those sections in greater depth. For example, module 1, "The Journal Article", includes an introductory chapter followed by individual chapters dedicated to the different sections of a journal article: the abstract, Introduction, Methods, Results, and Discussion (although not in that order). Most chapters begin with an authentic example of the targeted section from the chemistry literature, taken largely from **American Chemical Society (ACS)** journals, which you will become quite familiar with by the end of this book. You are asked to read the example (multiple times) and analyze it for its essential components (audience and purpose, organization, writing conventions, grammar and mechanics, and science content). The rest of the module includes excerpts from the chemistry literature, explanations, and exercises designed to strengthen your ability to read, analyze, and write in that genre. Interspersed throughout the chapters are "Writing on Your Own" step-by-step tasks that will guide you in writing in the target genre. While you write, we suggest ways for you to improve your writing through multiple revisions of your work.

ACS (American Chemical Society)

The premier American professional organization for chemists, chemical engineers, and other professionals interested in chemistry (http://www.acs.org).

Module 2 ("The Scientific Poster" and conference abstract) and module 3 ("The Research Proposal") are organized similarly to module 1; they include an introduction to the genre, authentic examples of the genre, a detailed discussion of each section composing the genre, and Writing on Your Own tasks. Through this combination of reading, analyzing, and writing, you will learn to recognize the defining characteristics of four important genres in the field of chemistry and to incorporate those characteristics into your own writing of those genres.

Section 2 of the textbook includes chapters that are relevant to the four genres covered in section 1. In these chapters, you will learn to format tables, figures, and schemes (chapter 16) as well as citations and references (chapter 17). In the last chapter (chapter 18), you will find useful hints for the final stages of revision for all your written work.

Additional language tips—related to audience and purpose, writing conventions, and grammar and mechanics—are included in appendix A. Each tip has explanatory notes, examples, exercises, and an answer key, making self-study easy. Appendix B repeats, for easy reference and accessibility, the move structures included in the textbook.

Chapter Review

As a review of what you've learned in this chapter, define each of the following terms for a friend or colleague who is new to the field of chemistry:

ACS	genre	optional move	word choice
audience	genre analysis	purpose	word usage
content	move	required move	writing conventions
formatting	move structure		

Similarly, explain the following to a friend or colleague who has not yet given much thought to chemistry genres geared toward expert and scientific audiences:

- Genres commonly used by chemists at various levels in their training
- Five components of genre analysis and how they facilitate the read-analyze-write approach to writing in chemistry
- Audiences that scientific writing typically addresses
- Common purposes of scientific writing
- Differences between broad and fine organization
- Relationship between a genre's move structure and its organization
- Examples of formatting, word choice, and word usage that are both appropriate and inappropriate in chemistry-specific writing genres
- Two different means of communicating science content

Additional Exercises

 **Exercise 1.12**

Use what you know about audience and purpose to place each of the genres introduced in this textbook (journal article, scientific poster, and research proposal) on the following continua.

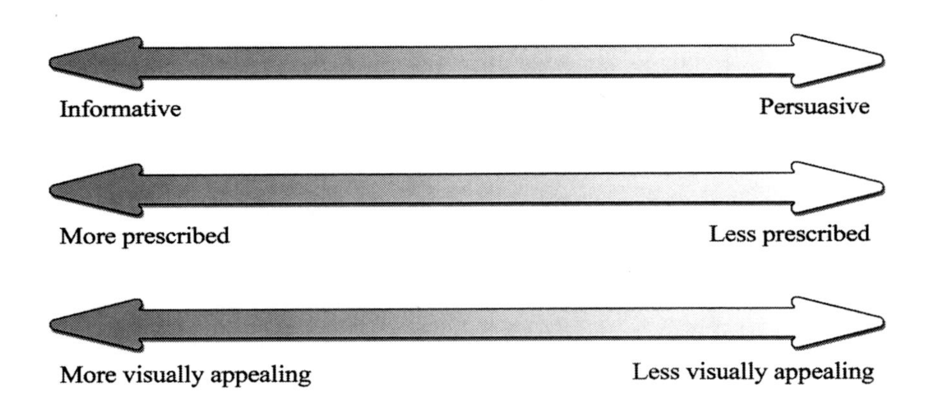

 **Exercise 1.13**

Skim the poem "Next Slide, Please," written by Roald Hoffmann, who in 1981 shared the Nobel Prize in Chemistry with Kenichi Fukui. In this poem, Hoffmann pokes fun at seminar presentations. The poem serves as an example of a genre that differs in many ways from the scientific papers written by this world-renowned chemist.

a. Who is Hoffmann's audience?

b. What was his purpose for writing this poem?

c. Although you may not have read many of Hoffmann's scientific papers, speculate on ways in which this poem is different in organization, writing conventions, grammar and mechanics, and science content from the many journal articles that he has written for expert audiences.

Next Slide, Please

there was no question that the reaction worked
but transient colors were seen
in the slurry of sodium methoxide in dichloromethane
and we got a whole lot of products
for which we can't sort out the kinetics

the next slide will show
the most important part
very rapidly
within two minutes
and I forgot to say on further warming
we get in fact the ketone
you can't read it on the slides
but I refer to the structure you saw before
the low temperature infrared spectrum
as I say
gives very direct evidence
so does the NMR
we calculated it
throwing away the geminal coupling
which is of course wrong
there's a difference of 0.9 parts per million
and it is a singlet
and sharp
which means two things
either
you're doing the NMR in excess methoxide
and it's exchanging
or
I would hazard a guess
that certainly in these nucleophilic conditions
there could well be
an alternative path
to the enone you see there
it's difficult to see
you could monitor this quite well in the infrared
I'm sorry in the NMR
my time is up I see
well this is a brief summary of our work
not all of which
I've had time to go into
in as much detail as I wanted
today.

(Hoffmann, 1987: pp 51–52)

Exercise 1.14

Science writers, working for magazines such as *Science News* and *Popular Science*, translate discoveries reported in journal articles (written for expert audiences) into articles that a more general audience can understand. Consider the following example, where we juxtapose an original passage from *Nature* with its translation in *Science News*. Both passages explore why staggered ethane is more stable than its eclipsed conformer. You may have learned in organic chemistry that the eclipsed conformation is higher in energy because of steric (crowding) effects, but computational results suggest that the real reason has to do with hyperconjugation.

a. Read and compare the titles of the *Nature* and *Science News* articles. What difference(s) do you notice?

   From *Nature*: Hyperconjugation Not Steric Repulsion Leads to the Staggered Structure of Ethane

   From *Science News*: Molecular Chemistry Takes a New Twist

b. Now read and compare the two passages below. Identify at least three differences in the writing styles for the two audiences.

   This structural preference is usually attributed to steric effects.[1–7] . . . Here, we report . . . that ethane's staggered conformation is the result of . . . hyperconjugation. (From Pophristic and Goodman, 2001)

   Textbooks . . . pin it on so-called steric effects . . . but . . . Pophristic looked at the other known influence on ethane's twisting—a quantum mechanical effect known as hyperconjugation. "The electrons of one methyl group jump over to the other methyl group," says Goodman. (From Gorman, 2001)

c. Select a concept that most chemistry majors are familiar with but that the general public is not (e.g., the resonance structures of benzene or the molecular shape of water). Write an explanation of the concept for a general audience.

Exercise 1.15

Below are five examples of a genre that you are likely familiar with, the Acknowledgments section. Acknowledgments are commonly included as a short section at the end of journal articles, just before the References section. Using the joke and used-car ad examples in this chapter as guides, conduct a full analysis of the genre of these Acknowledgments sections, using the five examples as representative samples of the genre. Specifically, identify the following:

a. The intended audience and purpose of the genre (consider level of detail, formality, conciseness, and word choice)
b. The writing conventions of the genre (e.g., abbreviations and acronyms, verb tense, voice)

c. The ways in which the content is communicated in the genre (topic, text, graphics)

d. The fine organizational structure of the genre. Do this by proposing a move structure (similar to figures 1.3 and 1.4) for the genre (not the individual acknowledgments). Assign move labels that reveal the actions taken by the writers. (Hint: There are some optional moves, i.e., moves that do not appear in all examples; be sure to indicate which moves are optional in your move structure.) Arrange the boxes so that they reflect the typical organization of journal-article Acknowledgments sections.

- (From Prevedouros et al., 2004) We are grateful to the UK Department of the Environment, Food and Rural Affairs (DEFRA) Air Quality Division for financial support. We also wish to thank Anna Palm of the Swedish Environmental Research Institute and Dr. Knut Breivik of the Norwegian Institute for Air Research (NILU) for their helpful comments.
- (From Huange et al., 2004) We gratefully acknowledge the support from the National Nature Science Foundation of China (20375005) and the Bilateral Scientific and Technological Cooperation Flanders Belgium-China (011S0503).
- (From Raczyńska and Darowska, 2004) E.D.R. and M.D. (SGGW) thank the Polish State Committee, the Conseil Général des Alpes Maritimes, and the French Ministry of Higher Education and Research for financial support and the Warsaw Agricultural University for the leave of absence. I.D. was financially supported by the U.S. DOEOBER Low Dose Radiation Research Program. Ab initio calculations were carried out at the Interdisciplinary Center for Molecular Modeling (ICM, Warsaw).
- (From Dick and McGown, 2004) This work was supported by the National Institutes of Health (Grant 1R03 AG21742–01).
- (From Vitòria et al., 2004) This study has been financed by CICYT Project REN2002–04288-C02–02 of the Spanish Government and partially by SGR01–00073 of the Catalonian Government. We would like to thank the Serveis Cientificotècnics of the University of Barcelona (Spain).

Exercise 1.16

Access the homepage for *Chemical and Engineering News* (*C&EN*) through the ACS Web site. On the *C&EN* homepage, find and select the last issue of the previous year, which will have a cover story titled "(Year) Chemistry Year in Review." Open the cover story and read several of the chemistry highlights. Describe the intended audience and purpose of these highlights.

Exercise 1.17

Reflect on what you have learned from this chapter. Select one of the reflection tasks below, and write a thoughtful and thorough response:

a. Reflect on the idea of audience in scientific writing.
   - What did you know about audience before reading this chapter? Where did you learn it?
   - What audiences have you written for in the past? What types of writing did you do for these audiences (e.g., lab reports, journal articles)?
   - Have you written for an expert audience before? What challenges do you think you'll encounter writing for an expert audience?

b. Reflect on the relationship between reading and writing in chemistry genres.
   - How might your writing improve by reading authentic examples of chemical writing (e.g., journal articles, posters, research proposals)?
   - How might your reading improve by learning to write in professional genres of chemistry?
   - What aspects of your reading and writing do you hope will improve? Why?

c. Reflect on the value of genre analysis activities and their role in the read-analyze-write approach to writing.
   - What are you likely to learn from engaging in genre analysis activities?
   - How might your reading and writing abilities improve as a result of genre analysis?
   - Which focal points of genre analysis (audience and purpose, organization, writing conventions, grammar and mechanics, and/or science content) do you think will be most useful to you when you attempt to write for an expert audience? Why?

d. Reflect on the value of move structures.
   - What is the value of depicting the fine organization of chemistry genres through move structures?
   - How might move structures help you with your writing?
   - Why might chemical writing be structured in such formulaic ways?

Module 1 *The Journal Article*

2 *Overview of the Journal Article*

An author should recognize that journal space is a precious resource created at considerable cost. An author therefore has an obligation to use it wisely and economically.

—American Chemical Society, *Ethical Guidelines to Publication in Chemical Research* (https://paragon.acs.org)

This chapter introduces the journal article module (comprising chapters 2–7). The chapter describes some of the defining characteristics of a journal article while emphasizing concise writing and organization. By the end of this chapter, you should be able to do the following:

- Recognize the importance of concise writing
- Identify the broad organizational structure of journal articles
- Explain what is meant by targeted reading and keywords

As you move through the chapter, you will begin to plan your own journal-quality paper. The Writing on Your Own tasks throughout the chapter will guide you in this process:

2A Get started

2B Select your topic

2C Conduct a literature search

2D Find additional resources

2E Decide on the broad organization of your paper

Module 1 focuses entirely on writing a journal-quality paper, a paper suitable for submission to a refereed chemistry journal. **Refereed journals** include only articles that have made it through a rigorous peer-review process. In this process, a submitted manuscript is critically reviewed by two or more anonymous reviewers. The reviewers are asked to judge both the scientific merit and writing quality of the manuscript. Authors are often required to revise their work before it can be accepted for publication. The entire review process can take six months or longer. An account of the review process typically appears in the published article, for example,

> Received for review March 9, 2008. Revised manuscript received August 3, 2008. Accepted August 5, 2008.

Once published, the journal article becomes part of the **primary literature** of chemistry. The primary literature is a permanent and public record of all scientific works, many of which are refereed journal articles.

Refereed Journals

Refereed journals publish only papers that have gone through a rigorous peer-review process.

Submitted manuscripts are evaluated by experts (peers) for quality and originality. Based on the reviewers' remarks, journal editors decide to accept, accept with revisions, or reject each submission.

Primary Literature

The primary literature comprises peer-reviewed publications that describe results of original research. In general, these publications are the first and most authoritative record of the work.

In this chapter, we take a bird's eye view of the journal article. We consider the journal article's audience and purpose, stress the importance of concise writing throughout the journal article, and examine the broad organizational structure of the journal article. In subsequent chapters, we examine sections of the journal article in more depth.

2A Writing on Your Own: Get Started

As you work through this module (chapters 2–7), you will be writing your own journal-quality paper. Your finished paper will be written for an expert audience, be organized into appropriate sections (title, abstract, Introduction, Methods, Results, and Discussion sections, and references), and contain at least one table or figure. The final paper must include experimental data; hence, a review of the literature or a summary of others' work will not suffice.

Step-by-step details on how to complete this assignment are provided in subsequent chapters in this module. In this chapter, you will prepare to write by selecting your topic, conducting a literature search, and determining the broad IMRD (Introduction, Methods, Results, Discussion) format for your paper.

Audience and Purpose

The major purpose for writing a chemistry journal article is to share the results of original research with other chemists. The primary audience for a journal article is an expert one; readers are typically well educated and highly experienced in the subfield of chemistry addressed in the article. Because journal articles are written largely for experts, newcomers to the field (e.g., students or chemists exploring a new research area) are often frustrated by the advanced level of these articles. Details are often omitted that the nonexpert reader would find useful. (If you find yourself in this situation, we recommend that you also consult related works written for a less sophisticated audience—textbooks, review articles, general science articles—to help you work your way through the journal article.)

Although the bulk of the journal article is written for experts, a few sections are often accessible to less sophisticated readers. For example, general or summative remarks in the abstract, Introduction section, and conclusions are often accessible to a scientific audience, allowing those readers to grasp the key concepts of the work. Similarly, many chemistry journals include features (e.g., book reviews, editorials, and news articles) that are written specifically for scientific and general audiences.

Exercise 2.1

Browse through a research article in three different ACS journals: *Analytical Chemistry*, *Biochemistry*, and *Environmental Science & Technology* and perform the following tasks:

a. Try to find two or three sentences that are easy to read and understand in each section of the article (abstract, Introduction, Experimental, Results, and Discussion). Next, find two or three sentences that are difficult to read in each section. What differences, if any, do you notice in the readability of these sections? Which sections are the easiest to read and understand? Why do you think some sections are easier to read than others?

b. Read one of your three articles more carefully. What makes the authors sound like experts? Jot down at least 10 examples of expert-like writing.

Exercise 2.2

Browse through the Table of Contents of several issues of *Analytical Chemistry* or *Environmental Science & Technology*. In most issues, you will see that, in addition to research articles, the journal also contains news articles, editorials, features, and/or book reviews. Glance through the pages of two such items. For each item, jot down its title and the name of the journal section in which it appears; identify the intended audience and purpose of the entry.

2B Writing on Your Own: Select Your Topic

Before you can begin to write, you must identify a topic for your paper. When selecting a topic, keep in mind that your project must be robust enough to result in a journal-quality paper (not a literature review). Minimally, you should be able to (1) introduce your topic, (2) provide background information about your topic, (3) describe the methods used to investigate your topic, (4) present your results (using at least one table or figure), and (5) discuss your results.

After you have selected a topic, write a two- to three-paragraph description about your selection. Briefly explain how you will meet the five criteria listed above, including what data you plan to present. Include a list of three to five keywords related to your topic. You will use these terms to conduct a literature search.

Conciseness

I find that there is nothing more tedious than papers that go on and on, with no obvious point.

—Richard Malkin, University of California–Berkeley

Vigorous writing is concise. A sentence should contain no unnecessary words, a paragraph no unnecessary sentences, for the same reason that a drawing should have no unnecessary lines and a machine no unnecessary parts.

—William Strunk, Jr., *Elements of Style*

Recall from table 1.1 that audience and purpose are communicated through four subcomponents (conciseness, level of detail, level of formality, and word choice). The first of these, conciseness, is a hallmark of writing in chemistry. Chemistry readers (experts and nonexperts alike) want crisp, clean sentences that say what needs to be said and no more. They do not want to be bogged down in words that fail to advance or, worse, confuse meaning. Because concise writing is important in every section of the journal article, we address it here, in chapter 2, as well as throughout this module.

Novice writers often equate wordy writing with expert writing. They adapt a wordy and pretentious writing style because (1) they want to make their papers longer (not at all a goal in scientific writing!) and (2) they want to sound more professional. However, most professionals (particularly scientists) prefer a more concise, direct style. As an example, scientists at a conference of the British Ecological Society were asked to read two texts that presented the same information, although one text was considerably wordier than the other (Turk, 1978). On the whole, the scientists rated the concise version as being easier to read and more appropriate than the wordy version. The scientists also asserted that the more concise author was more objective, had a more organized mind, and inspired more confidence in the work. Thus, wordy language, rather than making you sound more professional, can obscure your message and discredit your objectivity.

Wordiness

The ACS Style Guide recommends that the following phrases be omitted from papers because they are vacuous and contribute to wordiness:

- As already stated
- It has been found that
- It has long been known that
- It is interesting to note that
- It is worth mentioning at this point
- It may be said that
- It was demonstrated that

The ability to write concisely is a coveted skill among chemists and an important step toward sounding like an expert. The key is to say only what needs to be said, deleting unnecessary words (i.e., words that add little substance, state the obvious, or can be inferred by other words in the sentence). For example, compare the following two sentences. The wordy sentence contains so many unnecessary words that the authors' message is nearly lost.

Wordy	In a paper published by Bonderic et al.,[2] experiments were described that led to similar results.
Concise	Bonderic et al.[2] reported similar results.

(It is clear to the reader that Bonderic et al. published a paper because of the in-line citation. It is also obvious that "experiments were described" in that paper.)

Sound Like an Expert

Learning to write concisely will help you sound like an expert. This skill requires that you delete words, as you revise and edit your work, that add little substance or state the obvious.

Exercise 2.3

Read the following wordy passage (adapted from Liu et al., 2001) and identify five words and/or phrases that could be deleted to make the passage more concise:

After the mixture had been dried, the remaining residue (CD-capped gold nanoparticles + compound **5**) was found to express insolubility in dry $CHCl_3$ but the solubility was restored when water was used in the equilibration of the chloroform. In our judgment, this finding clearly makes it apparent that there must be some water necessary for the efficient phase transfer of the nanoparticles into $CHCl_3$. This finding leads to the conclusion that the idealized structure that has been proposed for the nanoparticles after they have been transferred to the chloroform phase (Scheme 2) has some aspects that must be similar to the structure of reverse micelles. We come to the conclusion that these nanoparticle-centered assemblies are similar in a conceptual way to gold-filled reverse micelles. (124 words)

Removing Unnecessary Words

There are several ways to make the adapted passage in exercise 2.3 more concise. One technique is to replace a group of words with a single word that has the same meaning. For example,

Wordy	This finding *makes it apparent* that . . . (6 words)
More concise	This finding *demonstrates* that . . . (4 words)
Wordy	We *come to the conclusion* that . . . (6 words)
More Concise	We *conclude* that . . . (3 words)

Contractions

Even though contractions seem to make writing more concise, do not use them in your papers; contractions are not appropriate in an academic style.

Table 2.1 Suggestions for concise writing (adapted from *The ACS Style Guide*: Coghill and Garson, 2006, pp 54–55).

Wordy Phrase	Concise Alternative
a number of	many, several
based on the fact that	because
by means of	by
despite the fact that	although
due to the fact that	because
if it is assumed that	if
in order to	to
in spite of the fact that	although
is/are known to be	is/are
it is clear that	clearly
reported in the literature	reported
subsequent to	after

The ACS Style Guide provides many useful tips for converting wordy multiple-word phrases into more concise alternatives (see table 2.1 for a few examples).

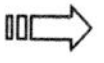

Concise Writing

See appendix A.

Exercise 2.4

Consider the suggestions for conciseness in table 2.1. Revise these passages by substituting the italicized phrases with more concise alternatives:

a. *Despite the fact that* the Lewis acid behavior of group 13 halides has been extensively studied,[8] the Lewis acid behavior of group 12 halides has not.[9] (Adapted from Borovik et al., 2001)

b. *In order to* explain this shift, they proposed that the smaller particles are more sensitive to UV curing *based on the fact that* their relative surface areas are larger. (Adapted from Bol and Meijerink, 2001)

Another way to be concise is to eliminate words that are redundant or that provide information the reader can assume. For example, the phrase "In our judgment" (at the beginning of the second sentence of the passage in exercise 2.3) can be removed because the reader can assume that the conclusions being drawn are based on the researchers' judgments:

Wordy	In our judgment, this finding clearly demonstrates... (7 words)
More Concise	This finding clearly demonstrates... (4 words)

(Some would argue that the word "clearly" could also be eliminated, but the word was included in the original article and is a favorite among chemists.)

Conciseness can also be achieved through the use of parentheses, as a way to eliminate information that is superfluous:

Wordy	The results, as illustrated in Table 3, suggest that... (9 words)
More Concise	The results (Table 3) suggest that... (6 words)
Wordy	The ethanol (research grade and purchased from Sigma-Aldrich, located in Milwaukee, Wisconsin) was added... (14 words)
More Concise	The ethanol (research grade, Sigma-Aldrich, Milwaukee, WI) was added... (9 words)

Exercise 2.5

Revise these sentences. Identify the parts that you consider to be too wordy. Make those parts more concise or delete them entirely.

Example

The lipid-binding potential was *observed to be* independent of pH; *as a consequence*, the results *presented in this paper* do not support hydrophobic interactions.

The lipid-binding potential was independent of pH; hence, the results do not support hydrophobic interactions.

a. Prior to irradiation, the sample chamber was thoroughly flushed with nitrogen, to be sure that air was absent during the irradiation. (Adapted from Bol and Meijerink, 2001)

b. It is possible that the products of the photochemical reaction that takes place upon irradiation in the presence of water passivate the surface better than the photooxidation products obtained during irradiation in dry air. (Adapted from Bol and Meijerink, 2001)

c. In consequence of this fact, it can be assumed that $ArCl^+$ does not interfere with the quantification of arsenic in the soil extracts.

d. In a large number of cases, cigarette smoke contributions could not be determined because the anteisoalkanes and isoalkanes that are used to trace cigarette smoke particles were below detection limits. (Adapted from Schauer et al., 2002)

e. Levels of all aldehydes increased during storage compared to the control sample, as exhibited in Figure 2. (Adapted from Vesely et al., 2003)

f. All of the chemicals were research grade, and they were all purchased from Fisher, which is located in Pittsburgh, PA.

Using Nominalizations

Another technique employed by chemists to achieve conciseness is to use **nominalizations**. Nominalizations are nouns that are formed from other parts of speech, usually by adding such endings as -tion, -sion, -ment, -ity, -sis, and -ence. For example,

solubility (noun, from the adjective *soluble*)
distillation (noun, from the verb *distill*)

Nominalizations often allow several words to be summarized in a single word.

Without a nominalization	After we distilled the product, it was a colorless liquid. (10 words)
With a nominalization	After distillation, the product was a colorless liquid. (8 words)

By using a nominalization in this last example, we could also remove the word "we" from the sentence, making it sound more objective. Table 2.2 lists common nominalizations used in chemistry journal articles.

Table 2.2 Common nominalizations used in chemistry writing.[a]

absorption	addition	agreement	calculation
activation	aggregation	analysis	comparison
concentration	emission	luminescence	reaction
conductivity	excitation	measurement	reactivity
conversion	extraction	oxidation	reduction
dependence	formation	preparation	synthesis
diffusion	intensity	presence	treatment
efficiency	interaction	purification	

a. Nominalizations determined through a computer-based search of 200 chemistry journal articles.

Nominalization

A noun formed from a verb or an adjective, usually by adding a word ending (e.g., -tion, -ment). For example,

Extraction (noun, from the verb *extract*)

Efficiency (noun, from the adjective *efficient*)

Nominalizations often result in more concise writing. (See appendix A.)

Keep words that convey important content; delete words with little substance.

Exercise 2.6

Consider these sentences taken from original sources in the chemistry literature. Identify instances of the common nominalizations listed in table 2.2.

Example

Potential *conversion* of the N1 adduct to its N^6 derivative was made possible through a Dimroth rearrangement, although the *efficiency* of this process is highly dependent on *reaction* conditions and adduct structure (Scheme 2).[26] (From Veldhuyzen et al., 2001)

a. The electronic spectra and structure of these systems clearly relate to the presence of closed-shell metal-metal bonding,[8,9] exciplex formation,[10] electron transfer,[11] energy transfer,[12–14] and chemical reactivity.[15] (From Rawashdeh-Omary et al., 2001)

b. The excitation spectra were corrected for the beam intensity variation in the Xe light-source used. (From Dhanaraj et al., 2001)

c. We have reported the preparation and characterization of deoxyuridine nucleosides and nucleotides where ferrocene was conjugated to the nucleobase through unsaturated bonds[5] and the preparation of adenosine and cytidine modified with ferrocene at the 2′-position through butoxy linkers.[6] (From Yu et al., 2001)

d. The relative intensity of this signal permits a comparison of the strength of the interaction between oxides in dependence on the preparation conditions (milling time, calcination temperature and time, presence or absence of water). (From Spengler et al., 2001)

Exercise 2.7

Try rewriting example sentence (d) in exercise 2.6 without using any nominalizations. How does your revision compare to the original in terms of conciseness and clarity?

Nominalizations usually make writing more concise. One important exception, however, is using nominalizations with forms of the verb *do* (i.e., *did* or *was done*). In such cases, nominalizations can make the sentence wordier and should be avoided. Consider the following examples:

Wordy   Komiyama et al.[2] did an analysis...
Concise   Komiyama et al.[2] analyzed...

Wordy   A synthesis was done by Martinez et al.[7]...
Concise   Martinez et al.[7] synthesized...

et al.

An abbreviation (for the Latin *et alia*) that means *and others.*

It is used when referring to authors of a publication that has three or more authors. The first author's last name is listed, followed by et al. (with the period).

Exercise 2.8

Rewrite the following sentences to make them more concise. (Do not delete original citations, indicated by superscript numbers.)

a. At this point in time, there exist only a small number of reported examples of the synthesis of carboranes from eneynes,[20] the first of these being the synthesis of 1-isopropenylcarborane.[21] (Adapted from Valliant et al., 2002)

b. Table 1 accurately summarizes the reaction products from the two reactions that were performed: the hydrogenation reaction and the oxidation reaction.

c. Polyadducts of C_{60} with well-defined three-dimensional structures are of great importance, based on the fact that they possess interesting biological[1a] and material properties.[1b,c] (Adapted from Mas-Torrent et al., 2002)

 **Exercise 2.9**

Consider the various ways in which you can make your writing more concise. Return to the wordy passage in exercise 2.3. Revise the passage so that it is more concise. As a guideline, consider that the original was only 90 words long!

2C Writing on Your Own: Conduct a Literature Search

In Writing on Your Own task 2B, you generated a list of three to five **keywords** related to your paper topic. Use these keywords to search for peer-reviewed journal articles about your topic. The goal is to find at least four articles relevant to your research area. Search scientific databases such as American Chemical Society Publications or ScienceDirect; be cautious with general Internet search engines, such as Google, because they do not limit searches to the primary literature.

When you find an article of potential relevance, read the title, abstract, and keywords. If this information appears relevant, skim the rest of the article (including tables and figures). Take notes on (1) what the study was about, (2) what methods were used, and (3) what conclusions were drawn.

Print out each article that you decide to use. Be sure to print the *entire* article. A common mistake is to print the text but not the references. Check to be sure the printout includes full bibliographic information (i.e., the full and abbreviated journal name and the article's title, authors, volume, year, and inclusive page numbers). For a Web-based article, write down the full Web address and the month and year that you accessed the site.

Keywords

Keywords can be used to search for articles in your field. Most journals require authors to include a list of keywords with their submissions. Oftentimes, the words are selected from a master list provided by the journal.

Organization

In chapter 1, we learned that genres have both broad and fine organizational structures. In this chapter, we focus on the *broad* organizational structure of the journal article, signaled by identifiable sections and headings. In general, journal articles are divided into four major sections. These sections have the familiar names **I**ntroduction, **M**ethods, **R**esults, and **D**iscussion; collectively, this organizational structure is referred to as the **IMRD format**. In addition to these four major divisions, journal articles also include a title, abstract, references, and often acknowledgments. (Acknowledgments are required for works supported by a funding agency; otherwise, this section is optional.) On occasion, journal articles also include a section for conclusions, but more often conclusions are included at the end of the Discussion section. Chapters 3–7 go into detail about all of these sections; in this chapter, we briefly highlight the IMRD sections.

IMRD Format

The typical broad organization of a journal article:

Introduction

Methods (or Experimental)

Results

Discussion

The Introduction section of a journal article identifies the research area, explains the importance of the research, provides background information, cites and summarizes key literature in the field, points out what still needs to be studied, and introduces the reader to the work presented in the article. The Methods section—formally known as Materials and Methods or Experimental (Section)—describes how the study was conducted. The Results section summarizes quantitative (and possibly qualitative) data collected during the study. In the Discussion section, authors interpret their data and suggest the larger implications and/or applications of their results. Each of these major sections can be further divided into moves, as we will see in subsequent chapters.

In recent years, variations have appeared in the traditional IMRD format. For example, some journals include explicit headings for all four divisions, some use fewer than four explicit headings, and some use no headings at all. The sequencing of the headings also varies. For example, *The Journal of Organic Chemistry* typically places the Experimental Section at the end of the article, while the *Journal of the American Chemical Society* places the Experimental Section in a footnote. Interestingly, in the past, chemists were discouraged from combining the Results section (presentation of data) with the Discussion section (interpretation of data), yet today a combined Results and Discussion section is commonplace. For the journal article that you write in this module, we recommend the traditional IMRD format. However, if you are submitting a paper to a journal for publication, you should follow the organizational structure recommended or required by that journal. Because every journal has slightly different requirements, it is important to read the "Information for Authors" for the particular journal to which you plan to submit your paper.

"Information for Authors"

A set of journal-specific instructions for hopeful authors that includes the following:

length specifications

section requirements

appropriate research areas

Exercise 2.10

Browse through *The Journal of Organic Chemistry*, *Journal of the American Chemical Society*, and *Organic Letters*. What variations in sectional divisions do you see? Are all headings shown explicitly? Repeat this exercise with two or three non-ACS chemistry journals (e.g., *Applied Surface Science*, *Chemical Physics*, and *Journal of Chromatography A*, all published by Elsevier).

Hourglass Structure

The IMRD format creates what is sometimes called the hourglass structure, a feature common to journal articles across many fields of academic research. The hourglass depicts the way in which the scope or specificity of the paper changes throughout its sections, as shown in figure 2.1. The Introduction section begins with a broad overview of the research area but narrows as the authors mention specifics about their presented work. This specificity is maintained throughout the Methods and Results sections and then broadens again at the end of the Discussion, where research findings are described in a broader context. An

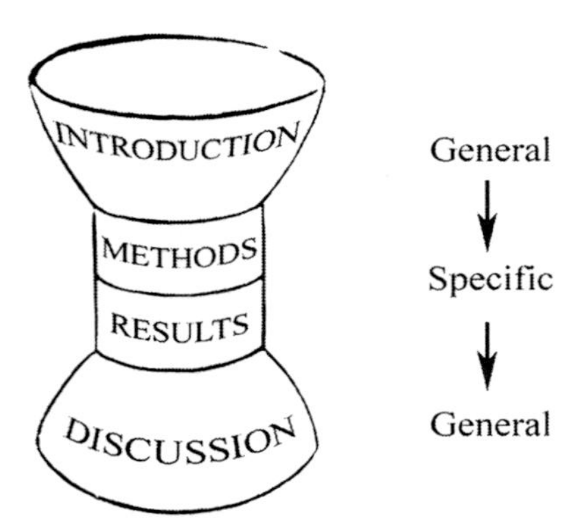

Figure 2.1 The hourglass structure of the IMRD format (adapted from Hill et al., 1982).

hourglass icon is used throughout the textbook to remind you where you are in the hourglass structure.

Hourglass Icon

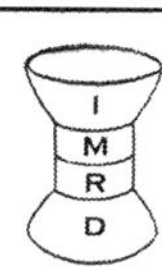

The hourglass icon is used in modules 1 and 2 as a visual reminder of the level of detail required at different points in your written work.

Exercise 2.11

Browse through three articles in *Chemical Research in Toxicology* or the *Journal of Agricultural and Food Chemistry*. For each article, copy and paste into a text document one sentence from each IMRD section that is consistent with the hourglass structure. Examine each group of four sentences. Is the hourglass structure apparent in these four sentences? Explain.

2D Writing on Your Own: Find Additional Resources

Use the articles that you found in Writing on Your Own task 2C to locate additional resources, as follows:

1. Begin with the most recently published articles. Browse through their reference lists to find additional sources that you may have missed in your literature search.
2. Read the Introduction sections of your articles. Because Introductions provide relevant background information and cite others' works, a targeted reading of the Introduction can help you identify commonly cited sources in your field (a technique known as *footnote chasing*).
3. Go to the library and/or the Web to obtain copies of these additional resources.
4. Read, sort, and code your articles by topic, methods, and/or results.

Targeted Reading

Before ending this chapter, one more purpose of the IMRD format is worth noting: the IMRD structure promotes targeted reading, allowing readers to locate science content in an expedient manner. You might be surprised to learn that chemists do not typically read every word of every article from start to finish. Rather, most chemists read selectively, targeting sections most pertinent to their interests. For example, most chemists glance first at an article's title, abstract, and keywords.

This cursory glance is often enough to decide if the article is sufficiently relevant to merit a closer look. If the content appears promising, they next target a specific section of the article. For example, a chemist interested in planning a synthesis will read the Methods section; a reader wanting to learn more about potential uses of a novel compound will read the Introduction and Discussion sections. Less relevant sections of the article are skimmed or skipped entirely. Only a few articles, those most pertinent to the reader's interests, will be read in their entirety (and then usually many times).

Targeted Reading

Students are generally accustomed to reading texts from start to finish, with the intent of understanding everything. You may have read a textbook chapter in this manner.

Scientists rarely read research articles like this. Generally, scientists *target* a given section to look for specific information. For example, a chemist might target the Methods section for the sole purpose of finding out what brand of instrument was used.

The IMRD format, together with the finer organizational structure (moves) and language of the journal article, helps readers quickly locate the content that they seek. As a developing writer, it is important that you learn to present your content in these expected ways and places. As readers scan your paper, they should be able to quickly locate your topic, the nature of your work, the methods you used, and your conclusions. The use of keywords in your title and abstract and throughout the body of your paper is especially important. Because so many scientists now turn to computer-based technologies and search engines to find pertinent and current information, if you fail to use expected keywords, your work may be missed even if relevant.

Exercise 2.12

Browse through three research articles in an ACS journal of your choice. With only a cursory glance at the title, abstract, and the IMRD sections, determine the topic of research, the methods used, and a brief summary of the conclusions. (Note: You should be able to identify the topic and methods, even if you do not understand them.)

2E Writing on Your Own: Decide on the Broad Organization of Your Paper

When you actually finish your journal article writing assignment, your paper will be written in either the IMRD format or a format specified by a particular journal. If you are writing your paper for publication, you must select the journal to which you will submit your manuscript and follow its submission guidelines (which may require a modified format). Most scientific journals (including most ACS journals) post their submission guidelines on their Web pages. In general, submission guidelines describe the type(s) of articles that the journal considers for publication and specify how manuscripts should be organized and formatted. For most ACS journals, you can also download a template to be used in your word-processing software that will help you follow the journal's specifications.

Decide on the broad organization of your paper. Will you follow the traditional IMRD format? Or will you follow a modified format specified by a particular journal?

Recall that the actual sequencing of sections in a completed journal article does not normally reflect the exact order in which most authors write their articles. In fact, writers often begin with the Methods section (as you will). As writers progress through the different sections of their papers, they go back and forth among the sections, revisiting previously drafted sections to modify them as needed.

Chapter Review

As a self-test of what you've learned in this chapter, define each of the following terms for a friend or colleague who is new to the field:

conciseness	IMRD format	primary literature
et al.	Information for Authors	refereed journal article
hourglass icon	keywords	submission guidelines
hourglass structure	nominalization	targeted reading

Also, explain the following to a friend or colleague who has not yet given much thought to writing a journal article:

- Main purpose of a journal article
- Audience of a journal article
- Broad organizational structure of a typical journal article and its variations
- Purpose(s) of each section (IMRD) of a journal article
- Importance of concise writing
- Techniques for writing more concisely
- Resources that can be found on a journal's Web site that can assist writers who want to prepare a paper for that journal

Additional Exercises

Exercise 2.13

Think about the purpose of each section in a journal article. For each of the sentences below, decide which section of a journal article (I, M, R, or D) it comes from and explain what led you to make your decision:

Example

When **13** and **17** were sialylated on a larger scale (2.0 and 3.5 mg, respectively) glycopeptides **20** and **21** could be isolated in 94 and 64% yields, respectively, after purification by reversed-phase HPLC. (From George et al., 2001)

Results. Data (specific numbers) are reported, along with a reminder of the methods used. The passage does not interpret the data. Therefore, it is probably not part of a Discussion section.

a. Most epithelial cells produce mucins, that is, glycoproteins in which the polypeptide backbone consists of highly conserved tandem repeats with complex carbohydrates linked to multiple serine and threonine residues.[1–4] (Adapted from George et al., 2001)

b. Studies directed toward chemoenzymatic synthesis of more complex mucin-derived glycopeptides, as well as attempts to use the glycopeptides described herein for development of cancer vaccines, are underway in our laboratories. (From George et al., 2001)

c. Tetrahydrofuran and diethyl ether were dried using sodium metal and then distilled, as required, from sodium benzophenone ketyl. (From Banwell and McRae, 2001)

d. Connectivities were observed between these protons and both C2 and C6 of dA. (From Veldhuyzen et al., 2001)

e. All aqueous solutions were made with water purified by standard filtration to yield a resistivity of 18.0 MΩ. (From Veldhuyzen et al., 2001)

f. Most every heteroatom of DNA exhibits at least some nucleophilic character, and each may be variably targeted for alkylation depending on the nature of the electrophile and the reaction conditions. (From Veldhuyzen et al., 2001)

g. Instead, it may have formed insoluble dimers, trimers, and higher-molecular-weight species as previously described for related structures.[39,40] (Adapted from Veldhuyzen et al., 2001)

h. This work has shown that CD-capped gold nanoparticles with average core diameters of ca. 3 nm act as effective hosts for cationic ferrocene derivatives, as evidenced by electrochemical and ^{1}H NMR spectroscopic data. (From Liu et al., 2001)

i. The fast development of methods for the preparation of metal and semiconductor nanoparticles capped with organic monolayers is opening interesting possibilities for the functionalization of their surfaces.[1] (From Liu et al., 2001)

Exercise 2.14

Familiarize yourself with the article by Boesten et al. (2001) about the Strecker reaction that appears at the end of this chapter. Look over the article as a professional chemist would to simply orient yourself to the article. Follow the steps below:

a. Read the title.
b. Read over the abstract. (This article does not include keywords.)
c. Skim the first sentences of most paragraphs.
d. Look over the tables and figures.
e. Answer these questions.
   1. What was the study about?
   2. What methods were used?
   3. What conclusions were drawn?
f. The Strecker synthesis article does not include section headings; however, it loosely follows the traditional IMRD format. Reread the article and indicate where each of the section headings could be placed. What difficulties did you encounter in placing the Results and Discussion headings?

Exercise 2.15

Compare the changes that you made for exercise 2.9 with the original Liu et al. (2001) passage below. Were you able to shorten the passage to the 90 words achieved by the authors? To what extent was the original meaning of the passage preserved in your version?

After drying, the residue (CD-capped gold nanoparticles + compound **5**) was found to be insoluble in dry $CHCl_3$, but the solubility was restored when the chloroform was equilibrated with water. This finding clearly demonstrates that some water is necessary for the efficient phase transfer of the nanoparticles into $CHCl_3$. Thus, the proposed idealized structure of the nanoparticles after their transfer to the chloroform phase (Scheme 2) has some similarities with the structure of reverse micelles. We conclude that these nanoparticle-centered assemblies are conceptually similar to gold-filled reverse micelles. (90 words)

Exercise 2.16

Revise the following sentences so that they are more concise (and more professional). If you need some help, consult the "Audience and Purpose" section above and the "Concise Writing" language tip in appendix A.

a. The energies of the associated LMCT transitions would be above the energy range that would appear to be accessible here (as shown in Table 6); as a consequence of this fact, all of the transitions that have been observed must be due to the axial (z-polarization) or equatorial (y-polarization) tyrosinates. (Adapted from Davis et al., 2002)

b. The present results are in agreement with the results that have been obtained from other spectroscopic studies that have been conducted on n-alkyl modified stationary phases.[27] (Adapted from Singh et al., 2002)

c. The reaction mixture was stirred for 1 h while allowing it to reach room temperature during that time.

Exercise 2.17

Reflect on what you have learned from this chapter. Select one of these reflection tasks and write a thoughtful and thorough response:

a. Reflect on the level of professionalism in published journal articles.
   - What are the predominant characteristics of published journal articles that make them appear so professional?
   - What aspects of this professional writing will you try to emulate?

b. Reflect on your own writing abilities.
   - What aspects of your writing will you need to improve to move toward more professional writing?
   - How will you go about making these improvements?
   - What do you think will be most challenging about learning to write for expert audiences?

c. Reflect on your reading habits.
   - Based on what you've learned in this chapter, how might you change the way in which you approach journal articles in the future to improve (1) your understanding of the articles and (2) your writing abilities?

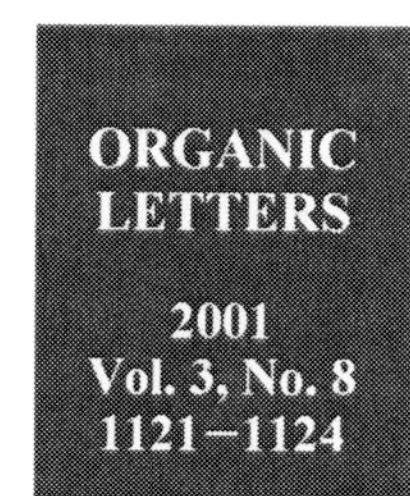

Asymmetric Strecker Synthesis of α-Amino Acids via a Crystallization-Induced Asymmetric Transformation Using (*R*)-Phenylglycine Amide as Chiral Auxiliary

Wilhelmus H. J. Boesten,† Jean-Paul G. Seerden,‡ Ben de Lange,*,† Hubertus J. A. Dielemans,† Henk L. M. Elsenberg,† Bernard Kaptein,† Harold M. Moody,† Richard M. Kellogg,‡ and Quirinus B. Broxterman*,†

DSM Research Life Sciences-Organic Chemistry and Biocatalysis, P.O. Box 18, 6160 MD Geleen, and Syncom B.V., Kadijk 3, 9747 AT Groningen, The Netherlands

rinus.broxterman@dsm-group.com

Received December 22, 2000 *(Revised Manuscript Received March 15, 2001)*

ABSTRACT

a) R_1R_2CO (*in situ*); b) NaCN, AcOH; solvent, time, temp

76-93% yield, dr > 99/1

3 steps

73% yield, > 98% ee

Diastereoselective Strecker reactions based on (*R*)-phenylglycine amide as chiral auxiliary are reported. The Strecker reaction is accompanied by an in situ crystallization-induced asymmetric transformation, whereby one diastereomer selectively precipitates and can be isolated in 76−93% yield and dr > 99/1. The diastereomerically pure α-amino nitrile obtained from pivaldehyde (R_1 = *t*-Bu, R_2 = H) was converted in three steps to (*S*)-*tert*-leucine in 73% yield and >98% ee.

The asymmetric synthesis of α-amino acids and derivatives is an important topic as a result of their extensive use in pharmaceuticals and agrochemicals and as chiral ligands. Many highly enantioselective approaches have been reported.[1] Industrial production of α-amino acids via the Strecker reaction is historically one of the most versatile methods to obtain these compounds in a cost-effective manner, making use of inexpensive and easily accessible starting materials.[2] The Strecker reaction is usually followed by resolution of the racemic amino acid or amino acid amide obtained after hydrolysis of the amino nitrile.[3] Either process leads to a maximum yield of 50% if the unwanted enantiomer is not racemized. In principle, asymmetric synthesis ap-

† DSM Research Life Sciences-Organic Chemistry & Biocatalysis. E-mail for Ben de Lange: Ben-B.Lange-de@dsm-group.com.
‡ Syncom B.V.

(1) (a) Calmes, M.; Daunis, J. *Amino Acids* **1999**, *16*, 215. (b) Cativiela, C.; Diaz-de-Villegas, M. D. *Tetrahedron: Asymmetry* **1998**, *9*, 3517 (c) Williams, R. M. *Synthesis of Optically Active α-Amino Acids*; Pergamon: Oxford, 1989.

(2) (a) Kunz, H. In *Houben-Weyl: Stereoselective Synthesis*; Helmchen, G., Hoffmann, R. W., Mulzer, J., Schaumann, E., Eds.; Thieme Verlag: Stuttgart, 1995, 1931; Vol. E 21, D.1.4.4. (b) Shafran, Y. M.; Bakulev, V. A.; Mokrushin, V. S. *Russ. Chem. Rev.* **1989**, *58*, 148 (c) Strecker, A. *Ann. Chem. Pharm.* **1850**, *75*, 27.

(3) For reviews, see: resolution by aminoamidases. (a) Sonke, T.; Kaptein, B.; Boesten, W. H. J.; Broxterman, Q. B.; Schoemaker, H. E.; Kamphuis, J.; Formaggio, F.; Toniolo, C.; Rutjes, F. P. J. T. In *Stereoselective Biocatalysis*; Patel, R., Ed.; Dekker: New York, 2000; Chapter 2. (b) Resolution by acylases. Drauz, K.; Waldmann, H. In *Enzyme Catalysis in Organic Synthesis*; VCH: Weinheim, 1995; Vol. 1, p 393.

10.1021/ol007042c CCC: $20.00
Published on Web 03/30/2001

Table 1. Asymmetric Strecker Reactions of (*R*)-Phenylglycine Amide **1** and Pivaldehyde **2**

(*R*)-**1** + **2** —(NaCN, HOAc; solvent, time, temp)→ (*R,S*)-**3** + (*R,R*)-**3**

entry	sovent	temp (°C)	time (h)	yield (%)[a]	dr (*R,S*)-**3**/(*R,R*)-**3**[b]
1	MeOH	rt	20	80	65/35
2	MeOH/2-PrOH, 1/9[c]	rt	22	51	99/1
3	2-PrOH	rt	22	84	88/12
4	2-PrOH/*t*-BuOH, 4/1[c]	rt	20	65	96/4
5	MeOH/H_2O, 35/1[c]	rt	20	69	81/19
6	H_2O	55	24	81	85/15
7	H_2O	60	24	84	96/4
8	H_2O	65	24	84	98/2
9	H_2O	70	24	93	>99/1

[a] Isolated yield after: evaporation of the solvent (entry 1) or filtration of precipitated amino nitrile **3** (entries 2–9). [b] The dr was determined by ^{1}H NMR spectroscopy. [c] Ratio in volume/volume.

proaches that lead to a maximum yield of 100% of a single enantiomer are more advantageous.

Recently several catalytic asymmetric Strecker reactions leading to N-protected amino nitriles in high ee's and high yields have been published.[4] Alternatively, diastereoselective Strecker syntheses using a broad variety of chiral inducing agents, like α-arylethylamines,[5] β-amino alcohols and derivatives,[6] amino diols,[7] sugar derivatives,[8] and sulfinates[9] have been reported to provide the α-amino nitriles with varying diastereoselectivities. A major drawback of these chiral auxiliaries can be cost and/or availability, because they are used in stoichiometric amounts and in principle lost during the conversion. Furthermore, in many cases the α-amino nitriles need to be purified in a separate step to obtain diastereomerically pure compounds. Purification requires, for example, crystallization or chromatography, which may lead to losses. An interesting solution to these problems would be a crystallization-induced asymmetric transformation,[10,11] in which one diastereomer precipitates and the other epimerizes in solution via the corresponding imine. This would lead both to high yield and high diastereoselectivity in a practical one-pot procedure.

Recently, optically pure (*R*)-phenylglycine amide **1** became readily accessible as a result of application on an industrial scale as key intermediate in the enzymatic synthesis of β-lactam antibiotics.[12] Either aminopeptidase-catalyzed hydrolysis of racemic phenylglycine amide[3] or asymmetric transformation of racemic phenylglycine amide with (*S*)-mandelic acid as resolving agent[13] can be used to prepare **1**. Because of its ready availability on a large scale and its anticipated easy removal via catalytic hydrogenolysis, we decided to investigate the application of (*R*)-phenylglycine amide **1** as chiral auxiliary in asymmetric synthesis.

In this paper, the first two examples of the use of (*R*)-phenylglycine amide in asymmetric Strecker reactions are presented. Pivaldehyde and 3,4-dimethoxyphenylacetone have been used as starting materials, which lead, respectively, to enantiomerically enriched *tert*-leucine and α-methyl-dopa, two important nonproteogenic α-amino acids for pharmaceutical applications. In addition, *tert*-leucine has considerable utility as a chiral building block.[14]

The asymmetric Strecker reaction of (*R*)-phenylglycine amide **1**, pivaldehyde **2** and HCN generated in situ from NaCN and AcOH was studied (Table 1). Amino nitriles (*R,S*)-**3** and (*R,R*)-**3** were obtained in 80% yield in a ratio of 65:35 by stirring an equimolar mixture of **1** (as AcOH salt)

(4) (a) Sigman, M. S.; Vachal, P.; Jacobsen, E. N. *Angew. Chem., Int. Ed. Engl.* **2000**, *39*, 1279. (b) Porter, J. R.; Wirschun, W. G.; Kuntz, K. W.; Snapper, M. L.; Hoveyda, A. H. *J. Am. Chem. Soc.* **2000**, *122*, 2657. (c) Ishitani, H.; Komiyama, S.; Hasegawa, Y.; Kobayashi, S. *J. Am. Chem. Soc.* **2000**, *122*, 762. (d) Vachal, P.; Jacobsen, E. N. *Org. Lett.* **2000**, *2*, 867. (e) Corey, E. J.; Grogan, M. *Org. Lett.* **1999**, *1*, 157.

(5) (a) Vincent, S. P.; Schleyer, A.; Wong, C.-H. *J. Org. Chem.* **2000**, *65*, 4440. (b) Wede, J.; Volk, F.-J.; Frahn, A. W. *Tetrahedron: Asymmetry* **2000**, *11*, 3231. (c) Juaristi, E.; Leon-Romo, J. L.; Reyes, A.; Escalante, J. *Tetrahedron: Asymmetry* **1999**, *10*, 2441. (d) Speelman, J. C.; Talma, A. G.; Kellogg, R. M.; Meetsma, A.; de Boer, A.; Beurskens, P. T.; Bosman, W. P. *J. Org. Chem.* **1989**, *54*, 1055. (e) Stout, D. M.; Black, L. A.; Matier, W. L. *J. Org. Chem.* **1983**, *48*, 5369.

(6) (a) Dave, R. H.; Hosangadi, B. D. *Tetrahedron* **1999**, *55*, 11295. (b) Ma, D.; Tian, H.; Zou, G. *J. Org. Chem.* **1999**, *64*, 120. (c) Chakraborty, T. K.; Hussain, K. A.; Reddy, G. V. *Tetrahedron* **1995**, *51*, 9179.

(7) Weinges, K.; Brachmann, H.; Stahnecker, P.; Rodewald, H.; Nixdorf, M.; Imgarter H. *Liebigs Ann. Chem.* **1985**, 566.

(8) Kunz, H.; Sager, W.; Schanzenbach D.; Decker, M. *Liebigs Ann. Chem.* **1991**, 649.

(9) Davis, F. A.; Fanelli, D. L. *J. Org. Chem.* **1998**, *63*, 1981.

(10) Only very few examples of crystallization-induced asymmetric transformations in Strecker reactions have been reported based on arylalkyl-methyl ketones: (a) Weinges, K.; Gries, K.; Stemmle, B.; Schrank, W. *Chem. Ber.* **1977**, *110*, 2098. (b) Weinges, K.; Klotz, K.-P.; Droste, H. *Chem. Ber.* **1980**, *113*, 710.

(11) For a broad discussion of crystallization-induced asymmetric transformation, see: Vedejs, E.; Chapman, R. W.; Lin, S.; Muller, M.; Powell, D. R. *J. Am. Chem. Soc.* **2000**, *122*, 3047 and references therein.

(12) Bruggink, A.; Roos, E. S.; de Vroom, E. *Org. Process Res. Dev.* **1998**, *2*, 128.

(13) (a) Boesten, W. H. J. European Patent Appl. EP 442584, 1991 (Chem. Abstr. **1992**, *116*, 42062r). (b) Boesten, W. H. J. European Patent Appl. EP 442585, 1991 (Chem Abstr. **1992**, *116*: 42063s).

(14) Bommarius, A. S.; Schwarm, M.; Stingl, K.; Kottenhahn, M.; Huthmacher, K.; Drauz, K. *Tetrahedron: Asymmetry* **1995**, *6*, 2851.

with **2** and NaCN in MeOH overnight at room temperature, followed by evaporation of the solvent (entry 1). The diastereomeric ratio of (*R,S*)-**3** and (*R,R*)-**3** was determined by ^{1}H NMR on the basis of the relative integration between the *t*-Bu signals at 1.05 ppm for (*R,S*)-**3** and 1.15 ppm for (*R,R*)-**3**. The assignments have been made on the basis of the absolute configuration as established by X-ray analysis and conversion to (*S*)-*tert*-leucine (vide infra).

Because in methanol crystallization of amino nitrile **3** did not take place, first the solvent was varied in order to attempt to find conditions for a crystallization-induced asymmetric transformation. At a MeOH/2-PrOH ratio of 1/9 amino nitrile (*R,S*)-**3** was isolated in 51% yield and dr 99/1 (entry 2). Other combinations of alcoholic solvents failed to lead to a higher yield of precipitated (*R,S*)-**3** in high dr (entries 3 and 4). On further screening of solvents it was observed that upon addition of H_2O to the methanol solution selective precipitation of amino nitrile (*R,S*)-**3** occurred giving (*R,S*)-**3** and (*R,R*)-**3** in a ratio of 81:19 and 69% yield (entry 5). The asymmetric Strecker reaction was further studied in H_2O alone using temperature as a variable. The results of these experiments are given in Table 1 (entries 6−9). After addition of NaCN/AcOH at 23−28 °C to (*R*)-phenylglycine amide **1** and pivaldehyde **2** in H_2O, the mixture was heated to the indicated temperatures.

After approximately 24 h of stirring, the mixture was cooled to 30 °C and the precipitated amino nitrile filtered and analyzed by ^{1}H NMR to determine the dr. The results in Table 1 show that optimal results were achieved after 24 h of stirring in water at 70 °C. The amino nitrile (*R,S*)-**3** was obtained in 93% yield and a dr > 99/1 via a crystallization-induced asymmetric transformation (entry 6). At lower temperatures the epimerization reaction is slower.[15]

The crystallization-induced asymmetric transformation in water at 70 °C is verified further by the observed increase of the dr of (*R,S*)-**3** as a function of the reaction time (Figure 1). After 30 h the precipitated (*R,S*)-**3** was obtained with a dr > 99/1.

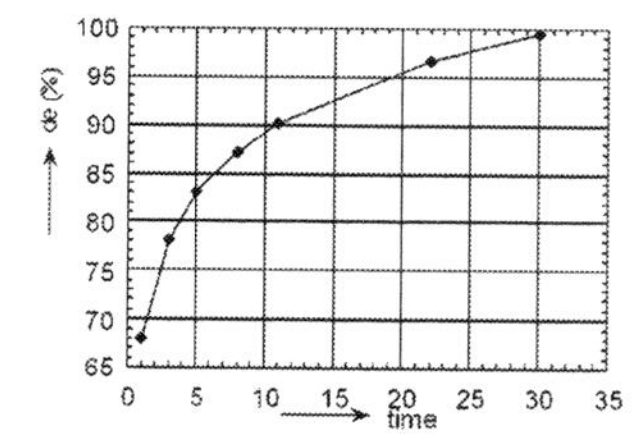

Figure 1. Crystallization-induced asymmetric transformation of amino nitrile **3** in water at 70 °C.

The observed diastereoselectivity in the asymmetric Strecker step via the crystallization-induced asymmetric transformation can be explained as shown in Figure 2. Apparently, the *re*-face addition of CN⁻ to the intermediate imine **4** is preferred at room temperature in methanol and results in a dr 65/35. At elevated temperatures in water the diastereomeric outcome and yield of the process is controlled by the reversible reaction of the amino nitriles **3** to the intermediate imine and by the difference in solubilities of both diastereomers under the applied conditions.[16,17]

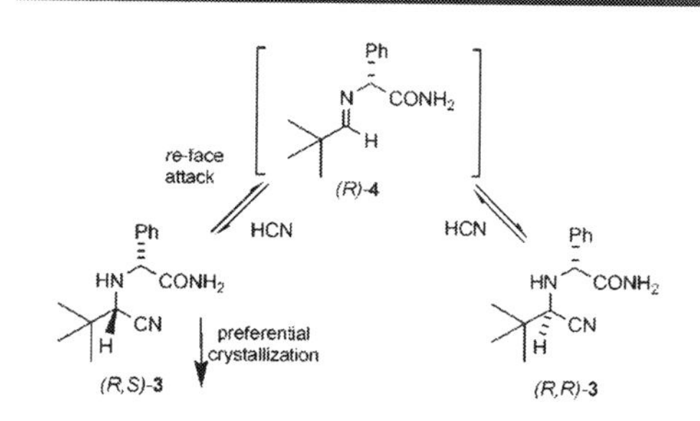

Figure 2. Crystallization-induced asymmetric transformation of amino nitrile **3**.

The absolute configuration of amino nitrile (*R,S*)-**3** was confirmed by X-ray analysis as shown in Figure 3[18] and by conversion to (*S*)-*tert*-leucine.

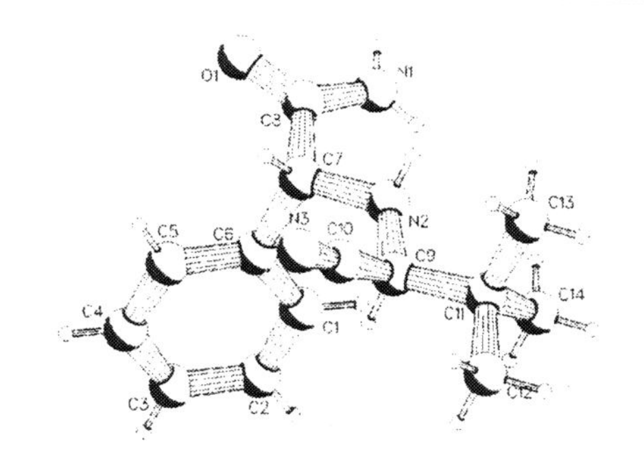

Figure 3. X-ray structure of amino nitrile (*R,S*)-**3**.

Conversion of the amino nitrile (*R,S*)-**3** to (*S*)-*tert*-leucine **7** was accomplished via the reaction sequence shown in Scheme 1. Hydrolysis of (*R,S*)-**3** to the diamide (*R,S*)-**5**

(15) At higher temperatures, lower yields of product were found, probably by degradation of amino nitrile.

(16) For example, in the case of phenylacetone (not illustrated) it was found that in solution the initially formed minor isomer preferentially precipitated under crystallization conditions.

(17) For a discussion of asymmetric transformation of α-amino nitriles with mandelic acid, see: Hassan, N. A.; Bayer, E.; Jochims, J. C. *J. Chem. Soc., Perkin Trans. 1* **1998**, 3747.

(18) The crystal structure of (*R,S*)-**3** has been deposited at the Cambridge Crystallographic Data Center and allocated the deposition number CCDC 154034.

Scheme 1

(*R,S*)-3 dr > 99/1 → H_2SO_4 (96%), CH_2Cl_2, 15°C → 40°C, 2h, 94% yield → (*R,S*)-5

→ H_2/Pd/C, EtOH, 20h, 90% yield → (*S*)-6 → 6N HCl, 100°C, 24 h, 86% yield → (*S*)-7 >98% ee

Scheme 2

8 → (*R*)-1.HCl, NaCN, MeOH/H_2O 6/1, RT, 96 h, 76% yield → (*R,S*)-9 dr 99/1

proceeded smoothly in concentrated H_2SO_4 in high yield and without racemization.

Removal of the phenylacetamide group under 2 atm of H_2 with catalytic Pd/C afforded (*S*)-*tert*-leucine amide **6** in 90% yield. Finally, hydrolysis of the amide was accomplished by heating in 6 N HCl at 100 °C to give (*S*)-*tert*-leucine **7** in 86% yield and >98% ee. The absolute configuration assignment, (*S*), was made by comparison with an authentic sample.[3] Obviously, other routes to convert the amino nitrile derivatives to the amino acid can be envisaged and are under investigation.

The crystallization-induced asymmetric transformation, using (*R*)-phenylglycine amide **1** as chiral auxiliary in diastereoselective Strecker reactions, was further explored with 3,4-dimethoxyphenylacetone **8** (Scheme 2).

The optimized asymmetric Strecker reaction of (*R*)-phenylglycine amide **1** (used as HCl salt) and an equimolar amount of 3,4-dimethoxyphenylacetone **8** in MeOH/H_2O (6/1 v/v) gave, after addition of NaCN (30% aqueous solution) and stirring for 96 h at room temperature, the nearly diastereomerically pure (dr > 99/1) amino nitrile **9** as a solid in 76% isolated yield. The dr could easily be determined by ^1H NMR analysis. It was found that in solution at room temperature an equilibrium of 55:45 exists between the two diastereomers (*R,S*)-**9** and (*R,R*)-**9**. Clearly, again a crystallization-induced asymmetric transformation has occurred.

In summary, (*R*)-phenylglycine amide **1** is an excellent chiral auxiliary in the asymmetric Strecker reaction with pivaldehyde or 3,4-dimethoxyphenylacetone. Nearly diastereomerically pure amino nitriles can be obtained via a crystallization-induced asymmetric transformation in water or water/methanol. This practical one-pot asymmetric Strecker synthesis of (*R,S*)-**3** in water leads to the straightforward synthesis of (*S*)-*tert*-leucine **7**. Since (*S*)-phenylglycine amide is also available, this can be used if the other enantiomer of a target molecule is required. More examples are currently under investigation to extend the scope of this procedure.[19]

Acknowledgment. Mr. A. Meetsma of the department of crystallography of the University of Groningen is acknowledged for the X-ray structure of (*R,S*)-**3**.

Supporting Information Available: Procedures and characterization data of all compounds. This material is available free of charge via the Internet at http://pubs.acs.org.

OL007042C

(19) Several other amino nitriles could be obtained as crystalline materials from H_2O/MeOH mixtures, e.g., R_1 = ^{i}Pr, R_2 = H; R_1 = Ph, R_2 = Me; R_1 = ^{i}Pr, R_2 = Me. Conditions are being sought to obtain also a crystallization-induced asymmetric transformation in these cases.

3 *Writing the Methods Section*

Tell me how you did it, but be concise. A long-winded step-by-step Methods section sidetracks and irritates the expert reader.

—Betty H. Stewart, Midwestern State University

In this chapter, we focus on writing a Methods section for a journal-quality paper. We begin with the Methods section because this is the section that many chemists write first, in part because this section describes what they know best: the procedures they have repeated (many times) to conduct their work. Moreover, most research groups use similar methodologies for several years; hence, previously written Methods sections can serve as models for writing new Methods sections. Together, these factors make the Methods section one of the easier sections to write and an excellent place to begin our writing instruction. By the end of this chapter, you will be able to do the following:

- Know how to address the intended audience of a Methods section
- Recognize which details to include and exclude from a Methods section
- Organize a Methods section following standard moves
- Use capitalization, abbreviations, and parentheses appropriately
- Format numbers and units correctly
- Use verb tense and voice in conventional manners

As you work through the chapter, you will write a Methods section for your own paper. The Writing on Your Own tasks throughout the chapter will guide you step by step as you do the following:

3A Read the literature

3B Describe materials

3C Describe experimental methods

3D Describe numerical methods

3E Practice peer review

3F Fine-tune your Methods section

The purpose of the Methods section is to address *how* a particular work was conducted. Relevant information about instrumentation and experimental and/or numerical procedures is described. The goal is to describe the information in enough detail that an expert (not a novice) could repeat the work. Usually, this section is formally called, for example, **Materials and Methods** or **Experimental Section**, but for brevity, we call it simply the Methods section.

Methods Section

What we call the Methods section is given a more formal name in journal articles, such as

- Materials and Methods
- Experimental Section
- Experimental Methods

Many of you have written a Methods section previously for a college-level chemistry course. Thus, we begin with an exercise to test your current knowledge about writing a Methods section.

Exercise 3.1

What have you learned about writing Methods sections from other writing courses and labs? Let's evaluate your knowledge with the following pretest. In column 1, place a "Yes" next to those items that you think should be included in a Methods section of a journal article. Place a "No" next to items that you think should be omitted. Use a question mark (?) if you are unsure. (You will use column 2 to repeat this exercise at the end of the chapter.)

1	2	Possible Items for a Methods Section
___	___	A table of chemicals with their physical properties (e.g., mp, MW, ρ)
___	___	Amounts of reagents used in a synthesis (e.g., mg, mmol)
___	___	Directions for preparing a stock solution
___	___	The quality (grade) of chemicals used
___	___	The names and locations of chemical vendors
___	___	The brand names of commercial instrumentation used
___	___	A diagram of a distillation apparatus
___	___	An illustration of a novel or custom-built apparatus
___	___	Operating conditions for a gas chromatograph

___	___	Equations used to calculate percent yield or dilution ratios
___	___	A list of disposable equipment (e.g., rubber gloves, Bunsen burners)
___	___	Step-by-step instructions of the procedure
___	___	Warnings to other scientists about unusual hazards
___	___	Quantitative statements of reaction times and temperatures
___	___	Descriptions of the physical appearances of synthesis products
___	___	IR or NMR data confirming product purity
___	___	Statistical packages used (including the name of the software)
___	___	Reports of other software used to keep track of data (e.g., Excel)

Reading and Analyzing Writing

We formally begin this chapter by asking you to read and analyze a Methods section taken from an article in the *Journal of Agricultural and Food Chemistry* pertaining to the chemical analysis of beer. (You will eventually read the entire article, section by section, throughout the module.) The study involves international collaboration (researchers from the Miller Brewing Company in the United States and the Institute of Chemical Technology in Prague, Czech Republic) and employs one familiar analytical technique, gas chromatography/mass spectrometry (GC/MS), and another worth learning more about, solid-phase microextraction (SPME). The study focuses on aldehydes in beer because of their potential role in affecting beer flavor and aroma. The authors developed a novel technique involving SPME to measure low-level concentrations of aldehyde in beer during storage.

Exercise 3.2

Read excerpt 3A below (a Methods section formally labeled Materials and Methods). Consider the audience, organization, writing conventions, and grammar and mechanics used in the excerpt as you answer the following questions:

a. Who is the authors' intended audience? How is this audience reflected in level of detail, level of formality, and word choice?

b. How do the authors make their writing concise?

c. How do the authors organize their information?

d. What writing conventions do you notice (i.e., formatting, abbreviations, verb tense, voice)?

e. Which passages correctly illustrate examples of parallelism (e.g., among section headings), the use of commas, and subject–verb agreement?

f. What instrumentation did the authors use, and how is it described?

g. Based on your response to exercise 3.1, what information did you expect to be in this Methods section that isn't included?

Excerpt 3A (adapted from Vesely et al., 2003)

Materials and Methods

Chemicals. The carbonyl compound standards 2-methylpropanal, 2-methylbutanal, 3-methylbutanal, pentanal, hexanal, furfural, methional, phenylacetaldehyde, and (*E*)-2-nonenal were purchased from Sigma-Aldrich (Milwaukee, WI). A stock solution containing a mixture of the standard compounds in ethanol was prepared daily in a concentration of 100 ppb each. An aqueous solution of the derivatization agent *O*-(2,3,4,5,6-pentafluorobenzyl)-hydroxylamine (PFBOA) (Sigma-Aldrich, Milwaukee, WI) was prepared at a concentration of 6 g/L. PFBOA solution was prepared every 3 months and kept refrigerated.

Beer Samples. American lager beer samples used for the aldehyde analysis were stored at 30 °C for 4, 8, or 12 weeks. Control samples were stored for 12 weeks at 0 °C.

The SO_2 level of the fresh beer was 3.4 ppm, a low level for beer. Knowledge of the SO_2 level in beer is important because SO_2 complexes with aldehydes and only "free" aldehydes are measured by the described method.

SPME Fiber. A 65 µm poly(dimethylsiloxane)/divinyl benzene (PDMS/DVB) fiber coating (Supelco, Bellefonte, PA) was used in this method. This fiber coating was selected for its ability to retain the derivatizing agent and for its affinity for the PFBOA-aldehyde oxime (7).

Derivatization Procedure. One hundred microliters of PFBOA solution and 10 mL of deionized water were placed in a 20 mL glass vial and sealed with a magnetic crimp cap (Gerstel, Baltimore, MD). Initially, the PDMS/DVB SPME fiber was placed in the headspace of the PFBOA solution for 10 min at 50 °C. The SPME fiber loaded with PFBOA was exposed to the headspace of 10 mL of beer placed in a 20 mL glass vial. Different derivatization times and temperatures as well as salt addition were tested in order to obtain the best results. To ensure the reproducibility of the method, an automated process using an MPS2 autosampler (Gerstel, Baltimore, MD) was employed.

GC Conditions. Aldehyde derivatives were analyzed using a HP6890 gas chromatograph equipped with a mass-selective detector (5972A, Agilent Technologies, Palo Alto, CA) and fitted with a DB-5 capillary column, 30 m × 0.25 mm × 0.50 µm (J&W Scientific, Folsom, CA). Helium was the carrier gas at a flow rate of 1.1 mL/min. The front inlet temperature was 250 °C. The injection was in the splitless mode with the purge valve set at 30 s. The oven temperature program used was 40 °C for 2 min, followed by an increase of 10 °C/min to 140 °C and 7 °C/min to 250 °C. The final temperature was held for 3 min.

Analyzing Audience and Purpose

The major purpose of a Methods section is to describe, for other chemists (the audience), the procedures that were used to obtain the results presented in the article. A well-written Methods section serves as a resource for expert chemists who wish to (1) develop similar procedures, (2) compare their own procedures with those presented, or (3) familiarize themselves with procedures in a branch of chemistry other than their own.

Because Methods sections of journal articles are written largely for experts, they are not at all like Methods sections of chemistry lab reports. Lab reports are written largely for instructional purposes, to reinforce new techniques and help students carry out experiments successfully. As such, they tend to include details (e.g., lists of equipment, safety precautions, and step-by-step directions) not needed (or wanted) by expert readers. Leaving out such details makes the writing more concise. Concise writing is important because, unlike a single lab experiment, methods sections in journal articles describe multifaceted works that took months or years to complete. Similarly, lab reports often include language inappropriate for journal articles. For example, in the Methods section of a lab report, a student might write "Stir mixture. Heat to reflux." In a journal article, however, this would be restated in past tense and passive voice as "The mixture was stirred and heated to reflux," making the writing more formal.

Experiment vs. Work

Because of the multifaceted nature of a research project, authors seldom use the word *experiment*, which connotes a simplistic activity that could be accomplished in a single day. Words such as *project* or *work* are preferred.

To see for yourself how chemists write Methods sections for journal articles, we have included excerpts from the published literature throughout this chapter. These excerpts illustrate appropriate levels of detail, formality, and conciseness when writing for expert chemists. We encourage you to use these excerpts (rather than lab reports or lab manuals) as models for your writing.

Exercise 3.3

Compare the following excerpts that describe the process of recrystallization for product purification. The first excerpt is adapted from an undergraduate laboratory experiment involving the extraction of caffeine from tea leaves. The last two, written for expert audiences, are taken from articles in *The Journal of Organic*

Chemistry. What details are included for the expert audience? What details are excluded?

a. The residue obtained in evaporation of the methylene chloride is next recrystallized by the mixed-solvent method. Using a steam bath or hotplate, dissolve the residue in a small quantity (about 2 mL) of hot acetone and add dropwise just enough low-boiling (bp 30–60 °C) petroleum ether to turn the solution faintly cloudy. Cool the solution and collect the crystalline product by vacuum filtration, using a small Buchner funnel. (From Pavia et al., 1998)

b. Further purification was obtained by recrystallization from ethyl acetate/hexane. (Adapted from Katritzky and Button, 2001)

c. **Bis(*p*-tolyl) Trisulfide (2b)**. Yield 97%; recrystallization from *n*-pentane at –15 °C afforded light yellow needles; mp 78–79 °C (lit. mp[24] 82–84 °C). (From Zysman-Colman and Harpp, 2003)

Analyzing Organization

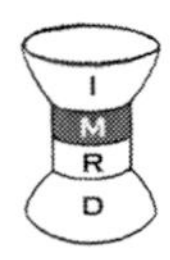

Most Methods sections follow a conventional organizational pattern. The pattern typically involves two or three separate steps, each of which corresponds to a move, as shown in figure 3.1. In accord with a common title for this section—Materials and Methods—the moves describe first the materials and then the methods (experimental and/or numerical) that were used in the work. Because these moves describe specific information, the Methods section is in the narrowest part of the IMRD hourglass structure.

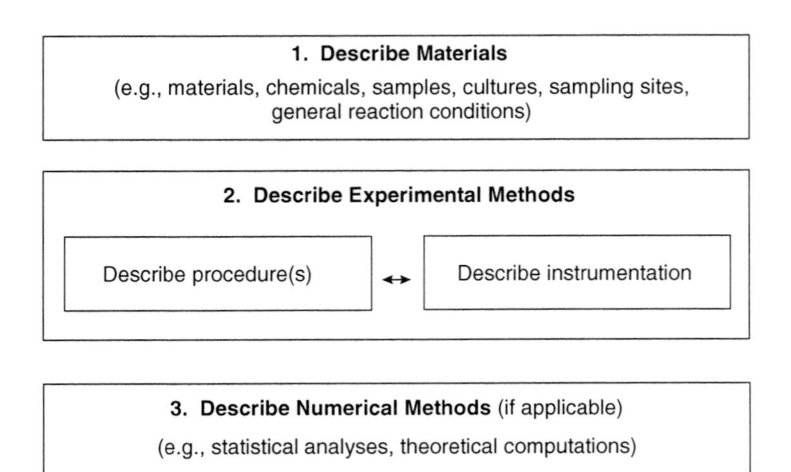

Figure 3.1 A visual representation of the move structure for a typical Methods section.

Move 1 tells readers about the materials used in the work. The term *materials* refers to chemicals (e.g., solvents and reagents), samples (e.g., soil, water, or food), biological media (e.g., bacterial cell cultures), and/or other tangible items used to conduct the work (e.g., the SPME fiber used in excerpt 3A). In field studies, the sampling site is also described in this move. When different types of materials are used in one work, subheadings (shown here in bolded font) are commonly used to set them off from one another (e.g., the subheadings **Chemicals**, **Beer Samples**, and **SPME Fibers** were used in excerpt 3A). Subheadings commonly used in move 1 include the following: **Reagents and Materials**, **Samples**, **Cell Cultures**, and **Site Description**.

Move 1 is approached slightly differently in synthesis papers (e.g., articles published in *The Journal of Organic Chemistry*). Such papers typically describe a series of related reactions, often totaling 10 or more individual syntheses. Rather than describe all of the chemicals used for these many reactions at the start of the Methods section, authors instead include only general information in move 1 (e.g., "All NMR studies were performed on a 500 mHZ instrument."). A common subheading for this move is **General**. Information about specific reagents and materials are included in move 2, where the individual syntheses are described.

In move 2, Describe Experimental Methods (figure 3.1), authors describe how they obtained their data. The move involves two submoves. The first submove, describe procedures, includes analytical procedures (e.g., the steps used to prepare, extract, concentrate, and/or derivatize a sample), field-collection procedures (e.g., the steps used to collect water samples from a polluted lake), and synthetic procedures (e.g., the steps used to synthesize target compounds), to name only a few. In some journals (particularly those describing analytical procedures), this submove also includes procedures used to ensure the accuracy and precision of the work. Such procedures are described as quality assurance/quality control (**QA/QC**).

QA/QC

This abbreviation refers to quality assurance/quality control. QA/QC procedures are standardized methods used to verify the quality (accuracy and precision) of data.

The second submove, describe instrumentation, describes the scientific apparatus used in the study. Both custom-built instruments (e.g., a high-vacuum chamber or a newly designed light source) and commercially available instruments (e.g., a gas chromatograph or an infrared spectrometer) are described. Ordinary lab equipment (e.g., a heating mantel or a rotary evaporator) is not described.

The submoves in move 2 are placed side by side in the move structure in figure 3.1 to indicate that authors may present the submoves in either order, to parallel the sequence of events in their study. For example, in excerpt 3A, the derivatization procedure is described before the GC instrumentation because the derivatization step was completed before the GC analysis. Alternatively, authors who employ instrumental analysis early on in the study may describe instrumentation first.

Subheadings are also common in move 2. As in move 1, subheadings help organize a paper that uses multiple methods. They also assist readers in quickly locating a method of interest. (Scientists often read papers selectively, looking only for information about a particular procedure or instrument.) The subheadings are often quite specific and include a name or description of the procedure (e.g., **HSSPME Extraction Procedure**, or **Fabrication of DNA Microarrays**). In synthesis papers, subheadings often name the compound synthesized (e.g., **2-(*p*-Toluenesulfonyl)-4′-methoxyacetophenone**). Similarly, for instrumentation, the subheadings specify the type of instrument (e.g., **FTIR-Raman Measurements**, **MALDI-TOF Mass Spectrometry**, or **Chromatographic Conditions**).

The last move of the Methods section, Describe Numerical Methods, is included only if numerical or mathematical procedures (e.g., statistical analyses) were used to analyze, derive, or model data presented in the paper. In such cases, the experimental methods are described first (move 2), and the numerical methods are described last (move 3). Subheadings used to demark move 3 include **Statistical Methods** or **Data Analysis**.

Exercise 3.4

Consider the following sets of subheadings as you perform the following tasks:

a. We have intentionally scrambled the order of these subheadings. Using the move structure in figure 3.1, arrange the subheadings in their correct order. (In some cases, more than one ordering is correct.)

b. Look carefully at the subheadings for formatting (e.g., bolding, italics), abbreviations and acronyms, capitalization, and punctuation. What do you notice?

Set 1 **Statistical Analysis.**
FTIR Measurements.
Starch Samples.
FT-Raman Measurements.

Set 2 **Time-of-Flight Mass Spectrometry (TOF).**
Chemicals.
SPE and Cleanup.
Sample Pretreatment.

Set 3 *Mutant Design.*
Structure Determination.
Enzyme Kinetics.
Statistics.

Exercise 3.5

Reconsider excerpt 3A in exercise 3.2. How well does the excerpt adhere to the move structure represented in figure 3.1? How do the authors use subheadings to help the reader locate the moves? Can you equate each subheading with one of the moves? Are any moves left out? If so, which one(s)?

Exercise 3.6

Examine how the move structure in figure 3.1 applies across different fields of chemistry.

a. Find two articles in *The Journal of Organic Chemistry* and two articles in *Organic Letters*. How are the methods presented in these articles? What subheadings are used? Do they adhere to the move structure suggested in figure 3.1? Explain.

b. Find two articles with the word "theoretical" or "computational" in their titles. How are the methods presented in these articles? What subheadings are used? Do they adhere to the move structure suggested in figure 3.1? Explain.

Upon completion of exercises 3.5 and 3.6, you probably noticed that not all written works in journals strictly adhere to the move structure in figure 3.1. Not surprisingly, the move structure does not apply to genres intended for a more general audience (e.g., news alerts, book reviews, editorial remarks), nor does it apply to all research-related works. For example, research articles published in *Organic Letters* omit a Methods section entirely; instead, the procedures are published on the Internet as supporting information.

Because of these variations, we end this section with a cautionary note: Although the move structure in figure 3.1 presents a common and effective way to organize your Methods section, it will not apply in all situations. Move structures vary from journal to journal and article to article; hence, ultimately you must model your organizational structure after an article similar to the one that you plan to write.

3A Writing on Your Own: Read the Literature

Read and review the Methods sections of the journal articles that you collected during your literature search (see chapter 2). As you read these articles, pay attention to how the authors organized their methods and what information they included. How much detail is included in descriptions of materials, instrumentation, procedures, and numerical methods? Are subheadings used? Do they help you navigate the section?

What ideas do these articles give you about ways to write your own Methods section?

Analyzing Excerpts

With general audience and organizational considerations in mind, let's next examine excerpts of Methods sections from the chemistry literature in more depth. We analyze excerpts in two parts:

- In part 1, we analyze excerpts move by move, focusing on levels of detail, formality, and conciseness (including the noticeable absence of ordinal language), writing conventions (including capitalization, abbreviations, numbers, and units), and grammar and mechanics.
- In part 2, we analyze excerpts for the purpose of examining the Methods section as a whole, focusing on verb tense and voice.

Part 1: Analyzing Writing Move by Move

There is no question that, in every part of my career, clear and simple writing has helped me communicate what I wanted said. In research, I often have had comments about the clarity of my writing. I cannot but think that this makes a favorable impression on reviewers. Although clarity cannot supplant content, it certainly "encourages" reviewers, and then readers, to spend some time with our work. Surely, that is what we want when we bring something to press.

—Robert Damrauer, University of Colorado–Denver

Move 1: Describe Materials

The first move of the Methods section provides a description of chemicals, materials, and/or samples. Beginning writers often wonder what to include in this section (level of detail) and how these details should be presented (level of formality), both issues related to audience. With respect to detail, it is customary to report the name, purity, and vendor for all essential chemicals and materials used in the work. (Incidental chemicals, e.g., solvents used to clean glassware, need not be reported.) Similarly, for samples, both how and where the samples were collected should be described. With respect to formality, the journal article requires complete sentences. A common mistake is to use lists; although commonplace in lab reports, lists should be avoided in journal articles because they often are fragments rather than complete sentences. Consider the following incorrect and correct examples:

Incorrect	**Chemicals.** Solvents (Aldrich): 99.8% purity methanol, 97% purity ethanol, 99% purity 1-pentanol, 99% purity 1-hexanol.
Correct	**Chemicals.** Methanol (99.8% purity), methyl formate (97% purity), 1-pentanol (99% purity), and 1-hexanol (99% purity) were purchased from Aldrich.
Correct	**Chemicals.** The following solvents were purchased from Aldrich: methanol (99.8% purity), methyl formate (97% purity), 1-pentanol (99% purity), and 1-hexanol (99% purity).

In the correct examples, complete sentences were used (achieved by adding the words "were purchased from Aldrich"). Also, parentheses were used for solvent purity, making the solvent names easier to read.

Fragments

Fragments are incomplete sentences. When writing your journal-quality paper, be sure to use complete sentences. Do not use lists or fragments.

Exceptions include (1) titles, (2) section headings, (3) figure captions, and (4) table titles.

When more than one vendor is involved, chemicals are typically grouped by vendor. It is customary to include the location of vendors (city, country) the first time a vendor is mentioned. Typically, the location is reported in parentheses at the end of the sentence, as shown in excerpts 3B and 3C.

Excerpt 3B (from Llompart et al., 2001)

Experimental Section

Reagents and Materials. The PCB congeners, 2,4,4′-trichlorobiphenyl (PCB-28), 2,2′,5,5′-tetrachlorobiphenyl (PCB-52), 2,2′,4,5,5′-pentachlorobiphenyl (PCB-101), 2,3,3′,4,4′-pentachlorobiphenyl (PCB-105), 2,3′,4,4′,5-pentachlorobiphenyl (PCB-118), 2,2′,3,4,4′,5′-hexachlorobiphenyl (PCB-138), 2,2′,3,4,4′,5′-hexachlorobiphenyl (PCB-153), 2,3,3′,4,4′,5′-hexachlorobiphenyl (PCB-156), and 2,2′,3,4,4′,5,5′-heptachlorobiphenyl (PCB-180) (PCB numbering according to IUPAC) were supplied by Ultra Scientific (North Kingstown, RI). Isooctane, acetone, and sodium hydroxide were obtained from Merck (Mollet del Valles, Barcelona, Spain). All the solvents and reagents were analytical grade.

Excerpt 3C (adapted from Plaper et al., 2002)

Experimental Procedures

Media, Chemicals, and Bacterial Strains. Proteinase K, ethylenediamine-tetraacetic acid (EDTA), tris(hydroxymethyl)aminomethane (Tris), *o*-nitrophenyl-β-D-galactopyranoside (ONPG) powder, tRNA, agarose, 3-4,5 dimethylthiazol-2,5 diphenyl tetrazolium bromide (MTT), ICR-191, and ethidium bromide were from Sigma (St. Louis, MO). Polymyxin B sulfate, Triton X-100, and 1,4-dithio-DL-threitol (DTT) were from Fluka Chemie (Buchs, Switzerland). Bacto yeast and Bacto trypton, used in LB medium, were from Difco (Detroit, MI). Chromium chloride ($CrCl_3 \cdot 6H_2O$) and chromium nitrate ($Cr(NO_3)_3 \cdot 9H_2O$) were from Merck (Darmstadt, Germany). Chromium oxalate ($KCr(C_2O_4)_2 \cdot 3H_2O$) was from Aldrich (Milwaukee, WI). Gyrase was from TopoGEN (Columbus, OH), and pUC19 plasmid DNA was from Promega (Madison, WI). All Cr^{3+} solutions were made fresh daily and diluted as required in sterile doubly distilled water immediately prior to use.

Excerpt 3D illustrates two additional points regarding naming chemicals in move 1: First, if all chemicals used were of the same grade and from the same company, state only the grade and company, not the names. The names will be mentioned in move 2, when the procedures in which the chemicals were used are described. Second, only those chemicals that were used as received, required minimal preparation (e.g., distilling or degassing), or were prepared according to literature methods (which should be cited in the text) are mentioned. Chemicals that require more detailed preparation steps are described in move 2.

Excerpt 3D (adapted from Ahmed et al., 2002)

Experimental Section

Reagents and Materials. All the chemicals used were of high grade purity and were obtained from E. Merck (Darmstadt, Germany). The standard reference material for Li was L-SVEC in Li_2CO_3 form. The ion exchangers, both anion and cation types, were prepared in our laboratory as described elsewhere.[8] Pure and Na-free SiO_2 gels were prepared in our laboratory by refinement of the procedure reported in the literature.[9,10]

Chemicals that are used as reference standards are also commonly described in move 1. As with other chemicals, the name, purity, and vendor should be included. Moreover, the final concentration of any stock standard solution should be mentioned (in addition to any other dilute solutions prepared from the stock solution). Do not explain how a stock solution (or any solution) was prepared; simply state the final concentration and solvent (e.g., 100 μg/L in ethanol), as illustrated below and in excerpt 3E.

Incorrect The internal standard phenanthrene-d10 was prepared by adding 1 mL of a 2000 μg/mL solution to a 50 mL volumetric flask and diluting to the mark with hexane (final concentration 40 ng/μL).

Correct Phenanthrene-d10 was used as the internal standard (40 ng/μL in hexane).

Excerpt 3E (adapted from Aguilera et al., 2003)

Reagents. (a) Pesticide standards of acephate, bromopropylate, chlorpyrifos, chlorpyrifos-methyl, chlorothalonil, diazinon, dichlorvos, endosulfan I, endosulfan II, endosulfan sulfate, lindane, methamidophos, phosalone, procymidone, pyrazophos, triazophos, and vinclozoline (purity >98%) were supplied by Riedel de Haen (Seelze, Germany). For each pesticide, a stock standard solution (about 500 mg/L) was prepared in acetone. Spiking standard solution, containing 50 mg/L of each pesticide, was prepared in acetone from the stock standard solutions.

Exercise 3.7

Imagine that you prepared a stock standard solution of arsenic in your research project. You purchased an arsenic concentrated standard (1000 μg/mL) from Spex Industries in Hoboken, New Jersey. You prepared a 100 μg/mL stock standard solution by adding 10 mL of the concentrated standard to a 100 mL volumetric flask and diluting to the mark with deionized water. How would you report this information in the Methods section of a journal article?

Deionized ≠ DI

Many beginning writers ask if they can use DI for deionized water or DD for double-distilled water. Using the ACS Journals Search, we found that both phrases appear more commonly in written-out form than in abbreviated form.

Samples are treated much like chemicals (excerpts 3F–3H). Complete sentences are used, and information is shared about the sample source and selection process. If sample collection follows an established procedure (e.g., a U.S. EPA protocol), that should be noted in the text (see excerpt 3H).

Excerpt 3F (adapted from Ozen and Mauer, 2002)

Materials and Methods

Samples. Eleven hazelnut oils, 25 olive oils, and 7 other types of oil (canola, soybean, corn, sunflower, sesame, walnut, and peanut) were purchased from local grocery stores and Internet suppliers. For the adulteration studies, 10 olive oil and 10 hazelnut oil brands were randomly chosen from the samples purchased, and blends of olive oil and hazelnut oil were prepared by mixing these oils. The hazelnut oil blend was adulterated with sunflower oil at 2–10% (v/v), and the blend of extra-virgin olive oils was adulterated with the hazelnut oil blend at 5–50% (v/v). Infrared spectra of pure oil samples (25 virgin olive oils, 11 hazelnut oils, and canola, soybean, corn, sunflower, sesame, walnut, and peanut oils) and adulterated samples then were obtained.

Excerpt 3G (adapted from Kunert et al., 1999)

Materials and Methods

Sampling. The moss samples were taken from a 30- to 40-year-old spruce forest (Hoerner Bruch) southeast of Osnabruck (F.R.G.). Starting in September 1985, samples were taken regularly (usually at weekly intervals) from 50–100 individual plants of *P. formosum* on an area of approximately 1 m^2 and made up into a mixed sample. The individual samples were picked about 1 cm above the soil. These samples represent a period of time of 2–3 years, and the content of metals reflected a measure of the atmosphere deposition during that period (*32*). Gloves were worn during sampling, and only synthetic materials were used. Attention was given to pick the moss samples without soil contamination. Markert and Weckert (*33–35*) have given detailed descriptions of the moss *P. formosum* in this area, which means that a large amount of background information was available. These investigations showed that it was sufficient to measure lead isotope ratios in four regularly selected samples per year (spring, summer, autumn, and winter).

Genus and Species Names

Genus names, as formal names, should be capitalized and italicized. Species should not be capitalized, but they should be italicized. Genus names are spelled out in full at first mention and abbreviated thereafter.

Seasons

Seasons of the year (fall, spring, summer, winter) should not be capitalized.

Excerpt 3H (adapted from Dellinger et al., 2001)

Experimental Section

Sample Collection. Multiple samples of $PM_{2.5}$ that were used in the mechanistic studies were collected at the Louisiana Department of Environmental Quality station 0.1 mi from Interstate highway I-10 and 1.5 mi east of the junction of I-10 and I-12 in Baton Rouge, LA, using the U.S. EPA protocol RFPS-0498–117 and a Rupprecht & Patashnick Partisol-FRM model 2000 air sampler. Samples of $PM_{2.5}$ from the other four sites were furnished by the EPA. They were collected using URG MASS 400 samplers as part of a sampler evaluation program conducted at EPA's "supersites" for ambient air pollution research (32). All samples were collected over a 24 h period except for 2 and 5 day samples from the Baton Rouge site, which were utilized in the mechanistic studies.

Unabbreviated Units of Time

The following units of time are not abbreviated: day, week, month, and year.

Exercise 3.8

Look over excerpts 3B–3H. Propose rules that describe the appropriate use of capitalization, abbreviations, parentheses, sentence fragments, numbers (spelled out or in number form), and units.

As mentioned above, organic synthesis papers seldom describe the chemicals in the first move; instead, the move is used to summarize general reaction conditions. Two examples are given in excerpts 3I and 3J:

Excerpt 3I (from Swenson et al., 2002)

Experimental Section

General. All reactions were performed under nitrogen. ^{1}H NMR and ^{13}C NMR spectra were recorded in ppm (δ) on a 300 MHz instrument using TMS as internal standard. Elemental analyses were performed by Robertson Micolit Laboratories. Anhydrous THF, toluene, and *tert*-butyllithium in pentane (1.7 M) were purchased. Flash chromatography was performed with silica gel 60 (230–400 mesh). Melting points were determined and are uncorrected.

Excerpt 3J (adapted from Demko and Sharpless, 2001)

Experimental Section

General. All ^{1}H NMR spectra were taken on a Bruker AMX-400 spectrometer in DMSO-d_6 with DMSO as a standard at 2.50 ppm. All ^{13}C NMR spectra were taken on the same machine at 100 MHz in DMSO-d_6 with DMSO as a standard at 39.50 ppm, unless otherwise noted. All melting points were taken on a Thomas-Hoover Uni-melt melting point apparatus. Reagents were used unpurified, and deionized water was used as the solvent.

Move 1 of the Methods section, because it describes compounds and materials, is an excellent place to examine writing conventions regarding capitalization. As you read excerpts 3B–3J, you probably noticed that chemists are a bit picky about what is capitalized and what is not. Most novice writers know (correctly) to capitalize the following:

Molecular formulas	F^-, $CrCl_3$, H_2O
Vendors and brand names	Alltech, Ultra Scientific
Select abbreviations	THF, PCB, NMR, EPA, ACS
Absolute configuration (*R* and *S*)	(*R*)-glyceraldehyde

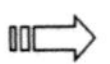

Capitalization

See appendix A.

Italics

Words (and their abbreviations) that indicate spatial orientation are italicized, for example, *cis*, *trans*, *o*, *m*, *p*, (*R*), and (*S*).

Except for (*R*) and (*S*), which are always uppercase, the others are always lowercase, even at the start of a sentence, for example,

o-Benzene...

trans-Butene...

(*S*)-Lactic acid...

However, novice writers also tend to capitalize words that should be written entirely in lowercase letters. For example, the names of molecules, compounds, and solvents are all written in lowercase:

The reaction of 6-octadecynoic acid (tariric acid) with potassium permanganate yields dodecanoic and 1,6-hexanedioic acid.

Acid-catalyzed dehydration of 2,2-dimethylcyclohexene in ether yields a mixture of 1,2-dimethylcyclohexene and isopropylidenecyclopentane.

The names of compounds are capitalized only at the start of a sentence, in which case only the first letter is capitalized. Even at the start of a sentence, words that signal stereochemistry—such as *ortho* (*o*), *meta* (*m*), *para* (*p*), *cis*, and *trans*—are lowercase (and italicized).

1,3-Dibromo-5-chlorobenzene was added to Br_2 and $FeBr_3$.

cis-2-Bromo-3-methylcyclohexane was added to (*R*)-lactic acid.

p-Toluenesulfonic acid reacts with NaOH and acid workup to form *p*-cresol.

Lowercase letters are also used for units derived from surnames when they appear in spelled-out form and do not follow a number. The abbreviated form, used as a unit after a number, is capitalized. Celsius and Fahrenheit are always capitalized because they refer to temperature scales (not actual units).

The temperature was measured in kelvins and converted to degrees Celsius.

The temperature was 298 K (25 °C).

Units of Measure from Surnames

Surnames that are used as units of measure, when not preceded by numbers, should be written in lowercase letters. For example,

ampere	joule
angstrom	kelvin
coulomb	newton
curie	ohm
einstein	pascal
hertz	watt

For more extensive guidelines, consult *The ACS Style Guide*.

Additional examples of appropriate and inappropriate uses of capitalization are shown in table 3.1.

Exercise 3.9

Reexamine excerpts 3B–3J. Which words are written in lowercase letters that you expected to be capitalized and vice versa?

In addition to capitalization, move 1 of the Methods section is also an excellent place to examine how chemists use abbreviations and acronyms in their writing. **Abbreviations** are short forms of words or phrases where each letter is often pronounced (e.g., DNA); acronyms are short forms of words or phrases that form pronounceable words (e.g., NASA). Both are common in chemical writing, in part because they make the writing more concise. A few abbreviations are so common that they can be used without ever introducing the full term (e.g., DNA, IR, NMR, UV, RNA). Most abbreviations, however, need to be defined before they can be used on their own; in such cases, abbreviations are placed in parentheses immediately *following* the full terms that they represent. For example,

Table 3.1 Capitalization rules followed in most journal articles.

Correct	Incorrect
As shown in Figure 2,	As shown in figure 2,
As shown in Scheme 1,	As shown in scheme 1,
As shown in Table 4,	As shown in table 4,
BF_3, a Lewis acid, was used to...	BF_3, a Lewis Acid, was used to...
The reaction was heated to 300 K.	The reaction was heated to 300 k.
The reaction was conducted under nitrogen.	The reaction was conducted under Nitrogen.
3-Bromobenzene and 3-chloropropane...	3-Bromobenzene and 3-Chloropropane...
The metals were lead, zinc, and tin.	The metals were Lead, Zinc, and Tin.
In the Southwest and in northern Arizona,	In the southwest and in Northern Arizona,
The disease is caused by *Salmonella typhimurium*.	The disease is caused by *Salmonella Typhimurium*.
UV–vis spectroscopy was used to...	UV–Vis spectroscopy was used to...
A differential scanning calorimeter (DSC)...	A Differential Scanning Calorimeter (DSC)...

- 2,4,4′-trichlorobiphenyl (PCB-28)
- 3-(4,5-dimethylthiazol-2-yl)-2,5-diphenyltetrazolium bromide (MTT)
- ethanol (EtOH), methanol (MeOH), and acetonitrile (ACN)

After an abbreviation or acronym has been defined, the abbreviated form may be used alone without parentheses. The full form may also be used again, if it seems more appropriate, but the abbreviated form should not be defined again. Note also that most abbreviations and acronyms are used without periods (e.g., NMR not N.M.R.).

Abbreviations

An abbreviation is a shortened form of a word. In some cases, the individual letters are pronounced, as in ACS.

The ACS Style Guide lists some abbreviations that need not be defined for expert audiences, (e.g., DNA, IR, RNA, and NMR).

Other abbreviations need to be written out in full form before using the abbreviated form, for example, tetrahydrofuran (THF). See *The ACS Style Guide* for a long list of accepted abbreviations.

The ACS Style Guide includes a long list of accepted abbreviations to discourage authors from creating their own. Moreover, authors should not use abbreviations that result from laboratory slang. For example, DI, DD, RT, and LN2 (slang for deionized water, double-distilled water, room temperature, and liquid nitrogen, respectively) should not be used in formal writing. A few incorrect and correct examples are illustrated below.

Incorrect The rxn proceeded for 3 h.
Correct The reaction proceeded for 3 h.

Incorrect The % yield was affected by adding NaBr.
Correct The percent yield was affected by adding NaBr.

Incorrect The product was washed 3x in DD water.
Correct The product was washed three times in double-distilled water.
Correct The produce was washed in double-distilled water (3 × 10 mL).

Incorrect Surrogate standards (SS) and internal standards (IS) were used.
Correct Surrogate standards and internal standards were used.

Some approved chemical abbreviations and acronyms, with their spelled-out equivalences, are given in table 3.2. When in doubt if an abbreviated form is correct, check *The ACS Style Guide* or search the literature (using, e.g., the search engine on the ACS Web site).

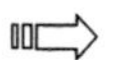

Abbreviations and Acronyms

See appendix A.

Another type of abbreviation—a bolded number, sometimes followed by a letter—is used to label reagents, products, or other compounds that are mentioned more than once in the text. (This is a useful convention because compounds can often be more than 100 characters in length!) Bolded numbers are also used to label species in reactions and schemes, so that the species can later be referred to by number. In the text, the bolded number is introduced, sometimes in parentheses, immediately after the first usage of the full name, like other types of abbreviations. Subsequent references to the compound are by bolded number (and possibly a letter) only, without the full name and without parentheses. Numbers are assigned sequentially if more than one compound is labeled in the text.

In the following example, compound **2a** is the second labeled compound and the first of several related compounds in the **2** series. The number is introduced in the subheading and used again in the subsequent text.

2-(*p*-Toluenesulfonyl)-4′-methoxyacetophenone (2a). A mixture of 2 bromo-4′-methoxyacetophenone (45.8 g, 200 mmol) and *p*-toluenesulfinic acid sodium hydrate

Table 3.2 Common abbreviations in chemistry (adapted from *The ACS Style Guide*: Coghill and Garson, 2006).

Instrumental Techniques	
atomic absorption spectroscopy	AAS
atomic force microscopy	AFM
electron-capture detector, detection	ECD
electron spin resonance	ESR
flame ionization detector, detection	FID
Fourier transform infrared	FTIR
high-performance liquid chromatography; high-pressure liquid chromatography	HPLC
inductively coupled plasma	ICP
mass spectrometry; mass spectrum	MS
scanning electron microscopy	SEM
secondary-ion mass spectrometry	SIMS
solid-phase microextraction	SPME
thin-layer chromatography	TLC
transmission electron microscopy	TEM
X-ray diffraction	XRD
X-ray fluorescence	XRF
Units of Measure	
atomic mass unit	amu
centimeters	cm
degrees Celsius	°C
degrees Kelvin	K
disintegrations per second	dps
foot/feet	ft
gram(s)	g
hour(s)	h
inch(es)	in.
liter(s)	L
liters per minute	lpm
meter	m
microgram(s)	µg
milligram(s)	mg
milliliter(s)	mL
millimeter(s)	mm
millimolar	mM
minute(s)	min
molar (mol L^{-1})	M
nanogram(s)	ng
parts per billion	ppb

Table 3.2 (continued)

parts per million	ppm
second(s)	s
volume per volume	v/v
Chemical Structures	
acetate	AcO
adenosine 5′-triphosphate	ATP
chlorofluorocarbon	CFC
dimethyl sulfoxide	DMSO
ethyl	Et
ethylenediaminetetraacectic acid	EDTA
inosine 5′-triphosphate	ITP
messenger RNA	mRNA
methyl	Me
methanol	MeOH
minimum Eagle's essential medium	MEM
polychlorinated biphenyl	PCB
tris(hydroxymethyl)aminomethane	Tris
Statistical Symbols	
correlation coefficient	r
degrees of freedom	df
probability	p, P
relative standard deviation	RSD
sample variance	s^2
standard deviation	σ, SD
standard error	SE
standard error of the mean	SEM
total number of individuals	n, N
Miscellaneous	
and others	et al.
anhydrous	anhyd
boiling point	bp
calculated	calcd
dose that is lethal to 50% of subjects	LD_{50}
enantiomeric excess	ee
inside diameter	i.d.
lethal dose	LD
mass-to-charge ratio	m/z
melting point	mp
molecular weight	M_r, MW
ultrahigh vacuum	UHV

(35.6 g, 200 mmol) in ethanol (1 L) was heated at reflux for 1.5 h. The mixture was stirred and cooled to room temperature, and the resulting solid was collected, washed with ethanol (2 × 50 mL), dried to give 54.6 g (90%) of pure **2a**: mp 126.0–127.0 °C; IR 2951, 2906, 1676, 1599, 1572 cm^{-1}; ^{1}H NMR ($CDCl_3$) δ 2.45 (s, 3H). (From Swenson et al., 2002)

In reactions and schemes, the bolded number is centered below the species and used without parentheses. For example, **1** and **2** represent bromomethane and methanol, respectively:

$$\underset{\mathbf{1}}{CH_3Br} + HO^- \rightarrow \underset{\mathbf{2}}{CH_3OH} + Br^-$$

Exercise 3.10

Look over excerpts 3B–3J again quickly. What types of information are presented in parentheses? For unfamiliar abbreviations in the excerpts, refer to table 3.2 for full terms.

Exercise 3.11

Consider the abbreviations used in the Methods sections of the journal articles that you have selected for your writing project. Make a list of these abbreviations and their definitions. (Note: Look for definitions earlier in the article if the abbreviations are used without definition in the Methods section.)

Exercise 3.12

Improve the following excerpt so that it uses capitalization, abbreviations, acronyms, parentheses, and complete sentences correctly:

Chemicals and Materials. Boric Acid and Methanol from Riedel-de Haën (Seelze, Germany). All peptides from Sigma (St. Louis, MO). 4-Amino-1-Naphthalenesulfonic acid (A.N.S.A.) from Aldrich (Steinheim, Germany). Sodium Nitrite and Cuprous Bromide, 98%, from Acros (Geel, Belgium) and HBr, 48%, from Fluka (Buchs, Switzerland). Used chemicals as received. (Adapted from Kuijt et al., 2001)

Exercise 3.13

Rewrite the following list of chemicals in a way that is appropriate for the Methods section of a journal article. Assume that all chemical compounds (i.e., reagents and solvents) were research grade and were purchased from Sigma-Aldrich in Milwaukee, Wisconsin.

- 2-Bromopropane
- Calcium chloride, anhydrous
- Magnesium sulfate, anhydrous
- Solvents: Distilled ethanol, acetonitrile, and dichloromethane

---

3B Writing on Your Own: Describe Materials

What chemicals, samples, and/or general conditions do you plan to describe in your Methods section? Prepare a list of these items, including any necessary supplementary information such as vendor, grade, and/or purity.

Convert your list into prose (using complete sentences), thereby writing the first move (Describe Materials) of your Methods section. Be sure to include an appropriate subheading for this section (e.g., **Chemicals**, **Materials**, or **Samples**).

Move 2: Describe Experimental Methods

Move 2 is typically the longest move in the Methods section. As shown in figure 3.1, move 2 involves two submoves (describe procedures and describe instrumentation), which can be addressed in either order. Each submove is addressed separately below.

Describe Procedures

Many of you have already described procedures in a lab report. Most likely, you included items such as the equation that you used to calculate percent yield or the step-by-step instructions that you followed to complete a synthesis (e.g., "Heat to reflux." or "Stir constantly for 10 min."). Are such items also appropriate in a journal article? To answer this question, we analyze several different excerpts. Each excerpt describes a common chemical procedure. Although by no means comprehensive, these few examples should get you started and help you understand what an expert audience expects in this move of the Methods section.

We begin with two excerpts that describe organic syntheses (excerpts 3K and 3L). Both excerpts begin with a subheading. In the first excerpt, the subheading is the name of the specific compound that was synthesized. In the second excerpt, the subheading refers to a general procedure for synthesizing a class of compounds (tetrazoles). The excerpts then go on to describe the steps in the synthesis.

Exercise 3.14

Read excerpts 3K and 3L and answer the following questions:

a. Different units are used to describe the amounts of reactants, solvents, and products. What are those units? Include correct punctuation and capitalization.

b. For what procedures are time and temperature reported?

c. What information is included about washing the products?

d. How is the writing style in excerpts 3K and 3L different from the writing style of most lab reports?

Excerpt 3K (adapted from Swenson et al., 2002)

2-(*p*-Toluenesulfonyl)-4′-methoxyacetophenone (2a). A mixture of 2-bromo-4′-methoxyacetophenone (45.8 g, 200 mmol) and *p*-toluenesulfinic acid sodium hydrate (35.6 g, 200 mmol) in ethanol (1 L) was heated at reflux for 1.5 h. The mixture was stirred and cooled to room temperature, and the resulting solid was collected, washed with ethanol (2 × 50 mL), and dried to give 54.6 g (90%) of pure **2a**: mp 126.0–127.0 °C; IR (cm^{-1}): 2951, 2906, 1676, 1599, 1572; ^1H NMR (400 MHz, $CDCl_3$): δ 2.45 (s, 3H), 3.90 (s, 3H), 4.67 (s, 2H), 6.95 (d, J = 8.8 Hz, 2H), 7.34 (d, J = 8.2 Hz, 2H), 7.76 (d, J = 8.2 Hz, 2H), 7.95 (d, J = 8.8 Hz, 2H); ^{13}C NMR ($CDCl_3$): δ 20.9, 55.1, 62.5, 113.4 (2C), 127.7 (2C), 128.3, 129.1 (2C), 131.1 (2C), 135.8, 144.3, 163.7, 186.0. Anal. Calcd for $C_{16}H_{16}O_4S$: C, 63.14; H 5.30; S, 10.54. Found: C, 63.49; H, 5.35; S, 10.33.

Excerpt 3L (adapted from Demko and Sharpless, 2001)

Large-Scale, Organic Solvent-Free Procedure for the Synthesis of Tetrazoles. To a three-necked 3 L round-bottomed flask equipped with a mechanical stirrer was added benzonitrile (103.1 g, 1.00 mol), 1 L water, sodium azide (68.2 g, 1.05 mol), and zinc chloride (68.1 g, 0.50 mol). The reaction was refluxed in a hood, but open to the atmosphere, for 24 h with vigorous stirring. After the mixture was cooled to room temperature, the pH was adjusted to 1.0 with concentrated HCl (~120 mL), and the reaction was stirred for 30 min to break up the solid precipitate, presumably $(PhCN_4)_2Zn$. The new precipitate was then filtered, washed with 1 N HCl (2 × 200 mL), and dried in a drying oven at 90 °C overnight to give 98.0 g of 5-phenyltetrazole as a white powder (67% yield, mp 211 °C (lit.[25] 216 °C)).

Reporting Analytical Data

In most synthesis papers, analytical data confirming product purity or composition are reported in the Methods section, not the Results section.

Excerpt 3K illustrates a conventional way to format mp, IR, NMR, and quantitative analysis. See *The ACS Style Guide* for other examples.

Units and Verb Agreement

Use a singular verb when one unit of measure is mentioned. Use a plural verb when two separate units of measure are mentioned.

Incorrect	To the mixture *were* added 8.5 g of X.
Correct	To the mixture *was* added 8.5 g of X.
Correct	To the mixture *was* added X (8.5 g).
Correct	To the mixture *were* added 8.5 g of X and 6.5 g of Y.

What details did you notice in excerpts 3K and 3L? Did you notice that both excerpts included the mass and moles of solid reagents, reaction times and temperatures, descriptions of the products (a solid, a white powder), and product yields? Both excerpts also included results from tests used to verify product purity and composition. The first included mp, IR, ^{1}H NMR, and mass spectral information; the second included only mp information. Perhaps you were surprised to see such "results" in the Methods sections. Synthetic chemists include such analytical information as part of the procedure, in the Methods section, rather than as a result, in the Results section. The formatting shown in excerpt 3K is typical for the presentation of such data.

There were also differences between the two excerpts. For example, no mention was made of a flask or beaker in excerpt 3K; however, the authors mention a three-necked 3 L round-bottomed flask in excerpt 3L. Similarly, no hood is mentioned in excerpt 3K, but the authors mention a hood, open to atmosphere, in excerpt 3L. What principles guided these decisions? Although there are no hard and fast rules, you should include common details (e.g., a flask or hood) only if you want to draw attention to those steps (as in excerpt 3L). (See figure 3.2.)

We have summarized a list of the most common details included in a synthesis in table 3.3, along with a list of what not to include. From these lists, it is clear that the journal article is not intended to teach a novice chemist how to conduct a first experiment. Instead, the journal article serves as a blueprint for

Figure 3.2 "Every little detail" is too much detail for experts (and Snoopy). PEANUTS: ©United Feature Syndicate, Inc.

Table 3.3 Details commonly included and omitted when describing a chemical synthesis.

Details Commonly Included	*Details Commonly Omitted*
• a specialized type of glassware • the mass and moles of solid reagents or volume of liquid reagents (often in parentheses) • information on the reaction time and temperature • information on stirring • information on how the product was filtered, washed, and dried (including solvents used and their amounts) • the final product mass and percent yield • a description of the final product (a solid, a white powder) • analytical information about the product (mp, IR, ^{1}H NMR, and mass analysis)	• the equation used to calculate yield • a list of disposables (gloves, pipettes) • names of common glassware (funnels, beakers) • steps taken in routine procedures (weighing, diluting, purifying) • mistakes made (e.g., spilling the product) (Note: If such mistakes are made, the synthesis must be repeated!)

the experienced chemist. In other words, the recipe is included, but basic cooking instructions are not.

We next focus on how authors describe a synthesis. As noted previously, command language should be avoided; moreover, the procedures should not read like a checklist of things to do. To get you started, we present "formulas" for three commonly described synthetic procedures. The formulas illustrate common organizational structures and phrases (e.g., the organic layer, the aqueous layer, the crude product) used to describe simple synthetic techniques such as refluxing, quenching, and purifying. By using these formulas and phrases, your writing will be more concise, and you will sound more like an expert. Note that the italicized passages in these examples will vary with each synthesis:

Describing a Synthesis

Phrases commonly used to describe a synthesis include the following:

- heated to reflux
- heated at reflux
- cooled to room temperature
- the organic layer
- the aqueous layer
- the/an aqueous solution

- the reaction mixture
- the resulting mixture
- the crude product

Formula 1: To describe how to prepare a mixture and heat at reflux

- A mixture of *X* (*g, mmol*), *Y* (*g, mmol*), and *Z* (*g, mmol*) in *A* (*x mL*) was heated at reflux for *1.5 h* and cooled to *room temperature*.
- To a mixture of *X* (*g, mmol*) and *Y* (*g, mmol*) in *A* (*x mL*) was added *Z* (*g, mmol*). The mixture was heated at reflux for *2 h* and cooled to *room temperature*.

Formula 2: To describe how to continue and stop a reaction

- After *x min*, a solution of *X* (*g, mmol*) and *Y* (*g, mmol*) in *A* (*x mL*) was added to the mixture. The mixture was stirred at *10 °C for 1 h*, warmed to *room temperature*, and quenched with *saturated aqueous ammonium chloride* (*5 mL*).

Formula 3: To describe how to extract, wash, and purify a product

- The water phase was extracted with *X* (*2 × 10 mL*). The combined organic layers were washed with *Y* (*2 × 10 mL*) and *Z* (*2 × 10 mL*), dried with $MgSO_4$, and evaporated in vacuo. The crude product was recrystallized from *ethanol* at *–10 °C* and further purified using *column chromatography*.

Exercise 3.15

Using excerpt 3K as a guide, rewrite the following ^1H NMR data in the appropriate format:

The ^1H NMR (400 MHz) solvent was $CDCl_3$. Two different sets of nonequivalent hydrogens were detected. The first set resonated at a chemical shift of 2.4 ppm (a singlet with an integrated area equal to 3H). The second set resonated at a chemical shift of 3.9 (also a singlet with an integrated area equal to 3H).

Exercise 3.16

Find three synthesis articles in *The Journal of Organic Chemistry*. Compare the Methods sections in these articles with excerpts 3K and 3L. Which of your three articles, if any, include the same types of information that you reported in parts a, b, and c of exercise 3.14?

Exercise 3.17

Rewrite the following synthetic procedure (adapted from D'hooghe et al., 2004) using language appropriate for a journal article. The five steps listed (a–e) are from a single procedure involving multiple steps. Note the use of bolded numbers in steps d and e.

a. Make a solution containing 15 mmol (3.60 g) of 2-(bromomethyl)-1-((4-methylphenyl)methyl)aziridine in 50 milliliters of acetonitrile.

b. Add 2.56 g (15 mmol) of benzyl bromide.

c. Heat the solution for 5 hours, refluxing.

d. Remove the solvent using a vacuum to produce *N*-benzyl-*N*-(2,3-dibromopropyl)-*N*-((4-methylphenyl)methyl)amine (**4**).

e. Purify the crude product using column chromatography on silica gel, eluting with a mixture of hexane and ethyl acetate. A typical yield of purified **4** is 85%.

We now consider procedures that do not involve synthesis (excerpts 3M–3O). There are literally hundreds of such procedures in chemistry; however, these few excerpts illustrate essential features of how such procedures are generally written. As in syntheses, nonsynthetic procedures begin with a descriptive subheading; following the subheading, the procedure is described in concise, complete sentences, including only those details needed by an expert audience. Such details are often quantitative in nature (e.g., the number of cells to plate). Information that is largely for students (e.g., "zero the balance") or available elsewhere (e.g., instructions from an operator's manual) is omitted.

Also like syntheses, a clear order of events is conveyed in the procedure. Novice writers inappropriately use words such as *first*, *second*, *next*, and *then* (examples of ordinal language) to convey the order of events; more experienced writers learn to omit most ordinal language. Consider the following example:

Inappropriate	Next, the extracts were combined, then they were reduced to 1 mL, and then they were frozen.
Inappropriate	First, the extracts were combined; second, they were reduced to 1 mL; and third, they were frozen.
Appropriate	The extracts were combined, reduced to 1 mL, and frozen.

The corrected version is uncluttered and more concise. The use of ordinal language, although grammatically correct, detracts from the flow of the text. It is unnecessary to state *first* or *next* because the sequencing of the procedure is implied through the order in which the steps are presented. Another example is presented in excerpt 3M (exercise 3.18). The authors describe a multistep cytotoxicity assay using very little ordinal language.

Ordinal Language

Ordinal language indicates order or position in a series. Examples include the following:

- first, second, third
- next, then

Ordinal language should be used sparingly in journal articles.

Exercise 3.18

Read excerpt 3M and answer the following questions:

a. What details do the authors include regarding the preparation of butachlor and incubation, harvesting, staining, and counting techniques?

b. How do the authors indicate the order of events? What did they do first, second, third, etc.? To what extent was ordinal language used to convey this sequence?

Excerpt 3M (adapted from Ou et al., 2000)

Cytotoxicity Assays. Cells were plated into 60 mm diameter dishes at several different densities (250–500 cells per dish, three dishes per density) to obtain more than one set in which the number of surviving colonies ranged from 100 to 200. The cells were treated with increasing doses (20–160 μM) of butachlor, previously dissolved in ethanol, and the treated cells were left overnight. The next day, the old medium was replaced with fresh medium. The cells were incubated for 10 days, harvested, fixed with methanol, and stained with Giemsa's solution. Colonies containing more than 50 cells per colony were counted as survivors, and survival rates for treatment groups relative to the control were calculated. The concentrations of butachlor used to produce a survival rate of >20% were used in further experiments.

Nonsynthetic procedures also require that authors express numbers and units in conventional ways. Consider excerpts 3N and 3O, which describe two extraction procedures. Note the use of numbers in each excerpt.

Exercise 3.19

Read excerpts 3N and 3O. Pay particular attention to (1) the form of numbers (numerical or word form) and (2) the formatting of units for time and measure (e.g., temperature, concentration, volume, area, and length). What do you notice?

Excerpt 3N (adapted from Tateo and Bononi, 2004)

Extraction Procedure from Hot Chilli. The sample was finely ground using an electric blender. Two grams were shaken for 20 min with 20 mL of ethanol (96%) and stirred in an ultrasonic bath for 10 min. The extraction was repeated three times, each time recovering the liquid phase after filtration on sodium sulfate anhydrous. The extract, collected in a sealed flask, was concentrated under vacuum in a rotary evaporator to about 10 mL, transferred into a 25 mL volumetric flask, and diluted to volume with ethanol (96%).

Excerpt 3O (adapted from Cabras et al., 2001)

Extraction Procedure from Powdery Stem Wood. A 0.1 g sample of *R. speciosa* ground powdery wood was weighed in a 40 mL screw-capped tube, and 10 mL of chloroform was added. Tubes were placed in an ultrasonic bath for 15 min at a temperature of 60 °C and then centrifuged for 5 min at 4500 rpm. A 1 mL aliquot was removed, and organic solvent was dried under a nitrogen stream, taken up with 1 mL of mobile phase (water/methanol, 75:25, v/v), and filtered with a 0.45 μm PTFE membrane filter. The resulting solution was analyzed by HPLC.

About Numbers

- Use numerals with units of time or measure (6 min, 5 g, 273 K).
- Use numerals for numbers greater than nine (10 samples).
- Use words for numbers less than 10 (nine samples), except for units of time or measure (9 min).

Some exceptions:

- Use all numerals in a series or range of values containing the number 10 or greater (e.g., 5, 8, and 12 samples).
- Use words for numbers that start a sentence, unless the number is part of a chemical name (e.g., "Nineteen samples were analyzed." "2-Butene was added.").

Consult *The ACS Style Guide* for more details.

As you read excerpts 3N and 3O, you probably noticed that scientists most often use the numerical form for numbers (e.g., 5) rather than the word form (e.g., five). Indeed, the numerical form is always used with units of time (e.g., s, h, min, days, weeks, years) and measure (e.g., mL, cm, m^3, g, K), unless the number starts the sentence. Note, too, that there is a space between the number and the unit.

Incorrect	2mL	0.6cm	4.2ft	0.015mg	8K	180°C	180° C
Correct	2 mL	0.6 cm	4.2 ft	0.015 mg	8 K	180 °C	

To remember the space, treat units like words. You would write "twelve compounds" not "twelvecompounds"; hence, you would write "12 mL" not "12mL." A notable exception to this rule occurs with percentages; in this case, there is no space between the number and the percent sign (%):

Incorrect	85 %	eighty-five %	eighty-five%
Correct	85%		

Units of Measure

Measure is a general term that implies units of

- volume (mL, cm^3)
- width or length (m, cm)
- mass (g, mg)
- temperature (°C, K)
- concentration (g/mL, M)

Occasionally, the word form of a number is preferred. For example, words are used for whole numbers less than 10 (e.g., nine flasks), except when the number refers to time or measure (7 s, 5 mL). The numerical form is used for numbers 10 or greater (e.g., 10 flasks, 10 samples, 25 trees, 100 cm).

Incorrect	five cm	6 fractions	3 samples	thirteen sites
Correct	5 cm	six fractions	three samples	13 sites

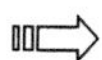

Numbers and Units

See appendix A.

The word form is also used for numbers that start a sentence, unless the number is part of a chemical name. Whenever possible, however, rewrite the sentence so that it does not begin with a number. Units are spelled out following a number in word form and the plural verb is used (e.g., Ten milliliters were . . .). A sentence that begins with a spelled-out number reverts back to numerical form, when appropriate, in the rest of the sentence. Consider the following examples:

Correct	2-Butene was purchased from Aldrich.
Correct	*Eleven* hazelnut oils, *25* olive oils, and *7* other types of oil were purchased. (Adapted from Ozen and Mauer, 2002)
Correct	*Two grams* of NaCl were shaken for *20* min with *20* mL of ethanol.
Better	NaCl (2 g) was shaken for *20* min with *20* mL of ethanol.

There are literally hundreds of units commonly used in chemistry; a few of them are listed in table 3.2, along with their recommended abbreviations. A more comprehensive listing is available in *The ACS Style Guide*. Some important rules about using numbers and units are summarized below:

- Abbreviate units of measure when they come after a numeral:
  A degassed solution of 312 mg (1.39 mmol)...
- Do not abbreviate units of measure that do not follow a numeral:
  ...several milligrams   Twenty percent...
- Leave a space between a numeral and its unit of measure, unless the unit of measure is a percent sign (%):
  30 in.   80 °C   600 g   95%
- Do not add "s" to make an abbreviated unit of measure plural:
  Incorrect   20 mgs
  Correct   20 mg
- Include a leading zero with numeric decimals:
  Incorrect   .6 mg
  Correct   0.6 mg
- When using symbols such as <, >, and ±, include spaces between the numbers and symbol if there are numbers on both sides of the symbol. Also include spaces if the symbol falls in between a variable and a number:
  Incorrect   > 60 g   35±2%   *P*<0.05
  Correct   <25 mL   80 ± 9%   ee > 99%
- Use numerals in a series or range containing numbers 10 or greater to maintain parallelism (even if smaller numbers would be written out in other circumstances):
  Incorrect   three, seven, and 14 samples
  Correct   3, 7, and 14 samples

Correctly formatted abbreviations, numbers, and units are an essential part of scientific writing. Formatting is the author's responsibility; it is not the responsibility of a faculty mentor, peer reviewer, or journal editor. If you are not sure how to format a number or word correctly, consult *The ACS Style Guide* and look for instructions to authors in your journal of interest.

Exercise 3.20

Correct the following (incorrect) uses of numbers and units. Assume that these numbers and units are not being used to start sentences.

7minutes   .15 mg   five percent yield   12 hrs.

10min.	0.175g	7 % recovery	13 hr.
15 mLs	5 sec.	100° C	300° K

Exercise 3.21

The paragraph below contains several errors in the use of numbers and units. Identify at least 10 errors and correct them.

3-Benzyloxy-6-bromo-4-methoxyphenethyliminophosphorane. To a solution of LAH (250mg) in THF (15ml) was added a solution of 3-benzyloxy-6-bromo-4-methoxy-β-nitrostyrene (500 mg, 1.37 mmol) in THF (15 mL), which was then boiled under reflux for eight hours. Addition of several mL of water followed by extraction with ether and drying the ether layer (K_2CO_3) gave the corresponding phenethylamine, which was purified by column chromatography on silica gel (hexane-CH_2Cl_2, 6:4 v/v): yellow oil (330 mg, 72 %); ^1H NMR (400 MHz, $CDCl_3$, δ): 7.33 (m, 6H), 6.77 (s, 1H), 5.05 (s, 2H), 4.62 (s, 2H), 3.82 (s, 3H), 2.94 (m, 4H). The amine (160 mg, .48 mmol) and triphenylphosphine (130 mg, .48 mmol) in 100 mgs of CCl_4 and CH_2Cl_2 (7 mL) were stirred at 40° C for 72 hours under dry N_2. (Adapted from Rodrigues et al., 2004)

We end our analysis of describing procedures (in submove 2) by examining ways in which authors describe QA/QC procedures in their Methods sections. In general, there are two basic approaches. The first approach embeds the QA/QC procedures in the procedure itself. For example, in excerpt 3P, the authors describe how they added a deuterated surrogate (recovery) standard to their samples at the start of their procedure and how they added a deuterated internal standard at the end of their procedure. The authors go on to describe the results of these procedures in their Results section.

QA/QC Standards and Blanks

Blanks, spikes, surrogates, and internal standards are all terms associated with QA/QC procedures. Collectively, they are used to measure sample contamination, analyte recovery, and analyte relative abundance.

Excerpt 3P (from Peck and Hornbuckle, 2005)

Sample Extraction and Cleanup. The sample extraction method has been described previously (27, 28). Prior to extraction, 100 µL of a d_{10}-fluoranthene surrogate standard solution (0.82 ng/µL) was added to each sample. The XAD-2 resin was extracted for 24 h in a Soxhlet apparatus with ~350 mL hexane/acetone (50/50 v/v). The extract volume was then reduced to ~100 µL using rotary evaporation followed by evaporation with nitrogen. Each sample extract was passed through a Pasteur pipet containing ~0.75 g 100–200 mesh Florisil with 4 mL ethyl acetate to provide cleanup as described by Foreman et al. (8). After cleanup, the sample extract was reduced to ~100 μL with nitrogen evaporation and 100 μL of an internal standard solution containing d_{10}-acenaphthene (2.5 ng/μL), d_{10}-phenanthrene (2.4 ng/μL), and d_{10}-pyrene (2.2 ng/μL).

In excerpts 3Q and 3R, the authors describe their QA/QC procedures in separate subsections, complete with their own subheadings (**Method Performance** and **Quality Assurance/Quality Control**). Results of the QA/QC procedures (e.g., relative standard deviation across replicate samples, recoveries, and accuracy) are commonly described in the Methods section, rather than in the Results section.

Excerpt 3Q (adapted from Meijer et al., 2001)

Method Performance. A blank sample, prepared using the same procedure as for the samples, was included with every five samples. PCB 28 and γ-HCH were the only compounds detected in the blanks. Detection limits, calculated as mean blank +3 SD, were typically 2.3–13.3 pg/μL = 0.02–0.12 ng/g soil. Results were not blank corrected. Replicate analysis (the same soil sample extracted three times) was done for several samples. The relative standard deviation (RSD) for replicate analysis was always less than ±20% (n = 3). Analytical recoveries were monitored with the aid of two recovery standards: mirex for F1 and δ-HCH for F2. The mean recovery for mirex was 100 ± 6% (range 89–115%); for δ-HCH, it was 80 ± 9% (range 62–94%). Data were not corrected for recoveries. Extraction procedures and recoveries of a standard containing all target compounds were assessed by spiking four soil samples; recoveries ranged from 74 to 135%.

Excerpt 3R (adapted from Grundl et al., 2003)

Quality Assurance/Quality Control. QA/QC measures included field blanks, solvent blanks, method blanks, matrix spikes, and surrogates. Percent recovery was determined using three surrogate compounds (nitrobenzene-d_5, 2-fluorobiphenyl, 4-terphenyl-d_{14}) and matrix spikes (naphthalene, pyrene, benzo[*ghi*]perylene); the recoveries ranged from 80 to 102%. Separate calibration models were built for each of the 16 PAHs using internal standards (naphthalene-d_8, phenanthrene-d_{10}, perylene-d_{12}). Validation was performed using a contaminated river sediment (SRM 1944) obtained from NIST (Gaithersburg, MD); accuracy was <20% for each of the 16 analytes.

Describe Instrumentation

Authors must also describe the instrumentation or scientific apparatus that they used in their work. (Note that ordinary equipment, e.g., a distillation apparatus or a rotary evaporator, should not be described.) Instrumentation generally falls into two categories: custom-built or commercial. Custom-built instrumentation includes novel or hand-built chambers, devices, or instruments. The first publication that describes a custom-built instrument offers the most detail and often includes a diagram. Subsequent publications briefly highlight the essential features of the apparatus and refer the reader to the original article for more information. Excerpt 3S, for example, includes a diagram of a home-built sonic-spray ionization (SSI) source used to optimize mass spectrometric conditions for generating amino acid clusters. When the authors of excerpt 3S make reference to the same sonic-spray source in a second article (see the short excerpt below), the authors refer back to the first article, rather than repeat a detailed description. You'll notice the superscript ([40]) that leads readers to the reference list at the end of the article. The reference list contains the full citation of the original article in which the SSI source is explained in more detail.

A home-built spray ion source,[40] operable in both the sonic-spray ionization (SSI) and electrospray ionization (ESI) modes, was used instead of the standard ESI source of the Finnigan LCQ instrument. (From Takats et al., 2003b)

Plural of Apparatus

The word *apparatus* has two plural forms:

apparatus

apparatuses

The ACS Style Guide (3rd ed.) recommends the former as the preferred form.

Descriptions of Apparatus

The ACS Style Guide (3rd ed.) specifies that an apparatus should be described only if it is not standard or not available commercially. With standard, commercially available apparatus, stating a company name and model number in parentheses is appropriate and adequate.

More than likely you will need to describe a commercially available instrument or standard piece of instrumentation in your Methods section. Several common types of instrumentation (and their abbreviations) are listed in table 3.2; these instruments are so common and standardized that no instructions or diagrams are needed to explain how they work. It is necessary, however, to report the **operational parameters** under which an instrument was operated. Parameters are selected and optimized for each particular application of an instrument and can vary among users, even for the same instrument. Moreover, parameters affect the outcome and reproducibility of an experiment; hence, they must be described. Characteristic ways to report parameters have been developed for many types of instruments. Excerpts 3I and 3K illustrate how to refer to a ^{1}H NMR spectrometer. Excerpts 3T–3V illustrate details that are typically included for FTIR and GC/MS measurements.

Operational Parameters

The conditions (settings) under which a particular instrument is operated.

Exercise 3.22

With which of the following instruments should you include a diagram of the instrumentation in a Methods section? Why?

a. a ^{1}H NMR instrument (400 MHz)

b. a reflux apparatus

c. a Soxhlet extraction apparatus

d. a Nicolet 870 FTIR spectrometer with an attenuated total reflectance (ATR) accessory

e. a new nozzle design for an ICP mass spectrometer

Excerpt 3S (adapted from Takats et al, 2003a)

Experimental Section

A sonic spray source was built following the Hirabayashi design but instead of an aluminum orifice of 0.4 mm diameter, a coaxial fused-silica capillary with internal diameter of 0.25 mm was used. A detailed cross-sectional view of the source is shown in Figure 1 [see p. 93]. The smaller difference between the o.d. (0.2 mm) of the sample capillary and the i.d. of gas capillary (0.25 mm) was expected to provide higher linear gas velocities at similar mass flow rates. The source was operated at a nitrogen nebulizing gas pressure of 1.2×10^6 Pa. Liquid sample was introduced at a flow rate in the range of 1–50 μL/min. Electrospray spectra were recorded using the same ion source operated in a pneumatically assisted electrospray mode. The nebulizing gas pressure was 1.0×10^5 Pa, and the sample flow rate was 1–3 μL/min. A high voltage of 2.3–2.7 kV was applied on the infusion syringe tip using a copper alligator clip. Experiments for the comparison of two techniques (ESI and SSI) were carried out without changing source geometry and instrumental settings, except for the spray high voltage and nebulizing gas pressure, both of which were optimized.

Excerpt 3T (adapted from Kizil et al., 2002)

FTIR Measurements. FTIR spectra were recorded using a Nicolet model 870 spectrometer (Madison, WI) equipped with a deuterated triglycine sulfate (DTGS) detector.

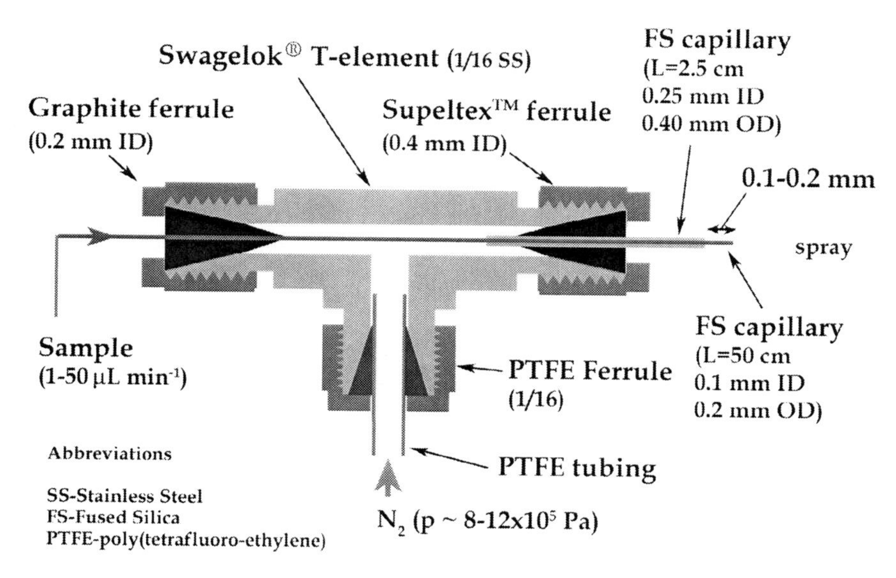

Figure 1. Schematic cross section of sonic spray ion source. Liquid sample is pneumatically sprayed by the coaxial gas flow. The cross section of the gas flow in this source is restricted to 0.017 mm^2.

The sampling station was equipped with an overhead DRIFTS accessory. The sample holder was used for the background spectra without KBr, and 256 coadded scans were taken for each sample from 4000 to 400 cm^{-1} at a resolution of 16 cm^{-1}. Single-beam spectra of the samples were obtained, and corrected against the background spectrum of the sample holder, to present the spectra in absorbance units. Spectra were collected in duplicate and used for multivariate analysis.

FT-Raman measurements. FT-Raman spectra were obtained using a Nicolet 870 spectrometer with the Raman module 32B (Madison, WI) and Nd:YAG laser operating at 1064 nm with a maximum power of 2 W. The system was equipped with an indium-gallium arsenide (InGaAs) detector, XT-KBr beam-splitter with 180° reflective optics, and a fully motorized sample position adjustment feature. A laser output power of 0.77 W was used, which was low enough to prevent possible laser-induced sample damage yet provided a high signal-to-noise ratio. Data were collected at 16 cm^{-1} resolution with 256 scans. Spectra were obtained in the Raman shift range between 400 and 4000 cm^{-1}. The system was operated with the OMNIC 5.1 software, and experiments were done in duplicate.

FTIR *What?*

Beginning writers often say "FTIR was used," but this is incorrect. FTIR is the abbreviation for "Fourier transform infrared" (all serving as adjectives). Thus, you need to include a noun following FTIR, such as the following: FTIR analysis, FTIR measurements, FTIR spectroscopy, an FTIR spectrum, or several FTIR spectra.

Excerpt 3U (adapted from Llompart et al., 2001)

Chromatographic Conditions. GC/MS–MS analyses were performed on a Varian 3800 gas chromatograph (Varian Chromatography Systems, Walnut Creek, CA) equipped with a 1079 split/splitless injector and a ion trap spectrometer (Varian Saturn 2000, Varian Chromatography Systems) with a waveboard for MS–MS analysis. The system was operated by Saturn GC/MS WorkStation v5.4 software. The MS–MS detection method was adapted from reference.[29] PCBs were separated on a 25 m length × 0.32 mm i.d., CPSil-8 column coated with a 0.25-µm film. The GC oven temperature program was as follows: 90 °C hold 2 min, ramp 30 °C/min to 170 °C, hold for 10 min, rate 3 °C/min to 250 °C, rate 20 °C/min to a final temperature of 280 °C, and hold for 5 min. Helium was employed as the carrier gas, with a constant column flow of 1.0 mL/min. The injector was programmed to return to the split mode after 2 min from the beginning of a run. Split flow was set at 50 mL/min. The injector temperature was held constant at 270 °C. Trap temperatures, manifold temperatures, and transfer line temperatures were 250, 50, and 280 °C, respectively.

Excerpt 3V (adapted from Pelander et al., 2003)

LC/TOFMS. The liquid chromatograph was an Agilent (Waldbronn, Germany) 1100 series system consisting of vacuum degasser, autosampler, binary pump, column oven, and diode array detector. Separation was performed in gradient mode with a Phenomenex (Torrance, CA) Luna C-18(2) 100 × 2 mm (3 µm) column and a 4 × 2 mm precolumn. The column oven was kept at 40 °C. Eluent components were 5 mM ammonium acetate in 0.1% formic acid and acetonitrile. Flow rate was 0.3 mL/min. The proportion of acetonitrile was increased from 10 to 40% in 10 min, to 75% in 13.50 min, to 80% in 16 min, and held at 80% for 3 min. Post-time was 5 min, and the injection volume was 10 µL.

The mass analyzer was an Applied Biosystems (Framingham, MA) Mariner TOF mass spectrometer equipped with a PE Sciex (Concord, ON, Canada) TurboIon Spray source and a 10-port switching valve. The instrument was operated in the positive ion mode. The eluent flow was carried to the ion source without splitting. The nebulizer gas (N_2) flow was 0.7 L/min, the curtain gas (N_2) flow 1.2 L/min, and the heater gas (N_2) flow 8 L/min. The spray tip potential of the ion source was 5.5 kV, and the heater temperature was 350 °C. Interface settings were as follows: nozzle potential 70 V, quadrupole rf voltage 800 V, and quadrupole temperature 140 °C. Skimmer 1 potential, quadrupole dc potential, deflection voltage, and Einzel lens potential varied depending on the daily tuning of the instrument. Analyzer settings were as follows: push pulse potential 492 V, pull pulse potential 225 V, acceleration potential 4.0 kV, reflector potential 1.55 kV, and detector voltage 1.9 kV. Pull bias potential varied depending on daily tuning. Spectrum acquisition time was 2 s, and a m/z range from 100 to 750 was recorded.

Exercise 3.23

Consider excerpts 3T–3V. What operational parameters should you report when FTIR spectroscopy is used in your research? What parameters should you include when GC is used in your research? (If unfamiliar with these techniques, ask your instructor what these parameters mean.)

Exercise 3.24

Consider the following sentence: "Samples were analyzed using UV–vis." Explain why such a sentence is inappropriate in a journal article, even if the instrumental parameters are specified in subsequent sentences. Rewrite the sentence so that it is appropriate for inclusion in a journal article.

Exercise 3.25

Find three articles that describe a single instrument that you have used. Compare the parameters reported in each article.

3C Writing on Your Own: Describe Experimental Methods

Procedures. Write the procedures section of your Methods section. Begin by creating an outline or flow chart that lists the steps that you followed in a given procedure. Cross out steps that are too basic for an expert audience. Organize the remaining steps into a concisely written paragraph, using the sequencing of your sentences to convey the order followed in your experiment. If more than one procedure was used, repeat this process as needed. Follow standard conventions for expressing numbers and units.

Instrumentation. Write the instrumentation section of your Methods section. When appropriate, be sure to include vendors, model numbers, and operating parameters. Use the literature to determine the operational parameters that you should include. Be sure to use parentheses appropriately.

Move 3: Describe Numerical Methods

The final move of the Methods section involves the description of statistical, computational, or other mathematical methods used to derive or analyze data. This move is required only if numerical methods were part of the work. Excerpts 3W and 3X demonstrate how some common types of statistical methods, including analysis of variance (ANOVA), are described. Once again, subheadings are used to guide the reader's attention to this information. Note that when specialized statistical software is used, the name and version number of the software package are reported. Important statistical parameters that affect the outcome of the statistical test (e.g., significance level) may also be reported, although these parameters may also be reported in the Results section. Note that routine software such as Microsoft Word or Excel should not be reported in this section (or anywhere else in the journal article).

Excerpt 3W (from Ye et al., 2000)

Statistical Analysis. Statistical analyses (two-way ANOVA) were performed by using the Statistical Analysis System (SAS, 1990). Means were compared by the least significant difference (LSD) test at $\alpha = 0.05$.

Excerpt 3X (from Besser et al., 2004)

Statistical Analysis. Analysis of variance (ANOVA) of toxicity data was conducted using SAS/STAT software (version 8.2; SAS Institute, Cary, NC). All toxicity data were transformed (square root, log, or rank) before ANOVA. Comparisons among multiple treatment means were made by Fisher's LSD procedure, and differences between individual treatments and controls were determined by one-tailed Dunnett's or Wilcoxon tests. Statements of statistical significance refer to a probability of type I error of 5% or less ($p \leq 0.05$). Median lethal concentrations (LC_{50}) were determined by the Trimmed Spearman-Karber method using TOXSTAT software (version 3.5; Lincoln Software Associates, Bisbee, AZ).

In excerpt 3Y, the authors refer to computational results performed with the Gaussian suite of programs, a computational package used to calculate molecular ab initio or semiempirical electronic structure theory. Computational parameters (e.g., the basis set and level of theory) are included in the description. Do not worry if you do not understand the content of excerpt 3Y; the language is intended for chemists with a computational or theoretical background.

Ab initio vs. Semiempirical

The Latin term "ab initio" means "from the beginning" or "from first principles." Ab initio calculations involve no experimental (empirical) data; they are derived solely from theory. (Note that ab initio should not be italicized or used with quotation marks.)

Semiempirical calculations combine both empirical (experimental) and nonempirical data.

Excerpt 3Y (from Kuwata et al., 2005)

II. Theoretical Methods

A. Quantum Chemistry Calculations. All electronic structure calculations were performed with the Gaussian 03 suite of programs.[31] The geometry, energy, and harmonic vibrational frequencies of each stationary point considered here were determined initially using the B3LYP functional[32,33] and the 6–31G(d,p) basis set.[34,35] Each reported minimum has all real frequencies, and each reported transition structure has one imaginary frequency. We determined the minima associated with each transition structure by animation of the imaginary frequency and, if necessary, with intrinsic reaction coordinate (IRC) calculations.[36,37]

3D Writing on Your Own: Describe Numerical Methods

If appropriate, write the numerical methods subsection of your Methods section. Depending on the type and extent of statistical, computational, or theoretical methods used, you may want to create a new subheading (e.g., Statistical Methods or Theoretical Methods) or simply add this information to the end of your experimental procedures.

If you are writing a theoretical paper, this section will be the bulk of your Methods section. Be sure to include the brand names and versions of specialized software packages used to analyze your data.

Part 2: Analyzing Writing across the Methods Section

Two writing conventions apply to the Methods section as a whole: the use of tense (past or present) and voice (passive or active). Past tense and passive voice predominate in the Methods section; however, in some cases, present tense and/or active voice are also used. Like other writing conventions, the proper use of tense and voice reveals authors' familiarity with the expectations of the field, their objectivity, and more expert-like writing abilities.

Past and Present Tense

The Methods section is largely written in the past tense. In general, the Methods section describes work that was done in the past, making the past tense the appropriate choice. This is different from a lab manual, which gives a set of instructions in the present tense.

Lab manual	Stir the mixture. Heat to reflux.
Journal article	The mixture was stirred and heated to reflux.

Although the Methods section is overwhelmingly written in the past tense, there are few correct instances of present tense. The general rule of thumb for deciding when to use past or present tense in the Methods section (and elsewhere in the journal article) is as follows:

Work was done in the past; knowledge exists in the present.

Work done in the past is described using past-tense verbs (e.g., analyzed, built, heated, investigated, isolated, measured, performed, synthesized, tested). Knowledge that exists in the present (and presumably into the future) is described using present-tense verbs (e.g., contains, defines, describes, explains, implies, is expected to, provides, suggests). Present tense is also used to describe fixed features of a custom-built instrument (e.g., length and width). Consider the following examples. In each case, the past tense describes actions taken by the researchers that led to their results; the present tense describes information that is expected to be true over time.

Past tense	The water *was* triply distilled.
Present tense	Triply distilled water *contains* less than 1 ppb of the impurity.
Past tense	Height measurements *were* taken using a nanoscope.
Present tense	Height data *provide* topographical information.
Past tense	The probe *was* modeled after work described elsewhere (4).
Present tense	The probe *projects* through the tee into the main chamber.
Past tense	Helium gas *was* used to purge the chamber.
Present tense	The outer diameter of the chamber *measures* 10 cm.

Exercise 3.26

Each of the following passages contains a present- and past-tense verb choice. For each italicized pair, select the correct verb tense:

a. Triply distilled water *has/had* a conductivity and surface tension lying within experimental error of the literature values for ultrapure water. (Adapted from Quickenden et al., 1996)

b. Phase data, which *measure/measured* the phase shift in the cantilever oscillation, *are/were* taken with an AFM operated in tapping mode. (Adapted from Clancy et al., 2000)

c. The main body of the probe system *is/was* a stainless steel tee. (Adapted from Van Berkel et al., 2002)

d. The tests in this study *were/are* conducted in TCE-contaminated groundwater in two distinct water-bearing zones, the A-zone and the C-zone. The A-zone *was/is* an unconfined shallow layer composed mainly of placed fill over Bay Mud. (Adapted from Hageman et al., 2001)

e. This peptide *was/is* chosen because it *was/is* predicted to form a loop structure on the surface of the folded protein. (Adapted from Stockton et al., 2003)

f. The lower spinning rate *was/is* chosen because higher rates routinely *interfere/interfered* with the cross polarization process. (Adapted from Vaisman et al., 2000)

Passive and Active Voice

The Methods section is also written largely in passive voice. Passive voice is most often combined with past tense:

Inappropriate	We heat the mixture to 80 °C.	[present tense, active voice]
	We heated the mixture to 80 °C.	[past tense, active voice]
	The mixture is heated to 80 °C.	[present tense, passive voice]
Appropriate	The mixture was heated to 80 °C.	[past tense, passive voice]

You may have been taught in other writing courses not to use passive voice because it is considered "weak." However, passive voice, when used appropriately, strengthens writing in chemistry journal articles (and other scientific genres). Figure 3.3 shows the frequency of passive voice in each section of a chemistry journal article. Note that passive voice is used in all sections, but it is most common in the Methods section.

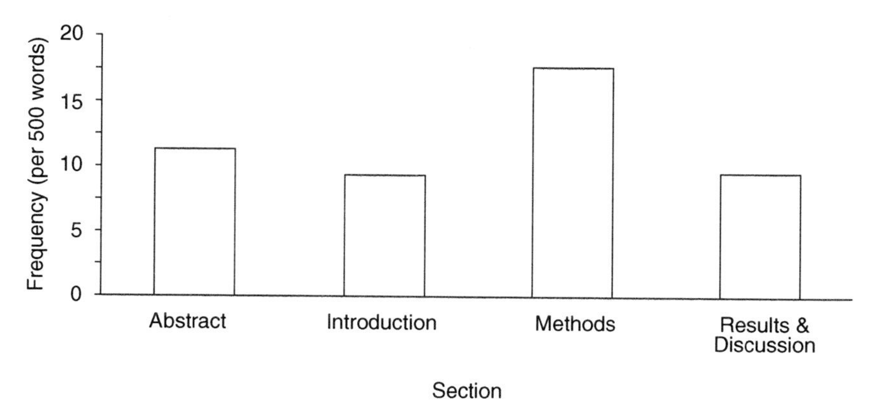

Figure 3.3 Frequencies of passive voice (expressed as the number of passive verbs per 500 words) in sections of chemistry journal articles, determined through a computer-based analysis of 60 journal articles (approximately 300,000 words).

Table 3.4 Passive-voice–past-tense combinations commonly used in Methods sections.[a]

was added	was determined	was maintained	was refluxed
was allowed	was dissolved	was performed	was removed
was assigned	was dried	was poured	was separated
was carried out	was evaporated	was prepared	was stirred
was collected	was extracted	was purified	was treated
was concentrated	was filtered	was quenched	was used
was cooled	was heated	was recorded	was washed

a. These passive-voice–past-tense combinations were identified through a computer-based analysis of passive voice in the Methods sections of 60 published chemistry research articles.

Passive voice is preferred because it sounds more objective. Passive voice essentially removes the human subject (i.e., the scientist) from the sentence so that the focus of the sentence is the object that was acted on.

Active We added solid Se (0.030 g) to the pale orange solution.
Passive To the pale orange solution was added solid Se (0.030 g).

Active We stirred the mixture for 10 min at room temperature.
Passive The mixture was stirred for 10 min at room temperature.

A list of passive-voice, past-tense combinations commonly used in Methods sections is provided in table 3.4.

Joining Sentences in Passive Voice

When you join two sentences that are in past-tense passive voice, each subject must have a verb that is preceded by "was" or "were."

Incorrect The eluant *was added* to the column, and the samples *collected* in 10 mL increments.

Correct The eluant *was added* to the column, and the samples *were collected* in 10 mL increments.

Passive Voice

See appendix A.

 **Exercise 3.27**

Rewrite these sentences so that they are more appropriate for the Methods section of a journal article; use passive voice and past tense:

a. We recrystallized the product from ethanol in a fume hood.

b. We measured the temperature with a K-type thermocouple located just above the catalyst bed.

c. Filter the precipitate. Wash three times with 10 mL of ethanol each time.

d. Add chlorosulfonic acid (0.350 mL) dropwise to a flask containing acetic acid in an ice bath.

e. We used a Nicolet model 590 FTIR spectrometer to analyze the water-ice films.

f. We collected all of our samples in amber glass bottles with Teflon-lined caps (EPA level 1, 33 mm, VWR).

g. We used the Box-Hunter program run under MAPLE computer algebra software (v. 5, Waterloo Maple, Inc.).

3E Writing on Your Own: Practice Peer Review

Before you engage in authentic peer review (when a classmate reviews your Methods section and you review a classmate's Methods section), practice the peer-review process. Imagine that a classmate or colleague has asked you for feedback on a draft of a Methods section. See "Peer Review Practice: Methods Section" at the end of the chapter for a copy of the draft, background information, and instructions for completing the task.

3F Writing on Your Own: Fine-Tune Your Methods Section

By now, you likely have a solid draft of your Methods section, including a description of materials, experimental methods, and numerical methods (if applicable); hence, it is time to revise and edit your Methods section as a whole. We recommend that you reread and edit your work, focusing on each of the following areas and using chapter 18 to guide you.

1. Audience and conciseness: Are you writing for an expert audience, leaving out unnecessary details? Try to find at least three sentences that can be written more clearly and concisely. Check for information that should be placed inside parentheses.
2. Organization of text: Check your overall organizational structure. Did you follow the move structure in figure 3.1 and include appropriate subheadings? Do your

experimental procedures clearly convey the order followed in your work (without using ordinal language)?

3. Writing conventions: Check to be sure that you have used voice (mostly passive) and tense (mostly past tense) correctly. Check your formatting of units and numbers, use of abbreviations and acronyms, and capitalization of compounds and vendors.
4. Grammar and mechanics: Check for typos and errors in spelling, subject–verb agreement, punctuation, and word usage (e.g., effect vs. affect, data).
5. Science content: Have you correctly conveyed the science in your work? Have you used words and units correctly? If asked, could you define all of the words that you have used? Do you understand, in principle, how the instruments described in your methods section work?

After thoroughly reviewing and revising your own work, it is common practice to have your work reviewed by a peer or colleague. A "new set of eyes" will pick up mistakes that you can no longer see because you are too familiar with your own writing. To facilitate the peer-review process, use the Peer Review Memo form (on the *Write Like a Chemist* Web site) to assist you and your peer reviewer. After your paper has been reviewed (and you have reviewed another's paper), consult the Peer Review Memo given to you by your partner to make final changes in your Methods section.

Finalizing Your Written Work

See chapter 18.

Chapter Review

Check your understanding of what you've learned in this chapter by defining each of the following terms for a friend or colleague who is new to the field:

abbreviation	numerical methods	passive voice
ab initio	operational parameters	QA/QC
acronym	ordinal language	semiempirical
active voice		

As a review, try explaining the following to a friend or colleague who has not yet tried to write a Methods section for a journal article:

- Main purpose of a Methods section
- Three typical moves of a Methods section

- Capitalization conventions for a Methods section
- Guidelines for spelling out abbreviations in a Methods section
- Guidelines for the inclusion of an illustration of an apparatus in a Methods section
- Guidelines for conveying the order of events in a Methods section
- Appropriate use of numbers and units in a Methods section
- Use of tense (past/present) and voice (active/passive) in a Methods section
- Reporting of quality control results in a Methods section versus a Results section

Additional Exercises

Exercise 3.28

Look back at exercise 3.1. Complete column 2 by placing a check next to items that you think should be included in a Methods section. Compare your answers with those in column 1. How have your ideas about Methods sections changed since the beginning of the chapter?

Exercise 3.29

Reread excerpt 3A and reexamine your answers to exercise 3.2. How would you modify your responses so that they are more accurate? Do you notice anything now that you did not notice earlier?

Exercise 3.30

Using the excerpts in this chapter as a guide, what are the correct abbreviations for the following units?

boiling point
centimeters
grams
hours
meters
micrograms
micrometers
milligrams
milliliters
millimeters
millimoles
minutes
moles
nanometers

Exercise 3.31

Rewrite the following lab manual passage so that it is appropriate for the Methods section of a journal article:

Clean the Erlenmeyer flask with deionized water and let it dry. Add 14.3 grams (or 0.25 moles) of activated zinc dust and 80 mL of HMPA to the dried flask. Stir to mix. Next, add 32 mL of chlorotrimethylsilane (equivalent to 0.24 mol). Stir the mixture for 90 minutes at room temperature. Cool the mixture on ice for 20 minutes.

Exercise 3.32

The passage below is the Methods section from a student paper. Considering what you have learned in this chapter, improve the paper through revision.

Methods

General Methods. Purity was determined using ^{1}H-NMR, IR, and TLC. Thin-layer chromatography was performed on a silica gel plate and developed in dichloromethane.

Preparation of isopropyl-MgBr (2). To ensure anhydrous conditions, we flame-dried the flask. Next, Mg (14.81 mmol) and three crystals of I_2 were added and heated until I_2 vapor filled the flask. Once the flask cooled to room temperature, 2-bromopropane **1** (1.13mL) in anhydrous diethyl ether (10 mL) was added, refluxed for 15 m at 40° C, then cooled to room temperature.

Preparation of 1-(4-methoxyphenyl)-2-methylpropan-1-ol (4). 4-methoxy-benzaldehyde(p-anisaldehyde) **3** (4.99 mmol) in anhydrous diethyl ether (10 mL) was gradually added to the Grignard reagent (0.5 mL increments), refluxed for 10 m at 40° C, and poured over ice water (50 mL). 1M H_3PO_4 was gradually added until the mixture became acidic. Then we rinsed the extracted ether layers with 5% aqueous NaOH (10 mL) and saturated NaCl (10 mL). It was dried with anhydrous magnesium sulfate and ether was extracted from the product using distillation. R_f = 0.45 (silica); IR (cm^{-1}) 3240 cm^{-1} (R-OH); ^{1}H NMR (400 MHz, $CDCl_3$, δ): 0.88 (d, J = 6.58 Hz, 3H), 0.91 (d, J = 6.58 Hz, 3H), 1.99 (s, 3H), 5.21 (m, 1H), 2.08 (s, 1H), 3.76 (s, 3H), 4.45 (d, J = 7.64 Hz, 1H), 6.85 (d, J = 8.18 Hz, 2H), 7.39 (d, J = 8.18 Hz, 2H).

Exercise 3.33

Rewrite the following wordy passages. Make them sufficiently concise so that they are appropriate for journal article Methods sections. Do not get distracted by the science; you do not need to understand the science fully to improve the passages with the conciseness techniques presented in chapter 2. Hints and word-count goals are provided for each passage to focus your efforts.

a. Prior to the reaction being started, the solution was evacuated to the point that gas evolution stopped and purged for a period of 10 min with Ar to free it from O_2. (32 words)

(Hint: Avoid using language that conveys the sequencing of steps and use nominalizations when appropriate. Goal: 11–12 words)

b. **Sample Preparation.** Two sets of coal samples, each prepared by mixing 0.5 g of liquid with 1.0 g of Pittsburgh No. 8 coal (used as received), were prepared. The coal used was 100-mesh and used as received, after which the liquid was added to the coal in drops. The resulting sample was then mixed by shaking. The samples that were used for the WAXRS experiments were then stored in 20 mL scintillation vials with screw-top caps. A portion of the sample was removed 1 day after its preparation and subsequently used as the subject of each wide-angle X-ray scattering experiment. (Adapted from Wertz and Smith, 2003) (103 words)

   (Hint: Combine several short sentences into fewer, more complex sentences, delete redundant or unnecessary information for an expert audience, and use parentheses to present information about materials and to define abbreviations. Goal: 50 words)

c. *Purification.* *E. coli* BL21 (DE3) containing the plasmid pMB1912 ($dadX_{PA}$) (6) was grown at 37 °C in Luria broth containing ampicillin (100 μg/mL). At OD_{600} = 0.5, IPTG (0.5 mM) was added, and afterwards cells were grown overnight at a temperature of 30 °C. Cell pellets were resuspended in 50 mM Tris, pH = 7.6, 0.5 mM PLP; next, 150 μg of purified *Serratia marcescens* nuclease was added. Cells were lysed using a Spectronic French Press at 16 000 psi, and cell debris was removed by centrifugation. $(NH_4)_2SO_4$ 20 and 60% cuts were done, and following the final cut, the protein pellet was dialyzed against 20 mM Tris, pH = 7.6 and filtered through a 0.45 μm syringe filter. After this preparation, the material was loaded on a Pharmacia Q-Sepharose HP column and eluted with a 0–0.5 M NaCl gradient. (Adapted from LeMagueres et al., 2003) (139 words)

   (Hint: Avoid using sequencing language and use parentheses where appropriate. Goal: 113 words)

Exercise 3.34

Rewrite the following sentences using the conventional language of a synthesis paper. Try to do it on your own first. Then review the excerpts, the Describing a Synthesis pointer (p. 82) and the three "formulas" (p. 83) to assist you with your final fine-tuning of these sentences.

a. To a beaker, which contained 30 mL of ether and 103.1 g benzonitrile (1.00 mol), was added 68.1 g (0.50 mol) zinc chloride.

b. A solution of water saturated with NaCl was used to wash the impure product that was produced at the end of the reaction.

c. The solution was heated until it began to boil and recondense. After 10 min of boiling, it was allowed to cool to the surrounding temperature.

d. The top (water-insoluble) layer was washed with a saturated solution of sodium chloride in water (2 × 25 mL).

Exercise 3.35

Reflect on what you have learned about writing a Methods section for a journal article. Select one of the reflection tasks below and write a thoughtful and thorough response:

a. Reflect on the differences among the ways in which methods are reported in lab manuals, lab reports, and journal articles.
   - What are the predominant differences between the ways in which methods are reported in lab manuals, lab reports, and journal articles?
   - Why do you think that the formats are so different?
   - What purposes do the different formats serve?

b. Reflect on the numerous writing conventions that are typical of a journal article Methods section.
   - Which writing conventions are relatively new to you?
   - Which writing conventions have you used before?
   - Which writing conventions do you have to make an effort to remember?
   - Why do you think expert readers and writers in chemistry take these conventions so seriously?

c. Reflect on the numerous excerpts that you have read in this chapter. Excerpts 3A–3Y come from different journals and report on different types of chemical research, but they have all been written for expert audiences.
   - What features do the excerpts have in common? Give specific examples in your response.
   - What features of this professional writing are most impressive to you?
   - What aspects of this writing do you think will be easiest to learn to use? Hardest to learn to use?
   - How might the reading of the chemical literature help you with your own writing?

d. Reflect on the ways in which tense and voice are used in a Methods section.
   - What rules have you created for yourself to remember when to use the present tense and past tense in a Methods section?
   - What rules have you created for yourself to remember when to use active voice and when to use passive voice in a Methods section?
   - How can the proper use of tense and voice help you achieve objectivity in your writing?
   - In what ways can the improper use of tense and voice cause miscommunication with your readers?

Peer Review Practice: Methods Section

Imagine that a friend has asked you to review a draft of a Methods section for a paper to be submitted to the *Journal of Agricultural and Food Chemistry.* Unfortunately, your friend has not had the benefit of a chemistry writing course; hence, a few writing tips would be appreciated. The project involves the identification of odor-active compounds in 19 California chardonnay wines. The steps involved in the study are as follows:

1. extracting the volatiles from the wine
2. concentrating the extracts (using distillation)
3. fractionating the concentrated extracts into three fractions (using silica gel chromatography)
4. screening "oral-active" fractions in the wine (using GC-olfactometry)
5. identifying/quantifying the compounds in the "oral-active" fractions (using GC-MS)

The first three steps are straightforward (extracting, concentrating, and fractionating the volatiles in the wine). In the fourth step, when the effluent comes off the GC column, half of it goes to a "sniffing port" (GC-olfactometry, step 4) and half of it goes to a mass spectrometer (GC-MS, step 5). Judges at the "sniffing port" are asked to indicate if the fraction is "odor-active" or not. If it is, they give a verbal description of the smell (using such words as glue, buttery, creamy, plastic, green grass, fruity, musty, and many others). The odor-active compounds are then identified and quantified by GC-MS. Eighty-one compounds were shown to be odor-active; of these, 74 were quantified and 61 were tentatively identified.

Using parts 2 and 3 of the Peer Review Memo form on the *Write Like a Chemist* Web site, provide feedback on the Materials and Methods draft. Give specific suggestions that can be used to improve the written work. (The Methods section below is adapted from an original source, cited in the Instructor's Answer Key.)

Materials and Methods

1 Purchased from EM Science, a division of EM Industries, Inc (New Jersey):
2 Diethyl Ether
3 Pentane
4 Silica Gel 60 (particle size 0.063–0.200 mm, 70–230 mesh)
5 Sigma-Aldrich Chemical Co. (St. Louis, MO):
6 Trichlorofluoromethane (Freon 11)
7 Absolute Ethanol
8 Compounds used as internal standards (IS)
9 Methyl Octanoate
10 2-Methyl-1-Pentanol

11 3-Methyl-3-Hydroxy-2-Butanone

12 GC-O/GC-MS Operating Parameters

13	Model:	Hewlett-Packard (HP) GC model 6890, Palo Alto, CA
14	Injector:	Split/splitless (operated in splitless time of 1 min.)
15	Column type:	DB-WAX bonded fused capillary column
16	Column size:	30 m x 0.25 mm i.d., film thickness—0.25 µm
17	Detector:	MS 6890 series mas selective dector, HP, Palo Alto, CA
18	Inlet temp:	220 °C
19	Carrier gas:	Helium gas at flow of 1.3 ML/min.
20	Program:	Oven temperature held at 40 °C for 4 min., ramped at 4 °C/min. to
21		185 °C, held for 20 min.

22 Nineteen 1997 Californian Chardonnay wines were analyzed in 2000, all of which had
23 been profiled by descriptive analysis (DA) 6–10 months before this study (*17*). All wines
24 were held at 10 °C during the studies.
25 All glassware was washed thoroughly with liquid soap and distilled water and
26 allowed to dry before use. The IS stock solution was prepared by adding 5 µg of each
27 internal standard to 100 ML of Absolute Ethanol.
28 Volatiles were extracted using a modification of a procedure described elsewhere (*18*).
29 Before extraction, 150 ML of wine was carefully poured into a 500 ML round-bottom
30 flask. First, 45 g of Sodium Chloride (NaCl) and 3 ML of IS stock solution were added
31 to the wine, which was then extracted 3 times with 50 ML of Trichlorofluoromethane
32 (Freon 11) using a liquid-liquid extractor at 28–30 °C. Next, the extract was concen-
33 trated to ~2 ML by distilling off the solvent on a Vigreux column (40 × 2 cm). The
34 solvent was further removed under a purified Nitrogen stream until the volume was
35 reduced to 1 ML. We fractionated the aroma extracts by Silica Gel chromatography to
36 provide better GC resolution, using a modification of Guth's method (*19*). The Freon
37 extract (1 ML) was placed in a glass column (30 × 1.9 cm i.d.) packed with Silica Gel
38 60. The sample was fractionated by elution with 200 ML of Pentane and Diethyl Ether
39 (Fraction 1, 85/15; Fraction 2, 70/30) and 200 ML of Diethyl Ether (Fraction 3). Finally,
40 the eluates were dried over Sodium Sulfate overnight and concentrated to a final volume
41 of 1 ML, as described above, and stored at -5 °C for subsequent analyses.
42 The recovery of internal standards after sample preparation (extraction, fraction-
43 ation, and GC analysis) was evaluated for 5 wines (JL, CDB-C, CAL, DEL and SH).
44 Recovery ranged from 82% for Methyl Octanoate in fraction 1 to 73% for 2-Methyl-1-
45 Pentanol in fraction 2 and to 61% for 3-Methyl-3-Hydroxy-2-Butanone in fraction 3.
46 Reproducibility of the sample preparation method was examined for 1 wine (SH). A
47 two-way analysis of variance for each peak showed no significant differences due to
48 extraction or injection.
49 A 1-µL sample of each concentrated wine fraction was analyzed by gas chromotog-
50 raphy-olfactometry (GC-O). GC operating parameters are listed above. As the effluent
51 came off the column, it was split 50:50 between a sniffing port (Gerstel, Germany) for
52 GC-O analysis and a mass spectrometer for GC-MS analysis (see below). The sniffing
53 port was held at 250 °C to prevent any condensation of volatile compounds. Humidified

54 air was added at 100 ML/min. in the sniffing cone to reduce fatigue and drying of the
55 judges' nasal passages. For determination of odor-active (OA) compounds, four judges
56 who had previous experience with GC-O were used. Assessors were seated in front of
57 the sniffing port and asked to smell the effluent off the column. An "olfactory button"
58 (Gerstel, Germany) was depressed when an aroma was detected. Judges also gave verbal
59 descriptions of perceived odors that the experimenter recorded.
60 Fractions identified as odor-active were analyzed using GC-MS. The column and
61 operating conditions were the same as those used for GC-O. The detector was a mass
62 spectrometer (MS 6890 series mass selective detector, Hewlett Packard, Palo Alto, CA).
63 Mass spectrum were taken over the *m/z* range 45–300. The total ion chromatogram
64 (TIC) acquired by GC-MS was used for peak area integration. HP MS chemstation soft-
65 ware G1701BA ver.B.01.00 was used for data acquisition.
66 We tentatively identified the oral-active compounds (screened by the GC-O) by
67 comparison of the Kovats retention index (KI) (*21*) and the MS fragmentation pat-
68 tern with those of reference compounds or with mass spectra in the Wiley 275 library
69 and previously reported Kovats retention indices. The Kovats retention indices (KI) of
70 unknown compounds were determined by injection of the sample with a homologous
71 series of alkanes (C_6-C_{28}).
72 The relative concentrations of the odor-active volatiles in all 19 wines were deter-
73 mined by GC-MS (TIC) by comparison with concentrations of internal standards,
74 assuming a response factor of 1. Methyl Octanoate, 2-Methyl-1-Pentanol, and 3-Methyl-
75 3-Hydroxy-2-Butanone were used as the internal standards for fractions 1, 2, and 3,
76 respectively.
77 GC data were first entered into an Excel (Microsoft) spreadsheet and later
78 imported into statistical analysis software.Analyses of variance were run on the GC
79 data using PROC GLM on Statistical Analysis Systems (SAS) for Windows, version 6.12
80 (Cary, NC).

4 *Writing the Results Section*

All sections of a journal article lead up to or away from the results section, and the results section may retain its value long after the methods and conclusions have become obsolete.
—Paradis and Zimmerman (1997)

This chapter focuses on the Results section of the journal article. The Results section makes use of both text and **graphics** to highlight the essential findings of a study and to tell the story of scientific discovery. In this chapter we focus on writing the text; we refer you to chapter 16 for information on formatting graphics (e.g., tables and figures). After reading this chapter (and chapter 16), you should be able to do the following:

- Distinguish between the description and interpretation of data
- Organize and present your results in a clear, logical manner
- Refer appropriately to a figure or graph in the text
- Use appropriate tense, voice, and word choice
- Prepare a properly formatted figure and table

Graphics

We use the term *graphics* to refer to figures, tables, and schemes.

Figures and tables are used to display, clarify, and summarize results, helping readers comprehend data more quickly.

Schemes are used to illustrate mechanisms (see chapter 5).

Formatting Graphics

Instructions for formatting figures, tables, and schemes are presented in chapter 16. Consult chapter 16 as you work through this chapter.

As you work through the chapter, you will write a Results section for your own paper. The Writing on Your Own tasks throughout the chapter will guide you step by step as you do the following:

4A Read the literature and review your results

4B Organize your results

4C Prepare figures and/or tables

4D Tell the story of scientific discovery

4E Practice peer review

4F Fine-tune your Results section

The purpose of a Results section (the third section in the standard IMRD format) is to present the most essential data collected during a research project. A well-written Results section guides the reader's attention back and forth between text and graphics while highlighting important features of the data and telling the story of scientific discovery. Months (possibly years) of accumulated knowledge and wisdom, and countless pages of data, are distilled into only a few pages; hence, only the essential threads of the story are included in the Results section.

In many journal articles, the Results section is actually a combined Results and Discussion (R&D) section. Combined R&D sections are preferred by many scientists who want to present and discuss results in an unbroken chain of thought. The combination is often more concise because less time is spent reminding the reader which results are being discussed. Combined R&D sections are not all alike; rather, they fall on a continuum with fully separated R&D sections at one end and fully integrated R&D sections at the other. Within this continuum, three patterns emerge: blocked R&D, iterative R&D, and integrated R&D.

In the **blocked R&D** pattern, a single block of results is followed by a single block of discussion. For example, for a set of three results, the pattern would be [results 1, results 2, results 3] [discussion 1, discussion 2, discussion 3]. In essence, the blocked R&D pattern is identical to that of fully separate sections but merged under a single "Results and Discussion" heading. In such papers, it is usually quite easy to determine where the Results section ends and the Discussion section begins.

Blocked R&D

An approach for combined R&D sections in which all results are presented first, followed by paragraphs dedicated to the discussion. For three sets of results, the pattern would be as follows:

[R1, R2, R3] [D1, D2, D3]

In the **iterative R&D** pattern (the most common pattern), authors alternate between presenting and discussing results. Thus, for three results, an iterative R&D pattern is achieved as follows: [results 1, discussion 1] [results 2, discussion 2] [results 3, discussion 3]. The story of scientific discovery is often easier to tell (and understand) if each finding is presented and discussed before moving on to the next.

Iterative R&D

An approach for combined R&D sections in which authors alternate back and forth between results and discussion. For three sets of results, the pattern would be as follows:

[R1 D1] [R2 D2] [R3 D3]

With the **integrated R&D** pattern, results are presented and discussed together, often in the same paragraph or even the same sentence. The text is organized in a way that best conveys the story of scientific discovery, with no obvious delineation between results and discussion. This pattern is less common, but when done well is quite effective.

Integrated R&D

An approach for combined R&D sections in which results and discussion are seamlessly integrated with no obvious pattern.

These patterns are intended to serve as guiding constructs only. In practice, most authors who use combined R&D sections will combine features of two or three patterns in their writing, making it difficult to find a pure example. For example, in some articles, the R&D section may generally follow the blocked R&D pattern, but authors may add some brief interpretative comments into their presentation of results. Some journals specify a required format for the R&D section; hence, it is always a good idea to refer to the "Information for Authors" section of a journal before beginning to write a manuscript for publication.

Despite the frequency with which combined R&D sections now appear in the chemical literature, we have chosen to address the sections separately in this textbook. The different purposes of Results and Discussion sections are important to understand and distinguish, even if you ultimately choose to write a combined R&D section. In this chapter, we focus on the Results section. The Discussion section and the integrated R&D approach are examined in chapter 5.

Reading and Analyzing Writing

In Results and Discussion sections, the reader should be led step-by-step through the subject, showing how conclusions unfold logically as the results accumulate.

—Charles H. DePuy, University of Colorado–Boulder

We formally begin this chapter by asking you to read and analyze a Results section on your own. Excerpt 4A is a continuation of excerpt 3A (in chapter 3), regarding the analysis of aldehydes in aged beer. The excerpt includes most of the original text, but, to conserve space, only one figure (Figure 3) and one table (Table 2) are included. Note that the excerpt is a combined R&D section.

Exercise 4.1

As you browse through excerpt 4A, consider the following questions:

a. Which organizational features and writing conventions (e.g., capitalization in abbreviations and subheadings, numerical formatting, and use of parentheses) appear to be similar to those used in the Methods section?

b. What formatting conventions do you notice in the figure and table?

c. Which sentences or paragraphs belong in the Results section? Which belong in the Discussion section? How does the language help you differentiate between the two?

d. What did the authors do to make their writing concise?

Excerpt 4A (adapted from Vesely et al., 2003)

Results and Discussion

Identification. Identification of the carbonyl PFBOA derivatives was performed by mass spectrometry using electron impact ionization running in the scan mode. It was confirmed that fragment *m/z* 181 was the main fragment of all analyzed aldehydes (6). Figure 1 shows as an example the mass spectrum of the PFBOA derivative of methional. To increase the selectivity of the method, all aldehyde analyses were run in the single-ion monitoring (SIM) mode with monitoring for *m/z* 181. Beer was also analyzed by GC/MS without being derivatized by PFBOA in order to ensure that there were no other sources of *m/z* 181 besides the derivatization agent.

Optimization of Derivatization Procedure. Three parameters that may affect the partition of aldehydes between the headspace and the solution were tested: derivatization time, temperature, and ionic strength. The effect of pH was not examined because it was previously shown that the natural pH of beer, 4.5, is sufficiently low for the derivatization reaction (6). Therefore, the pH of standard mixtures was adjusted to 4.5 using 0.1% phosphoric acid. **Because** methional appeared to be the most problematic aldehyde to detect, optimization was carried out in a 5% ethanol (pH 4.5) solution spiked with 5 ppb of methional.

The effect of temperature on the extraction of methional from ethanol solution and its derivatization on a PFBOA-loaded fiber was examined for 35 and 50 °C (Figure 2). Increasing the extraction temperature caused an increase in the peak area of the derivatized methional. Based on this result, subsequent derivatizations were conducted at 50 °C.

The optimal derivatization time was also tested. The ethanol solution spiked with 5 ppb of methional was exposed for 15, 30, 60, 90, and 120 min at 50 °C. It was determined that the time to reach equilibrium between stationary phase and sample headspace was 90 min (Figure 3). A derivatization time of 60 min at 50 °C appeared to be a good compromise between the time of reaction and analyte response.

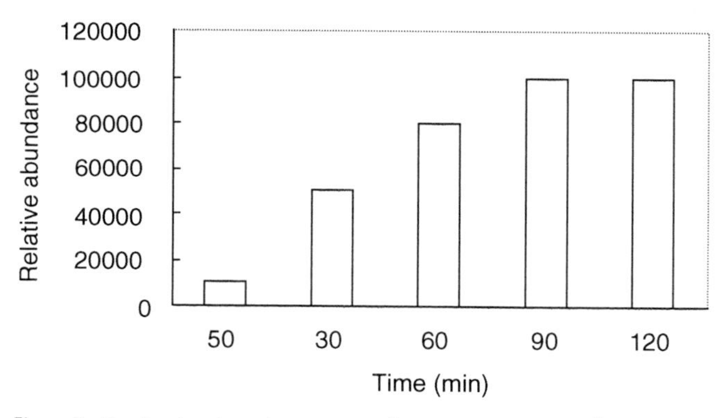

Figure 3. Derivatization time versus detector response of PFBOA derivative of methional.

Figure 4 shows that addition of salt (2 g of NaCl in 10 mL of methional solution) did not have any effect on the extraction and derivatization procedure (60 min, 50 °C).

Calibration. Most aldehydes, except formaldehyde, form two geometrical isomers of the derivatives and appear as two peaks in the chromatogram. The sum of these two peak areas was used in the calibration measurements. A six-point calibration curve for nine carbonyl compounds was measured. The calibration range was 0.1–50 ppb, except for (*E*)-2-nonenal, where the calibration range was 0.01–5 ppb. The matrix used for calibration solutions was 5% ethanol solution, pH 4.5. Correlation coefficient (R^2) values indicate that this method can be used for analysis of aldehydes in a wide range of concentrations (Table 1).

Method Validation. Reproducibility of the method was determined by analyzing one beer sample 10 times. Table 1 shows that the method provides very good reproducibility, with coefficients of variations for monitored aldehydes below 5.5%, except for (*E*)-2-nonenal. The higher coefficient of variation for (*E*)-2-nonenal may be due to extremely low levels of this aldehyde in the analyzed beer.

Beer Analysis. Nine aldehydes were detected in analyzed beer (Figure 5). The resolution of two peaks, representing two geometrical isomers of each aldehyde, was good, except for furfural, where the first peak was clustered with a peak of an uncharacterized compound.

The aldehydes 2-methylpropanal, 2-methylbutanal, 3-methylbutanal, methional, and phenylacetaldehyde are so-called Strecker aldehydes, formed as a result of a reaction between dicarbonyl products of the Amadori pathway and amino acids, having one less carbon atom than the amino acid (*1*). According to Schieberle and Komarek (*8*), the increase of Strecker aldehydes and some esters might play a central role in flavor changes during beer aging. The same authors exclude (*E*)-2-nonenal, a degradation product of linoleic acid, as a key contributor to the stale flavor of beer. Other aldehydes related to the autoxidation of linoleic acid are pentanal and hexanal (*1*). Furfural, a product of the Maillard reaction, is a known heat exposure indicator that does not impact beer flavor due to its high flavor threshold (*9*).

During long-term storage at elevated temperatures, American-style beers develop a stale flavor (*10*). Analyzed beer samples were stored at 30 °C for 4, 8, and 12 weeks. Levels of all aldehydes increased during beer storage compared to the control sample (Table 2). Although the increase after 12 weeks at 30 °C was significant (16-fold increase for furfural, 7-fold increase for 2-methylpropanal), none of the analyzed aldehydes exceeded their flavor threshold in beer (*11*). However, it is probable that additive or synergistic effects take place when aldehydes contribute to the stale flavor of aged beer.

Table 2. Aldehyde Level Changes (ppb) in Beer during 12 Weeks Storage at 0 and 30 °C

	0 °C	30 °C			FT[a]
	12	4	8	12 weeks	
2-methylpropanal	6.1	20	30.6	42.4	1000
2-methylbutanal	1.8	3.1	4.2	5.2	1250
3-methylbutanal	12.2	17.2	20.7	24.4	600
pentanal	0.3	0.6	0.7	0.8	500
hexanal	1.0	1.8	20.1	2.5	350
furfural	28.8	202.8	362	458.3	150000
methional	2.8	3.6	4.1	4.6	250
phenylacetaldehyde	6.6	9.9	10.1	12.7	1600
(*E*)-2-nonenal	0.01	0.02	0.02	0.03	0.11

[a] Flavor threshold in American-style beer (*11*).

Because and Since

See appendix A for more information on these commonly confused words.

Exercise 4.2

Because excerpt 4A uses a combined R&D section, you likely found sentences that are clearly results and sentences that are clearly discussion (see exercise 4.1c). Given this observation, which combined R&D pattern do you think best characterizes this excerpt: blocked, iterative, or integrated? Of course, the match may not be perfect.

Exercise 4.3

Compare this figure to Figure 3 in excerpt 4A. Using Figure 3 as a guide, identify five formatting mistakes in the bar graph below (we found seven). (See chapter 16 for more information on formatting bar graphs.)

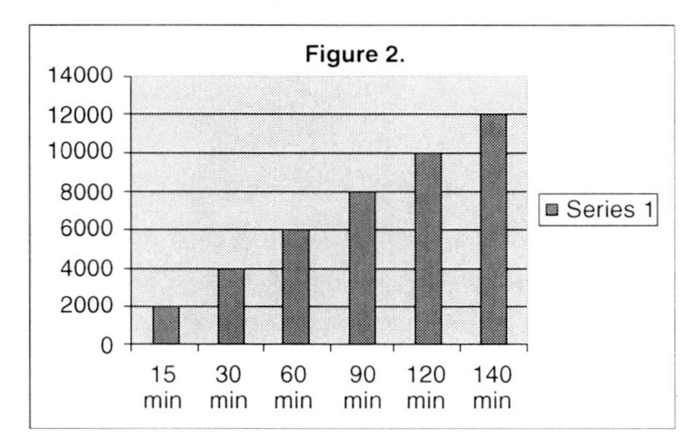

Exercise 4.4

Compare this table (adapted from Vesely et al., 2003) with Table 2 in excerpt 4A. Using Table 2 as a guide, identify five formatting mistakes in the table below (we found six). (See chapter 16 for more information on formatting tables.)

Table 1. Correlation Coefficient (R^2), Coefficient of Variations (CV), and Relative Recovery (RR) of Analyzed Aldehydes

	R^2	CV	RR
2-Methylpropanal	.9639	4.7%	110%
2-Methylbutanal	.9723	4.6%	104%
3-Methylbutanal	.9706	4.0%	109%

continued

Table 1 (continued)

	R^2	CV	RR
Pentanal	.9951	3.9%	114%
Hexanal	.9925	4.3%	103%
Furfural	.9892	5.1%	99%
Methional	.9983	2.4%	90%
Phenylacetaldehyde	.9839	5.3%	98%
(E)-2-Nonenal	.9944	8.0%	89%

Analyzing Audience and Purpose

An author's central obligation is to present an accurate account of the research performed as well as an objective discussion of its significance.

—American Chemical Society, *Ethical Guidelines to Publication in Chemical Research* (https://paragon.acs.org/)

The central purpose of the Results section is to describe your research findings to other scientists (an expert audience) in a clear and concise manner. As you will see in chapter 5, the central purpose of the Discussion section is to interpret those findings. The distinction between description and interpretation is not always clear-cut. The following rule of thumb helps to distinguish between the two:

Description (Results) answers the question *What did you find?*
Interpretation (Discussion) answers the question *What do your findings mean?*

Truth?

Words such as *truth* and *prove* seldom appear in scientific writing. In a computer-based analysis of 180 journal articles, *prove* was found only twice, and *truth* never occurred.

Hedging words are used instead. For example, data *suggest* (not prove), results offer *evidence* (not proof), and *findings* (not truths) are reported.

Hedging is discussed in more detail in chapter 5.

An objective description of results allows readers to examine the data unbiased by interpretation. Results are sometimes viewed as a glimpse at the "truth"; alternatively, interpretations are educated opinions that are likely to change over time. As a writer, you must learn to distinguish between description and interpretation, especially if you write a combined R&D section. The following sentences from excerpt 4A (referring to Figure 3) help to clarify the difference:

Description	The ethanol solution spiked with 5 ppb of methional was exposed for 15, 30, 60, 90, and 120 min at 50 °C (Figure 3).
Interpretation	The higher coefficient of variation for (*E*)-2-nonenal may be due to extremely low levels of this aldehyde in the analyzed beer.

That the spiked ethanol solution was exposed for different lengths of time at 50 °C (presumably) conveys a "truth" or "fact" that will not change; thus, the statement is descriptive. However, the higher coefficient of variation for (*E*)-2-nonenal may not necessarily be due to the low levels of aldehydes in the analyzed beer; hence, the statement is interpretive.

Exercise 4.5

The following statements are taken from the R&D section of an article that reports on a study of tartary buckwheat as a source of dietary rutin (adapted from Fabjan et al., 2003). For each sentence, decide whether its primary purpose is to describe, interpret, or both:

a. The highest content of rutin, 2.5–3% dry weight, was observed as sampling started.

b. Trends in rutin content were rather similar in all of the buckwheat varieties, indicating a more important influence of environment than genotype on the rutin content of the buckwheat herb.

c. Later sowing had essentially no impact on rutin content in herb, as shown in Figure 2.

d. On the basis of this study, it is clear that buckwheat herb production is feasible and that it could readily be produced as a nutritionally rich food, a rutin-rich herb tea, or food additive.

4A Writing on Your Own: Read the Literature and Review Your Results

Read the Results sections of the journal articles that you collected during your literature search (starting with Writing on Your Own task 2C). As you read these articles, pay attention to how the authors organized their results and what results they chose to emphasize in both text and graphics. What ideas do these articles give you about ways to write your own Results section?

Now is also a good time to review your results. What have you learned from your data? What do you want your readers to learn? You will not be able to share all of your results with your readers, so begin to think about the most important points that you want to communicate.

Analyzing Organization

The purpose of the Results section is to present—without interpretation—the results of the study. Two moves are suggested to accomplish this task (figure 4.1). The first move, Set the Stage, serves to transition the reader from the Methods to the Results section. Two submoves are involved: in submove 1.1, the reader is briefly reminded how a particular set of results was obtained; in submove 1.2, the reader is referred to a graphic (a table or figure) that displays those results. These complementary submoves are often accomplished in a single sentence (a poignant reminder of the conciseness in chemistry writing). After the graphic has been introduced, the authors shift to the second move, Tell the Story of Scientific Discovery, where important findings are identified, trends are highlighted, and unexpected results are underscored. Importantly, the story is rarely told in the way that it actually occurred (chronologically); rather, it is told in a way that logically leads the reader to the conclusions of the paper. These two moves are repeated, as needed, for each set of results. Because these moves describe quite specific information, the Results section is in the narrowest part of the hourglass.

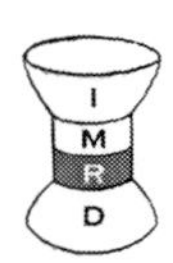

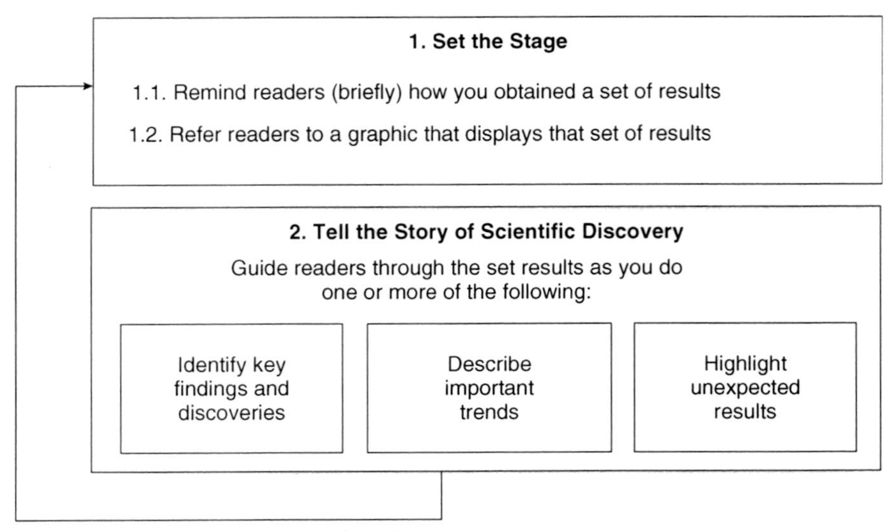

Figure 4.1 A visual representation of the move structure for a typical Results section.

Sets of Results

Results sections often include multiple sets of results. Each set presents a different piece of evidence or a different part of the project. The various sets are linked to lead logically to the conclusions of the paper.

Exercise 4.6

The authors of excerpt 4A present several sets of results using both text and graphics (only two graphics are included here). These results include the optimization of temperature, time, and ionic strength. Reread excerpt 4A and consider the third and fourth paragraphs, which present temperature and time results, respectively. For each paragraph, determine how well the authors adhere to the move structure in figure 4.1. Explain.

Exercise 4.7

Browse through three articles in one of the following journals: *The Journal of Organic Chemistry*, *Analytical Chemistry*, *Environmental Science & Technology*, or *Chemical Research in Toxicology*. How well do the articles adhere to the move structure illustrated in figure 4.1?

4B Writing on Your Own: Organize Your Results

Organize your data into one or more sets of results. (Omit results that led to false starts or dead ends or that were preliminary in nature.) What evidence does each set of results reveal? Organize the sets of results in a logical sequence so that the pieces of evidence lead ultimately to the conclusion(s) of your work. (Remember that you do not need to follow the actual order in which the data were collected.)

Analyzing Excerpts

Because we do not work in a vacuum in the academic world, we must learn to communicate with others. The written word remains the foundation of this communication, and I would hope that some of the capstones in my career have come from being able to communicate in a clear and controlled manner.

—Richard Malkin, University of California–Berkeley

We are now ready to read and analyze excerpts of Results sections in more depth, one move at a time. After examining the individual moves in part 1, we look holistically at the Results section in part 2.

Part 1: Analyzing Writing Move by Move

Move 1: Set the Stage

The goal of move 1 is to transition the reader from the Methods section to the Results section. The move begins with a brief reminder of how a set of results was obtained and then refers the reader to a graphic that displays these results. Consider the following passage:

> P1 The EPR spectra of samples of $PM_{2.5}$ from five different sites in the U.S. are shown in traces A-E of Figure 1. (From Dellinger et al., 2001)

In one sentence, the authors remind us that $PM_{2.5}$ was collected in five sites and analyzed with EPR, and they refer us to Figure 1, which displays the EPR spectra. (Details about $PM_{2.5}$ and EPR were included in the Methods section.)

Referring to Graphics

Remember to refer to graphics before actually guiding readers through the results in the text.

Avoid "dangling graphics," that is, graphics that are included in a paper but never mentioned in the text.

Note that the reference to the graphic (move 1) comes before guiding the reader through the graphic (move 2). In this way, the reader has the opportunity to view the data before reading the associated prose. Often, the graphic is introduced in parentheses, as shown in the next two examples. Note, too, the use of subheadings in these passages; subheadings are a particularly effective way to direct the reader's attention to each set of results.

> P2 **Chromium Accumulation in *E. coli* Cells.** The cellular uptake of chromium is presented as milligrams of chromium per dry weight of treated *E. coli* cells (Table 1). (From Plaper et al., 2002)

> P3 **Experiment 1.** When fed at a concentration of 3 μg of Se/g of diet, high-Se broccoli significantly reduced the incidence and total number of mammary tumors as compared to rats fed 0.1 μg of Se/g of diet as

either selenite alone or 0.1 μg of Se/g of diet in combination with low-Se broccoli (Table 1). (From Finley et al., 2001)

Another common way to refer to a graphic in the text is with the phrase "as shown in Figure__" or "are shown in Figure___." Consider the following example:

P4 The first part of this study involved structurally differentiating between hazelnut oil and other oils and then detecting adulterating oils in hazelnut oil. For this purpose, the spectra of pure hazelnut oil were compared with the spectra of seven other oil types, *as shown in Figure 1.* (Adapted from Ozen and Mauer, 2002)

"As shown in Figure __"

What's the most common four-word combination in a chemistry journal article?

"as shown in Figure"

(based on a computer-based analysis of 200 journal articles)

Exercise 4.8

How well do passages P1–P4 set the stage (move 1)? For each passage, predict what the results will be about, even though you have not read the Introduction or Methods section.

Exercise 4.9

Both past tense and present tense are used in passages P1–P4. Find instances of both and explain when each is used. Based on your answer, select the right tense for this sentence: The data *are/were* shown in Table 1.

Exercise 4.10

Based on passages P1–P4, describe how the words "table" and "figure" should be formatted in the text. Identify three ways in which you can refer to a table or figure in the body of your text.

Move 2: Tell the Story of Scientific Discovery

After the stage is set, you are ready to tell the story behind the data. The story is told using both text and graphics (typically tables and figures). Most authors determine the order of their graphics first (Table 1, Figure 1, Figure 2, Table 2, etc.), as well as the order of the data within each graphic (entry 1, entry 2, etc.), and then write the prose to complement the graphics. Ultimately, the graphics and prose should work together, reinforcing but not duplicating one another; the reader's attention should naturally be shifted back and forth between the two. That said, we point out that graphics are not required in all instances, and occasionally an article will be published with no figures or tables. Indeed, graphics should only be included if there is sufficient data (see chapter 16 for guidelines on how much data is needed for tables and figures), and if the graphics make the data more accessible to the readers. Graphics that are superfluous or that repeat content that is easily described in the text will detract from rather than enhance the Results section.

The Bare Minimum

See chapter 16 for guidelines on how much data you need for a table or figure.

The best way to learn how to write a Results section is to read and analyze Results sections from the literature. To this end, we examine excerpts (prose and graphics) from six published Results sections. We guide you through the content of these articles, elucidate what the authors have "discovered," and analyze how the authors have organized their results. (In each case, the original articles had multiple sets of results, but for brevity, we include only a few.) As you read these excerpts, pay particular attention to how the authors use text, in combination with graphics, to describe their data (i.e., to identify important findings, describe trends, and highlight unexpected results). In subsequent chapters, we read additional excerpts from these articles, taken from the Discussion sections (chapter 5), Introduction sections (chapter 6), and abstracts (chapter 7).

4C Writing on Your Own: Prepare Figures and/or Tables

Select the data that you plan to present in a figure or table. We recommend that you include at least one figure or table in your paper. Check to be sure that you have enough data for the graphic (if not, ask your instructor for possible sources of additional data). Organize the information in the graphic in a logical sequence. You will follow this sequence as you write the text that accompanies your graphic.

Create the figure and/or table using guidelines in this chapter and in chapter 16.

We begin with an excerpt from *Environmental Science & Technology* (excerpt 4B). In a combined R&D section, the authors tell us what happened when they coated different types of soil with randomly methylated β-cyclodextrins (RAMEB). Cyclodextrins are highly water-soluble, crystalline sugars; their shape (referred to as toroidal) resembles a water pail without a bottom. The outer surfaces of the pail are hydrophilic (water-loving), which accounts for their solubility in water and their ability to attract water molecules. RAMEB alone adsorbs water molecules; hence, the authors predicted that RAMEB-coated soils would adsorb more water than their noncoated counterparts.

To test this theory, the authors measured water vapor adsorption isotherms for RAMEB-treated soils. The amount of water adsorbed (kg water/kg soil) was monitored as a function of the partial pressure of water (p/p_0), the dose of RAMEB in the soil (0, 1, or 9%), and the type of soil. Seven soils were studied and arranged in order of increasing clay content (3, 8, 11, 16, 25, 36, and 49% clay content for S1, S2, S3, . . . S7, respectively).

With that background, let's look at excerpt 4B. The authors use subheadings to present each set of results (only the first set of results is included in excerpt 4B). The opening sentence (in accord with move 1) reminds us how this set of results was obtained and refers us to Figure 1. Recall that seven soils were studied, but only three are plotted in Figure 1 (a low-, medium-, and high-clay-content soil). The authors realized that the trends in their data would be clearer if they graphed representative data only, not all of the data. Novice writers might have been tempted to include three figures, one for low-, one for medium-, and one for high-clay-content soils. Such an approach, however, would have made the trends more difficult to discern.

Representative Data

Avoid the temptation to plot all of your data. Instead, plot only representative data, the data needed to illustrate important trends in your study.

In the corresponding text, the authors emphasize the important trends: clay-rich soils show lower adsorption, clay-poor (sandy) soils show higher adsorption, and soils with medium clay content show intermediate adsorption, when compared to their untreated counterparts. Moreover, the unexpected behavior of the clay-rich soils is highlighted. The authors first state the expected behavior ("An increase in water sorption was expected after RAMEB addition to all soils.") and then point out the unexpected behavior ("However, the isotherms for RAMEB-treated clay-rich S6 and S7 soils show lower adsorption than the original soils."). Because this is a combined R&D section, the authors also offer a tentative interpretation for the unexpected finding ("that RAMEB decreases the amount of water-available surfaces in clay-rich soils").

Exercise 4.11

Read and analyze excerpt 4B (text and figure) to determine what the adsorption isotherm measurements reveal and then answer the following questions:

a. How quickly do the authors refer to Figure 1?

b. What trends revealed in the graph do the authors describe in the text? Are there any unexpected findings? If so, how are they highlighted?

c. Explain why the authors included only three of the seven soils studied (S2, S5, S7) in Figure 1.

d. Excerpt 4B is an example of an iterative R&D section. In the full journal article, the authors report numerous sets of results; for brevity, only the first set of results is included here. In excerpt 4B, how do the authors present their results regarding sandy soil? How are these results discussed?

e. Examine the symbols the authors used to graph the data. What pattern guided their approach? Comment on the advantages and disadvantages of the approach that they selected. (See chapter 16 for recommendations for symbol selection.)

Excerpt 4B (adapted from Jozefaciuk et al., 2003)

Results and Discussion

Effect of RAMEB on Water Vapor Adsorption on Soils. Experimental adsorption isotherms for the RAMEB-treated soils are presented in Figure 1 [see p. 127]. As pure RAMEB sorbs a very high amount of water (ca. 1 g g^{-1} at $p/p_0 = 0.99$), an increase in water sorption was expected after RAMEB addition to all soils. However, the isotherms for RAMEB-treated clay-rich S6 and S7 soils showed lower adsorption than the original soils, which is illustrated for S7 soil with 49% clay. This potentially indicates that RAMEB decreases the amount of water-available surfaces in clay-rich soils, similar to what was observed for pure clay minerals (*20*). In sandy soils (S1–S4), the water sorption markedly increased, particularly at higher RAMEB doses, as is illustrated for S2 soil. This may be attributed to water sorption by free RAMEB, that is, RAMEB molecules that did not interact with the sandy soils. For soil S5 of medium clay content (25%), the effect of RAMEB on water sorption was small, which can reflect a balance between the two tendencies described above.

Excerpt 4C is taken from an article in *Analytical Chemistry*. Headspace solid-phase microextraction (HSSPME) is coupled with GC to quantify polychlorinated biphenyls (PCBs) in milk. The PCBs are volatilized out of the liquid phase (milk) into the gas phase (headspace) and concentrated on an SPME fiber. The concentrated PCBs on the fiber are then injected into the GC.

This excerpt offers an excellent example of what we mean by the "story of scientific discovery." The authors guide us through their discovery process,

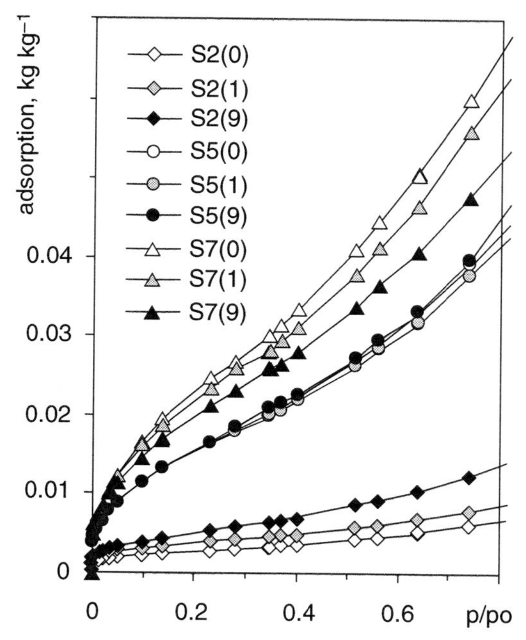

Figure 1. Water vapor adsorption isotherms for RAMEB-treated soils. Soil symbols are as in Table 1. The number in parentheses following the soil symbol is the dose of RAMEB (%).

highlighting first the successful analysis of PCBs in skim milk, then the unsuccessful analysis of PCBs in full-fat milk, and ultimately, the successful analysis of PCBs in full-fat milk. The story appears to follow chronological order, but this need not be the case. What's important is that the authors purposefully sequenced their results to make the story easy to follow. Their logical presentation makes it clear that the problem involves full-fat milk, and this problem can be solved with saponification.

Scientific Terms (excerpt 4C)

HSSPME	Headspace solid-phase microextraction; a preconcentration technique that concentrates volatile analytes on a fiber than can be inserted directly into a GC
ECD	Electron-capture detector; a detector that is very sensitive to halogenated compounds
Saponification	The process of converting a fat (RCOOR′) to its corresponding carboxylate anion ($RCOO^-$) and alcohol (R′OH) by reaction with NaOH

Chronological Order?

Results may appear to be presented in chronological order, but usually they are not. Actual chronological order is often quite messy because of false starts, dead ends, and "wrong turns." True chronological order only confuses the reader.

Instead, effective writers use hindsight to intentionally rearrange their results in a logical sequence of events.

Exercise 4.12

As you read excerpt 4C, examine the figure and text carefully to determine what the HSSMPE-GC measurements revealed. Answer the following questions:

a. What sentences accomplish the goals of move 1?
b. List the sequence of events portrayed by the authors in graphics and text to tell their story of scientific discovery.
c. The figure contains a lot of information. Was all of the information described in the text? What aspects of the graphic were most important to the authors?
d. The authors use a combined R&D section. Find examples of both description and interpretation.
e. Comment on why you think that the authors included panel B in Figure 1 of their article instead of simply reporting a more concise story (e.g., "Method A works for skim milk" and "Method B works for full-fat milk").

Excerpt 4C (adapted from Llompart et al., 2001)

Results and Discussion

Preliminary Experiments. *Influence of Fat Contents.* Initial HSSPME experiments were performed using spiked skimmed and full-fat milk samples.... Figure 1A shows the ECD chromatogram obtained for the skim milk. When the same experiments were performed on full-fat milk, the results were considerably lower, as can be seen in the chromatogram shown in Figure 1B. Also, in the full-fat milk sample, the background appeared higher, and it increased after each SPME injection. This indicates that this simple procedure might be adequate for the analysis of PCBs in milk samples having low fat content; however, it is not adequate when the percentage of fat increases. This is quite logical because PCBs are more strongly retained in the sample matrix as the fat content increases....

Taking into account these results, our objective was to develop a SPME procedure that improved the release of PCBs from the sample to the fiber coating irrespective of the fat content of the samples. Saponification of fats to their corresponding glycerols and carboxylates facilitates the release of PCBs from fatty matrixes and also can

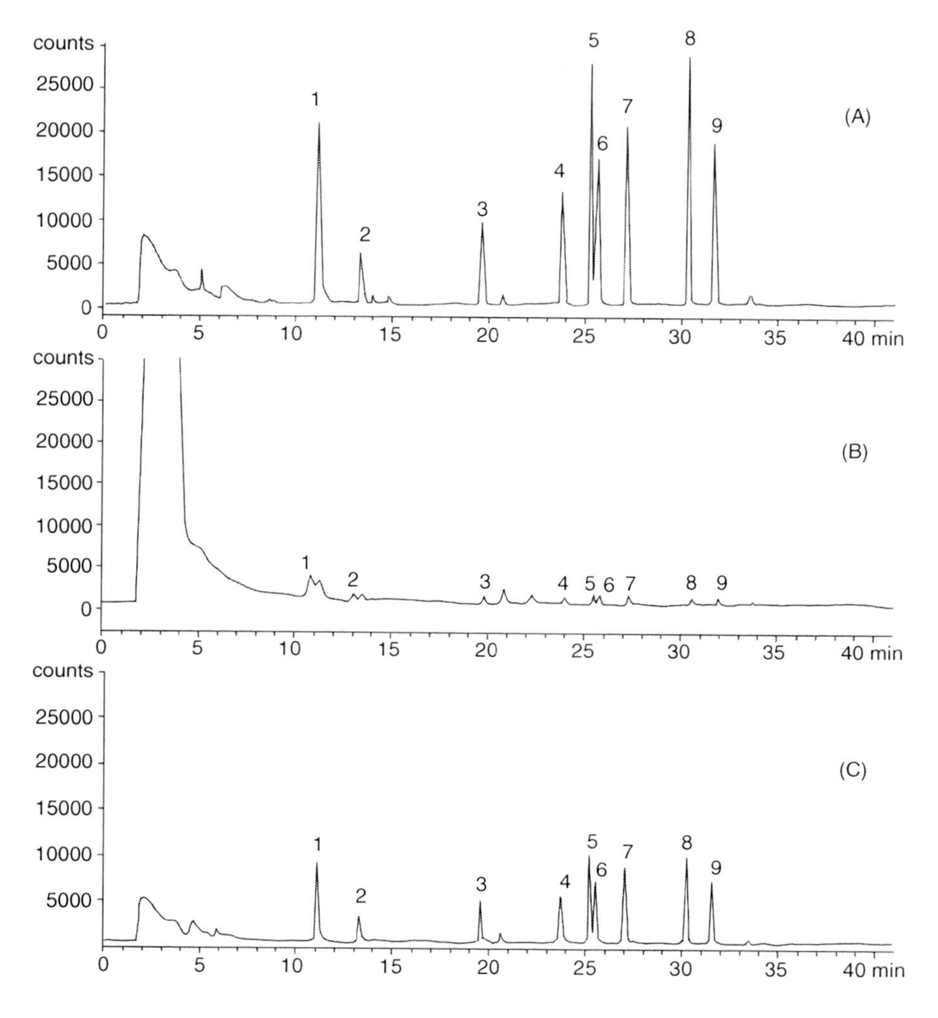

Figure 1. HSSPME-GC-ECD chromatograms of spiked milk samples: (A) skim milk; (B) full-fat milk; (C) full-fat milk after saponification. Peak identification: (1) PCB-28, (2) PCB-52, (3) PCB-101, (4) PCB-118, (5) PCB-105, (6) PCB-153, (7) PCB-138, (8) PCB-156, and (9) PCB-180.

selectively degrade many other interfering substances without affecting the PCBs.[1] Sets of preliminary HSSPME experiments were run after 2 mL of 20% NaOH was added to the samples. Figure 1C shows the chromatogram obtained by HSSPME for the full-fat milk with the addition of NaOH solution. When comparing chromatograms B and C in Figure 1, we can see the increase in response, as well as the lower background obtained, after saponification.

Excerpt 4D is from *Chemical Research in Toxicology*. The authors investigate the genotoxicity of three different chromium (III) compounds (chromium chloride, chromium nitrate, and chromium oxalate) and link several sets of results (we consider only three) to make their case. As we will see, the organization of these results is important for convincing the readers of their conclusions.

The first set of results (not included in excerpt 4D) presents data from a Pro-Tox (C) assay, a test that looks at 13 possible stress promoters that can be induced in bacteria by the chromium compound(s) under investigation. According to this test, both chromium chloride and chromium nitrate induced similar stress promoters, producing profiles indicative of DNA damage. Alternatively, chromium oxalate induced very few stress promoters, and its profile was not indicative of DNA damage. As the authors state (italics added): *Interestingly, none of the 13 stress promoters were induced when bacteria were treated with chromium oxalate* (Figure 2). The authors use the word *interestingly* to call the reader's attention to the observation that chromium oxalate is somehow different than chromium chloride and chromium nitrate. As you will see, this statement foreshadows results that are presented in excerpt 4D.

Interestingly, ...

Authors sometimes use the word *interestingly* to foreshadow a result that will be explained in more detail elsewhere in the paper. The word is most effective when used only once in a paper.

Presumably, for DNA damage to occur, chromium must enter the cell. The second set of results (excerpt 4D) addresses this issue. The authors look for chromium uptake inside *E. coli* cells using a technique known as flame atomic absorption spectroscopy (FAAS). These results build on the authors' first set of results, providing additional evidence that chromium oxalate is somehow different than the other chromium compounds studied.

Exercise 4.13

Read and analyze excerpt 4D to understand what the table and text reveal. Then answer the following questions:

a. What data in the table do the authors choose to emphasize in the text?

b. How does the information presented in excerpt 4D build on the results of the Pro-Tox (C) assay (see description of full article above)? What new insights do you have into the genotoxicity of Cr^{3+}?

c. How well do the authors adhere to table-formatting conventions (see chapter 16)?

Scientific Terms (excerpt 4D)

E. coli — *Escherichia coli*, bacteria that live in the human intestinal tract. Some strains of *E. coli* are harmless; others cause diarrhea-like symptoms. The Pro-Tox (C) assay and the Cr^{3+} tests use *E. coli* K-12 strains. (Note that the genus and species names are italicized and the *E.* is capitalized. See chapter 3.)

FAAS — Flame atomic absorption spectroscopy; the flame atomizes metals in solutions. Once in the gas phase, the atoms absorb UV–vis light, exciting electrons to higher energy levels. The amount of light absorbed is used to determine the metal concentration.

Excerpt 4D (from Plaper et al., 2002)

Chromium Accumulation in *E. coli* Cells. The cellular uptake of chromium is presented as milligrams of chromium per dry weight of treated *E. coli* cells (Table 1). FAAS measurements of total chromium concentrations in *E. coli* cells showed that only chromium chloride and chromium nitrate accumulate intracellularly but not chromium oxalate. As shown in Table 1, at all concentrations of chromium oxalate used, the quantity of chromium was below the detection limit.

Table 1. Uptake of Chromium by *E. coli* Cells Treated with Different Cr^{3+} Compounds

Cr^{3+} compound	concentration of added compound (mM)	intracellular content of Cr (mg of Cr/g of dry weight)
$CrCl_3 \times 6H_2O$	0	<0.021
	0.63	1.8
	1.25	6.7
$Cr(NO_3)_3 \times 9H_2O$	0	<0.021
	0.63	2.4
	1.25	5.7
$KCr(C_2O_4)_2 \times 3H_2O$	0	<0.025
	0.63	<0.022
	1.25	<0.02

In the third set of results (not included in excerpt 4D), the authors examine how Cr^{3+} affects gyrase, an enzyme that regulates the ability of supercoiled DNA to relax. Results, however, are reported only for chromium chloride, not for chromium oxalate or chromium nitrate. By omitting these latter two compounds, the authors illustrate what we call a broad-to-narrow approach. At the start of a research project, there are typically many variables; however, as knowledge is gained, some of these variables can be eliminated. In this case, chromium oxalate was eliminated because it does not enter the cell; chromium nitrate was eliminated because it mimics chromium chloride. Even if the authors had conducted the third test on all three compounds, the Results section is best written without this information. Authors risk diluting relevant results by including extraneous data that fail to advance their story. We will see the broad-to-narrow approach again in excerpt 4E.

Broad-to-Narrow Approach

Chemists often limit the number of variables in later experiments, based on information acquired in earlier work.

This approach also applies to writing. Authors begin by sharing results for many variables, then limit their focus to only those that advance their story.

Before moving on to excerpt 4E, we call your attention to two ways in which the concept of zero is addressed in excerpt 4D. First, we consider the concept of zero in measured concentrations (i.e., the concentrations reported in the last column of Table 1). Recall that no chromium oxalate was detected in the cells; however, the authors do not report this with a zero. Rather, they use the phrase "below the detection limit" in the text and the less-than symbol (e.g., <0.025 mg/g) in Table 1, which puts an upper limit on the amount of chromium oxalate present. Novice writers might (incorrectly) suggest that "no chromium was present" in the text and use a zero in the table (0 mg/g). Such uses of zero, however, are incorrect, because (for measured concentrations) zero varies with the sensitivity of the detecting instrument. For example, on one instrument, zero will be less than one part per million; on a more sensitive instrument, zero will be less than one part per billion. Instead of reporting zero, authors report that the measurement was below the detection limit for that instrument. Some common ways to express this concept in the text and table are as follows:

In the text	X was not detected
	X could not be detected
	We did not detect X
	X was below detection limits
In a table	<0.02
	bdl (below detection limits)
	ND (not detected)

Alternatively, it is correct to use zero to indicate added amounts of compounds. You may say "that no chromium oxalate was added" and use a "0" to indicate this in a table. For example, the second column of Table 1 includes three zeroes, indicating that no chromium compounds were added to the cells in these three experiments.

Zero?

Measured concentrations should not be reported as zero, even if the substance is not detected. A more sensitive instrument (one with a lower detection limit) might be able to detect it. In other words, even if you don't "see" the substance, it might be there!

Excerpt 4E is taken from an article in *Chemical Research in Toxicology* and involves the toxicity of fine particulate matter, airborne particles with effective diameters ≤2.5 μm (also known as $PM_{2.5}$). The fine particulate was collected using a $PM_{2.5}$ monitor. Ambient air is pulled through the monitor, diverting the larger particles (≥2.5 μm) and capturing only the smaller ones onto a filter. Such fine particles arise from a number of sources including industrial emissions, vehicle exhaust, and forest fires and may lead to asthma, bronchitis, and possibly cancer.

Like excerpt 4D, the authors of excerpt 4E present multiple sets of results and use a broad-to-narrow approach. In the first set of results (not shown), the authors display electron paramagnetic resonance (EPR) spectra of $PM_{2.5}$ collected at five U.S. cities. The EPR data indicate that free radical concentrations in the $PM_{2.5}$ samples are high and at times exceed the free radical concentrations in cigarette smoke. In the second set of results (not shown), the authors present results from a comet assay, an electrophoresis technique. Cells without DNA damage migrate together as a group, forming what looks like a comet's head; cells with DNA damage migrate faster and at different rates, forming what looks like a comet's tail. The tail moment, a measurement of the tail length, is used to assess the amount of DNA damage. Two types of human cells (K562 myeloid leukemia cells and IB3–1 lung epithelial cells) were treated with extracts from the $PM_{2.5}$ filters and from "blank" (clean) filters. Cells treated with $PM_{2.5}$ extracts showed significantly greater tail moments and more DNA damage (74–90%) than cells treated with blank extracts (0–14%).

Thus far, the results have highlighted two findings: (1) $PM_{2.5}$ contains free radicals and (2) $PM_{2.5}$ causes DNA damage. The next logical step is to link the free radicals to the DNA damage. As yet, the authors have not done this. Perhaps other toxins in the $PM_{2.5}$, such as metals, are causing the DNA damage. Can the authors strengthen the case for the free radicals?

To this end, the authors perform a second comet assay. Only limited results, however, are shared; $PM_{2.5}$ from only one city (Baton Rouge) was tested using only one cell line (K562 myeloid leukemia cells), thereby following a broad-to-narrow approach. The results are presented in excerpt 4E. As shown in Figure 3 (excerpt 4E), cells were exposed to (A) a blank extract, (B) a $PM_{2.5}$ extract, (C, D, E) a $PM_{2.5}$ extract mixed with one of three different free radical scavengers, (F) a positive control, and (G, H) a $PM_{2.5}$ extract mixed with one of two different metal-ion chelators. The free radical scavengers remove free radicals; the metal-ion chelators remove metal ions, thereby preventing them from causing DNA damage. If damage occurs without the additive, but disappears or is reduced with the additive, then the radical (or metal) is likely causing the DNA damage.

With this background in mind, complete exercise 4.14 as you read excerpt 4E. Can you figure out what is causing the damage in the cell DNA? What is the conclusion of the authors' story?

Scientific Terms (excerpt 4E)

Chelators	Organic compounds to which metal ions (e.g., Fe^{3+}) bind (or chelate) to form a complex, where the metal ion is in the center of a ring, coordinated to two or more organic species
Electrophoresis	The movement of particles in a gel due to electrodes (positive and negative) applied to opposite ends of the gel
Free radical	A highly reactive species with an unpaired electron and no charge (e.g., the hydroxyl radical, $HO^{\cdot}$)
In vivo	In the living cell (written without italics)
In vitro	In glass, like a test tube (written without italics)
Positive or negative controls	Substances that are known to give a positive or negative result; controls are used to check that the experimental design and instrumentation are working correctly.

Exercise 4.14

Use the following questions to unravel the authors' story of scientific discovery in excerpt 4E:

a. Explain why the authors treated the cells with a blank filter extract and H_2O_2. What do these cell treatments tell us?

b. Using Figure 3 and Table 2 in excerpt 4E, decide which of the following is (are) most likely causing DNA damage (there may be more than one correct answer):

- free radicals
- Cu^{2+}/Cu^{+}
- Fe^{3+}/Cu^{2+}

c. The authors use "0" in Table 2 to report the percentage of cells with DNA damage. Explain why the use of "0" is correct in this instance. (Hint: "0" is used in a relative way and is related to the percent damage observed in the blank filter extracts.)

d. Consider the order in which the authors reveal their results, beginning with the EPR and comet assay tests described in text (before the excerpt), and ending with the comet assay test on the K562 cells treated with the Baton Rouge extract (Figure 3, in excerpt 4E).
   1. List the order in which the authors reveal their results.
   2. Offer a rationale for why the authors followed this order.
   3. Do you think the order is chronological? Explain.

e. The authors of excerpts 4D and 4E begin their Results sections by reporting findings that are quite broad in scope (i.e., the toxicity of Cr^{3+} from three compounds and the toxicity of $PM_{2.5}$ from five cities), but they end with findings with a more narrow focus (i.e., Cr^{3+} from only $CrCl_3$ and $PM_{2.5}$ from only Baton Rouge). Why do you think the authors shift from a broad to a specific focus? How might this approach make the story of discovery easier to follow?

Excerpt 4E (adapted from Dellinger et al., 2001)

Results

To understand the mechanistic basis for the DNA damage, additional in vitro and in vivo experiments were undertaken that involved the use of free-radical scavengers and metal chelators to determine what effect they may have on $PM_{2.5}$-mediated DNA damage. Because of their greater sample masses and availability, these experiments were performed using the 5-day samples collected from the Baton Rouge site.

These studies were performed in vivo using human myeloid leukemia K562 cells. This cell line was chosen because it lacks p53-induced apoptosis that causes double-strand breaks and whose presence could therefore complicate the interpretation of our results (*40*). As shown in panel A of Figure 3 [see p. 136], extracts from a blank, unloaded filter left 72% of the DNA from K562 cells undamaged, whereas extracts from filters containing $PM_{2.5}$ left just 24% undamaged DNA (panel B). Superoxide dismutase (SOD) (panel C), catalase (panel D), and catalase plus SOD (panel E) all provided complete protection of the DNA. Exposure of the cells to 100 μM hydrogen peroxide (panel F) was used as a positive control (35) and produced 100% damaged DNA. As shown in panel H, the Fe^{3+} and Cu^{2+} chelator deferoxamine provided almost complete protection, but the Cu^{2+}/Cu^{+} chelator bathocuproine, as shown in panel G, provided only partial protection.

The fitted means and standard errors for log-transformed comet tail moments, as well as the percentage of cells exhibiting extensive DNA damage (e.g., cells labeled 3 and 4) are reported in Table 2 [see p. 137]. An adjusted *p* value indicated no differences existed between cells treated with extracts from exposed filters or with hydrogen peroxide. Cellular responses were significantly different ($P < 0.05$) between unloaded $PM_{2.5}$ filter extracts and loaded $PM_{2.5}$ extracts as well as extracts containing deferoxamine.

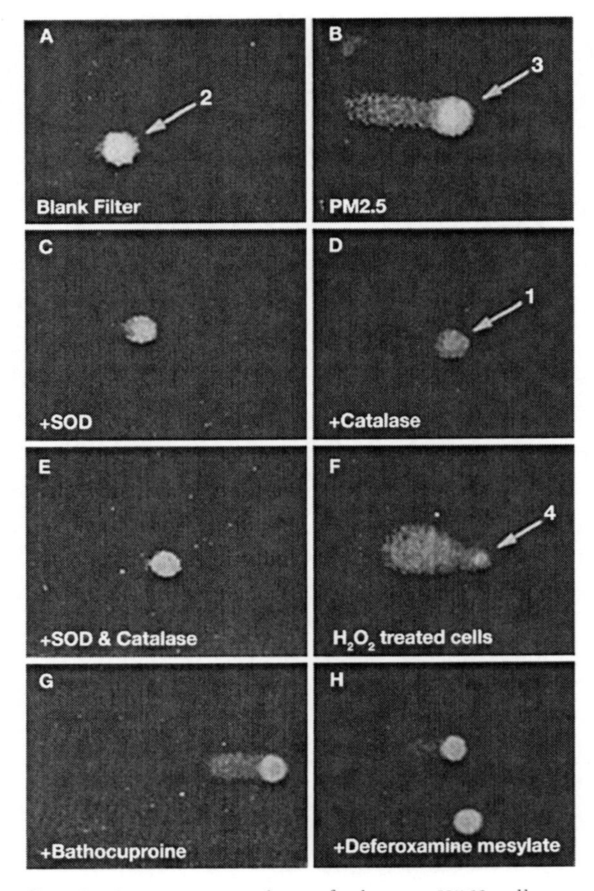

Figure 3. Comet assay is shown for human K562 cells exposed to an extract produced from particulate matter released from sample $PM_{2.5}$ filters, to an extract derived from an unloaded filter, or to hydrogen peroxide (100 μM) as a control. Cells with little or no DNA damage are labeled 1 and 2, and those with extensive damage are labeled 3 and 4.

In the articles from which excerpts 4D and 4E were taken, reported results represent only a small fraction of the data actually collected. The authors found ways to condense their data as they wrote their papers. In excerpt 4D, the authors condensed their data by reporting only representative results (i.e., results from three soils instead of all seven). In excerpts 4D and 4E, the authors initially reported multiple variables (i.e., three chromium compounds and five cities) but ended with a narrower focus (i.e., one chromium compound and one city). In each case, the readers benefited from the researchers' hindsight. Learning to tell your story of scientific discovery in retrospect, by reorganizing your data and highlighting only the most illustrative pieces, is an essential skill in effective writing.

Table 2. Results of Comet Assays for K562 Cells Treated with Baton Rouge $PM_{2.5}$ Extract

cell treatment (label in Figure 3)	adjusted tail moment (±SE)	cells with DNA damage (%)
blank filter extract (A)	16.63 (0.19)	28
$PM_{2.5}$ extract alone (B)	18.02 (0.20)	76
$PM_{2.5}$ extract plus free radical scavenger:		
SOD (C)	15.25 (0.19)	0
catalase (D)	13.60 (0.19)	0
SOD & catalase (E)	15.36 (0.19)	0
positive control: H_2O_2 (100 μm) (F)	18.03 (0.20)	100
$PM_{2.5}$ extract plus metal chelator:		
bathocuproine (Cu^{2+}/Cu^{+}) (G)	15.80 (0.78)	72
deforoxamine (Fe^{3+}/Cu^{2+}) (H)	14.18 (0.14)	4

The final two excerpts (excerpts 4F and 4G) illustrate a few ways in which results from a chemical synthesis can be described. The Results sections of synthesis papers often include a hint of what was tried and failed as well as what ultimately succeeded. Reading these excerpts will give you a feel for the trial-and-error process of science and the "aha" experience that sometimes occurs when a successful approach is discovered.

Excerpt 4F is taken from an article written by Demko and Sharpless. (Barry Sharpless was a co-recipient of the Nobel Prize in Chemistry in 2001 for his work on chirally catalyzed oxidation reactions.) In this article, the authors propose a way to synthesize aromatic tetrazoles from nitriles in water, using only sodium azide and a zinc salt. Water, despite its obvious advantages (i.e., safe and inexpensive), rarely succeeds as a solvent in organic synthesis. Thus, a synthesis that uses water successfully is an important scientific accomplishment.

We begin our analysis of excerpt 4F by examining its use of compound labels. You have seen compound labels before. The ubiquitous R group (R, R′, R″,... or R_1, R_2, R_3,...) in organic chemistry, used to connote radical or residue (e.g., CH_3– or CH_3CH_2– in R–Br), is one example. Another example is when authors include a bolded number (or bolded number and letter) immediately after the name of a compound in the text or table. Consider the following example, where **1** and **2** are the compound labels:

> A mixture of 2-bromo-4-methoxyphenone **1** was reacted with *p*-toluenesulfinic acid sodium hydrate **2**.

After being introduced, **1** and **2** can be used instead of the compound names, thereby saving considerable space. The numbers are bolded so that they can be easily differentiated from other numbers in the sentence, such as citation numbers, which are never bolded. In some journals, the bolded number is placed inside parentheses.

Compound Labels

Compound labels, in addition to being concise, focus attention on important features of a reaction. They allow authors to communicate the versatility and generality of a reaction, mechanism, or scheme efficiently. (See appendix A for more details.)

Let's examine the use of compound labels in excerpt 4F. The authors first use compound labels in their eq 1 (our eq 4.1) to illustrate the conversion of an unspecified nitrile **1** to an unspecified tetrazole **2**:

$$\underset{\mathbf{1}}{R{-}C{\equiv}N} \xrightarrow[\text{water, reflux}]{1.1\ \text{equiv}\ NaN_3,\ 1.0\ \text{equiv}\ ZnBr_2} \underset{\mathbf{2}}{\text{R-tetrazole (N–NH, N=N)}} \qquad (4.1)$$

In Table 2 of excerpt 4F, labels for specific tetrazoles (**2a**, **2b**, **2c**,... **2j**) are introduced. Table 2 also provides the information needed to unlock the specific structures of **1** and **2** in the authors' eq 1. For example, in table entry **2a**, the R group on the tetrazole is phenyl; hence, R is also phenyl on the nitrile, and their eq 1 becomes explicitly (our eq 4.2)

$$\underset{\mathbf{1a}}{C_6H_5{-}C{\equiv}N} \xrightarrow[\text{water, reflux}]{1.1\ \text{equiv}\ NaN_3,\ 1.0\ \text{equiv}\ ZnBr_2} \underset{\mathbf{2a}}{\text{C}_6\text{H}_5\text{-tetrazole (N–NH, N=N)}} \qquad (4.2)$$

Exercise 4.15

Glance ahead to excerpt 4F. Examine the connections between the compound labels used in eq 1 and Table 2. Use these connections to complete eq 1 when **2c** is the product formed.

In journals such as *The Journal of Organic Chemistry* and *Organic Letters*, equations are often included in tables, and the equation and table entries are linked by compound labels. Exercises 4.16 and 4.17 illustrate this convention.

Exercise 4.16

Consider the following table, adapted from Usugi et al. (2004):

Table 1. $GaCl_3$-Mediated Reaction of Various Alkynes with Disulfide[a]

R—≡—H (**1**) + R′SSR′ (**2**) —($GaCl_3$, PhH, 0°C, 30 min)→ (R)(R′S)C=C(SR′)(H) (**3**)

entry	R	1	R′	2	3	yield (%)
1	*n*-Bu	**1a**	Ph	**2a**	**3a**	83
2	*i*-Pr	**1b**	Ph	**2a**	**3b**	58
3	*t*-Bu	**1c**	Ph	**2a**	**3c**	50
4	Ph	**1d**	Ph	**2a**	**3d**	87
5	Ph	**1d**	*p*-Tol	**2b**	**3e**	84

[a] Disulfide (0.50 mmol), alkyne (0.75 mmol), $GaCl_3$ (0.50 mmol), and PhH (2 mL) were employed.

The table includes an equation depicting a general chemical reaction (1 + 2 → 3). The table entries describe five specific reactions by defining R and R′ in each case. Using Table 1, write out the specific reactions for entries 2–5. Entry 1 has been completed for you (R = *n*-Bu and R′ = Ph):

n-Bu—≡—H (**1**) + PhSSPh (**2**) —($GaCl_3$, PhH, 0 °C, 30 min)→ (*n*-Bu)(PhS)C=C(SPh)(H) (**3**)

Exercise 4.17

Imagine that you are in a research group that measures the relative rates of nucleophilic substitution (S_N2) reactions. The reactions that you have investigated are listed below. Prepare a table, like Table 1 in exercise 4.16, to summarize these reactions. Include the following equation at the top of the table: R–Br + Nu^- → R–Nu + Br^-. Use compound labels to link the equation and table entries.

Reaction						Relative Rate
t-BuBr	+	Cl^-	→ *t*-BuCl	+	Br^-	<1
*neopentyl*Br	+	Cl^-	→ *neopentyl*Cl	+	Br^-	1
i-PrBr	+	Cl^-	→ *i*-PrCl	+	Br^-	500
EtBr	+	Cl^-	→ EtCl	+	Br^-	40,000
MeBr	+	Cl^-	→ MeCl	+	Br^-	2,000,000
MeBr	+	CH_3COO^-	→ $MeOOCH_3$	+	Br^-	1,000,000
MeBr	+	HO^-	→ MeOH	+	Br^-	32,000,000
MeBr	+	I^-	→ MeI	+	Br^-	200,000,000

Excerpt 4F does a good job of moving the reader's attention back and forth between the table and text, highlighting important results, but not repeating the data. As you read the text and table together, it becomes clear that the authors have organized their table not in an arbitrary way, but according to nitrile type (aromatic, electron-poor aromatic, electron-rich aromatic, etc.). This organization helps the reader see trends in the data.

Excerpt 4F (adapted from Demko and Sharpless, 2001)

[From Introduction section]

We report here a safer and exceptionally efficient process for transforming nitriles **1** into tetrazoles **2** in water; the only other reagents are sodium azide and a zinc salt (eq 1).

$$\underset{\mathbf{1}}{R-C\equiv N} \xrightarrow[\text{water, reflux}]{\text{1.1 equiv NaN}_3\text{, 1.0 equiv ZnBr}_2} \underset{\mathbf{2}}{\text{5-R-1}H\text{-tetrazole}} \qquad (1)$$

[From a combined R&D section, although no heading was included]

A wide variety of nitriles were converted to tetrazoles on a 20 mmol scale. Other things being equal, the more electron-poor a nitrile, the faster it reacts. Aromatic nitriles (see Table 2) with a variety of substituents (**2a**,**b**,**c**,**i**) reach completion within several days at reflux. Electron-poor aromatic and heteroaromatic nitriles, such as 2-cyanopyridine and cyanopyrazine (**2d**,**e**), are complete within a few hours. Some electron-rich aromatic nitriles (**2f**,**g**) require higher temperatures, which are achieved using a sealed glass pressure reactor. Ortho-substituted aromatic nitriles are the most challenging, sometimes proceeding at reflux (**2h**), but often requiring much higher temperatures (**2j**). We have not been able to achieve significant conversion of any aromatic nitriles bearing an sp^3-hybridized substituent in the ortho position.[18]

Table 2. Aromatic Tetrazoles

entry	tetrazole[a]	temp/time	yield (%)	mp (°C)
2a		reflux/24 h	76	215–216
2b		reflux/24 h	94	220
2c		reflux/48 h	86	231–232
2d		reflux/6 h	79	211
2e		reflux/2 h	83	193–195
2f		140 °C/24 h	96	234–236
2g		140 °C/48 h	73	205–207
2h		reflux/48 h	64	228–230
2i		reflux/12 h	67	158–160
2j		170 °C/48 h	67	150

[a] These reactions were run on 20 mmol scale.

Excerpt 4G involves a variation of a bioorganic reaction known as the Strecker synthesis. The Strecker synthesis is a two-step process that leads to the formation of an α-amino acid, the building block of proteins. The general structure of an α-amino acid involves a tetrahedral carbon (the α-carbon) bonded to an amino group ($-NH_2$), a carboxylic acid (–COOH), a hydrogen (–H), and a variable side chain (–R). Of the 20 naturally occurring α-amino acids found in proteins, 19 are chiral, all with a counterclockwise (*S*) configuration. This makes the (*S*) amino acid a desirable synthetic target and motivates the asymmetric Strecker synthesis. Unlike the Strecker reaction, which results in a racemic (50:50) mixture of the (*S*) and (*R*) α-amino acid, the asymmetric Strecker reaction leads to an enantiomeric excess (ee) of the (*S*) α-amino acid.

R vs. (*R*)

R refers to a variable side chain, radical, or residue, for example,

$R-CH_2Br$

(*R*) and (*S*) are stereochemical terms, referring to clockwise and counterclockwise orientations, respectively. According to *The ACS Style Guide*, (*R*) and (*S*) are italicized and placed inside parentheses, for example,

The reaction leads to the (*R*)-amide.

Before we consider excerpt 4G, we first walk you through the asymmetric Strecker reaction (scheme 4.1). (The authors did not include a similar scheme in their article because the reactions depicted in scheme 4.1 are familiar to their audience.) First, note the use of compound labels in scheme 4.1. For clarity, we use the same labels that are used in excerpt 4G. The synthesis begins by reacting a chiral primary amine, (*R*)-phenylglycine amide (*R*)-**1**, with an aldehyde **2**. (We use a generic aldehyde in scheme 4.1; excerpt 4G uses pivaldehyde.) The product yields an imine that retains the chiral carbon (shown with an asterisk). The imine is reacted with NaCN in acetic acid (HOAc) to form a pair of diastereomeric amino nitriles, each with two chiral carbons. Following hydrolysis, the nitriles are converted to a pair of diasteromeric α-amino acids, (*R*,*S*)-**3** and (*R*,*R*)-**3**. Recall that diastereomers, unlike enantiomers, have different physical properties and can be separated based on these differences. (*R*,*S*)-**3** is ultimately separated from (*R*,*R*)-**3** and converted to the (*S*)-α-amino acid (not shown).

With this in mind, consider Table 1 in excerpt 4G. The authors include only the initial reactants (*R*)-**1** and **2** and the final products (*R*,*S*)-**3** and (*R*,*R*)-**3** in their

Scheme 4.1

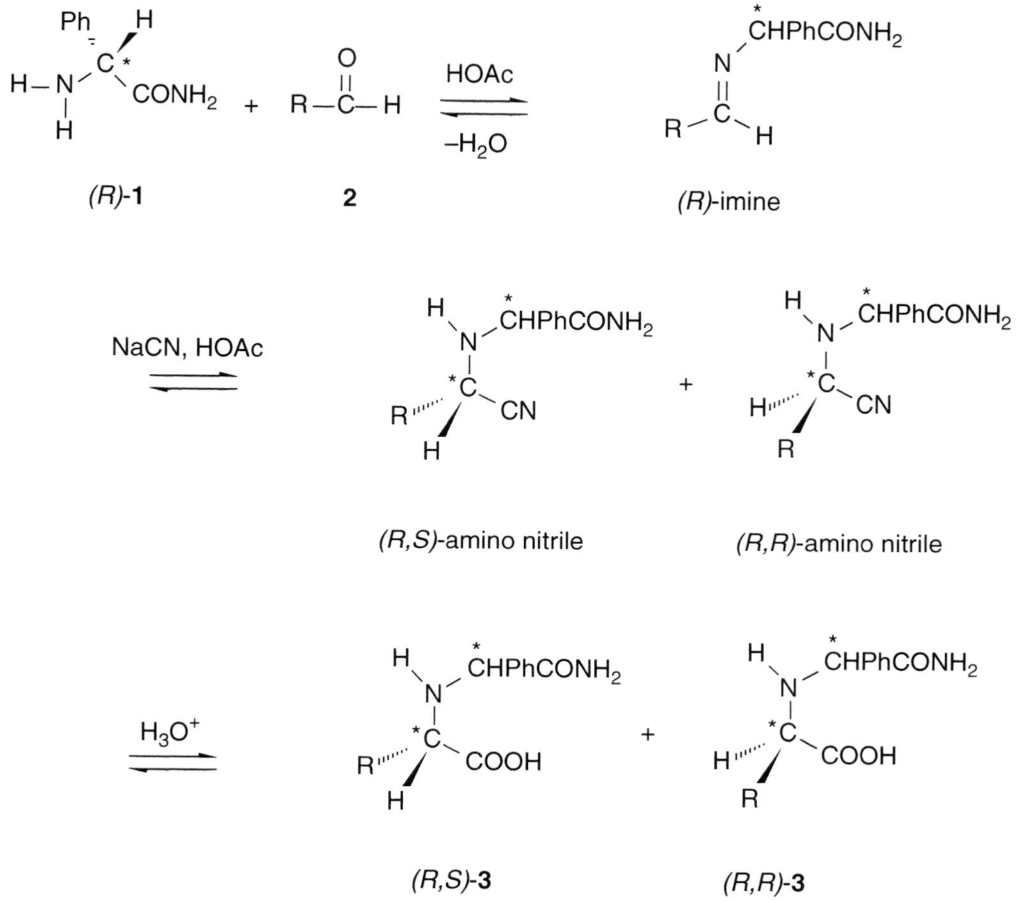

(*R*)-**1** + **2** $\xrightleftharpoons[-H_2O]{HOAc}$ (*R*)-imine

$\xrightleftharpoons{NaCN,\ HOAc}$ (*R,S*)-amino nitrile + (*R,R*)-amino nitrile

$\xrightleftharpoons{H_3O^+}$ (*R,S*)-**3** + (*R,R*)-**3**

table. The arrow (↓) following (*R,S*)-**3** indicates that (*R,S*)-**3** preferentially precipitates out of solution, leaving (*R,R*)-**3** in solution, and successfully separating the two diastereomers.

Schemes

Schemes are used to illustrate progress in a chemical reaction (see chapter 16).

As you read the rest of the table and text in excerpt 4G, you will see that the authors tried a number of different reaction conditions to maximize the yield of (*R,S*)-**3**, the diastereomer that is ultimately converted to the (*S*) α-amino acid.

(These final steps are not included in excerpt 4G, but if you are interested, the full article is included at the end of chapter 2.) The text describes these efforts, following the order of the entries in Table 1. First, different solvents were tried beginning with methanol (entry 1). Next, various alcohol mixtures were tried, but yields dropped (entries 2–4). The yield improved slightly when water was mixed with methanol (entry 5); hence, water alone was tried at different temperatures (entries 6–9). This truly is a story of scientific discovery! The readers learn both what did not work and what did. This approach is quite common in papers describing organic synthesis.

Scientific Terms (excerpt 4G)

Diastereomer	A compound with two stereocenters where one stereocenter is the same and one is different from its isomer (e.g., (*2R*, *3R*)-**1** and (*2R*, *3S*)-**1** are diasteomers); diastereomers have different physical properties
Diastereomeric excess (dr)	An excess of one diastereomer in a reaction that leads to a pair of diastereomers (e.g., 90% (*2R*, *3R*)-**1** and 10% (*2R*, *3S*)-**1**)
Enantiomeric excess (ee)	An excess of one enantiomer in a reaction that leads to a pair of enantiomers (e.g., 90% (*2R*, *3R*)-**1** and 10% (*2S*, *3S*)-**1**)
Enantiomer	A nonsuperimposable mirror image of a molecule (e.g., (*2S*, *3S*)-**1** and (*2R*, 3*R*)-**1** are enantiomers); enantiomers have the same physical properties
Racemic mixture	An equal (50:50) mixture of two enantiomers

Exercise 4.18

Given the background provided on the asymmetric Strecker reaction, complete the following tasks as you read excerpt 4G:

a. The goal of this reaction is to maximize the preferential crystallization of (*R*,*S*)-**3**. Several different attempts were tried; some worked and some did not. How do the authors share this process of discovery with the reader? Include what was tried first and what ultimately succeeded in your answer.

b. Let's pretend you are a member of Boesten's research group. You were the first to tackle this reaction. You intuitively chose water as the solvent (at 70 °C) and let your reaction run for 24 h; these turned out to be the ideal conditions. Over the next several months, you tried other solvents and reaction conditions, but never achieved better results. Would you still write the story as it appears in excerpt 4G? Explain why or why not.

c. How do the authors organize the results in Table 1? Argue for or against reorganizing Table 1 by percent yield (lowest to highest).

d. How do the authors highlight trends in their text without repeating the data in Table 1? Give an example.

e. Excerpt 4G describes the synthesis of (*R*,*S*)-**3**, a critical intermediate in the synthesis of the α-amino acid (*S*)-*tert*-leucine. The article goes on to present several additional sets of results. We have summarized these below, but in a scrambled order. Propose a logical order of presentation for these findings. Justify your proposed order.

   1. An X-ray structure of (*R*,*S*)-**3**, confirming its absolute configuration
   2. A statement that points out that further examples of this reaction are under investigation using (*R*)-phenylglycine amide with different aldehydes
   3. A graph that shows how the formation of (*R*,*S*)-**3** depends on reaction time (0.5–30 h)
   4. A description of the synthetic steps taken to convert (*R*,*S*)-**3** to the desired product (an α-amino acid)
   5. Results from an alternate synthesis of an intermediate amino nitrile (similar to (*R*,*S*)-**3**) that uses a ketone as a reactant instead of the aldehyde **2**

Excerpt 4G (adapted from Boesten et al., 2001)

The asymmetric Strecker reaction of (*R*)-phenylglycine amide **1**, pivaldehyde **2**, and HCN generated in situ from NaCN and AcOH was studied (Table 1). Amino nitriles (*R*,*S*)-**3** and (*R*,*R*)-**3** were obtained in 80% yield in a ratio of 65:35 by stirring an equimolar mixture of **1** (as AcOH salt) with **2** and NaCN in MeOH overnight at room temperature, followed by evaporation of the solvent (entry 1). The diastereomeric ratio (dr) of (*R*,*S*)-**3** and (*R*,*R*)-**3** was determined by ^{1}H NMR on the basis of the relative integration between the *t*-Bu signals at 1.05 ppm for (*R*,*S*)-**3** and 1.15 ppm for (*R*,*R*)-**3**. . . .

Because in methanol crystallization of amino nitrile **3** did not take place, first the solvent was varied in order to attempt to find conditions for a crystallization-induced asymmetric transformation. At a MeOH/2-PrOH ratio of 1/9, the amino nitrile (*R*,*S*)-**3** was isolated in 51% yield and dr 99/1 (entry 2). Other combinations of alcoholic solvents failed to lead to a higher yield of precipitated (*R*,*S*)-**3** in high dr (entries 3 and 4). On further screening of solvents, it was observed that upon addition of H_2O to the methanol solution selective precipitation of amino nitrile (*R*,*S*)-**3** occurred giving (*R*,*S*)-**3** and (*R*,*R*)-**3** in a ratio of 81:19 and 69% yield (entry 5). The asymmetric Strecker reaction was further studied in H_2O alone using temperature as a variable. The results of these experiments are given in Table 1 (entries 6–9). After addition of NaCN/AcOH at 23–28 °C to (*R*)-phenylglycine amide **1** and pivaldehyde **2** in H_2O, the mixture was heated to the indicated temperatures.

Table 1. Asymmetric Strecker Reactions of (*R*)-Phenylglycine Amide **1** and Pivaldehyde **2**

(*R*)-**1** + **2** → (NaCN, HOAc; solvent, time, temp) → (*R*,*S*)-**3** + (*R*,*R*)-**3**

entry	solvent	temp (°C)	time (h)	yield (%)[a]	dr (*R*,*S*)-**3**/(*R*,*R*)-**3**[b]
1	MeOH	rt	20	80	65/35
2	MeOH/2-PrOH, 1/9[c]	rt	22	51	99/1
3	2-PrOH	rt	22	84	88/12
4	2-PrOH/*t*-BuOH, 4/1[c]	rt	20	65	96/4
5	MeOH/H_2O, 35/1[c]	rt	20	69	81/19
6	H_2O	55	24	81	85/15
7	H_2O	60	24	84	96/4
8	H_2O	65	24	84	98/2
9	H_2O	70	24	93	>99/1

[a] Isolated yield after evaporation of the solvent (entry 1) or filtration of precipitated amino nitrile **3** (entries 2–9).

[b] The diastereomeric ratio was determined by ^1H NMR spectroscopy.

[c] Ratio in v/v.

4D Writing on Your Own: Tell the Story of Scientific Discovery

Identify the major trends that you will highlight in the text of your Results section. If applicable, decide how you will highlight unexpected results and/or compare important findings.

Using as guides (1) your gathered data and notes, (2) your sequenced sets of results and graphics, (3) figure 4.1, and (4) chapter 16, write the text for your Results section. Start with the first move (set the stage by transitioning from the Methods section and referring to a graphic) and then continue with the second move (tell your story of scientific discovery). Be sure to follow a logical sequence of events as you tell your story.

Part 2: Analyzing Writing across the Results Section

In examining Results sections move by move, we looked at how authors refer to figures and tables, how they use compound labeling, and how they highlight trends in the data. We examined how to report values below detection limits and how to use R to consolidate reactions in a synthesis paper. In this part of the chapter, we analyze a few writing conventions that are characteristic of the entire Results section, including verb tense, voice, and word choice.

Past and Present Tense

Unlike the Methods section, which is written primarily in past tense, both past and present tense are used in the Results section. In general, present tense is used (1) to refer the reader to a figure or a graph and (2) to make statements of general knowledge expected to be true over time. Consider the following examples:

Present Tense Used to Refer to a Figure

The experimental desorption isotherms... *are* presented in Figure 1. (From Jozefaciuk et al., 2003)

Figure 1A *shows* the ECD chromatogram obtained for the skim milk. (From Llompart et al., 2001)

Present Tense Used to Indicate Knowledge Thought To Be True over Time

Saponification of fats... *facilitates* the release of PCBs from fatty matrixes. (From Llompart et al., 2001).

Pure crystals of 2,2-diphenyl-1-picrylhydrazyl, the stable, low-molecular-weight free radical, *contain* about 2×10^{21} radicals/g. (Adapted from Dellinger et al., 2001)

Present Tense and Past Tense

Use the present tense in the Results section to refer to a graphic and to make statements about knowledge expected to be true over time; in other cases, use the past tense.

Past-Tense Verbs in Results Sections

The 15 most frequent past-tense verbs in Results sections, in order of frequency (based on an analysis of 60 Results sections from ACS journals), are as follows:

1. observed
2. obtained
3. showed
4. found
5. gave
6. revealed

7. produced	8. formed	9. prepared
10. studied	11. reported	12. catalyzed
13. used	14. led	15. resulted

Passive and Active Voice

Recall from chapter 3 that passive voice allows writers to remove the human subject from a sentence, allowing the writer to focus on the science rather than the scientists. One way to test if a sentence is in passive voice is to see whether you can add "by someone" to the end of it:

Passive	The mixture was stirred (by someone).
Not passive (not correct)	We stirred the mixture (by someone).

In figure 3.3 (chapter 3), we reported the frequencies of passive voice in each section of a journal article. If you look back at figure 3.3, you will see that passive voice is used more frequently in Methods sections than in Results (or Discussion) sections. This distribution suggests that both active voice and passive voice are used in Results sections. Past and present tense, when combined with active and passive voice, form four different tense–voice combinations. Each combination has its own function, several of which are illustrated in table 4.1.

Table 4.1 Common functions of different verb tense–voice combinations in Results sections.

Function	Tense–Voice Combination	Example
To describe specific results in your work	Past–active	Other combinations of alcoholic solvents *failed* to lead to a higher yield. (From Boesten et al., 2001)
To describe specific steps in your work	Past–passive	Initial HSSPME experiments *were performed* using spiked skimmed and full-fat milk samples. (From Llompart et al., 2001)
To state scientific "truths" or knowledge	Present–active	PCBs *are* more strongly retained in the sample matrix as the fat content *increases*. (From Llompart et al., 2001)
To refer to a figure or table	Present–active	Figure 1 *shows* as an example the mass spectrum of the PFBOA derivative of methional. (From Vesely et al., 2003)
	Present–passive	Experimental adsorption isotherms for the RAMEB-treated soils *are presented* in Figure 1. (From Jozefaciuk et al., 2003)

Exercise 4.19

Follow steps a–c as you examine sentences 1–5 below (adapted from Weston et al., 2004):

a. Using table 4.1 as a guide, determine the function of each sentence (1–5).
b. Identify the verb tense and voice used in each sentence.
c. Decide whether the verb tense used is appropriate given the function of the sentence. If the tense used is inappropriate, rewrite the sentence so that it is more appropriate for a Results section.

1. This assumption is reasonable because the toxicity of pyrethroids to benthic organisms is predictable from the equilibrium partitioning-derived pore water concentration (*8*), and the pyrethroids in this study have K_{oc} values comparable to those of cypermethrin (*10*).
2. A toxicity unit (TU) approach was used to identify pesticides potentially responsible for observed toxicity.
3. Esfenvalerate concentrations are ≥0.5 TU in five samples.
4. Sediments of the tailwater ponds not only have the highest concentrations of many pesticides but also prove to be highly toxic.
5. TU calculations for samples not toxic to *C. tentans* are shown in Table 3.

Use of "We"

Historically, the use of *we* (and other personal pronouns, e.g., *I* and *our*) in scientific writing has been controversial. Those opposed to the use of *we* argue that it makes the writing sound less objective; hence, many scientists (particularly

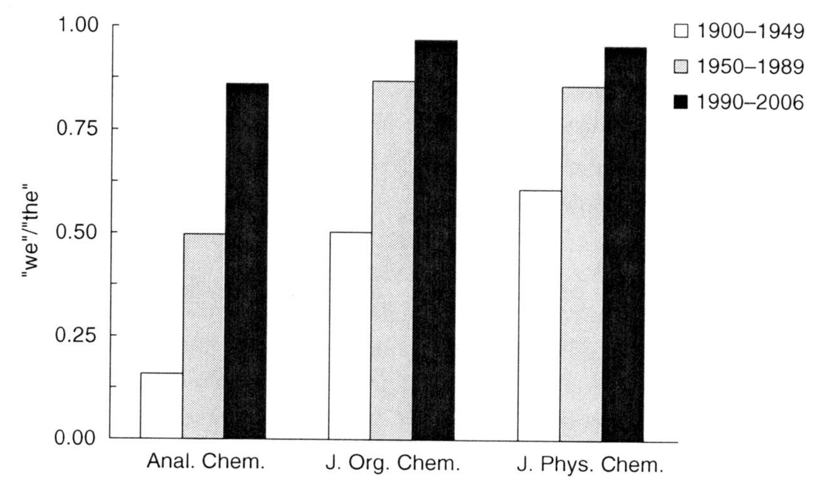

Figure 4.2 The number of documents using *we* at least once (relative to the number using *the*) over three time periods, determined using the ACS Journals Search. (Note: *J. Phys. Chem.* includes *J. Phys. Chem. A* and *B* after 1996.)

analytical chemists) have been taught to avoid *we* entirely. This trend is reflected in the published literature. For example, figure 4.2 shows the number of articles that included the word *we* (normalized against the number of articles that included the word *the*) in three different chemistry journals during three time periods. Until 1990, analytical chemists used *we* far less frequently than their organic- and physical-chemistry colleagues. Today, however, *we* appears at least once in more than 85% of the documents published in these three journals.

"We" in the R&D Section

In a computer-based analysis of 60 Results and Discussion sections, the word *we* occurred only 3 times per 1000 words. Thus, its use is quite rare.

The most compelling reason to use *we* in the Results section is to highlight a decision or choice made while conducting your work.

The ACS Style Guide advises against using phrases such as "we believe", "we feel", and "we can see".

Despite its increased frequency over time, the use of *we* is still restricted. In the Results section, where data are to be presented as objectively as possible, *we* is generally not used in the first move (e.g., see excerpts 4D and 4E). Recall that the purpose of the opening move is to remind readers of research methods, not to draw attention to the researchers themselves; thus, *we* should be avoided when describing work done in the past (e.g., *X was measured* is preferred over *We measured X*). Alternatively, *we* is used (sparingly) in the second move of the Results section, where authors tell their story of scientific discovery. *We* can be used to highlight a (human) decision or choice made during the course of the work. Consider the following four examples:

- In MALDI-MS, this challenge is overcome by coadding individual spectra. *We* have adopted a similar approach to achieve reproducible SERS spectra. (Adapted from Jarvis and Goodacre, 2004)
- After injection of the methanol/water extract of kelp powder and kelp powder spiked with 0.5 μg mL^{-1} of As(III), DMA, MMA, and As(V), *we* observed an overlap of DMA and phosphate ribose. *We* therefore decided to change the mobile phase. (From Almela et al., 2005)
- Modification typically takes advantage of electrostatic interactions between charges on the surface of the macromolecules and the polar headgroups of surfactants. *We* reasoned that the host-guest interactions at the nanoparticle-solution interface investigated in this work could be used for similar purposes. (From Liu et al., 2001)
- *We* have defined our sets of compounds for cross-comparison more broadly than in previous studies. (From Vieth et al., 2004)

The ability to use *we* appropriately comes only from reading the literature and growing accustomed to the convention. As we've said before, many aspects of scientific writing are not right or wrong; they are simply conventional or unconventional. It is not incorrect to use *we* frequently in the Results section; it is simply not customary for experts to do so. To sound like an expert, you must learn the convention.

Exercise 4.20

Consider the most compelling reasons for using *we* in a Results section. For each passage below, decide whether the use of *we* is appropriate. Explain each decision.

a. The results for the As and Pb concentrations *we* obtained for the 83 samples are reported in Table 3. *We* note from Table 3 that the range of As and Pb concentrations in the two populations is quite distinct.

b. *We* use the term K_D, the distribution coefficient, in the following discussion, although equilibrium may not have been achieved in all cases.

c. To reduce the problems of ligand-specific bias, *we* developed a modified rating for each molecule. *We* have called this corrected score the multiple active site correction rating (MASC).

d. Within the reference group, *we* found that the mean oxidative damage for smokers was significantly higher than that for nonsmokers.

e. At the end of each experiment, *we* measured the release of adsorbed alkenes from air–water interfaces.

Use of "Respectively"

The word *respectively* (meaning "separately, in the order specified") often appears in science writing and can be used to make your writing more concise. Generally, *respectively* appears at the end of the sentence; on rare occasions, however, it appears within the sentence. Note, too, that when two or more items have the same unit, the unit is stated only once.

- The concentrations of 2-ABP, 3-ABP, and 4-ABP in PPD were estimated at 70, 310, and 500 ppb, *respectively.*
- The turnover of 2-propanol and tosylate must exclusively take place via equilibria 5 and 6, *respectively.*
- Assuming that the sample volume of seawater is 5 L, and the chemical recovery is 70%, the ^{239}Pu and ^{240}Pu concentrations in the final solution (~0.7 mL) could be approximately 10.5 and 1.9 fg/mL, and U and Pb concentrations would be approximately 15 and 1 µg/mL, *respectively.*

- Curves 1–3 correspond to film thicknesses of 10, 25, and 50 nm, *respectively*, on a water–ice substrate that is 100 nm thick.

Compare the following sentences with and without the word *respectively*.

Less concise	A was measured at X °C, B was measured at Y °C, and C was measured at Z °C.
More concise	A, B, and C were measured at X, Y, and Z °C, respectively.

In these examples, you can see how the word *respectively* helps achieve conciseness. More important, it aids clarity. By grouping values together, trends in the data are easier to discern. To ensure that the correct meaning is conveyed when using *respectively*, it is crucial that the order of the first set of items (e.g., A, B, C) parallels the order of the second (e.g., X, Y, Z).

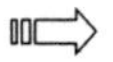

Respectively

See appendix A.

Exercise 4.21

Practice using *respectively* by rewriting each of the following passages to include the word, when appropriate. If *respectively* cannot be appropriately introduced into the passage, indicate "no change needed".

a. The extrapolated FH parameters for infinite molar volume are plotted in Figure 7, while the coefficients A are plotted in Figure 8. (Adapted from Schwahn and Willner, 2002)

b. The conductivity increased by 0.5% in trial 1, 5.6% in trial 2, and 10.1% in trial 3.

c. Ion intensities for *m/z* 29, 45, and 83 indicate that the Nafion membrane discriminates in favor of methanol by a factor of 67 relative to chloroform and ethanol by a factor of 55 relative to chloroform, assuming equal analyte responses. (Adapted from Creaser et al., 2002)

d. For the low-energy transition, the origins are located at 626.6 nm in **1**, 626.1 nm in **2**, and 627.6 nm in **3**. (Adapted from Spanget-Larsen et al., 2001)

e. The reacting system as a solute was solvated in boxes containing 396 molecules of H_2O (approximate dimension of 20 Å × 20 Å × 30 Å), CH_3OH (approximate dimension of 27 Å × 27 Å × 40 Å), and THF (approximate dimension of 33 Å × 33 Å × 49 Å). (Adapted from Xue and Kim, 2003)

Quantitative Language

As mentioned above, a Results section is descriptive, not interpretive. At times, the difference between description and interpretation can be subtle; this difference is often a matter of word choice. One way to keep your Results section descriptive is to use precise language. By avoiding overly positive or negative words that are not particularly precise, such as *excellent*, *very good*, or *poor*, and using more neutral terms instead, such as *high* or *low*, you can maintain a descriptive tone in your Results section. Even better, you can replace qualitative terms with more precise, quantitative values. Consider the following examples:

Vague	Heating the mixture to 93 °C gave very good yields.
Better	Heating the mixture to 93 °C gave high yields.
Even Better	Heating the mixture to 93 °C gave a 98% yield.

Vague	The solution was very acidic.
Better	The pH of the solution was 1.2.

 **Exercise 4.22**

Rewrite the following passages so that the language is more descriptive than interpretive. Feel free to "invent" measured data or details if you think that they will help.

a. Because of the high acidity of the water, samples were collected in an appropriate container.

b. The cells were suspended in MSM, which gave good results.

c. Although GC/MS has been used on similar samples before, our preconcentration technique afforded a significant increase in sensitivity for the brominated compounds.

Use of "Very"

One of the most overused words by inexperienced writers is the word *very*. Some scientists would argue that all instances of the word *very* should be eliminated from journal articles because its use contributes to wordiness, minimizes objectivity, and indicates a lack of precision on the part of the writer. Nevertheless, *very* is observed in the chemical literature, although infrequently. In excerpts 4A–4G, it appears only twice:

> Table 1 shows that the method provides *very* good reproducibility, with coefficients of variations for monitored aldehydes below 5.5%, except for (*E*)-2-nonenal. (From excerpt 4A)

As pure RAMEB sorbs a *very* high amount of water (ca. 1 g g^{-1} at $p/p_0 = 0.99$), an increase in water sorption was expected after RAMEB addition to all soils. (From excerpt 4D)

Very Rare

Be careful not to overuse the word *very* in your own writing.

Exercise 4.23

Consider the use of *very* in the following passages. In which passages is *very* necessary? In which passages is *very* most appropriate? What word substitutions could be made for more precise writing?

a. Consequently, the voltage required was *very* high and as a result…
b. The spectrum of maltohexaose is *very* similar to that of…
c. Additional experiments were carried out, some of which yielded *very* surprising results.
d. Alkanes are also *very* inert to alcohols and ketones.
e. Because low-molecular-weight hydrocarbons are volatile and *very* poorly soluble in water…
f. Because the rates of reaction were *very* slow…
g. The gel filtration experiments confirmed that *very* little Zn^{2+} was released…
h. The rate constants for the slowly and *very* slowly desorbing fractions were…
i. When BINAP or DPPF as a ligand was used, the yields were *very* low.

Scientific Plurals

The word *data* is commonly misused by writers; the mistake involves using *data* as a singular noun. In nearly all instances, the word *data* is plural and should be used with a plural verb:

Incorrect Data is…
Correct Data are…

Until you become more familiar with *data* being plural, you may read over a passage using *data* with a singular verb and not even notice the error (e.g., "data shows" sounds correct to many native English speakers). It is difficult to catch a mistake that does not sound wrong. In fact, many chemists use the word *data* as singular when speaking; it is only in writing, when there is more time for reflection and revision, that they correct themselves. A trick that may help you catch the mistake in your own writing is to mentally replace the word *data* in your sentence with a common plural word. For example, "The *data* shows a strong trend" may not sound wrong to you, but "The *values* shows a strong trend" likely does. (In both cases, "shows" should be "show.") The singular form of *data* is *datum*, but the word *datum* is rarely used; a single point of data is typically referred to as a *data point* (rather than a datum). If the term *data set* is used instead of *data*, the singular verb form is correct (e.g., "The second *data set* confirms this trend."). Some correct uses of data, with plural and singular verb forms, are shown in table 4.2.

Data

In nearly all instances, the word *data* is plural. The singular form, *datum*, is rarely used. The word *data* is almost always used with a plural verb:

Incorrect	Data is...
Correct	Data are...

In addition to the word *data*, other scientific plurals are often problematic for novice writers. Table 4.3 includes a list of confusing singular and plural word forms.

Table 4.2 Uses of *data* in the literature (identified using the ACS Journals Search).

Data *with a Plural Verb*

These data show...

The data imply...

The data were biased by...

These data are supported by...

The data suggest...

The data reveal...

Data *with a Singular Verb*

The profile suggests that the data (*set*) is well converged.

Data *Followed by a Singular Verb That Agrees with a Different Noun*

Inspection of the data *reveals*...

A complete *set* of data *is* available.

A key *feature* of the bond length data *is*...

Table 4.3 Singular and plural forms of common scientific words (adapted from *The ACS Style Guide*: Coghill and Garson, 2006, p 128).

Singular Form	Plural Form[a]
apparatus	apparatus, apparatuses
appendix	appendixes, appendices
bacterium	bacteria
basis	bases
criterion	criteria, criterions
formula	formulas, formulae
fungus	fungi, funguses
index	indexes (indices if mathematical)
matrix	matrixes (matrices if mathematical)
medium	media, mediums
spectrum	spectra, spectrums

a. When more than one plural form is recognized, the preferred plural form is given first.

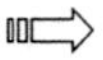

Scientific Plurals

See appendix A.

Exercise 4.24

The following sentences may contain one or two errors with regard to singular and plural forms of words. Decide whether each sentence contains errors. If it does, correct the sentence. If the sentence is correct as written, indicate "no change needed".

a. The HSSPME technique is most applicable for volatile analytes contained in complex matrices.

b. As shown in Table 3, the data was highly reproducible with relative standard deviations of less than 3% in all cases.

c. The mass spectrum (Figures 2 and 3) was collected in the selective ion monitoring (SIM) mode.

d. The criteria used to determine when trichloroethylene (TCE) had equilibrated between the solution and headspace was a change in headspace concentration of less than 5% over a 30 min period.

e. The basis for our decision was the demonstrated inertness of PTFE toward uptake of organics.

f. More data are needed to determine whether microbes significantly influence the fate of selenium in this system.

4E Writing on Your Own: Practice Peer Review

Before you engage in authentic peer review, practice the peer review process. Imagine that a colleague has asked you for feedback on a draft of a Results section. See "Peer Review Practice: Results Section" at the end of this chapter for a copy of the draft, background information, and instructions.

4F Writing on Your Own: Fine-Tune Your Results Section

By now, you should have made progress writing your own Results section by completing the preceding Writing on Your Own tasks. When you have a good draft of your Results section (having set the stage, move 1, and told your story of scientific discovery, move 2), it is time to revise and edit your Results section as a whole. Focus on each of the areas specified below while you reread your written work. Refer to chapter 18 to guide you in the revision process.

1. Organization of text: Check your overall organizational structure. Did you follow the move structure outlined in figure 4.1 and include appropriate subheadings? Have you referred the reader to a figure or table at the end of move 1? Have you included at least one of the following: important findings, trends, and unexpected results?
2. Audience and conciseness: Are you writing for an expert audience, leaving out unnecessary details? Have you answered the question "What did you find?" as a way to focus on description (rather than interpretation)? Find at least three sentences that can be written more clearly and concisely. Are there sentences that could be made more concise by using the word *respectively*? If the word *we* is used, check to see if it is used correctly. Replace such words as *excellent*, *very good*, or *poor* with more precise words or phrases.
3. Writing conventions: Check to be sure you have used voice and tense correctly (see table 4.1). Are your graphics formatted correctly? Refer to chapter 16 to review formatting conventions.
4. Grammar and mechanics: Check for typos and errors in spelling, subject–verb agreement, and punctuation. Be sure that you have used troublesome scientific plurals (e.g., data) correctly.
5. Science content: Have you correctly conveyed the science in your work? Have you used words and units correctly? If asked, could you define all of the words that you have used

in this section? Do you understand the results of your work? Include only the most relevant, representative data. If possible, use the broad-to-narrow approach to limit the amount of data that you present to tell your story.

After thoroughly reviewing your own work, it is a common procedure to have your work reviewed by a peer or colleague. A "new set of eyes" will pick up mistakes that you can no longer see because you are too familiar with your own writing. To facilitate the peer review process, use the Peer Review Memo (on the *Write Like a Chemist* Web site) to assist you and your peer reviewer. After your paper has been reviewed (and you have reviewed another's paper), consult the Peer Review Memo given to you by your peer reviewer to make final changes in your Results section.

Finalizing Your Written Work

See chapter 18.

Chapter Review

As a self-test of what you've learned in chapter 4, define each of the following terms, in the context of this chapter, for a friend or colleague who is new to the field:

blocked R&D
broad-to-narrow approach
compound labels
detection limit
graphics
integrated R&D
iterative R&D
representative data
sets of results

Also explain the following to a friend who hasn't yet given much thought to writing a Results section for a journal article:

- Main purpose of a Results section
- Moves of a Results section
- Relationship between text and graphics in a Results section
- Differences between description (for a Results section) and interpretation (for a Discussion section)
- Circumstances in which the use of *we* could be acceptable in a Results section
- Uses of the past tense and present tense in a Results section
- Common irregular plural words used in scientific writing
- Uses of the word *respectively*
- Use of the word *very* in a Results section

Additional Exercises

Exercise 4.25

Rewrite the wordy sentences below to make them more concise:

a. As can be seen from Table 1, any of the samples that had been passivated with PVB, MA, and PVA were seen to exhibit a substantial increase in quantum efficiency upon UV irradiation. (33 words; goal = 22 words)

b. It is worth mentioning that, in some experiments, partial racemization of the alcohol was detected. (15 words; goal = 11 words)

c. In order to obtain further insight into the ways that the polymer may have an influence on the UV enhancement, several samples that had previously been coated with different polymers were irradiated. (32 words; goal = 15 words)

Exercise 4.26

Reword the following passages using more precise language. Feel free to "invent" measured data or details if you think that they will help.

a. Among methanolic extracts from four specialty mushrooms, only *Dictyophora indusiata* (basket stinkhorn) showed very good antioxidant activity (2.26% of lipid peroxidation).

b. During photo-Fenton treatment, diuron degradation is extremely quick, but final byproducts are formed only slowly.

c. All chlorinated hydrocarbons were observed at similar concentrations.

d. The second fertilizer showed more promise for increasing crop yield than the first.

Exercise 4.27

Look at the Results sections of three journal articles. Examine the authors' use of tense, voice, *we*, *respectively*, neutral and precise language, *very*, and scientific plurals. Are their uses consistent with your expectations? Explain.

Exercise 4.28

How might the following text and table from a Results section be improved? Modify both to be more in line with journal article expectations. Consult chapter 16 if necessary.

Results

4-Methoxyacetophenone **3** crystals were obtained in 70% yield by reacting equimolar amounts of **1** and **2** at 35 °C for 5 min (entry 1). Temperature was varied in 5 °C increments between 35–70 °C (entries 1–8), with the greatest yield occurring between 45–70 °C (entries 3–7). At 35 °C, the yield was 70% and at 40 °C it was 80%. From 45–70 °C, yield remained at 85%. The reaction was then studied at 50 °C using time as a variable (entries 4, 9–11). The greatest yield was achieved at 50 °C and 10–15 min reaction time (entries 9, 10).

Table 2.

1 + **2** $\xrightarrow[CH_2Cl_2]{AlCl_3}$ **3** + OH

entry	temp (°C)[a]	time (min)	yield (%)
1	35	5	70
2	40	5	80
3	45	5	85
4	50	5	85
5	55	5	85
6	60	5	80
7	65	5	80
8	70	5	85
9	50	10	92
10	50	15	92
11	50	20	90

[a] The water bath was kept within 5 °C of the specified temperature for the specified time.

Exercise 4.29

Reflect on what you have learned about writing a Results section for a journal article. Select one of the reflection tasks below and write a thoughtful and thorough response:

a. Reflect on how reading the excerpts in this chapter has influenced your scientific reading and writing abilities.
   - Give three examples of how reading and analyzing excerpts of Results sections has (1) improved your ability to write your own Results section and/or (2) changed your approach to writing your own Results section. Explain.
   - If your research advisor were to give you a paper to read, how confident are you that you would be able to read and understand the Results section

of the paper? How has your experience reading and analyzing the excerpts in this chapter made this task easier? Explain.

b. Reflect on the move structures of Methods and Results sections. (If necessary, consult figures 3.1 and 4.1.)
   - How do the different move structures reflect the vastly different purposes of the Methods and Results sections?
   - Which writing conventions do you associate with the different moves and the different sections?
   - How do those writing conventions assist writers in achieving their purposes?
   - Which moves were easier for you to write? Why?

c. Reflect on the numerous excerpts that you have read in this chapter.
   - Although excerpts 4A–4G come from different journals and report on different types of chemical research, they share some common attributes. What are their commonalities?
   - Which excerpt assisted you the most in writing your own Results section? Why was it so helpful?
   - What new chemistry content have you learned as a result of reading these excerpts?

Peer Review Practice: Results Section

Imagine that a friend has asked you to review a draft of a Results section written for a paper about the decomposition of solid biowastes (yard and food wastes). Such biowastes typically decompose through a process known as anaerobic (i.e., without oxygen) biodegradation. However, anaerobic biodegradation generally results in low methane yields (50–60% of the theoretical maximum). Your friend's research group is trying to boost methane production during degradation by first exposing the waste to a process known as wet oxidation.

The research involves many steps. Your friend's part in the project is to examine how wet oxidation (WO) affects the amount of volatile suspended solids in the waste. Volatile suspended solids (VSS) are solids that can be converted to gases during wet oxidation, thereby affecting the total mass of waste leftover to undergo anaerobic biodegradation. The Results section draft below describes this part of the research project.

Using parts 2 and 3 of the Peer Review Memo on the *Write Like a Chemist* Web site, review the Results section draft. Provide specific suggestions in your memo that can be used to improve the Results section. (The Results section below is adapted from an original source, noted in the Instructor's Answer Key.)

Results

1 **Wet Oxidation (WO) Treatment.** As described in the Experimental section above,
2 WO experiments were carried out in a high-pressure autoclave with a tubular loop and
3 an impeller constructed at Risø National Laboratory (*17*). The autoclave was designed as
4 a cylindrical vessel (V = 1890 mL) made of Sandvik Sanicro 28 (27% Cr, 31% Ni, 3.5%
5 Mo, and 1% Cu) with an impeller that continuously pumped the liquid through the
6 tubular loop.
7 Raw yard waste and food waste were oxidized under the same conditions (except
8 for WO time) (Table 1), while the digested biowaste was oxidized under 4 different
9 conditions (A–D). Yard waste underwent WO at 185 °C for 15 min at 12 bar. Food waste
10 underwent WO at 185 °C for 10 min. The WO conditions for digested biowaste treat-
11 ments A, B, C, and D were 185 °C for 15 min at 0 bar, 185 °C for 15 min at 3 bar, 185 °C
12 for 15 min at 12 bar, and 220 °C for 15 min at 12 bar, respectively.

			digested biowaste			
	yard waste	food waste	A	B	C	D
temperature	185 °C	185 °C	185 °C	185 °C	185 °C	220 °C
time	15 min	10 min	15 min	15 min	15 min	15 min
oxygen pressure	12 bar	12 bar	0 bar	3 bar	12 bar	12 bar
pH before WO	9.5	7.2	8.3	8.3	8.3	10.1
pH after WO	3.7	4.6	7.3	6.6	4.4	6.4
VSS before WO (g/L)	33	41	21	21	21	21
VSS after WO (g/L)	30	27	19	18	17	13
VSS oxidized (%)	9 %	10 %	9 %	14 %	20 %	32 %

Table 1. Wet Oxidation (WO) Conditions and Volatile Suspended Solid (VSS) Losses for Raw Yard Waste, Raw Food Waste, and Digested Biowaste (Conditions A–D)

13 For all cases, the WO treatment will cause a pH drop from 1 to 5.8 units, with the
14 most pronounced decrease in pH at the highest oxygen pressure. Furthermore, table 1
15 shows generally that 9–20% of the VSS contained in the waste will be oxidized during
16 wet oxidation at a WO temperature of 185 °C. At a WO temperature of 220 °C, approxi-
17 mately 32% of the organic content will be oxidized during WO.
18 In light of the Kyoto agreements and the EU green electricity certificates, additional
19 technologies to enhance the methane yield from various wastes and to ensure a biologi-
20 cally safe digested product are needed. Wet oxidation has a higher techno-economical
21 feasibility as compared to other pretreatment technologies for anaerobic digestion due
22 to the low oxygen consumption for the presented WO conditions, the self-sustaining
23 character of the WO reaction, and the opportunity for heat and oxygen recovery. There
24 is still a need to establish the technical and economical benefits of the WO technology in
25 addition to methane and ethanol recovery from various biomasses and waste.

5 *Writing the Discussion Section*

A Discussion section should be as satisfying to read as the last chapter of a mystery novel. The groundwork is laid in the Introduction section, technological tools are described in the Methods section, evidence is revealed in the Results section, but it is in the Discussion section where the mystery is solved.

This chapter focuses on the Discussion section, the last part of the standard IMRD structure for a journal article. The Discussion section, as mentioned in chapter 4, can stand alone or can be part of a combined Results and Discussion (R&D) section. In either case, it serves the same major purpose: to interpret the results of the study. In this chapter, we analyze excerpts from various Discussion sections, including those that accompany results presented in chapter 4 (excerpts 4B–4G). Upon completion of this chapter, you should be able to do the following:

- Organize a Discussion section following the major moves
- Interpret your results (but avoid overinterpretation)
- Describe the greater importance of your findings
- Follow appropriate writing conventions

As you work through this chapter, you will write a Discussion section for your own paper. The Writing on Your Own tasks throughout the chapter will guide you step by step as you do the following:

5A Read the literature

5B Prepare to write

5C Draft your Discussion section

5D Practice peer review

5E Fine-tune your Discussion section

In the Discussion section of a journal article, authors interpret their data, address *why* and *how* questions (e.g., Why was the reaction faster? How did the mechanism proceed?), and, ultimately, extend their findings to a larger context (e.g., What value will these findings have to the scientific community?). Ideally, the Discussion section explains the story revealed by the data, postulates reasons for the observed behaviors, and furthers our fundamental understanding of the underlying science.

Although interpretation is the primary goal of the Discussion section, authors must be careful not to overinterpret their data, misinterpret their results, overstate their assumptions, or stray too far from scientific evidence. The excerpts selected for this chapter illustrate ways to avoid these pitfalls. Similarly, the excerpts illustrate that the language of the Discussion section is typified by restraint and understatement. Such words as *fact*, *truth*, and *prove* are rarely used in a Discussion section. **Hedging words**, such as *theory* and *evidence*, are much more common, as are such verbs as *appear*, *indicate*, *seem*, and *suggest*. By hedging, writers acknowledge that their knowledge is limited and will be subjected to scientific scrutiny over time.

Hedging Words

Words that soften interpretations and suggest that interpretations are not absolute facts. Common hedges in scientific papers include the following:

apparently	largely	possibly	should
appear	likely	potentially	suggest
can	mainly	presumably	support
could	may	probable	typically
generally	might	probably	would
indicate	possible	seem	

Reading and Analyzing Writing

Chemists should seek to advance chemical science, understand the limitations of their knowledge, and respect the truth. Chemists should ensure that their scientific contributions, and those of the collaborators, are thorough, accurate, and unbiased in design, implementation, and presentation.

—The Chemist's Code of Conduct (www.chemistry.org)

To begin the analysis of the Discussion section, we ask you to read and analyze the Discussion section from the article on the analysis of aldehydes in beer.

Because this article uses a combined R&D section, we refer you to chapter 4 for this excerpt (excerpt 4A). The bulk of the Discussion section begins after the subheading "Beer Analysis", although a few discussion-like sentences appear before this subheading.

Exercise 5.1

Read the "Beer Analysis" section in excerpt 4A and answer the following questions:

a. What moves do you see in this excerpt? Propose a move structure for the Discussion section, with at least one move for each paragraph.

b. Find the two instances of hedging in this section. Explain the purpose of each.

c. What is the purpose of the concluding paragraph?

d. Why was this study conducted? Where in the Discussion section is this reason stated?

5A Writing on Your Own: Read the Literature

Read and review the Discussion sections of the journal articles that you collected during your literature search (begun with Writing on Your Own task 2C). Read these Discussion sections to learn more about your topic and to find ways to describe and interpret your findings. Examine how the authors applied their findings to a broader research context. Identify articles that you want to cite in your paper, such as works that offer supporting or conflicting evidence. Jot down careful notes as you read.

Analyzing Audience and Purpose

Two major purposes of the Discussion section are to interpret or explain results presented in the paper and to propose broader implications of these findings. Not surprisingly, each purpose is associated with a slightly different audience. The interpretation of results, like the Results section itself, is typically written for an expert audience. (Hence, the novice organic chemist is likely to have trouble understanding the references made to "Strecker aldehydes" and the "Amadori pathway" in excerpt 4A.) Alternatively, the broader implications of the work are typically accessible to a scientific audience or even a general audience. This shift in audience completes the hourglass structure of the IMRD format. The Discussion section forms the bottom of the hourglass; it begins with a specific focus but ultimately expands to offer a more general perspective.

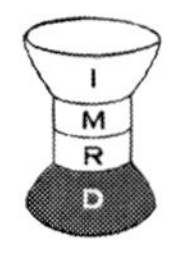

Discussion Section

Two major purposes of a Discussion section are to *interpret* results and *suggest* broader implications of findings.

Exercise 5.2

Reread the Discussion section of excerpt 4A.

a. Find two sentences that are accessible to a scientific (or even general) audience and two sentences that are geared toward a more expert audience.

b. What are the purposes of these different sentences?

Analyzing Organization

As shown in figure 5.1, the Discussion section is organized around two major moves: Discuss Specific Results and Conclude the Paper. The first move is divided into two submoves. Submove 1.1 reminds readers about the result that will be discussed, serving as a transition between the Results and Discussion sections. Such a reminder is often not needed in a combined R&D section but is necessary in a stand-alone Discussion section. Its purpose is to draw the reader's attention to a particular finding, not to restate all of the results. This submove is often accomplished in only a few sentences.

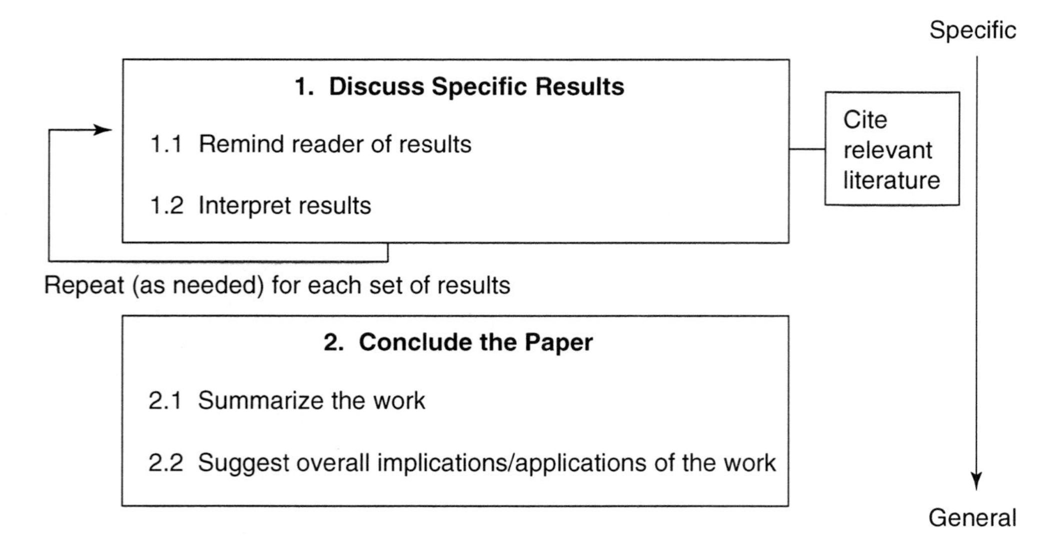

Figure 5.1 A visual representation of the move structure for a typical Discussion section.

Submove 1.2 is the heart of the Discussion section; mechanisms are proposed, results are elaborated, and/or the authors postulate why or how a particular behavior was observed. Whenever possible, references to relevant literature should be included as part of this submove. In particular, references that provide additional insights, refute an argument, or **corroborate** the findings at hand should be cited. In this way, your work can be connected to a larger body of evidence, moving toward the ultimate goal of scientific consensus.

Corroborate

To strengthen and support results or interpretations, using evidence from the literature.

The first move is reiterated as needed for multiple sets of results, ideally paralleling the order that was used in the Results section. As in the Results section, you can use subheadings to help the reader locate the discussion for each result. If, while writing the discussion, you question the logic of your sequencing, you need to revise both the Results and Discussion sections to align them.

Align Results and Discussion

The order in which you present findings (in Results) and interpret findings (in Discussion) should be parallel.

The second and last move of the Discussion section signals the conclusion of the paper. A heading (**Conclusions** or **Summary**) can be used to demark this section, or it can be identified by the phrase *In conclusion* or *In summary* at the start of a paragraph. Move 2 is also divided into two submoves. Submove 2.1 provides a brief summary of the work, highlighting the **take-home message(s)** of the paper. This is followed by a brief narrative (submove 2.2) that suggests implications and/or applications of the work and addresses at least one of the following questions:

- What are the implications of the work?
- What new insights were gained?
- How has this work increased our fundamental understanding of the research area?
- What are the practical applications of this work?
- How will the work affect society (e.g., industry, medicine, technology, the environment)?

To answer these questions, authors must look beyond the specific details of their own work and focus instead on the broader goals of the research project. Attaining this broader outlook can be challenging, especially for students who spend most of their time focused on only a small part of a larger project. Over time, however, your grasp of the broader picture will improve as you continue to read the literature, attend seminars and conferences, and read and write research proposals in your area of research.

Signaling the Conclusion

The last part of the Discussion section is often signaled with phrases such as

> In conclusion,
>
> In summary,

Note that these phrases are always followed by a comma.

If a conclusion is included in a section of its own, it is often marked with a **Conclusions** heading.

Take-Home Message

A sentence or two that summarizes the essential features of the work, that is, the information that you want your readers to "take home with them."

Exercise 5.3

Consider the following sentences taken from Discussion sections. Although these sentences are presented out of context, specify which submove you think is accomplished in each. Refer to figure 5.1.

a. The stress promoters induced in the Pro-Tox (C) test indicated that Cr^{3+} could cause changes in DNA topology. (From Plaper et al., 2002)

b. In summary, (*R*)-phenylglycine amide **1** is an excellent chiral auxiliary in the asymmetric Strecker reaction with pivaldehyde or 3,4-dimethoxyphenylacetone. Nearly diastereomerically pure amino nitriles can be obtained via a crystallization-induced asymmetric transformation in water or water/methanol. (From Boesten et al., 2001)

c. However, available data for size-fractionated fine particulate matter indicate that PAH quinones, including 1,4-naphthoquinone, 5,12-naphthacenequinone, benz[*a*]anthracene-7,12-dione, and anthracene-9,10-dione, are important organic components (*41*, *42*). The detection of these molecular species that are similar in structure to semiquinone-type radicals supports the assignment of our EPR signals. (Adapted from Dellinger et al., 2001)

d. Our findings suggest that the surface properties and pore structure of minerals change dramatically upon RAMEB addition. (From Jozefaciuk et al., 2003)

e. A possible explanation for this effect can be attributed to an increased density and viscosity of the milk–NaOH phase when NaOH concentration increases. (From Llompart et al., 2001)

f. Because it is known that humans exposed to different Cr^{3+} species accumulate high levels of Cr^{3+} intracellularly (*17*), presented results may have an impact on human intake of Cr^{3+} as a nutrition additive. (From Plaper et al., 2002)

g. In conclusion, we have shown that $PM_{2.5}$ contains stable radicals that can be detected by EPR. The EPR parameters, persistence in air, and DNA-damaging capacity of the $PM_{2.5}$ radicals are similar to those of the radicals in cigarette tar. (From Dellinger et al., 2001)

h. The data in this article are consistent with a wealth of evidence showing that dietary Se consumed in excess of the Recommended Dietary Allowance lowers the risk of several important cancers (*2*, *14*, *26*, *27*). (Adapted from Finley et al., 2001)

Before we continue, we point out two additional submoves, associated with move 2, that are less common today than in the past: (1) indicate limitations of the work and (2) suggest ideas for future work. Although neither submove is addressed in this textbook, you will see both submoves in the literature, and, in many disciplines, they are still quite common. Thus, as you prepare your paper, check to see if these submoves are expected in Discussion sections in your targeted journal.

5B Writing on Your Own: Prepare to Write

In accord with the major purposes of a Discussion section, determine if you have enough information to (1) interpret your results, (2) explain the broader implications of your work, and/or (3) apply your results to a larger, broader context.

Similarly, review the move structure of a typical Discussion section (figure 5.1). Determine if you have enough information to develop all moves and submoves.

Gather together this information, in outline form or as notes, to prepare to write your Discussion section.

Analyzing Excerpts

With the audience, purpose, and organization of the Discussion section in mind, we are ready to analyze excerpts of Discussion sections from the chemistry literature. We begin by analyzing the excerpts move by move and then examine a few writing conventions common to the entire section.

Part 1: Analyzing Writing Move by Move

In this part of the chapter, we revisit the journal articles that were introduced in chapter 4, where we focused on Results. Here we focus on the Discussion sections of these articles. We examine how well the authors follow the move structure in figure 5.1, how they interpret their results, and how they conclude their work.

Let's begin with the excerpt on randomly methylated β-cyclodextrin (RAMEB)-enriched soils in chapter 4 (excerpt 4B). The authors use an iterative R&D approach: They state their first result (R1), pertaining to clay-rich soils, and then immediately offer an interpretation (D1) of that result. This is followed by a result and interpretation for clay-poor soils (R2D2) and a result and interpretation for medium-clay-content soils (R3D3). Thus, the result–discussion sequence is iterated three times. In each case, the discussion immediately follows the result; hence, submove 1.1 (which reminds readers of the result) is not needed.

Iterative R&D

The iterative R&D approach and other combined R&D approaches are described in chapter 4.

R1D1 focuses on clay-rich soils (for the full excerpt, see excerpt 4B). Recall that the clay-rich soils did not show the expected behavior. All RAMEB-treated soils were expected to adsorb more water than their untreated counterparts, but such was not the case with the clay-rich soils. Let's see how the authors interpret this unexpected result.

R1 However, the isotherms for RAMEB-treated clay-rich S6 and S7 soils showed lower adsorption than the original soils, which is illustrated for S7 soil with 49% clay.

D1 This potentially indicates that RAMEB decreases the amount of water-available surfaces in clay-rich soils, similar to what was observed for pure clay minerals (*20*).

The authors offer a single sentence to explain the result: RAMEB decreases the number of water-binding sites available in clay-rich soils. The authors offer this interpretation cautiously (using the hedging phrase "potentially indicates") and provide additional support for their position by citing an earlier publication. Moreover, the authors do not overinterpret this finding. They present general evidence that RAMEB physically interacts with the particle surface but, for example, do not go on to speculate about specific RAMEB–soil binding interactions. To do so would constitute a **hand-waving argument**, an interpretation that lacks sufficient empirical evidence. Although hand-waving arguments serve as useful constructs for scientists to think and talk about their results, they should be avoided in scientific papers.

Hand-Waving Argument

An expression used among scientists that implies that a researcher is advocating a theory or belief with little evidence to support it.

Hand-waving arguments should be avoided. Interpretations included in Discussion sections should be supported by data.

In R2D2, the authors focus on clay-poor soils. For continuity, the authors must explain why, in contrast to clay-rich soils, clay-poor (sandy) soils do take up more water when treated with RAMEB. Importantly, the explanation for less water uptake in clay-rich soils and more in clay-poor soils must be consistent with each other. They cannot argue, for example, that RAMEB both increases and decreases the number of water-available sites on the soil. With that in mind, let's see how they explain the clay-poor soil results:

R2 In sandy soils (S1–S4), the water sorption markedly increased, particularly at higher RAMEB doses, as is illustrated for S2 soil.

D2 This may be attributed to water sorption by free RAMEB, that is, RAMEB molecules that did not interact with the sandy soils.

Restated, sandy soils have smaller surface areas; hence, fewer RAMEB molecules interact with their surfaces. This leaves more free RAMEB molecules, which adsorb the extra water. Aha! The authors have painted a picture of what may be happening that is consistent with the results for both soil types. This logic is continued in R3D3, the results and discussion for the intermediate-clay-content soils, as illustrated in the following exercise:

Exercise 5.4

R3D3 offers the results and interpretation for the water-sorption behavior of the medium-clay-content soil (S5). We include R3 below (slightly modified):

R3 For soil S5 of medium clay content (25%), the effect of RAMEB on water sorption was small.

a. Compose your own D3 in a way that is consistent with D1 and D2 above.

b. When you are done, check to see the authors' D3 in excerpt 4B. How similar is your interpretation (D3) to that of the authors? Explain.

Jozefaciuk et al. (2003) go on to present and interpret several other sets of results that are not included here (e.g., the effects of RAMEB addition on surface area and porosity). Then, they conclude their article (move 2) as follows:

Our results demonstrated that RAMEB strongly interacts with soils, modifying their surface, pore, and aggregate properties. These **effects** can **affect** soil remediation technologies

In these brief two sentences, the authors provide a take-home message (that RAMEB strongly interacts with soils) and suggest an application of their work (soil remediation). In our estimation, they clearly win points for conciseness!

Affect vs. Effect

See appendix A for more information on these easily confused words.

The next example (excerpt 5A) returns us to the analysis of PCBs in milk. Recall that Llompart et al. (2001) (excerpt 4C) used HSSPME techniques to detect PCBs in milk. Their stumbling block was the fat content. The original method worked fine on skim milk, but the fat contained in nonskim milk trapped the PCBs in the liquid phase (those pesky matrix effects!), rendering headspace (i.e., gas-phase) analysis of the PCBs disappointing at best. More promising results were obtained when the milk fat was first saponified with base (NaOH), a process that helps to release the PCBs from the fatty matrix.

Llompart et al. (2001), like Jozefaciuk et al. (2003), use a combined R&D section (the preferred format in *Analytical Chemistry*, the journal that published this article). Their R&D section describes both preliminary tests and optimization procedures. Results and discussion of the preliminary tests were presented in excerpt 4C; results and discussion of the optimization procedures are presented in excerpt 5A. The optimization process used a factorial design in which five experimental parameters were systematically varied and tested to improve the saponification technique. These variables included the concentration of NaOH, the volume of NaOH, the extraction and stirring times, and the kind of SPME fiber used.

Preference for Combined R&D Sections

The ACS journal *Analytical Chemistry* prefers a combined R&D section. As stated in its authors' guide, "In most cases, combining results and discussion in a single section will give a clearer, more compact presentation" (*Anal. Chem.* **2007,** *79,* 390).

As you read through the R&D section in excerpt 5A, you will notice that equal time is not given to the Results and Discussion sections; more emphasis (and text) is given to the Results. In a few instances, the authors offer an explanation (e.g., why the filter type was important only for the lighter PCBs and why, for some PCBs, the concentration of NaOH showed a negative effect), but for the most part, the focus is on the results of the optimization tests. Such an approach is not unusual for an analytical paper whose major purpose is to improve an analytical method. Results that demonstrate increased efficiency, detection limits, and/or accuracy are the major reasons for doing such a study; hence, evidence that the new technique works is emphasized over why it works.

Exercise 5.5

Read excerpt 5A. Which factors were the most important in the optimization procedure? How are these factors emphasized in the text?

Excerpt 5A (adapted from Llompart et al., 2001)

Results and Discussion

[Continuation of excerpt 4C]

Optimization of the Saponification-HSSPME Process: Factorial Design. [*The first three paragraphs are omitted*]

As can be seen in Figures 2 and 3, the agitation of the sample was the most important factor for almost all the analytes. In all cases, this factor has a positive effect, and it appears to be of increasing importance as the degree of chlorination of the PCBs increases. Also, the extraction time was a significant factor for all the compounds, as might be expected. On the other hand, the type of fiber was only important for the lighter PCBs, mainly for PCB-28 and PCB-52. For these compounds, the PDMS-DVB fiber is more efficient than the PDMS fiber. The effect of the fiber factor appeared negative (Figure 2) because PDMS-DVB was selected as **its** low level (Table 1). For the highly chlorinated PCBs, the two fibers tested seem to have similar performance, and this factor lacks statistical significance.

The volume of NaOH was also a significant factor for most of the compounds but especially for the high-molecular-weight PCBs. The plot curvature of this factor (Figure 3) shows optimum experimental settings that vary (2–3.5 mL) depending on the PCBs to be extracted. The concentration of NaOH was only statistically significant for PCB-105 and PCB-180. This factor showed a negative effect, which means that the extraction efficiency decreases when the concentration of NaOH was at the high level. A possible explanation for this effect can be attributed to an increased density and viscosity of the milk–NaOH phase when NaOH concentration increases. This retards the kinetics of the HSSPME process, and consequently, the extraction efficiency decreases.

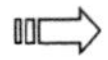

Its vs. It's

See appendix A for more information on these easily confused words.

Llompart et al. (2001) conclude their article with a table (not shown) that compares the amount of PCBs absorbed by the filter with and without saponification; results from performance evaluations and validation procedures are also presented. Because the article does not end with a summary statement or broader applications, it varies slightly from the move structure presented in figure 5.1. In general, however, articles published in *Analytical Chemistry* follow the move structure shown in figure 5.1. As indicated in the authors' guide to *Analytical Chemistry*, a separately demarked Conclusions section is preferred. The purpose of the Conclusions section is to offer interpretative remarks and present broader implications of the work; the journal editors caution against repeating information that is presented elsewhere in the article or abstract. As an example, consider excerpt 5B, a Conclusions section from a different *Analytical Chemistry* article. The conclusion follows the submoves in figure 5.1 by beginning with a summary of the major achievement (improvements in the sensitivity of an electrochemical DNA sensor) and then going on to suggest possible applications of this technique.

Excerpt 5B (adapted from Wong and Gooding, 2006)

Conclusion

Following our initial investigation on the transduction of DNA hybridization via long-range charge transfer conducted in sequential steps, we have made significant improvements with regard to sensitivity and ease of use via a single-step in situ electrochemical approach. Using the in situ approach, the DNA biosensor is able to detect target DNA in the subnanomolar range within 1 h. Furthermore, the in situ detection scheme is also able to differentiate between complementary, noncomplementary target DNA, and even target DNA with single-base pair mismatches, including the most thermodynamically stable G-A mismatch, without requiring any additional stringency steps. This new approach also has the ability of studying biological processes in real time and thus allows the kinetic processes to be monitored. The good sensitivity, excellent selectivity, and simplicity of use of the DNA biosensor make it more compatible for integrating with on-chip PCR reactors than other DNA biosensors of which we are aware.

Exercise 5.6

Glance through six articles in a current issue of *Analytical Chemistry*.

a. How many articles use a combined R&D section rather than separated Results and Discussion sections?

b. How many articles include a separate Conclusions section?

c. Select two Conclusions sections in *Analytical Chemistry*. How well do they adhere to the second move represented in figure 5.1?

We next consider the Discussion section that follows excerpt 4D concerning the genotoxicity of Cr^{3+} in bacterial cells (Plaper et al., 2002). Recall from excerpt 4D that three forms of Cr^{3+} were initially investigated: chromium chloride, chromium nitrate, and chromium oxalate. Chromium chloride and chromium nitrate both induced stress promoters indicative of DNA damage, but chromium oxalate did not. The authors also measured chromium accumulation in *E. coli* cells. Chromium from chromium chloride and chromium nitrate was taken up by the cells, but chromium from chromium oxalate was not. Finally, the authors examined how Cr^{3+} (from chromium chloride) affects gyrase, the enzyme that regulates the relaxation of supercoiled DNA. Gyrase inhibition was observed at $CrCl_3$ concentrations >10 μM.

Plaper et al. (2002) use a stand-alone Discussion section, rather than a combined R&D section. According to the move structure in figure 5.1, a separate Discussion section should begin with a brief reminder of a specific result or set of results; this is essentially how Plaper et al. (2002) begin. The authors first remind the readers about the study in general (in the first two sentences); then, in the third sentence, they remind the readers specifically about the Pro-Tox (C) assay with chromium chloride.

> P1 Many studies have shown that not only Cr^{6+} but also Cr^{3+} cause damage inside the cells. In this work, three Cr^{3+} compounds were examined for their impact on genotoxicity and cell proliferation in vitro. Chromium chloride added to *E. coli* strains in the Pro-Tox (C) test system induced *lacZ* gene transcription from several stress promoters. (Adapted from Plaper et al., 2002)

The paragraph goes on to interpret these findings, indicating that DNA is a target for Cr^{3+} inside the cell. The authors describe some of the stress promoters that were induced and frequently cite the literature to offer additional insights and corroborating evidence. At the end of the paragraph, the authors suggest possible causes for these inductions. Two references to the literature support the suggestion that the hydroxyl radical may be the cause of the DNA damage.

> P2 Most of the latter were associated with different types of DNA damage, indicating that DNA is one of the main targets for Cr^{3+} inside the cell. The most prominent effect was on *lacZ* transcription from *dinD* promoter, which responded to all types of DNA lesion. *RecA* promoter

was also induced, indicating problems with DNA replication (*37*). Additional evidence of DNA damage was the induction of *merR* and *osmY* promoters. The induction of *lacZ* transcription from *merR* is not surprising as *merR* is the promoter that responds to the presence of the heavy metals such as cadmium and mercury (*24*). Moreover, it has been shown that *merR* is induced by changes in DNA topology (*26*). Changes in DNA supercoiling could also account for strong induction of *lacZ* transcription from *osmY* promoter. The *osmY* stress gene usually responds to hyperosmotic environmental conditions, but it can also be regulated by changes in DNA topology, similarly to the *proU* operon (*10, 16, 39*). In our test, there was also 5-fold induction from *soi28*, an oxidative stress promoter responding to oxidative damage in the cells. This induction could be caused by the hydroxyl radical produced in the reaction **between** Cr^{3+} and hydrogen peroxide (22). The hydroxyl radical is a known mediator of DNA damage, causing lesions to DNA bases and to the phosphodiester sugar backbone (*13*). (From Plaper et al., 2002)

Between vs. Among

See appendix A for more information on these easily confused words.

The first move is reiterated in each of the next two paragraphs. The second paragraph initially reminds the reader about the negative results obtained with chromium oxalate, both in the Pro-Tox (C) test and in the FAAS measurements of *E. coli* cells. The paragraph goes on to interpret these results, suggesting that the lack of induction by chromium oxalate is "probably due to the inability of that compound to enter the bacterial cells." Note that the authors do not speculate on how chromium chloride and chromium nitrate enter the cells, or on how chromium oxalate might be excluded. Again, to do so would be hand-waving. Their data show only that chromium concentrations are negligible in test cells when chromium oxalate is used; their analytical probe (FAAS) does not provide insights into the mechanisms of chromium transport.

P3 When *E. coli* test strains were treated with chromium nitrate, the induction profile obtained was very similar to chromium chloride profile (data not shown). The promoters that were induced indicate that the action of Cr^{3+} was on DNA. However, none of the 13 stress promoters responded to chromium oxalate at any of the concentrations tested.... The lack of *lacZ* induction in the case of chromium oxalate is probably due to the inability of that compound to enter the bacterial cells. This was confirmed with FAAS measurements of total chromium concentrations in *E. coli* (Table 1). In cells treated with chromium oxalate, concentrations of chromium were negligible. (Adapted from Plaper et al., 2002)

The third paragraph focuses on the influence of Cr^{3+} on gyrase. The paragraph begins by providing motivation for the gyrase experiments and reminds the reader of the experimental results (that Cr^{3+} binds to gyrase and DNA and may inhibit gyrase activity). Citing references to the literature for supporting evidence, the authors suggest that Cr^{3+} may be binding to the OH group in the active site of gyrase, thereby causing the inhibition. An alternative explanation is also provided.

P4 The stress promoters induced in the Pro-Tox (C) test indicated that Cr^{3+} could cause changes in DNA topology and in this way also affect proper DNA replication and transcription. For this reason, we examined the influence of Cr^{3+} on gyrase, an essential enzyme that relaxes and supercoils double-stranded DNA in bacterial cells.... We showed that Cr^{3+} binds to gyrase as well as to DNA. Additionally, we showed that Cr^{3+} may inhibit the enzyme's activity.... [T]he inhibition of gyrase activity may well be a consequence of Cr^{3+} binding to the OH group of the Tyr in the active site, preventing the enzyme's interaction with DNA phosphate groups (27). It is also possible that Cr^{3+} binds to some other part of the enzyme and causes changes in the gyrase conformation so that Tyr122 is no longer available for interaction with DNA. (Adapted from Plaper et al., 2002)

The conclusion of the work (move 2) is accomplished in the final paragraph of the article. The conclusion is broad in scope, reiterating the take-home message that Cr^{3+} can affect DNA. Implications of these findings are also suggested.

P5 Our results suggest that Cr^{3+} has an impact on DNA, DNA topology, and consequently processes leading to cell growth and proliferation. This could ultimately lead to the mutagenic and carcinogenic potential of Cr^{3+}. Because it is known that humans exposed to different Cr^{3+} species accumulate high levels of Cr^{3+} intracellularly (*17*), presented results may have an impact on human intake of Cr^{3+} as a nutrition additive. (Adapted from Plaper et al., 2002)

Exercise 5.7

Refer to passages P1–P5 as you complete the following tasks:

a. Read over passages P1–P4. Find groups of two to three sentences that together remind the reader of and interpret specific results.

b. Do the authors present their interpretations as facts, or do they use more cautious language? Support your answer with words from the text.

c. Is the last paragraph (P5) accessible to a scientific audience (as opposed to an expert audience), thereby completing the hourglass structure? What larger implications do the authors present?

d. Let's imagine that several studies (7–9) suggest that chromium oxalate cannot enter bacteria cells because of its size. (It is too big; the oxalate anion is $^{-}OOC–COO^{-}$, significantly larger than either NO_3^- or Cl^-.) Add a sentence or two to the end of the second paragraph (P3) to relate the current work to these studies.

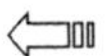

Audience

See chapter 1 for distinctions among expert, scientific, student, and general audiences.

Next, we analyze the Discussion section from Dellinger et al. (2001) regarding the toxicity of fine particulate. Remember that the authors postulated that free radicals in $PM_{2.5}$ contribute to its toxicity (excerpt 4E). The authors provided evidence that $PM_{2.5}$ contains free radicals and that untreated extracts of $PM_{2.5}$ induce DNA damage. However, DNA damage was not observed when free radical scavengers or Fe^{3+} chelators were added to the $PM_{2.5}$ extracts. Taken together, the data suggest that both free radicals and Fe^{3+} are involved in the toxicity of $PM_{2.5}$.

Dellinger et al. (2001) go a bit further and suggest that the free radicals in $PM_{2.5}$ are also the ones in cigarette tar (semiquinones), and they propose a mechanism by which this free radical can induce DNA damage (excerpt 5C). These suggestions are not mere speculation (which would constitute hand-waving) but are corroborated by the literature. The literature on the health effects of cigarette tar is extensive; hence, the authors build on this knowledge base. In this way, excerpt 5C represents a good example of how the literature can be used to provide deeper insights into data.

Excerpt 5C also provides a good example of how prose can be used to guide readers through a reaction cycle or mechanism. In this excerpt, a redox cycle is described (reactions 1–5); the text accompanying the cycle points out which reactions lead to the biologically damaging hydroxyl radical (reactions 1–3), how Fe^{2+}/Fe^{3+} may also be involved (reaction 3), and how the semiquinone radical is regenerated in the cycle (reaction 4). The authors conclude their article by reiterating their findings and suggesting that their results may apply, more generally, to the deleterious health effects associated with combustion-generated particulate matter.

Exercise 5.8

As you read through excerpt 5C, answer the following questions. To guide your reading, we include structures for quinone, the semiquinone radical, and hydroquinone:

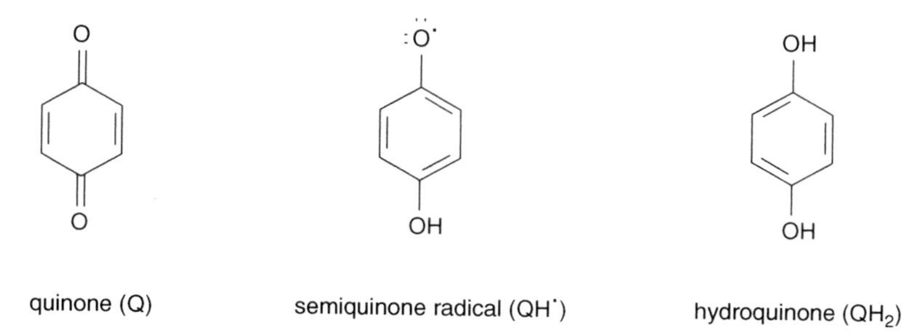

a. What is the purpose of the first sentence in excerpt 5C? Is this first sentence consistent with the general move structure presented in figure 5.1?

b. In the first paragraph, the authors make the case that the radicals in $PM_{2.5}$ are the same as those in cigarette tar (semiquinones). What evidence do they cite for this interpretation? Do you find the evidence convincing? What if they are wrong? Should the publisher issue a correction to the article?

c. The authors include a redox cycle in their Discussion section. The cycle shows a way in which the hydroxyl radical (HO·) can be continuously produced from the semiquinone radical (QH·). Is this an original mechanism, or one that the authors have included from the literature? How could you tell?

d. Recall that the authors point out, in the Results section, that both free radicals and Fe^{3+} may be the cause of DNA damage. Are both species implicated in the proposed mechanism and accompanying text? In other words, is the proposed mechanism consistent with their data? Explain.

e. In which paragraph does the second (and last) move begin? How can you tell? The move is divided into two paragraphs. Which submove is accomplished in each paragraph?

f. The word *we* occurs two times in the penultimate paragraph. Rewrite these sentences without *we*. Which do you like better? Explain.

Excerpt 5C (adapted from Dellinger et al., 2001)

[Continuation of excerpt 4E]

Discussion

The radicals in $PM_{2.5}$ have similar EPR g-values and line shapes to the radicals found in cigarette tar (*12, 18, 31, 33, 37*). The cigarette tar radicals produce DNA damage that is similar to that produced by $PM_{2.5}$. The cigarette tar radicals are a family of semiquinone radicals that are present in a dynamic hydroquinone (QH_2), semiquinone ($QH\cdot$), quinone (Q) system (*12, 18, 31, 33, 37*). Our body of data leads us to propose that the radical signals we report are also due to **semiquinone-type** radicals.

The sample sizes collected by our $PM_{2.5}$ sampling systems are insufficient to conduct detailed chemical analyses. However, available data for size-fractionated fine particulate matter indicates that PAH quinones, including 1,4-naphthoquinone, 5,12-naphthacenequinone, benz[*a*]anthracene-7,12-dione, and anthracene-9,10-dione, are important organic components (*41, 42*). The detection of these molecular species that are similar in structure to semiquinone-type radicals supports the assignment of our EPR signals.

A semiquinone radical ($QH\cdot$) can lead to the production of the hydroxyl radical ($HO\cdot$), as shown in reactions 1–3. $QH\cdot$ can reduce oxygen to form superoxide ($O_2\cdot^-$), reaction 1. Superoxide production leads to the formation of hydrogen peroxide (H_2O_2), reaction 2, which, in turn, yields the biologically damaging hydroxyl radical in a metal ion-dependent reaction, reaction 3. In addition, the quinone (Q) produced in reaction 1 can be reduced back to $QH\cdot$ and further to the hydroquinone (QH_2) by reducing agents present in biological systems, as shown in reactions 4 and 5 (*43*).

$$QH^{\cdot} + O_2 \longrightarrow Q + O_2^{\cdot -} + H^+ \quad (1)$$

$$2O_2^{\cdot -} + 2H^+ \longrightarrow H_2O_2 + O_2 \quad (2)$$

$$H_2O_2 + Fe^{2+} \longrightarrow HO^{\cdot} + HO^- + Fe^{3+} \quad (3)$$

$$Q + e^- + H^+ \longrightarrow QH^{\cdot} \quad (4)$$

$$QH^{\cdot} + e^- + H^+ \longrightarrow QH_2 \quad (5)$$

Reactions 1–5 establish a redox cycle (*43*) in which the hydroxyl radical is continuously produced from the radicals in $PM_{2.5}$. It is well documented that the hydroxyl radical causes DNA strand breaks (*33*).

The data shown in Figures 1–4 support the suggestion that $PM_{2.5}$ contains radicals that, like those in cigarette tar, can reduce oxygen to superoxide, which then forms hydrogen peroxide and, ultimately, the hydroxyl radical, as shown in reactions 1–3. Iron and copper ions, which are the transition metals most frequently found in combustion-generated particles (*44*) and also are ubiquitous in biological systems, could be involved in reaction 3. . . .

In conclusion, we have shown that $PM_{2.5}$ contains stable radicals that can be detected by EPR. The EPR parameters, persistence in air, and DNA-damaging capacity of the $PM_{2.5}$ radicals are similar to those of the radicals in cigarette tar. Therefore, we propose that the radicals associated with $PM_{2.5}$ include semiquinone-type radicals that, like the cigarette tar radical, can reduce oxygen to produce superoxide and ultimately produce the DNA-damaging hydroxyl radical.

The results presented here suggest a new mechanism of toxicity for $PM_{2.5}$ based on sustained hydroxyl radical generation by the semiquinone radicals present in $PM_{2.5}$. Because a substantial fraction of the fine particles in the atmosphere arises from combustion sources (9), it is possible that the deleterious health effects associated with $PM_{2.5}$ can be at least partially ascribed to radicals associated with combustion-generated particulate matter.

Hyphenated Modifiers

Chemistry writing includes many two-word modifiers that describe nouns. Consider these examples:

- low-energy process
- high-energy sites
- bioremediation-accelerating effect
- semiquinone-type radicals
- 3-fold induction
- DNA-damaging capacity
- combustion-generated particulate matter

(See appendix A.)

We complete our analysis of the Discussion section by examining two representative organic synthesis articles that appeared in *Organic Letters* and *The Journal of Organic Chemistry*. These journals typically omit section headings entirely (*Organic Letters*) or include only an Experimental Section heading at the end of the article (*The Journal of Organic Chemistry*). Nevertheless, we can still find the typical moves of a Discussion section in these seemingly "discussion-less" articles.

We first consider the Demko and Sharpless article (2001) on the synthesis of substituted tetrazoles from nitriles in water (excerpt 5D). This excerpt is particularly useful because it illustrates several types of content that authors typically discuss in synthesis papers. The authors begin by proposing two possible mechanisms for the tetrazole reaction, a two-step mechanism and a concerted mechanism. The mechanisms are presented in a **scheme** (Scheme 1). In the accompanying text, the authors cite evidence for both mechanisms, highlight salient features of the mechanisms, mention the results of kinetic studies, and point out that the role of zinc metal is as yet unclear.

Schemes

Schemes, like figures and tables, are a form of graphics. They are often used to depict proposed reaction mechanisms. A proposed mechanism, by its very nature, is interpretative; hence, schemes are commonly found in Discussion sections.

For information on formatting schemes, see chapter 16. Also see exercises 5.9 and 5.10.

The authors go on to discuss several additional factors related to their synthesis, such as stoichiometric considerations (molar equivalents needed to ensure that the reaction goes to completion), chief competing reactions and how they might be minimized, and other reagents that might be successful substitutes in the reaction. Finally, they discuss their efforts to scale up the reaction, taking it from bench-scale to large-scale applications. Although some of these discussion points are not interpretive in a formal sense, they are representative of common topics in the Discussion section of a synthesis paper. Following typical organizational conventions, the authors conclude with a brief summary of the article and suggest larger implications and applications of their findings (following the move structure represented in figure 5.1).

Exercise 5.9

As you read through excerpt 5D, notice that the authors use a scheme to present two proposed mechanisms for the reactions. Using this excerpt as an example, propose a set of rules for formatting a scheme (see also chapter 16).

Excerpt 5D (adapted from Demko and Sharpless, 2001)

[Continuation of excerpt 4F]

Kinetic studies using the water-soluble nitrile **1i** revealed first-order dependence in both nitrile and azide and one-half order dependence for zinc bromide. The mechanism of the addition of hydrazoic acid/azide ion to a nitrile to give a tetrazole has been debated, with evidence supporting both a two-step mechanism[8b,21] (Scheme 1, eq 2) and a concerted [2 + 3] cycloaddition[22] (Scheme 1, eq 3). Our mechanistic studies to date imply that the role of zinc is not simply that of a Lewis acid; a number of other Lewis acids were tested and caused little to no acceleration of the reaction.[23] In contrast, Zn^{2+} exhibited a 10-fold rate acceleration at 0.03 M, **which** corresponds to a rate acceleration of approximately 300 at the concentrations typically used. The exact role of zinc is not yet clear.

Empirically, we found that to ensure complete reaction one needs a 0.5 molar equiv of the zinc salt (ZnX_2); however, in many cases, lower loadings of zinc may be used.[24] The chief competing reaction is hydrolysis of the nitrile to the primary amide; therefore, in cases where the tetrazole-forming reaction is sufficiently fast, namely, with electron-poor nitriles, lower zinc loadings did not entail significant formation of the amide byproduct. Other zinc salts such as zinc perchlorate and zinc triflate also work; zinc chloride, while less expensive, led to more of the amide byproduct. Zinc bromide was chosen as the best compromise between cost, selectivity, and reactivity (see Table 3).

Scheme 1

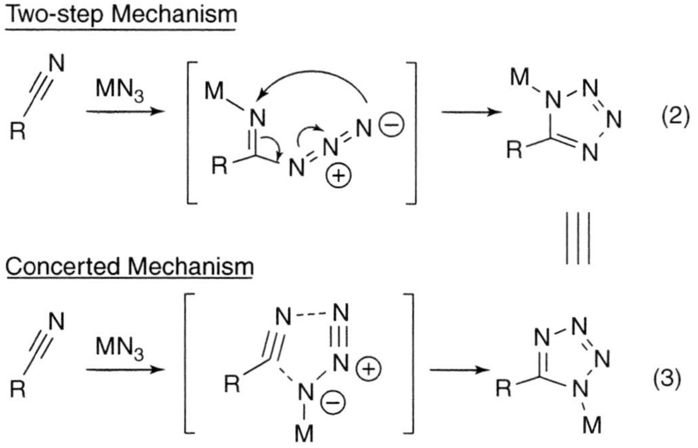

Two-step Mechanism

MN_3 (2)

Concerted Mechanism

MN_3 (3)

M = H, L_nZn, other metals

The process became even more attractive for large-scale applications when we found that it could be run at higher concentration without sacrificing yield and without the use of organic solvents in the workup or isolation phases. The resulting products were spectroscopically identical by ^{1}H NMR and ^{13}C NMR to those synthesized by the general method outlined below; however, the melting points were slightly lower.

In summary, we have demonstrated an exceedingly simple protocol for transforming a wide variety of nitriles into the corresponding 1*H*-tetrazoles. By using zinc salts as catalysts, we showed that water can be used as the solvent despite the relative insolubility of the starting materials. This discovery should facilitate the preparation of tetrazoles in the laboratory.

Topics of Discussion in a Synthesis Paper

- a mechanism
- kinetic (rate) vs. thermodynamic (energy) considerations
- role of a catalyst
- stoichiometry
- competing reactions (and how to minimize them)
- alternative reagents
- efforts to scale up the reaction

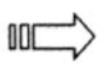

Which vs. That

See appendix A for more information on these easily confused words.

Exercise 5.10

Find out if your set of rules for schemes (generated as part of exercise 5.9) applies to schemes in other articles. Check three articles in *The Journal of Organic Chemistry* or *Organic Letters* to see if your rules hold true. In these articles, how do the authors use compound labels and accompanying text to walk the readers through the schemes?

Exercise 5.11

Browse through three Discussion sections of synthesis articles in *The Journal of Organic Chemistry*. In addition to proposed mechanisms, make a list of other topics routinely addressed in Discussion sections.

Lastly, we consider Boesten et al. (2001) (excerpt 5E), which describes an asymmetric Strecker synthesis. Recall that the synthesis results in two diastereomers, which can be separated based on their different solubilities in water. In exercise 2.14 (chapter 2), you were asked to decide where the Discussion section began in this article. That exercise was more challenging than you may have realized. The article presents numerous results, with only a few sections of integrated discussion.

Exercise 5.12

With these comments about the Boesten et al. (2001) article in mind, go back and try exercise 2.14(f) again. Reexamine the Strecker synthesis article (at the end of chapter 2) and assign R or D to the appropriate sentences and/or paragraphs.

Consider your answers to exercise 5.12. You will probably agree with us that Figure 2 (in the Boesten et al. article), along with the paragraph that describes it, is part of the discussion. We have reproduced the figure and its accompanying text in excerpt 5E. Once again, the discussion focuses on a mechanism, although in this case, the mechanism is presented in a figure rather than a scheme. (The choice to use a figure or scheme is usually left to the authors.) The authors propose a mechanism that includes a reaction intermediate (*R*)-**4** and shows how (*R*,*S*)-**3** is produced preferentially over (*R*,*R*)-**3**. The text walks the reader through important aspects of the mechanism, pointing out that, at room temperature, CN^- attacks the *re* face of (*R*)-**4** preferentially, producing (*R*,*S*)-**3** in greatest yield. At elevated temperatures, because (*R*,*S*)-**3** crystallizes out of solution more readily than (*R*,*R*)-**3**, equilibrium favors the (*R*,*S*)-**3** product. In a separate graphic (Scheme 1), the authors go on to describe how (*R*,*S*)-**3** is ultimately converted to the target molecule, (*S*)-*tert*-leucine. The article ends with a short paragraph that summarizes the work, underscores its usefulness ("a practical one-pot" synthesis), and suggests ways in which the synthesis can be applied to other target molecules.

The *re* Face

An sp^2 carbon in which the three substituents are oriented clockwise according to Cahn-Ingold-Prelog sequence rules.

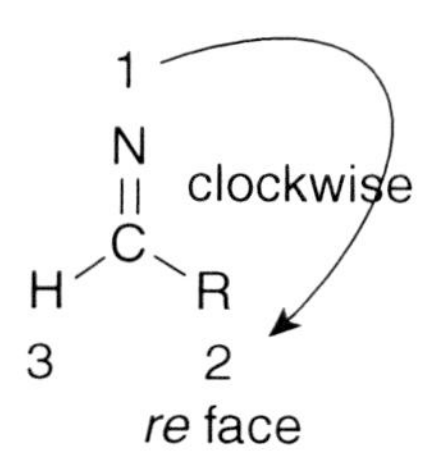

Exercise 5.13

As you read excerpt 5E, complete the following tasks:

a. Explain why the imine intermediate in Figure 2 is labeled (*R*)-**4** even though it leads to the formation of (*R*,*S*)-**3** and (*R*,*R*)-**3**. (Hint: Look at Table 1 of Boesten et al. (2001) at the end of chapter 2.)

b. Typically, it is sufficient (and recommended) that you refer to a compound by number only, after the number has been introduced. However, Boesten et al. (2001) do not follow this rule in their final paragraph. Both (*R*)-phenylglycine amide **1** and (*S*)-tert-leucine **7** are mentioned a second time using name and number. Why you think that the authors do this, rather than simply using **1** and **7**?

c. Imagine that you are a synthetic chemist and that you have successfully synthesized the compound ABC **5**. Write a one- to two-sentence summary statement for your paper on the synthesis of **5** based on the following information:

   1. the reaction is easy and involves only three steps
   2. the reaction is inexpensive
   3. the synthesis is a general one and can be applied to other compounds
   4. the yields are good (>95%) with few side reactions

Excerpt 5E (adapted from Boesten et al., 2001)

[Continuation of excerpt 4G]

Ph
N CONH$_2$
H
re-face attack
(*R*)-**4**
HCN
HCN
Ph
HN CONH$_2$
CN
H
preferential crystallization
(*R,S*)-**3**
Ph
HN CONH$_2$
CN
H
(*R,R*)-**3**

Figure 2. Crystallization-induced asymmetric transformation of amino nitrile **3**.

The observed diastereoselectivity in the asymmetric Strecker step via the crystallization-induced asymmetric transformation can be explained as shown in Figure 2. Apparently, the *re* face addition of CN^- to the intermediate imine **4** is preferred at room temperature in methanol and results in a dr 65/35. At elevated temperatures in water, the diastereomeric outcome and yield of the process are controlled by the reversible reaction of the amino nitriles **3** to the intermediate imine and by the difference in solubilities of both diastereomers under the applied conditions.[16,17] . . .

In summary, (*R*)-phenylglycine amide **1** is an excellent chiral auxiliary in the asymmetric Strecker reaction with pivaldehyde or 3,4-dimethoxyphenylacetone. Nearly diastereomerically pure amino nitriles can be obtained via a crystallization-induced asymmetric transformation in water or water/methanol. This practical one-pot asymmetric Strecker synthesis of (*R,S*)-**3** in water leads to the straightforward synthesis of (*S*)-*tert*-leucine **7**. Because (*S*)-phenylglycine amide is also available, this can be used if the other enantiomer of a target molecule is required. More examples are currently under investigation to extend the scope of this procedure.[19]

5C Writing on Your Own: Draft Your Discussion Section

Write a complete draft of your Discussion section. Begin with an outline of your results and interpretations, and then convert the outline to full sentences. Sketch out any schemes (or figures) you plan to include, and draft the text that will accompany those schemes. End your draft with a paragraph that concludes your paper and suggests the broader applications of your work.

Part 2: Analyzing Writing across the Discussion Section

Here, we examine writing conventions that are common throughout the Discussion section. We focus on tense and voice in addition to two word-choice issues, the use of *we* and hedging words.

Tense and Voice

Both past tense and present tense are common in the Discussion section. The same rule of thumb introduced in the Methods section applies to the Discussion section:

The work was done in the past, but knowledge exists in the present.

For example, Plaper et al. (2002) state that "three Cr^{3+} compounds *were examined*" (work done in the past) but that the "hydroxyl radical *is* a known mediator of DNA damage" (knowledge that exists in the present). Similarly, the present-tense, active-voice combination is used in the Discussion section to state scientific "truths" (knowledge expected to be true over time), just as it was in the Results section. Note that interpretations and/or mechanisms put forth in a Discussion section are often considered to be "truths" and therefore are stated in present-tense active voice. Table 5.1 summarizes common verb tense–voice combinations and their functions, with example sentences.

Exercise 5.14

Examine sentences 1–6 below (adapted from Wu et al., 2004). Using table 4.1, table 5.1, and the information presented in this chapter, do the following:

a. Determine the function(s) of each sentence.
b. Identify the tense–voice combination(s) used in each sentence.
c. Decide whether the tense–voice combinations are appropriate, given the function(s) of the sentence. Rewrite the sentence, if necessary, so that it is more appropriate for a Discussion section.

1. Because fruits and vegetables are the major antioxidant sources in the daily diet of humans, we calculated estimated daily antioxidant capacity intake from these foods.
2. Our data suggested that foods with active polyphenolic flavonoids were more resistant than foods with vitamins and related compounds.
3. Removal of the peel is one factor that influences antioxidant capacity (*28*, *66*) as indicated by lower values in apples compared to that of the intact apple.

4. In our study, cooked tomatoes have a significantly higher H-$ORAC_{FL}$ and L-$ORAC_{FL}$ compared to uncooked samples, which agrees with observations in previous studies (*42*, *68*).
5. Under normal reaction conditions, phenolic compounds are the predominant antioxidants in hydrophilic extracts of samples that easily transfer one hydrogen to the peroxy radical ($ROO^{\cdot}$).
6. In summary, the lipophilic and hydrophilic $ORAC_{FL}$ values for more than 100 common foods in U.S. markets are obtained for the first time.

Use of "We"

The word *we* (used to refer to the authors of the work) is commonly used to achieve the following purposes in the Discussion section:

To indicate a decision or course of action

P6 To correct this fluctuation, *we* incorporated caffeine standard measurements. (Yan et al., 2007)

P7 To further validate the suitability of using caffeine as a standard, *we* tested 15 additional compounds. (Adapted from Yan et al., 2007)

To compare findings with previous works

P8 However, like others,[28] *we* often observed quantification inconsistencies when compounds with adjacent nitrogen atoms were analyzed. (Adapted from Yan et al., 2007)

P9 Previously, *we* reported that oral intake of BCAs has many beneficial physiological effects (2–6). (From Matsumoto et al., 2006)

To offer an interpretation

P10 *We* speculate that, due to structural variations, some unavoidable signal loss may be structure dependent and an error range of 10–20% is possible. (Yan et al., 2007)

P11 Therefore, *we* assumed the decrease of mitochondrial membrane potential and release of cytochrome *c* to cytosol did not happen in the PDTC-pretreated cells. (Cheng et al., 2006)

To report or summarize findings

P12 *We* found that sulfite alters the pattern of expression of the main aldehyde dehydrogenase gene, reinforcing the link at the molecular level between both metabolites. (Adapted from Aranda et al., 2006)

Table 5.1 Common functions of different verb tense–voice combinations in Discussion sections.

Function	Tense–Voice Combination	Example
To remind readers about what was studied in the current work	Past (active or passive)	In this work, three Cr^{3+} compounds *were examined* for their impact on genotoxicity and cell proliferation in vitro (From Plaper et al., 2002) (past–passive)
To remind readers about/ summarize specific result(s) in the current work	Past (active or passive)	Similar to regular garlic, regular broccoli florets *reduced* the incidence of mammary tumors (From Finley et al., 2001) (past–active)
To share corroborating or conflicting results from others' works	Past (active or passive)	Uptake of dissolved organic carbon by zebra mussels *was* also *reported* by Roditi et al. (*30*). (Adapted from Voets et al., 2004) (past–passive)
To interpret results presented in the current work	Present–active	The 0.5 TU threshold *is* arbitrary but *suggests* a strong likelihood that the analyte *makes* a substantial contribution to the observed mortality. (From Weston et al., 2004)
To propose "truths" based on the current work and others' works	Present–active	The data in this paper *are* consistent with a wealth of evidence showing that dietary Se consumed in excess of the Recommended Dietary Allowance *lowers* the risk of several important cancers (*2*, *13*, *26*, *27*). (From Finley et al., 2001)
To present the take-home message of the current work	Present–active	In summary, (*R*)-phenylglycine amide **1** *is* an excellent chiral auxiliary in the asymmetric Strecker reaction with pivaldehyde or 3,4-dimethoxyphenyl-acetone. (From Boesten et al., 2001)
To suggest overall implications and/or applications of the current work	Present–active	The good sensitivity, excellent selectivity, and simplicity of use of the DNA biosensor *make* it more compatible for integrating with on-chip PCR reactors than other DNA biosensors of which we are aware. (Adapted from Wong and Gooding, 2006)

P13 In conclusion, *we* have demonstrated that PDTC inhibits luteolin-induced apoptosis, which it might do by causing the phosphorylation of caspase-9 in human leukemia HL-60 cells. (Cheng et al., 2006)

Exercise 5.15

Consider the sentences below, all taken from Discussion sections. What are the authors trying to achieve in each sentence: (1) indicate a decision or course of action, (2) compare findings with previous works, (3) offer an interpretation, or (4) report or summarize findings?

a. *We* used this value as a correction factor for caffeine calibration, and all compound analyses were adjusted accordingly. (Adapted from Yan et al., 2007)
b. *We* rationalize that this is likely the root cause for the structure dependence. (From Yan et al., 2007)
c. In conclusion, *we* have shown that $PM_{2.5}$ contains stable radicals that can be detected by EPR. (From Dellinger et al., 2001)
d. In our previous study, *we* observed that luteolin could trigger cytochrome *c* released to cytosol. (Cheng et al., 2006)
e. *We* confirmed that, when compounds contain isolated nitrogen atoms, the response is close to quantitative with a variation about 10–20% depending on structures. (Adapted from Yan et al., 2007)

Hedging Words

Hedging is the mark of a professional scientist, one who acknowledges the caution with which he or she does science and writes on science.

—Crismore and Farnsworth (1990)

As mentioned at the beginning of the chapter, the language of a Discussion section is typified by restraint and understatement. The goal is to let the science speak for itself. Words should be used to clarify, not convince. By overselling a point, even one that is well substantiated, you will appear to be biased. For this reason, chemists soften their interpretations and claims to "truth" by using hedging words. The bar below lists verbs that are commonly used in combination with hedging words, such as *can*, *may*, and *might*.

Verbs Used with Hedging Words

The hedging words *can, could, may, might, should,* and *would* commonly occur with verbs such as these:

assign	conclude	expect	make	rationalize
attribute	deduce	explain	observe	relate
cause	determine	form	obtain	use

Exercise 5.16

Consider the passages below. Find words that indicate that the authors are hedging, rather than offering proof. Then read over excerpts 5A–5E to identify at least five additional examples of hedging.

a. Our results suggest that Cr^{3+} has an impact on DNA, DNA topology, and consequently processes leading to cell growth and proliferation. This could ultimately lead to the mutagenic and carcinogenic potential of Cr^{3+}. Because it is known that humans exposed to different Cr^{3+} species accumulate high levels of Cr^{3+} intracellularly (*17*), presented results may have an impact on human intake of Cr^{3+} as a nutrition additive. (Passage P5)

b. A possible explanation for this effect can be attributed to an increased density and viscosity of the milk–NaOH phase when NaOH concentration increases. (Excerpt 5A)

c. In all cases, this factor has a positive effect and it appears to be of increasing importance as the degree of chlorination of the PCBs increases. (Excerpt 5A)

d. For the highly chlorinated PCBs, the two fibers tested seem to have similar performance, and this factor lacks statistical significance. (Excerpt 5A)

e. The results presented here suggest a new mechanism of toxicity for $PM_{2.5}$ based on sustained hydroxyl radical generation by the semiquinone radicals present in $PM_{2.5}$. Because a substantial fraction of the fine particles in the atmosphere arise from combustion sources (*9*), it is possible that the deleterious health effects associated with $PM_{2.5}$ can be at least partially ascribed to radicals associated with combustion-generated particulate matter. (Excerpt 5C)

f. By using zinc salts as catalysts, we showed that water can be used as the solvent despite the relative insolubility of the starting materials. This discovery should facilitate the preparation of tetrazoles in the laboratory. (Excerpt 5D)

Exercise 5.17

All hedges have been removed from the paragraph below (adapted from Lissens et al., 2004). Rewrite this paragraph so that it better conforms to the conventions of a journal article. Remember that not all sentences may require hedging. (Note that when you add hedging words, you may have to make other minor changes in the sentence.)

Effects of WO on Anaerobic Biodegradability of Raw Waste. Figure 1 shows the effect of the composition of a waste stream (Figure 1A) as well as the effect of the applied wet oxidation conditions (Figure 1B) on the anaerobic biodegradability of raw and digested waste after assessing wet oxidation. Although a doubling of the methane yield was achieved for wet oxidation yard waste compared to the reference, a minor increase

(7%) in methane yield was observed when raw food waste was subjected to wet oxidation. This was due to inherent differences in lignocellulose composition and characteristics of the lignin fraction of both wastes. Although it was previously shown that both wastes have a similar lignin content (21–22 g/100 g) and also rather similar cellulose and hemicellulose content (Table 2) (*18*, *19*), the amount of readily biodegradable and soluble organics in the food waste must be much higher than that in the woody yard waste. Hence, the wet oxidation pretreatment leads to a substantial beneficial effect on the biodegradability of the fibrous yard waste, although this was not the case for food waste.

5D Writing on Your Own: Practice Peer Review

Imagine that a colleague has asked you to review a draft of a Discussion section. Based on what you have learned in this chapter, read the draft, background information, and instructions at the end of this chapter under "Peer Review Practice: Discussion Section" and provide written feedback.

5E Writing on Your Own: Fine-Tune Your Discussion Section

By now, you should have made good progress writing your own Discussion section; thus, it is time to refine and edit your work. Focus on each of the areas specified below. Refer to chapter 18 to guide you in the revision process.

1. Organization of text: Check your overall organizational structure. Did you follow the move structure outlined in figure 5.1 and include appropriate subheadings?
2. Audience and conciseness: Are you writing for an expert audience, leaving out unnecessary details? Try to find at least three sentences that can be written more clearly and concisely. Have you directed your closing comments (e.g., applications and/or implications) to a more scientific audience? Check that you used *we* and hedging words appropriately.
3. Writing conventions: Check to be sure that you have used voice and tense correctly.
4. Grammar and mechanics: Check for typos and errors in spelling, subject–verb agreement, and punctuation. Be sure you have used troublesome words such as *effect*, *affect*, and *data* correctly.
5. Scientific content: Have you correctly conveyed the science in your work? Have you used words and units correctly? If asked, could you define all of the words you have used in this section? Do you understand the implications of your work?

After thoroughly reviewing your own Discussion section, ask a colleague to review your work. A "new set of eyes" will pick up mistakes that you can no longer see because you are too familiar with your own writing. To facilitate this process, use the Peer Review Memo on the *Write Like a Chemist* Web site. After your paper has been reviewed (and you have reviewed another's paper), use the feedback provided to make final changes to your Discussion section.

Finalizing Your Written Work

See chapter 18.

Chapter Review

As a self-assessment of what you've learned in this chapter, define each of the following terms for a friend or colleague who is new to the field:

hand-waving argument	implications	scheme
hedging	interpretation	take-home message
hyphenated modifier		

Also, explain the following to a friend or colleague who has not yet given much thought to writing a journal article Discussion section:

- Main purpose(s) of a Discussion section
- Standard moves of a Discussion section
- Use of present-tense active voice in a Discussion section
- Use of *we* in a Discussion section
- Purpose(s) of hedging and common hedges

Additional Exercises

Exercise 5.18

Revise the following sentences (intended for a Discussion section) to make them more concise:

a. It is likely that both specific and nonspecific effects add up to lead to the production of the observed differences in CO stretching frequencies.

b. In the case of the photobleaching experiment, the analysis is relatively straightforward, and it is further simplified by the molecular symmetry of **1**.

c. Conversion trends a lot like those shown in Figure 12 have been reported in an article by Al-Dhabi et al.[3]

Exercise 5.19

Excerpt 5F includes the Discussion section from a 2001 article in the *Journal of Agricultural and Food Chemistry*. Read the excerpt and then complete the following tasks:

a. Identify each of the moves and submoves of the Discussion section. Does this Discussion section follow the typical move structure presented in figure 5.1, or does it vary somewhat? If it varies, explain the deviation.
b. Find instances of hedging in the excerpt. What purposes do these examples of hedging serve?
c. Do the authors of this Discussion section both interpret results and apply them to a wider context? Do the authors explore the broader implications of their work? If so, what applications are mentioned? What implications are mentioned?
d. Does this excerpt include any two-word modifiers? If so, list them and determine if they are hyphenated properly.
e. Consider the authors' use of verb tense and voice. Find at least two sentences in which present tense–active voice is used to state interpretations.

Excerpt 5F (from Finley et al., 2001)

Discussion

The data in this paper are consistent with a wealth of evidence showing that dietary Se consumed in excess of the Recommended Dietary Allowance lowers the risk of several important cancers (*2*, *14*, *26*, *27*). Previous research has established a strong association between the dietary form of Se and the cancer-preventive properties of this element (*5*, *28*). This paper extends the evidence that Se in chemical forms known to accumulate in garlic and some *Brassica* species is especially effective in the prevention of chemically induced carcinogenesis (*14*, *17*).

Similar to regular garlic, regular broccoli florets reduced the incidence of mammary tumors (Table 1) (*3*); this indicates that there are components in addition to Se in each of these plants that have anticarcinogenic activity. Se-enriched broccoli was not more effective (Table 1) than enriched garlic (*3*) in reducing the number of tumors; this suggests that the combination of sulforaphane, indole carbinol, and chlorophyll with Se did not provide additional protection against mammary tumors. However, firm conclusions cannot be made because the concentrations of these compounds were not determined in the broccoli used in this experiment and because a direct comparison of high-Se garlic and high-Se broccoli was not made.

Results of the second experiment (Table 3) show that Se-enriched broccoli sprouts have properties similar to enriched broccoli florets that contain SeMSC as the predominant form of Se (*13*). Consumption of Se from high-Se broccoli sprouts, as compared to Se from selenite, resulted in a significant decrease in the number of aberrant crypts. Additional experimentation is needed to determine whether the decrease in carcinogenesis is a result primarily of the presence of SeMSC, and if there is a correlation between SeMSC content in enriched plants and the reduction of carcinogenesis. If such a correlation is established, then the SeMSC content of various enriched plants could be used to screen for the greatest efficacy in tumor reduction. Se-enriched broccoli appears to be similar to enriched broccoli florets, for which the predominant form of selenium was also shown to be SeMSC (*13*).

High-Se broccoli sprouts were not more effective than high-Se broccoli florets for the prevention of DMH-induced aberrant colon crypts (Table 3). This result, in conjunction with the finding that low-Se broccoli was not more effective than regular garlic for prevention of MNU-induced mammary tumors (experiment 1, Table 1), provides strong evidence that the cancer-preventive qualities of secondary plant compounds found in broccoli but not garlic, such as sulforaphane, indole carbinol, and chlorophyll, are masked by a much stronger protective effect of Se in broccoli.

The present results also point to differences between the mammary tumor model and the ACF model for evaluating the potential cancer protective effects of Se in broccoli. A previous study (*17*) showed that high-Se broccoli florets decreased the number of DMH-induced ACF. Similarly, in the present study high-Se broccoli sprouts decreased DMH-induced ACF, but low-Se broccoli sprouts alone did not have any effect (Table 3). In the mammary tumor model, however, broccoli alone, similar to garlic alone (*3*), reduced the number of tumors (Table 1). This contrasting effect could be the result of a difference between tumor and preneoplastic lesion models, a difference between carcinogens, or a difference between mammary and colonic tissues.

In response to the findings of the Se-responsive reduction of cancer risk in humans, many nutritionists and other health professionals have begun to suggest supplemental intakes of as much as 200 μg of Se/day. However, the present results provide evidence that the total Se intake is not the only factor to consider for the reduction of carcinogenesis. Similar to our previous findings concerning the ability of high-Se broccoli to reduce the incidence of colon cancer (*17*), an equal amount of Se supplied as selenite did not significantly reduce the incidence of ACF. This means that in addition to total Se intake, the form of Se in a particular food or supplement must be taken into consideration. Grains and meat supply a major portion of dietary Se (*29*), and the Se in grains and meat is very effective for increasing tissue Se concentrations and GSH-Px activities (*30–32*). However, the form of Se in meat and grain is greatly different from the form in broccoli and garlic (*13*), foods that seem to provide superior anticarcinogenic properties. Consequently, more work needs to be conducted before concrete recommendations of the optimum forms of supplemental Se can be made.

Exercise 5.20

Excerpt 5G showcases the last move of a Discussion section.

a. As you read the excerpt, identify the goals of the work, the methodology used, the major findings, and the interpretation of findings.
b. In your opinion, is this a typical conclusion for a Discussion section? Why or why not?

Excerpt 5G (adapted from Yu et al., 2001)

In summary, we have designed and synthesized a new ferrocene-modified phosphoramidite **9** for the electronic detection of single-base mismatches in an array format. By employing automated DNA/RNA synthesis techniques the ferrocenyl complexes have been inserted into oligonucleotides at various positions. The thermal stability of the metal-containing DNA oligonucleotides has been investigated and indicates that the incorporation of **9** into DNA oligonucleotides causes little or no destabilization of the duplex. Electrochemical analysis of oligonucleotides containing **9** reveals that the derivative can function as a signaling probe for the electronic detection of nucleic acids. When incorporated into a CMS-DNA chip, results clearly show that dual-signaling oligonucleotide probes containing **9** and the phosphoramidite **1** detect single-base mismatches.

Exercise 5.21

Reflect on what you have learned about writing a Discussion section for a journal article. Select one of the reflection questions below and write a thoughtful and thorough response:

a. Reflect on the move structure of a typical Discussion section.
   - How does knowing the typical move structure of a Discussion section assist you with reading authentic excerpts from the chemical literature?
   - How has knowing the typical move structure of a Discussion section helped you write your own Discussion section?
   - Which parts of the typical move structure of a Discussion section are easiest to write? Most difficult to write? Why?
b. Reflect on the different ways in which authors connect their Results and Discussion sections (i.e., stand-alone Discussion, blocked R&D, iterative R&D, integrated R&D).
   - Which format has the greatest appeal to you? Why?
   - Which format is the easiest to read? The most difficult to read? Why?
   - Can you explain the logic of each formatting convention? What might the benefits of each format be?
c. Reflect on your experience writing your own Discussion section.

- What problems did you encounter writing your Discussion section? How did you resolve your problems?
- What parts of this chapter have helped you the most in writing your Discussion? How did you use chapter information to assist you in your writing?
- What have you learned from the experience of writing your Discussion that will help you in the future?

Peer Review Practice: Discussion Section

Imagine that a classmate has asked you to review and provide feedback on a draft of a Discussion section. Your valuable feedback will be used to improve the written work.

Using parts 2 and 3 of the Peer Review Memo on the *Write Like a Chemist* Web site, review the Discussion section below. We have included the last paragraph from the Introduction section to help you better understand the Discussion. You do not need to review the Introduction. Provide specific suggestions that your classmate can use to improve the Discussion section. (The Discussion section below is adapted from an original source, noted in the Instructor's Answer Key.)

[Excerpt from the Introduction]

The aim of this study was to investigate the effect of high but environmentally realistic concentrations of humic acid on the relative long-term accumulation of Cd in the freshwater mussel *D. polymorpha* under controlled laboratory conditions. We assessed whether the long-term uptake of Cd is in agreement with the free-ion activity model and if Cd accumulation is related to the Cd^{2+}-ion activity in the water.

Discussion

1 As stated previously in the Methods section, zebra mussels were exposed to varying
2 concentrations of cadmium in 80 L of water for 31 days. Each aquarium contained 150
3 individuals at the start of the experiment. Five experimental aquaria and one control
4 aquarium were run. We found that zebra mussels efficiently accumulate cadmium and
5 other heavy metals. Even at low environmental Cd concentrations or short exposure
6 periods cadmium concentrations in the tissues are significantly elevated. Our data
7 shows that the cadmium concentration in zebra mussel tissue increases from 0.054 ±
8 0.004 to 0.214 ± 0.031 μmol/g dry wt at an exposure concentration of 0.020 μM after
9 31 days. The uptake rate of cadmium is strongly correlated with the free Cd-ion activi-
10 ties in the water, and the Cd concentrations in the tissues are still increasing at the end
11 of the 31-day exposure period.
12 The presence of humic acid decreases the cadmium accumulation by the zebra mus-
13 sels. Although this reduced accumulation can mainly be explained by the decrease in

14 free cadmium-ion activity in the exposure water, cadmium accumulation was higher
15 than expected based on cadmium-ion activity alone. This proves that zebra mussels
16 must accumulate cadmium complexed to humic acid. The Cd^{2+} binding to humic acid
17 must occur through interactions between the cadmium ion and negatively charged sur-
18 face functional groups.

6 *Writing the Introduction Section*

In writing an article about my chemical research, I like to use as a model a travel or travel-adventure article that might appear in a magazine or in the travel section of a newspaper. Because, after all, what is research but a voyage to a completely unknown place where everything is new. So the first few paragraphs of the article [the Introduction] should tell the potential reader where we are going and why.

—Charles H. DePuy, University of Colorado–Boulder

This chapter focuses on the Introduction, the first formal section of the journal article. The Introduction is often the first section to be read (by readers) but the last section to be written (by writers). This is because the Introduction must tell readers "where the article is going and why", a mission that is most easily accomplished after the rest of the sections have been completed. By the end of this chapter, you should be able to do following:

- Write an Introduction following its conventional organizational structure
- Compose the all-important opening sentence of an Introduction
- Cite and summarize others' works in concise and appropriate ways
- Conclude your Introduction in an effective manner

As you work through the chapter, you will write an Introduction section for your own paper. The Writing on Your Own tasks throughout the chapter guide you step-by step as you do the following:

6A Read and paraphrase the literature

6B Prepare to write

6C Draft your opening paragraph

6D Identify a gap

6E Draft your full Introduction

6F Practice peer review

6G Fine-tune your Introduction

The Introduction, as its name implies, sets the stage for the rest of the journal article by introducing the research area, describing its importance, and hinting at what new knowledge and insights the authors have gained. The Introduction is also where authors summarize others' works; this involves several important writing skills such as paraphrasing, writing concisely, and correctly citing the literature. Paraphrasing and writing concisely are addressed in this chapter; citing the literature is addressed in chapter 17.

Citing the Literature

See chapter 17 for information on how and what to cite from the primary literature.

Reading and Analyzing Writing

At long last, we ask you to read the Introduction to the aldehydes-in-beer article (excerpt 6A). If you have progressed through these textbook chapters sequentially, you have already read the Methods, Results, and Discussion sections. Admittedly, this order may seem a bit unusual. Remember, however, we want you to read the Introduction through the eyes of the writer, not the reader. As authors write their Introduction, they already know what unfolds in the rest of their paper; now you, too, have this perspective. (If necessary, refer back to chapters 3–5, and excerpts 3A and 4A, to refresh your memory.) As you read the Introduction, consider how the authors introduce their story of scientific discovery.

Exercise 6.1

As you read through excerpt 6A, complete the following tasks:

a. Identify the major purpose of each paragraph. Use these purposes to propose a move structure for the Introduction.
b. How detailed are the authors' descriptions of others' works?
c. What do you notice about the language and writing conventions used by the authors?
d. Refer back to excerpts 3A (Methods) and 4A (Results and Discussion). Jot down two to three key concepts for each excerpt. Where, and to what extent, is this information shared in the Introduction?

Excerpt 6A (from Vesely et al., 2003)

Introduction

Carbonyl compounds, particularly aldehydes, are considered to play an important role in the deterioration of beer flavor and aroma during storage. Strecker degradation of amino acids, melanoidin-mediated oxidation of higher alcohols, oxidative degradation of lipids, aldol condensation of short-chain aldehydes, and secondary oxidation of long-chain unsaturated aldehydes are mechanisms implicated in their formation (*1*). Their levels in beer are usually very low, and therefore it has always been a challenge for brewing chemists to develop an analytical method that would enable routine analysis of aldehydes.

Several analytical methods for the determination of aldehydes in beer have been developed, and good results have been obtained using liquid–liquid extraction (*2*), distillation (*3*), or sorbent extraction (*4*). However, these methods are rather complicated and not highly selective.

A simple way to increase the selectivity of extraction techniques is to derivatize the carbonyl compounds. *O*-(2,3,4,5,6-Pentafluorobenzyl)hydroxylamine (PFBOA) is commonly used as a derivatization agent in gas chromatography (*5*). This technique has been applied to the analysis of carbonyl compounds in water and also in beer (*6*). Although these methods provide good reproducibility, they are time-consuming and require use of solvents, materials for the derivatization, and isolation steps. Martos and Pawliszyn (*7*) developed an original extraction technique based on PFBOA on-fiber derivatization of gaseous formaldehyde followed by gas chromatography with flame ionization detection.

In this work, we adapted a method for the analysis of beer aldehydes using solid-phase microextraction (SPME) with on-fiber derivatization. This extraction technique does not require solvents, consists of a one-step sample preparation procedure, and provides high sensitivity and reproducibility. It enabled a detailed study of aldehyde level changes during packaged beer storage.

Exercise 6.2

Examine the language in excerpt 6A as you answer these questions:

a. Both present and past tenses are used. Find two examples of each. Based on this Introduction, what general trends might you suggest about the use of tense in the Introduction of a journal article?

b. Both active voice and passive voice are used. Find two examples of each. Which voice appears to be used more often?

Exercise 6.3

A central purpose of the Introduction is to help readers understand why the targeted area of research is important. Reread the first and last paragraphs of excerpt 6A and then restate the importance of the work in your own words.

Exercise 6.4

Examine the ways in which the authors of excerpt 6A cite others' works. Propose two citation rules that the authors appear to be following.

6A Writing on Your Own: Read and Paraphrase the Literature

In chapters 3–5, we suggested that you begin Writing on Your Own by reviewing the targeted section (i.e., Methods, Results, and Discussion) in each of the articles that you collected during your literature search. To prepare to write the Introduction, however, we recommend that you review each article in its entirety, in order to summarize the major findings of each work.

Take careful notes as you review each article, looking for key ideas and themes that will help you organize your notes into categories. Use your own words while taking notes; avoid the temptation to copy exact words from original sources. In this way, others' ideas will be easier to paraphrase when you write your Introduction. Consider these note-taking guidelines for paraphrasing the works of others:

1. Carefully read the passage that you wish to paraphrase, making sure that you understand what you have read. (This often requires consulting textbooks, reference materials, and other publications cited in the article.)
2. Without looking at the original passage, jot down notes about the passage; try to capture the ideas most relevant to your study.
3. Summarize your notes into a concise, well-worded statement, without looking back at the original. (Note cards are ideal for this step.)
4. Check your summary against the original to ensure that you have not plagiarized but have still captured the ideas most relevant to your study.
5. Add full bibliographical information to the summary. The names of all authors, article title, journal name, volume number, year, and pages (first and last) may eventually be required.
6. Label your summary with a keyword, creating a master list of keywords as you review the literature. Use these keywords to organize your summaries into categories.
7. Repeat these steps for additional passages in the same article or new passages in other articles.

Analyzing Audience and Purpose

You have probably noticed that the Introduction is one of the most easily understood sections of a journal article. This is because the Introduction is written for a more general audience than the rest of the paper. A wide range of readers, from students to experts, should be able to read at least parts of the Introduction. The purposes of the Introduction are also more general than other sections of the paper:

- To introduce the area of research
- To explain the importance of the research area
- To highlight relevant, precedent works
- To justify the need for the current work
- To introduce the current work

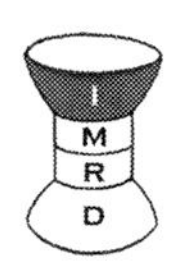

The Introduction begins with the most general information (the research area) and gradually shifts to a more specific focus (the current work), preparing the reader for the highly specific focus of the Methods section. This transition from general to specific is apparent in the now-familiar hourglass shape of the journal article.

Exercise 6.5

Each of the following passages is about a Grignard reaction. Four are from journal articles (two from Introduction sections and two from Discussion sections), and one is from a textbook. Based on your understanding of the purpose and intended audience of each section and genre, read each passage and decide where it comes from. Explain what information you used to make your decisions.

a. Entirely different regioselectivity to that observed with the BuMgX/CuCN reagents was obtained with $Bu_2Cu(CN)Li_2$ in THF[1] and BuLi/CuCN (cat.) in both THF and Et_2O . . . , while other combinations of the reagents and the solvent showed a similar efficiency. (Adapted from Ito et al., 2001)

b. The Grignard reagent has probably been the most widely used intermediate in organic chemistry since its introduction by Victor Grignard in 1900.[1] Despite this wide use, it is not possible to assign a specific structure for a particular reagent. This is because both RMgX and R_2Mg are formed during preparation of the reagent and are connected by the equilibrium described by Schlenk and Schlenk.[2] (Adapted from Walter, 2000)

c. Organohalides (RX) react with magnesium metal in ether or tetrahydrofuran (THF) solvent to yield organomagnesium halides, RMgX. The products, called Grignard reagents after their discoverer, Victor Grignard, are examples of *organometallic* compounds because they contain a carbon–metal bond. (Adapted from McMurry, 2004)

d. The mechanism of the Grignard reaction with chlorosilanes is different from that with alkoxysilanes. An S_Ei mechanism, common for electrophilic substitution reaction with organometallic compounds,[1–3] can be assumed. (Adapted from Tuulmets et al., 2003)

e. One of the most important methods for forming carbon–carbon bonds is through the nucleophilic addition of an organometallic reagent to a carbonyl derivative. Such reactions are exemplified by the Barbier-Grignard type reactions.[1–3] (Adapted from Li and Meng, 2000)

Analyzing Organization

The Introduction, like other sections of a journal article, follows a conventional set of moves. In fact, the move structure for the Introduction section is likely the most consistently followed move structure presented in this textbook (figure 6.1).

The first move, with three submoves, has the broad purpose of describing the general research area. Submove 1.1 identifies the research topic, and submove 1.2 stresses its importance. Together, these two submoves are frequently accomplished in the first few sentences of the paper. Note that only the general topic is mentioned at this point, not the specific work that is presented in the paper. (We refer to the specific work presented in the paper as the *current work.*) Submove 1.3 is where authors summarize essential works in the field and situate the current work in its appropriate context. This submove does not provide an exhaustive review of the literature but rather includes "sound bites" that alert readers to works that critically influenced the current work or led to fundamental knowledge in the field. The entire first move is usually accomplished in a few

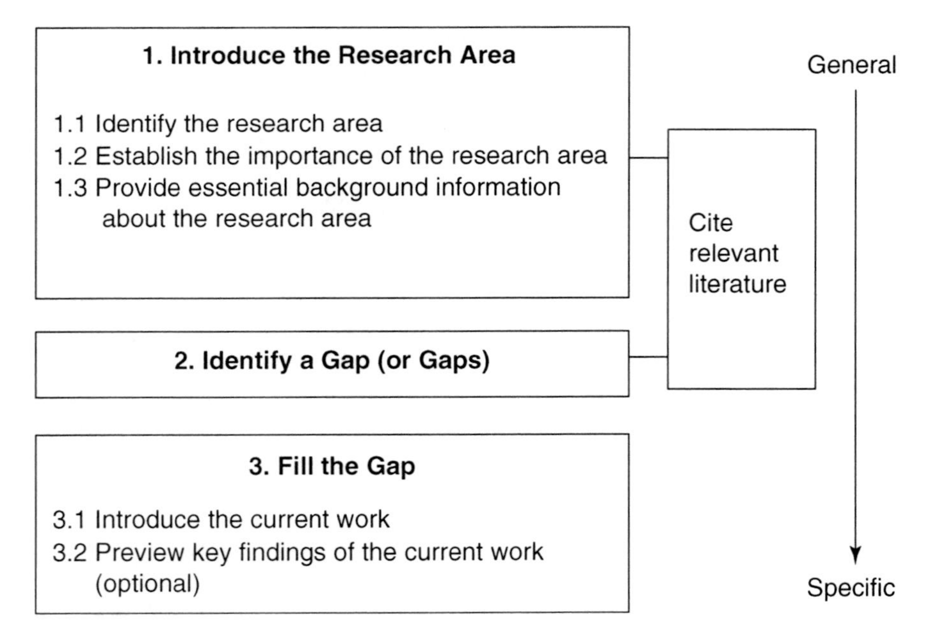

Figure 6.1 A visual representation of the move structure for a typical Introduction section.

paragraphs, with most attention devoted to submove 1.3. All three submoves are strengthened by citations to the literature. Works by other authors are cited as well as previous, relevant works by the current authors.

Current Work

This term is reserved for the specific work presented in the journal article. The phrase should not be used to refer to others' works or even past works by the authors.

With a few exceptions, most Introduction sections refrain from mentioning the current work in the first move.

See table 6.3 for other phrases used to refer to the current work.

The second move in the Introduction section (Identify a Gap) shifts the reader's attention from what has been done (or learned or understood) to what still needs to be done (or learned or understood). The essence of this move is captured in the sentence, "Although much is known about X, little is known about Y." Gap statements come in various forms; a few possibilities are listed in table 6.1. Of course, to correctly identify a gap, the authors must have thoroughly reviewed the literature; hence, citations to the literature are common in this move, too.

After a gap has been identified, the third and final move of the Introduction section is to fill the gap. This move typically comprises a short paragraph at the end of the Introduction and begins with a phrase like "In this paper, we...." At last,

Table 6.1 Examples of gap statements. (X represents what has been done, learned, or understood; Y represents what needs to be done, learned, or understood.)

Type	Example
A question that remains unanswered	Numerous questions remain unanswered about Y.
A research area that remains poorly understood	Although much has been learned about X, Y remains poorly understood.
A next step that needs to be taken	The next step is to apply X to the study of Y.
An area that has yet to be studied	X has been the subject of several studies; however, to our knowledge, no studies on Y have been reported.
A procedure that needs to be improved (made less expensive, simpler, more efficient, etc.)	Although X achieves the desired detection limits, the method is costly and time-consuming.
A new hypothesis or observation that needs to be validated	Additional studies are needed to corroborate these findings.

the authors can refer to the current work, and many use the personal pronoun *we* to accomplish this task. The authors give a short description of the current work (typically a few sentences), highlighting how the work fills the identified gap (submove 3.1). The Introduction can end here, or the authors can elect to preview their principal findings (submove 3.2). If the authors do the latter, care must be taken not to repeat sentences verbatim that appear elsewhere in the paper.

Exercise 6.6

Reread excerpt 6A with the move structure of the Introduction section in mind. Which moves are present? Which sentences are associated with which moves? Support your answers with specific examples from the text.

As you might suspect, not all authors or journals adhere strictly to the move structure depicted in figure 6.1. One variation (employed commonly by organic chemists) is to mention the gap and the current work in the first paragraph of the Introduction. We will encounter one such variation later in this chapter, when we examine the Introduction section of a journal article from *The Journal of Organic Chemistry*.

6B Writing on Your Own: Prepare to Write

With knowledge of the move structure for the Introduction section in mind, look through your notes and be sure that you have sufficient information to address each move. How will you introduce your topic? How will you justify the importance of your research area? What essential works will you describe to provide relevant background information? What gap does your current work fill? If you cannot address each of these questions, you may need to extend your review of the literature.

Analyzing Excerpts

If a chemical article is going to describe something new about a well-known reaction, the introduction might say, "Although the study of substitution reactions occupies a large portion of undergraduate chemistry and has been extensively studied for decades, no one has examined how the reactions change when the reagents are in the gas phase instead of in solution."

—Charles H. DePuy, University of Colorado–Boulder

We now read and analyze excerpts of Introduction sections, including those from articles examined in chapters 3–5. In part 1, we examine the excerpts one move at a time; in part 2, we look specifically at ways to make your writing more concise and fluid in the Introduction.

Part 1: Analyzing Writing Move by Move

Move 1: Introduce the Research Area

As revealed in figure 6.1, the research area is introduced in three submoves. In the first submove, the research area is described broadly. This step is initiated and often accomplished in the opening sentence of the paper. This all-important first sentence, in many genres, is used to set the tone for the work (see figure 6.2); however, in chemistry journal articles, the first sentence tells the reader, with a broad stroke, what the story is about. Consider the following examples, in which the general topic is mentioned at the start of each sentence (our bolding):

P1 **Chromium** is a metal widely distributed in soil and plants (*1*). (From Plaper et al., 2002)

P2 **Polychlorinated biphenyls** (PCBs) are a group of pollutants widely distributed in the environment due to their generous use in the past,

Figure 6.2 Even Snoopy recognizes the importance of the opening sentence. In a novel, the first sentence sets the tone; in a journal article, it identifies the topic. PEANUTS: ©United Feature Syndicate, Inc.

their lipophilic character, and their chemical stability.[1,2] (From Llompart et al., 2001)

The specific content of each article (Cr^{3+} toxicity and the detection of PCBs in full-fat milk, respectively) is stated much later in the Introduction section. Four other examples of opening sentences are included in exercise 6.7.

All-Important Opening Sentence

The first sentence in the Introduction section should convey the general topic of the paper (i.e., the research area), not the specific work to be reported.

Exercise 6.7

Passages P3–P6 open the Introduction sections to four key articles cited throughout the textbook. Briefly state the general topic of each article based on only the first sentence.

P3 **Since** the possibility of using cyclodextrins (CDs) for soil remediation was first mentioned in 1992 (*1*), two main soil treatment technologies have been developed: washing of contaminated soils with a relatively concentrated CD solution (sugar flushing) (*2*) and using small amounts of CDs as a bioavailability-enhancing additive to accelerate the biodegradation of organic pollutants (*3*). (From Jozefaciuk et al., 2003)

P4 Epidemiological studies indicate increases in human mortality and morbidity due to exposure to airborne fine particulate matter (*1*). (From Dellinger et al., 2001)

P5 The asymmetric synthesis of α-amino acids and derivatives is an important topic as a result of their extensive use in pharmaceuticals and agrochemicals and as chiral ligands. (From Boesten et al., 2001)

P6 The literature on tetrazoles is expanding rapidly.[1] (From Demko and Sharpless, 2001)

Since vs. Because

See appendix A for more information on these easily confused words.

Exercise 6.8

Select a chemistry journal, and jot down its title. Select five research articles from the journal, and read the first sentence of each article. Based only on the first sentence, identify the general topic of each paper. Summarize the topic in a word or two. Repeat this exercise for a second journal. Did you run across any articles in which the general topic is not identified in the first sentence? Explain.

Two writing features are worth pointing out regarding the opening sentence of a journal article. First, the topic is usually introduced in the present tense; only one passage above (P3) did not use present tense. Second, citations to the literature are quite common in the first sentence. Once again, all passages but one (P5) included at least one citation in the first sentence.

Equally important is what *not* to do in the opening sentence. First, we caution against using catchy language. Unlike many forms of writing (e.g., Snoopy's novel in figure 6.2), when writing for expert chemists, catchy language must be avoided, not only in the opening sentence but throughout the work. Consider the following two sets of opening sentences; the first is written for a general audience and the second for an expert audience:

Written for a general audience	Beer foam. Most beer drinkers try to minimize it when doing a pour. (McCue, 2002)
Written for an expert audience	Foam and flavor stability are important considerations for a brewer as it is through these that the consumer judges the quality of the beer. However, these foaming and flavor properties are seriously damaged by lipids... (Cooper et al., 2002)

Second, we caution against mentioning the current work in the opening sentence. As stated above, the conventional place to mention the current work is toward the end of the Introduction section. Third, it is uncommon to mention scientists by name in the opening sentence. Although the opening sentence often includes citations to the literature, names are normally not included.

Exercise 6.9

The following sentences appear in the Introduction section of an article on asbestos (Webber et al., 2004). Based on content and language, which is likely the first sentence of the Introduction? Based on the move structure in figure 6.1, predict the order of all three sentences:

a. This paper demonstrates the first reconstruction of airborne asbestos concentrations from the last century, including the periods of highest exposures.
b. Asbestos fibers are naturally occurring hydrated silicate mineral fibers that have found myriad uses in the 20th century.
c. However, airborne asbestos fibers became the well-recognized cause of asbestosis, bronchogenic carcinoma, and mesothelioma during the latter half of that century (*1–3*).

Exercise 6.10

Rewrite the following sentences so that they conform more closely to the conventions of an opening sentence for the Introduction section of a journal article:

a. We used NMR spectroscopy to show that we had successfully synthesized a novel arylated quinoline, a molecule belonging to a class known to be important in biologically active compounds.[1–3]
b. Smith et al.[1] reviewed carbon nanotubes (CNTs) in their recent publications and showed that CNTs have novel electrical properties[1–3] and many possible applications in electronic and sensing devices.[2,4]
c. Believe it or not, with the wonders of modern science, we can now manipulate single biomolecules such as DNA and proteins, and by using highly focused laser light, we can grab them and move them to new positions in new orientations.
d. You already know that tobacco causes cancer, but did you know that it's the nitrosamines in the tobacco that lead to carcinogenesis?

After the general topic has been identified in the Introduction section, the next step is to describe the importance of the research area (submove 1.2). The authors must explain why the subject is important and refer to key articles in the permanent literature to substantiate this importance. Many authors begin to establish the importance of the research area in the very first sentence, along with identifying the topic of the paper. Thus, for example, in P4 (exercise 6.7), we learn both that fine particulate matter (PM) is the topic of the article and that studying PM is important because exposure can lead to human mortality and morbidity. The authors go on to stress this importance even more strongly in the rest of the first full paragraph of the article:

P7 Epidemiological studies indicate increases in human mortality and morbidity due to exposure to airborne fine particulate matter (*1*). This has led to the promulgation of stringent new air pollution regulations that limit the atmospheric concentration of particles with a mean aerodynamic diameter of less than 2.5 μm, $PM_{2.5}$. The link between exposure of $PM_{2.5}$ and excess mortality is well established (*1, 3, 4*), although the causative agent(s) have not been conclusively identified, and the confounding of effects due to co-exposure to gaseous pollutants is inherently difficult to analyze (*5, 6*). The **principal** source of airborne $PM_{2.5}$ is combustion emissions, either primary particulate emissions or particulate matter formed from atmospheric reactions of gaseous combustion emission (*7–9*). Thus, it is likely that species directly emitted from combustion sources or their atmospheric reaction products play a role in the health effects of airborne $PM_{2.5}$. (Adapted from Dellinger et al., 2001)

Principal vs. Principle

See appendix A for more information on these easily confused words.

Impact on human health is a common theme used by authors to justify the importance of their research. This theme also appears in the Introduction section of the chromium article. As illustrated in P8, the authors first highlight the well-documented health risks of hexavalent chromium (Cr^{6+}) and then indicate why it is also important to study trivalent chromium (Cr^{3+}).

P8 In numerous studies, Cr^{6+} compounds have been shown to be carcinogenic in vivo and mutagenic in vitro. Cr^{6+} can induce tumors in experimental animals and can neoplastically transform cells in culture (*2*). In cultured cells, Cr^{6+} induces DNA single-strand breaks, binding of amino acids and proteins to DNA, DNA–DNA cross-links, and Cr–DNA adducts (*2, 4–8*).... In contrast to Cr^{6+}, trivalent chromium (Cr^{3+}) is actually an essential nutrient, needed for the expression of glucose tolerance (*1*). Because the normal human dietary intake of this element is less than 60% of the minimum suggested intake (*1*), many people, particularly in developed countries, supplement their diets with trivalent chromium. The use of chromium nutrition additives is widespread and unsupervised, and there is no awareness of possible side effects. It is therefore important that Cr^{3+} toxicity be investigated in **further** detail, to ensure safer application of this additive. (Adapted from Plaper et al., 2002)

Farther vs. Further

See appendix A for more information on these easily confused words.

Another common theme that authors use to establish importance involves environmental impacts. For example, an environmental slant is used in the first sentence of the cyclodextrin article (P3, exercise 6.7), where the study of cyclodextrins is justified based on their role in soil remediation. The importance of work that benefits air or water quality and/or promotes **green chemistry** can also be stressed. Work is also viewed as important if it has cross-disciplinary applications. For example, in the Introduction section of the tetrazole article, the authors stress the importance of tetrazoles in coordination chemistry, medicinal chemistry, and in various materials science applications and point out their role as useful intermediates in the preparation of substituted tetrazoles:

Green Chemistry

Green chemistry refers to practices designed to prevent pollution and promote the sustainable use of natural resources.

P9 The literature on tetrazoles is expanding rapidly.[1] This functional group has roles in coordination chemistry as a ligand, in medicinal chemistry as a metabolically stable surrogate for a carboxylic acid group,[2] and in various materials science applications, including specialty explosives.[3] Less appreciated, but of enormous potential, are the many useful transformations that make tetrazoles versatile intermediates en route to substituted tetrazoles, and especially to other 5-ring heterocycles via the Huisgen rearrangement.[4] (From Demko and Sharpless, 2001)

Note that when establishing the importance of a research area, it is appropriate to cite works that substantiate this importance. For example, 22 citations were included in P7–P9.

Exercise 6.11

Read through the first sentences in P1–P6 again. In addition to P4, which other passages both identify the topic and hint at its importance in the first sentence?

Exercise 6.12

Read through P10 and P11. In each case, do the following:

a. Identify the topic of the article and why the topic is important.

b. State how many citations to the literature are used to substantiate this importance.

c. State how often the science is emphasized, rather than the scientist.

P10 Polychlorinated biphenyls (PCBs) are a group of pollutants widely distributed in the environment due to their generous use in the past, their lipophilic character, and their chemical stability.[1,2] Thus, PCBs have a long environmental half-life and tend to accumulate in the food chains; the highest concentrations were usually found in human beings and higher animals at the top of the food chain.[3,4]

Food, and especially fatty food, has been widely recognized as the main source of intake of toxic chemicals such as PCBs.[5] Dairy products, and milk in particular, have received special interest due to their extensive and elevated consumption by the population.[6] Several countries have established levels (recommended maximum limits, RMLs) for PCBs in dietary products such as fish (~2000 ng/g), meats (ranging from 200 to 2000 ng/g), and eggs (100–300 ng/g). For milk and dairy products, RMLs range from 200 (Canada) to 1500 ng/g (Thailand). Germany has established RMLs for some congeners (PCBs 28, 52, 101, 180) in 8 ng/g of fat and 10 ng/g of food, for the food with more and less than 10% fat, respectively.[7] (From Llompart et al., 2001)

P11 The asymmetric synthesis of α-amino acids and derivatives is an important topic as a result of their extensive use in pharmaceuticals and agrochemicals and as chiral ligands. Many highly enantioselective approaches have been reported.[1] Industrial production of α-amino acids via the Strecker reaction is historically one of the most versatile methods to obtain these compounds in a cost-effective manner, making use of inexpensive and easily accessible starting materials.[2] (From Boesten et al., 2001)

Exercise 6.13

Browse through three different chemistry journals. To make this exercise more relevant, choose journals related to your own field of study. For each journal, write down the name of the journal and three themes used to justify the importance of works in the journal.

When authors cite others' works to establish the importance of their own work (discussed above), or to provide background information (discussed below), they frequently use present tense. You might find this surprising because, after all, the cited works were done in the past; yet, the importance of the work is expected to be true today and into the future. Consider the following example:

P12 This functional group *has* roles in coordination chemistry as a ligand, in medicinal chemistry as a metabolically stable surrogate for a carboxylic acid group,[2] and in various materials science applications, including specialty explosives.[3] (From Demko and Sharpless, 2001)

It would sound funny to say, in the past tense, that the functional group "had" roles in coordination chemistry, implying that those roles are no longer important. In addition to present tense, another verb construction is commonly used when citing others' works. Consider the following sentence:

P13 In numerous studies, Cr^{6+} compounds *have been shown* to be carcinogenic in vivo and mutagenic in vitro. (From Plaper et al., 2002)

The construction "have been shown" (in P13) is an example of a verb form known as present perfect. There are two forms of the present perfect:

Present perfect–active   has shown, have shown
Present perfect–passive   has been shown, have been shown

Present perfect is typically used to signal that the knowledge gained from work completed in the past is still believed to be true in the present. Present perfect combines *has* or *have* with a past-participle verb form, which is usually (but not always) the same as the past tense form (see table 6.2). Note that the use of the present perfect is not limited to this submove; you will see it used in other sections of the journal article, as well.

Table 6.2 Examples of active and passive constructions in present perfect.

Active Voice: has/have + past participle[a]	Passive Voice: has/have + been + past participle[a]
has demonstrated	has been demonstrated
has recognized	has been recognized
has shown	has been shown
have discovered	have been discovered
have observed	have been observed
have received	have been received

a. Present perfect and passive voice constructions require the use of the past participle. For most regular verbs, the past tense and the past participle verb forms are the same (e.g., for the verb "demonstrate", both the past tense and past participle are "demonstrated"). With many irregular verbs, however, the past tense and past participle verb forms are different (e.g., for the verb "show", the past tense is "showed" but the past participle is "shown"). Common past participles include the following: become, chosen, given, grown, seen, shown, taken, undergone, undertaken.

Present Tense and Present Perfect

When citing others' works to establish importance or provide background information, authors often use present tense or present perfect in either active or passive voice.

Present tense–active	Nagy et al. (*1*) propose...
Present tense–passive	Arsenic is recognized as...(*1*).
Present perfect–active	Nagy et al. (*1*) have proposed...
Present perfect–passive	Arsenic has been recognized as...(*1*)

Verb Forms

Native speakers of English will likely hear the difference between the past tense and past participle forms of the verb and use them appropriately, even if they do not know the terminology used to distinguish one from the other. Hence, "it has been showed" will sound wrong and "it has been shown" will sound right.

For nonnative speakers, however, it is helpful to know the different verb forms and the rules behind what "sounds right" to the native ear.

 **Exercise 6.14**

The following passages comprise the first several sentences of two articles from the same issue of the *Journal of Agricultural and Food Chemistry*. Make a list of the present tense and present perfect verb constructions used in each passage. Which construction do the authors seem to prefer?

a. Progress in plant lectin biochemistry has exploded during the past decade (*1*). New lectins with interesting properties have been isolated and characterized; a great many lectins have been cloned, and homologies in their amino acid sequences and similarities in their molecular structures have been established. The X-ray crystallographic structure of a host of these lectins has been solved at high atomic resolution. The biosynthesis of many lectins has been elucidated. Many lectins have been expressed in bacteria and eukaryotic cells and their mutant forms studied to determine changes, if any, in their carbohydrate-binding specificity. A number of lectins have also been transfected into food crops with the intention of conferring resistance to various insect vectors. (Adapted from Goldstein, 2002)

b. *Allium fistulosum* L. (Liliaceae) is a perennial herb that is widely cultivated throughout the world, ranging from Siberia to tropical Asia. China, Japan, and Korea grow most of the world production. The common name "Welsh onion" derives from the German *welshche*, meaning foreign. Other local names include the following: in China, *Cong*; in English-speaking countries, Japanese bunching onion, Spanish onion, two-bladed onion, spring onion, green bunching onion, scallion, green trail, and Chinese small onion; in Japan, *negi*; and in Korea, *pa*. It is believed to have originated in northwestern China (*1*). Both the leaves and the bulbs are edible. It has also been used as an herbal medicine for many diseases. According to the dictionary of Chinese drugs (*2*), the bulbs and roots of this plant have been used for treatment of febrile disease, headache, abdominal pain, diarrhea, snakebite, ocular disorders, and habitual abortion, as well as having antifungal and antibacterial effects. The seeds are used as a tonic and an aphrodisiac. (From Sang et al., 2002)

After the importance of the area has been established in the Introduction section, the next step (submove 1.3) is to provide readers with relevant background information. The goal is to alert readers to essential works in the field, not to review the literature exhaustively. This point is often emphasized in the Information for Authors section provided by journals, as illustrated below for three ACS journals:

Background material should be brief and relevant to the research described. Detailed or lengthy reviews of the literature should be avoided. (Scope, Editorial Policy, and Preparation of Manuscripts, *Chemical Research in Toxicology* **2007**, *20*, 12A)

The introduction should state the purpose of the investigation and must include appropriate citations of relevant, precedent work but should not include an extensive review of marginally related literature. (Authors' Guide, *Analytical Chemistry* **2007**, *79*, 389)

Do not attempt a complete survey of the literature. . . . In general, the introduction should be no more than 2 double-spaced pages without figures or tables and should include fewer than 20 references. (Instructions to Authors, *Environmental Science & Technology* **2007**, 29)

Despite the label assigned to submove 1.3, it is not the only submove (or move) that presents background information; indeed, relevant background information is (and should be) integrated throughout the Introduction section, first to introduce the topic and establish its importance, and subsequently to identify the gap (discussed below).

To see how authors report relevant background information in their Introductions, we consider five examples (P14–P18). As you read these passages, notice how concisely they are written. The authors do not summarize one work at a time, in whatever order they choose; instead, multiple references are grouped together in a logical sequence, ultimately leading up to the current work. (Hints for how to achieve such conciseness in your writing are included in part 2 of this chapter.)

Also, note that no direct quotes are used. Although common in other genres, direct quotes are exceedingly rare in chemistry journal articles, in part, because space is so limited. Instead, chemistry authors must capture the essence of others' works in only a few words or sentences. For example, consider P14; in a single sentence, the authors summarize background information and cite nine other works. Imagine the space it would take to directly quote from each of these works!

> P14 The most common cleanup methods are as follows: shaking the extract with concentrated sulfuric acid,[9–12] Florisil,[9–13] alumina,[14,15] and silica gel[8,13] and using size exclusion chromatography (SEC).[15,16] (From Llompart et al., 2001)

Avoid Direct Quotes

Although common in other forms of writing, chemists almost never use direct quotes in journal articles.

We next consider a passage from the article on Cr^{3+} toxicity. The authors have already established the importance of chromium research in general and Cr^{3+} specifically (P1, P8, P13); now the authors provide background information on what is known about Cr^{3+} toxicity. The authors begin by acknowledging that, so far, Cr^{3+} appears to be nonmutagenic on cellular test systems; however, recent studies indicate that Cr^{3+} can cause DNA damage, a finding that underscores the need for their work.

> P15 So far Cr^{3+} has been shown to be nonmutagenic in most of the studies performed on cellular test systems, probably due to the inability of the hydrated Cr^{3+} complexes to cross plasma membranes (*10*). However, in some recent studies, Cr^{3+} was shown to cause Cr–DNA adducts and DNA–DNA cross-links (*13, 14*). Cr^{3+} also increased DNA polymerase processivity and decreased its fidelity during DNA replication in vitro (*15–17*) besides causing formation of mutagenic adducts of amino acids to the DNA phosphate backbone (*18*). (From Plaper et al., 2002)

As a third example, consider a passage from the cyclodextrin article (P16). The authors provide useful background information on two types of cyclodextrins used in bioremediation: HPBCD and RAMEB. The two cyclodextrins are compared, and the authors explain why RAMEB was chosen over HPBCD, thereby leading up to the current work.

> P16 Two CD derivatives are most commonly used for soil remediation: hydroxypropyl and random methylated β-cyclodextrins (HPBCD and

RAMEB, respectively), both being mixtures of isomers with different degree and pattern of substitution, extremely soluble in water (more than 50%) (*13*), and nonvolatile. Although RAMEB has a higher solubilizing effect, HPBCD is more feasible for the "sugar flushing" technology because of lower surface activity (*5*). An aqueous solution of RAMEB is not the best choice for soil washing because the mobility of the nonaqueous phase liquids (NAPLs) might be increased (*2*). The other unfavorable property of RAMEB may be its slight adsorption on clay minerals (illite), whereas HPBCD is not adsorbed (*14*). The high solubilizing effect of RAMEB is utilized in bioremediation techniques requiring only low amounts (0.1–1% w/w soil) of the additive (*16*, *17*). In this case RAMEB acts as a catalyst improving the transfer of pollutants from the solid phase to the aqueous phase of the soil. β-CD was also reported to improve the biodegradation of a single hydrocarbon (dodecane) (*15*). γ-CD, HPBCD, and RAMEB were effective in the intensification of PCB biodegradation in soils (*10*, *16*). Especially remarkable bioavailability-enhancing properties were exhibited by RAMEB in hydrocarbon-polluted soils (*17*). Because soil bioremediation needs months to years depending on type and concentration of the contaminants, soil properties, and microflora, an additive that degrades slowly in the soil is required. RAMEB meets this requirement. Its half-life time is about 1 year in a soil contaminated with motor oil (*18*), while HPBCD is decomposed rapidly (*19*). (Adapted from Jozefaciuk et al., 2003)

Numbering Citations

The citation numbers may appear to be out of sequence in P15 (because they skip from 10 to 13) and in P16 (because 13 is cited before 5 and 2), but this sequencing just means that the missing citations were mentioned earlier in the Introduction sections. (See chapter 17.)

A fourth example (P17) is from the Introduction section of the article that examines PCBs in full-fat milk. For background information, the authors outline the general four-step procedure used to determine PCBs in full-fat milk. Conventional methods used to accomplish two of these steps, extraction and cleanup, are also described. In a new paragraph, the authors introduce solid-phase microextraction (SPME), a technique that greatly simplifies this four-step process. But SPME is not recommended for complex matrixes; hence, the authors motivate the topic of their current paper, headspace mode SPME (HSSPME).

P17 The general analytical procedure for the determination of PCBs in full-fat milk includes four main steps: extraction from the matrix, preconcentration and cleanup steps, gas chromatographic separation, and detection. Conventional methods of extraction are liquid-liquid extraction and Soxhlet extraction.[1] Also, solid-phase extraction (SPE) has been employed.[8] The most common cleanup methods are as follows: shaking the extract with concentrated sulfuric acid,[9–12] Florisil,[9–13] alumina,[14,15] and silica gel[8,13] and using size exclusion chromatography (SEC).[15,16]

Arthur and Pawliszyn[19] introduced solid-phase microextraction (SPME) in 1990 as a solvent-free sampling technique that reduces the steps of extraction, cleanup, and concentration to a unique step. SPME utilizes a small segment of fused-silica fiber coated with a polymeric phase to extract the analytes from the sample and to introduce them into a chromatographic system. Initially, SPME was used to analyze pollutants in water[20,21] via direct extraction. Subsequently, SPME was applied to more complex matrixes, such as solid samples or biological fluids. With these types of samples, direct SPME is not recommended; nevertheless, the headspace mode (HSSPME) is an effective alternative to extracting volatile and semivolatile compounds from complex matrixes. (Adapted from Llompart et al., 2001)

As a final example, consider passage P18 from the Introduction section of the article on substituted tetrazoles. As background information, the authors examine previous efforts to synthesize 5-substituted 1*H*-tetrazoles and categorize these methods into three main groups.

P18 The most convenient route to 5-substituted 1*H*-tetrazoles **2** is the addition of azide ion to nitriles **1**.[5] The literature is replete with methods to perform this transformation; they fall into three main categories: those that make use of tin or silicon azides,[6] those that use strong Lewis acids,[7] and those that are run in acidic media.[8] (Adapted from Demko and Sharpless, 2001)

Exercise 6.15

Consider passages P19 and P20, adapted from the Introduction sections of articles on particulate matter and the asymmetric Strecker reaction:

a. What background information is provided in each passage?
b. Identify the verb constructions italicized in each passage as past, present, or present perfect.

P19 It *has become* evident that fine particulate matter has the ability to generate reactive oxygen species (ROS) (*10, 11*). This *is* striking because generation of ROS is intimately linked to the genesis of pulmonary and cardiovascular injury (*12*). Proposed sources of ROS *include* generation by particle-activated polymorphonuclear leukocytes, and direct ROS generation by the particles themselves and/or their constituents (*13*). For example, it *has been reported* that iron released from airborne fine particles or present on their surfaces plays a role in the generation of ROS, and this *is* also true for coal fly ash particles (*14–16*). . . .

These pathways are thought to result in the production of superoxide (*13*) or in the release of superoxide directly from the particles themselves. Superoxide production *leads* to the formation of hydrogen peroxide, and metal ions such as Fe^{2+} react with hydrogen peroxide to produce the hydroxyl radical. It *is* well documented that the hydroxyl radical can damage DNA as well as lipids and proteins (*18, 19*). Some of the health effects of cigarette tar and smoke are attributed to free radicals that can initiate production of superoxide and hydroxyl radical (*3, 10, 11, 20, 21*). (Adapted from Dellinger et al., 2001)

P20 Several catalytic asymmetric Strecker reactions leading to N-protected amino nitriles in high ee and high yields *have been published*.[4] Alternatively, diastereoselective Strecker syntheses using a broad variety of chiral inducing agents, such as α-arylethylamines,[5] β-amino alcohols and derivatives,[6] amino diols,[7] sugar derivatives,[8] and sulfinates,[9] *have been reported* to provide the α-amino nitriles with varying diastereoselectivities. (Adapted from Boesten et al., 2001)

Exercise 6.16

Consider the following direct quotation that describes lead concentrations in ice cores and firn (loosely compacted granular snow) taken from a glacier at the Swiss-Italian border. Convert the direct quote into a summary of the work as it might appear in the Introduction section of a journal article on lead emissions in the last decade.

Lead concentrations in firn dated from the 1970s are ~25 times higher than in ice dated from the 17th century, confirming the massive rise in lead pollution in Europe during the last few centuries. A decline of the lead concentration is then observed during the last two decades, that is, from 1975 to 1994. . . . These variations are in good agreement with available information on variations in anthropogenic lead emissions from West European countries, especially from the use of lead additives in gasoline. (Adapted from Schwikowski et al., 2004)

6C Writing on Your Own: Draft Your Opening Paragraph

Reread the notes that you have taken on the importance of your topic and relevant background information. If you used note cards, sort them by keywords to identify multiple works that can be grouped and cited together in your Introduction. If you did not use note cards, figure out other ways to organize your notes by keywords or key concepts.

Begin writing your opening paragraph, including the all-important first sentence (making sure the topic is clear). Stress the importance of your research area and summarize relevant, precedent works, remembering that the key is to (1) alert your readers to works that laid the groundwork for your study, (2) illustrate key findings in the field, and (3) summarize essential knowledge related to the current work.

Move 2: Identify a Gap

After the importance of a research area has been identified and the relevant background information has been summarized, the Introduction section shifts from a focus on what has been done (or learned) to an emphasis on what remains to be done (or learned). This change in emphasis is signaled with a gap statement (table 6.1). The gap statement points out what is lacking in the field and, in so doing, infers the next step that needs to be taken. Consider the following examples:

P21 Persistent electron paramagnetic resonance (EPR) signals have been reported in coals, chars, and soots (26–29), but $PM_{2.5}$ has not been studied by EPR. (From Dellinger et al., 2001)
[Next step: Use EPR to study free radicals in $PM_{2.5}$.]

P22 Although HSSPME has been applied to an enormous variety of matrixes,[22,24] to date, the number of publications that apply SPME technology to milk samples is relatively low,[25,26] and most of them are related to the determination of flavors.[27,28] (Adapted from Llompart et al., 2001)
[Next step: Use HSSPME to study milk.]

P23 A major drawback of these chiral auxiliaries can be cost and/or availability, because they are used in stoichiometric amounts and in principle lost during the conversion. Furthermore, in many cases the α-amino nitriles need to be purified in a separate step to obtain diastereomerically pure compounds. Purification requires, for example, crystallization or chromatography, which may lead to losses. (From Boesten et al., 2001)
[Next step: Develop better chiral auxiliaries.]

P24 Because soil bioremediation needs months to years depending on type and concentration of the contaminants, soil properties, and microflora,

an additive that degrades slowly in the soil is required. (From Jozefaciuk et al., 2003)
[Next step: Find slowly degrading bioremediation additives.]

Exercise 6.17

Reread P8, a passage selected to demonstrate how authors emphasize the importance of their research area (in this case, the toxicity of Cr^{3+}). A gap statement is also present in the passage. Find the gap statement and state it. What next step or steps are suggested by the gap statement?

Exercise 6.18

Reread P18 and then read P25 below, which immediately follows P18 in the published article. A gap is expressed in P25. State the gap in your own words, and identify the next step that is implied.

P25 Each of these [previous methods] involves one or more of the following drawbacks: uses expensive and toxic metals, demonstrates severe water sensitivity, or produces hydrazoic acid, which is highly toxic and explosive as well as volatile. The few methods that seek to avoid hydrazoic acid liberation during the reaction, by avoiding acidic conditions, require a very large excess of sodium azide.[9] In addition, all of the known methods use organic solvents, in particular, dipolar aprotic solvents such as DMF. This is one of the solvent classes that process chemists would rather not use.[10] (Adapted from Demko and Sharpless, 2001)

Exercise 6.19

Select three different chemistry-specific journals. In each journal, find two articles that include gap statements in their Introduction sections. Write down the name of the journal, the gap statement (either restated or verbatim), and the location of the gap statement (near the beginning, middle, or end of the Introduction). Comment on whether the gap statement serves as a transition between describing previous work and the current work.

6D Writing on Your Own: Identify a Gap

Make a list of the possible gaps that your work fills. For each gap, consider the conclusions that might be drawn by your readers. Identify references that will support your claims.

Move 3: Fill the Gap

After the gap has been established, the last move of the Introduction section is to fill the gap. The authors must show how the current work takes at least a small step forward toward addressing the specified need, problem, or lack of knowledge in the field. The start of the third move is commonly signaled with a new paragraph and a phrase such as "In this paper," or "In this work, we . . . " (see table 6.3). Following this phrase, the authors go on, typically in a sentence or two, to tell the readers about the current work. Four examples are given below:

P26 In this article, a simple and rapid saponification-HSSPME procedure has been developed for the extraction of PCBs from different milk samples. Saponification of the fats helps the transference of the PCBs from the sample to the microextraction fiber. Moreover, saponification acts as a cleanup step, thereby improving selectivity and reliability in peak identification. (Adapted from Llompart et al., 2001)

P27 In this article, we report the results of studies that indicate that $PM_{2.5}$ does indeed contain semiquinone-type radicals, and these radicals can initiate damage to DNA through a catalytic cycle involving ROS. (Adapted from Dellinger et al., 2001)

P28 In this article, the first two examples of the use of (*R*)-phenylglycine amide in asymmetric Strecker reactions are presented. (Adapted from Boesten et al., 2001)

P29 To assess the influence of Cr^{3+} on the eukaryotic cells, its effect on the viability and proliferation rate of murine B16 melanoma cells, and . . . human epithelial cells was tested. (From Plaper et al., 2002)

The phrases in table 6.3 are often followed by the personal pronoun *we* (e.g., In the present study, *we* . . .). In such instances, *we* is used to signal the beginning of the authors' presented work in the journal article. (Recall that *we* is also used in Results sections to signal human choice and in Discussion sections to signal interpretative remarks.) Table 6.4 lists some verbs that typically follow *we* in the fill-the-gap statement. Note that the verbs are in present tense when they refer to what is presented in the paper (e.g., "we present"); they are in past tense when they refer to work done in the past (e.g., "we measured"). (See table 6.5 for a summary of common functions of verb tense–voice combinations in Introductions.)

Table 6.3 Common phrases used to transition from the second to the third move of the Introduction section.[a]

In the present study,	In this context,	In this study,	In this paper,
In the present work,	In this investigation,	In this work,	Herein,

a. These phrases were identified through a computer-based analysis of Introduction sections from 60 published chemistry journal articles.

Table 6.4 Common verbs that follow *we* in the fill-the-gap statement of the Introduction section.[a]

In this work, we (present tense verbs)	In this work, we (past tense verbs)
carry out	analyzed
demonstrate	calculated
describe	chose
develop	determined
employ	employed
present	examined
propose	focused on
provide	found
report	investigated
show	measured
use	solved
	studied
	synthesized

a. These phrases were identified, in part, through a computer-based analysis of Introductions from 60 published chemistry journal articles.

Words to Avoid

When introducing the current work, avoid informal phrases such as "we looked into", "we looked at", "we saw if". See Formal Vocabulary in appendix A.

On occasion, a fill-the-gap statement will appear in the opening paragraph of the Introduction section. This deviation from the conventional move structure in figure 6.1 is particularly common in organic journals such as *The Journal of Organic Chemistry*. Moreover, in such journals, the authors often include more than one fill-the-gap statement. For example, in the tetrazole article, Demko and Sharpless (2001) include fill-the-gap statements at the end of their first (P30), second (P31), and fourth paragraphs (P32). (The final paragraph of their Introduction section previews principal findings, as shown in P36 later in this chapter.) Each fill-the-gap statement emphasizes steps taken (in this case, to advance the use of water as a solvent in organic reactions).

P30 (Fill-the-gap statement in first paragraph) We report here a safer and exceptionally efficient process for transforming nitriles into tetrazoles in water; the only other reagents are sodium azide and a zinc salt.

Table 6.5 Common functions of different verb tense–voice combinations in Introduction sections.

Function	Tense–Voice Combination	Example
To introduce research area	Present–active	Chromium *is* a metal widely distributed in soil and plants (*1*). (From Plaper et al., 2002)
To describe importance of the research area	Present–active	PCBs *have* a long environmental half-life and tend to accumulate in the food chains. (From Llompart et al., 2001)
	Present perfect (active and passive)	Many highly enantioselective approaches *have been reported*.[1] (From Boesten et al., 2001) (present perfect–passive)
To provide relevant background information	Present active Present perfect (active or passive)	Conventional methods of extraction *are* liquid-liquid extraction and Soxhlet extraction.[1] Also, solid-phase extraction (SPE) *has been employed*.[8] (From Llompart et al., 2001) (present active; present perfect–passive)
	Past–active	In some recent studies, ... Cr^{3+} *increased* DNA polymerase processivity and *decreased* its fidelity during DNA replication in vitro (*15–17*). (Adapted from Plaper et al., 2002)
To identify a gap	Present–active	[A]n additive that degrades slowly in the soil *is* required. (From Jozefaciuk et al., 2003)
	Present perfect–passive	$PM_{2.5}$ *has* not *been studied* by EPR. (From Dellinger et al., 2001)
To introduce the current work (as a means to fill the gap)	Present–active	In this paper, we *report* the results of studies that ... (Adapted from Dellinger et al., 2001)
	Past (active or passive)	In this work, we *adapted* a method for the analysis of ... (Vesely et al., 2003) (past–active)
To hint at findings, focusing on work done in the past and/or "truths" gleaned from the research	Past (active and passive) Present–active	During the study, it *was discovered* that Cr^{3+} *causes* DNA damage, *has* influence on DNA topology, most probably via effects on DNA gyrase, and also *reduces* the proliferation rate of certain eukaryotic cells. (From Plaper et al., 2002) (past–passive; present–active)

P31 (Fill-the-gap statement in second paragraph) We sought a method that avoided these drawbacks and was easy to use on both a laboratory and industrial scale.

P32 (Fill-the-gap statement in fourth paragraph) Thus encouraged, we envision a special style of organic synthesis, one based on an entire family of reactions for which water is the best "solvent."

The fill-the-gap statement is an appropriate way to end the Introduction section. For example, Dellinger et al. (2001) conclude their Introduction with the fill-the-gap statement in P27. Alternatively, some authors elect to end their Introductions by previewing a principal finding (optional submove 3.2 in figure 6.1). We consider four examples, each only a sentence or two in length. The first (P33) previews an experimental method, because the focus of this work was to develop a more accurate and sensitive technique. The others preview major findings. Note also the use of present and past tense in these passages. Past-tense verbs (italicized) are used to refer to work done in the past; present-tense verbs (bolded) are used to describe "truths" gleaned from the research. (See table 6.5 for a summary of common functions of verb tense–voice combinations in Introduction sections.)

P33 Analyses *were performed* on a gas chromatograph equipped with an electron capture detector (ECD) and a gas chromatograph coupled to a mass-selective detector working in mass spectrometry-mass spectrometry (MS-MS) mode, to achieve better limits of detection and selectivity. The proposed method **yields** high sensitivity, good linearity, precision, and accuracy. (From Dellinger et al., 2001)

P34 During the study, it *was discovered* that Cr^{3+} **causes** DNA damage, **has** influence on DNA topology, most probably via effects on DNA gyrase, and also **reduces** the proliferation rate of certain eukaryotic cells. (From Plaper et al., 2002)

P35 Pivaldehyde and 3,4-dimethoxyphenylacetone *were used* as starting materials, which **lead**, respectively, to enantiomerically enriched *tert*-leucine and α-methyl-dopa, two important nonproteogenic α-amino acids for pharmaceutical applications. In addition, *tert*-leucine **has** considerable utility as a chiral building block.[14] (From Boesten et al., 2001)

P36 As a result of these endeavors, we *found* that in the presence of zinc salts[9b,c] tetrazole formation **proceeds** with excellent yields and scope in refluxing water.[16] Thanks to the low pK_a of 1*H*-tetrazoles (ca. 3–5) and their highly crystalline nature, a simple acidification *is* usually sufficient to provide the pure tetrazoles. (Adapted from Demko and Sharpless, 2001)

Exercise 6.20

Table 6.3 lists common phrases used to transition from the second to the third move of an Introduction section. Select two or three journals of your choice; browse through the Introductions in these journals until you find three new ways to introduce the third move of the Introduction. Remember that you will most often see these phrases toward the end of the Introduction section. Add these new phrases to table 6.3 to make it more complete.

Exercise 6.21

Select two journals of your choice and then select two Introduction sections in each journal. Examine how the authors end their Introductions.

a. Do the Introduction sections conclude with submove 3.1, 3.2, or a variation? Support your answer with excerpts from the Introductions.

b. Comment on how the authors use past tense and present tense in these excerpts. (Refer to table 6.5, if needed.)

6E Writing on Your Own: Draft Your Full Introduction

Following the move structure of an Introduction section, as shown in figure 6.1, write a draft of your full Introduction.

Check the all-important opening sentence that you drafted as part of Writing on Your Own task 6C. Did you identify the topic of your paper?

When describing the importance of your research area, focus on the research rather than the researchers. When presenting background information, paraphrase the work of others; do not use direct quotations. When filling the gap, remember to introduce your work by using one of the phrases in table 6.3. Decide whether you want to report principal findings.

After you have completed a good draft, insert citations and begin your reference list (if you have not started it already). Refer to chapter 17 for guidelines on formatting citations and references.

Part 2: Writing Concisely and Fluidly

I always take my writing seriously. Although I don't know my readers, they all meet me through my writing. I want my words to communicate my commitment to good science, my professionalism, and my desire to engage in a dialogue with the larger scientific community.

—David B. Knaff, Texas Tech University

It is not easy to write a clear and concise Introduction. Before pen is put to paper (or fingers to keyboard), authors must first find, read, and understand appropriate literature and consolidate key concepts, trends, and findings. Next, they must organize this information in a logical order, linking like concepts in fluent prose, using language that is neither repetitive nor choppy. These are not trivial writing skills, even for an experienced writer. Here we focus on ways to help you develop these writing abilities, targeting conciseness and fluidity. These practices apply not only to the Introduction section but to other parts of the journal article, and to other genres as well.

Be Concise

When beginning writers first attempt to summarize others' works, they often do so using a wordy and repetitive writing style. For example, when multiple works are cited, novice writers often resort to the following monotonous pattern:

Garcia et al.[1] showed . . . Daloğlu et al.[2] showed . . . Manygoats et al.[3] showed . . .

Or worse (because it is wordier), inexperienced writers might compose the following:

In a study by Garcia et al.[1], it was shown that. . . . Another study was conducted by Daloğlu et al.[2] to show that. . . . In a more recent study conducted by the scientists Manygoats et al.,[3] additional evidence was provided to show that. . . . Finally, other researchers[4–6] have shown that. . . .

Although nothing is wrong grammatically with the sentences above (even et al. is used correctly), they signal a novice writer because they are wordy and include unconventional words such as *researcher* and *scientist.* More important, the wordiness of the sentences interferes with clarity. To enhance the clarity and conciseness of your writing, consider the following two suggestions (discussed in more detail below): (1) focus on the science, not the scientist(s), and (2) group related ideas.

Concise Writing

See appendix A.

Researchers and Research

In a computer-based analysis of Introduction sections from 60 published chemistry journal articles, there were no occurrences of the term *researchers.*

The term *research,* used infrequently, rarely refers to authors' own work or the work of others. Rather, the term *research,* when used, usually has a more generic sense:

A key element in genetics *research* is...

Many theoretical and experimental *research* studies on optical nonlinearity of fullerene...

Surprisingly little *research* has been reported on SBA-1...

Research in this area has involved multicomponent molecules...

Recommendations:

Use the term *research* sparingly.

Do not use the term *researchers*.

The first suggestion to make your writing more concise is to eliminate the names of scientists and the titles of their works from your sentences. (If relevant, this information is included in the references, so it is redundant to repeat it in the text.) The authors of the six key articles in this textbook followed this advice; more than 125 articles were cited in their Introduction sections, but authors' names appeared only once! What this means, then, is to make the science the subject of your sentences. Consider the following examples:

Very wordy	In their article titled "Preparation of 5-Substituted 1*H*-Tetrazoles from Nitriles in Water," Demko and Sharpless[5] propose a way to synthesize 1*H*-tetrazoles using nitriles and sodium azide in water.
Wordy	Demko and Sharpless[5] recently proposed a way to synthesize 1*H*-tetrazoles using nitriles and sodium azide in water.
Concise	1*H*-tetrazoles have been prepared using nitriles and sodium azide in water.[5]

Exercise 6.22

Revise the following sentences so that the names of all scientists, article titles, and book titles have been removed. How many fewer words are needed? Does the conciseness aid or detract from the readability of the sentences?

a. Molnar et al. (*17*) were among the first scientists to point out that remarkable bioavailability-enhancing properties are exhibited by RAMEB in hydrocarbon-polluted soils. (Adapted from Jozefaciuk et al., 2003) (24 words)

b. Fatty foods, as first suggested by the researchers Gallo et al.,[5] are widely recognized as a main source of intake of toxic chemicals such as PCBs. (Adapted from Llompart et al., 2001) (26 words)

c. Reichardt,[11] in his important book titled *Solvents and Solvent Effects in Organic Chemistry*, asserted that water has extraordinary physical properties as a solvent, making its use as a solvent widely appreciated. (Adapted from Demko and Sharpless, 2001) (31 words)

A second strategy for achieving conciseness requires that you group related ideas and use punctuation appropriately. For example, when summarizing the literature, introduce related ideas with the phrase *such as*. Consider the following examples, juxtaposing a fabricated wordy passage with its more concise (authentic) counterpart:

Wordy Several countries have established levels (recommended maximum limits, RMLs) for PCBs in dietary products. Levels of ~2000 ng/g have been established for fish. Levels ranging from 200 to 2000 ng/g have been established for meats. Levels between 100 and 300 ng/g have been established for eggs. (4 sentences, 46 words)

Concise Several countries have established levels (recommended maximum limits, RMLs) for PCBs in dietary products, *such as fish* (~2000 ng/g), *meats* (200–2000 ng/g), *and eggs* (100–300 ng/g). (adapted from Llompart et al., 2001) (1 sentence, 28 words)

An alternative way to achieve conciseness is to use the word *respectively*:

Concise Several countries have established levels (recommended maximum limits, RMLs) for PCBs in dietary products, such as fish, meats, and eggs, with RMLs of ~2000, 200–2000, and 100–300 ng/g, *respectively*. (1 sentence, 31 words)

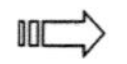

Respectively

See appendix A and chapter 4.

The use of colons, with grouped ideas, can also contribute to conciseness. In the following examples, the fabricated wordy passage uses no colon; the more concise (authentic) passage lists the related items after a colon:

Wordy The general analytical procedure for the determination of PCBs in full-fat milk includes four main steps. This first step involves extraction of the PCBs from the matrix. The second step involves preconcentration and cleanup. The third step uses gas chromatographic separation. Finally, the last step involves detection. (5 sentences, 47 words)

Concise The general analytical procedure for the determination of PCBs in full-fat milk includes four main steps: extraction from the matrix, preconcentration and cleanup, gas chromatographic separation, and detection. (From Llompart et al., 2001) (1 sentence, 28 words)

Note that when you use a colon, it must be preceded by a complete sentence:

Incorrect The procedure includes: extraction, cleanup, and detection.
Correct The procedure includes three steps: extraction, cleanup, and detection.

Commas, Colons, and Semicolons

See appendix A.

Items grouped in a series may also be numbered (with numbers enclosed in parentheses) to achieve conciseness.

Wordy Over the past few years, we have encountered numerous examples of water as the "perfect" solvent. We observed this first in osmium-catalyzed dihydroxylation reactions[12] and also in nucleophilic ring-opening reactions of epoxides.[13] We also observed this in cycloaddition reactions[13] and in most oxime ether, hydrazone, and aromatic heterocycle condensation processes.[14] Finally, we observed it in formation reactions of an amide from a primary amine and an acid chloride using aqueous Schotten-Baumann conditions.[15] (4 sentences, 72 words)

Concise Over the past few years, we have encountered numerous examples of water as the "perfect" solvent in, for example, (1) osmium-catalyzed dihydroxylation reactions;[12] (2) nucleophilic ring-opening reactions of epoxides;[13] (3) cycloaddition reactions;[13] (4) most oxime ether, hydrazone, and aromatic heterocycle condensation processes;[14] and (5) formation reactions of an amide from a primary amine and an acid chloride using aqueous Schotten-Baumann conditions.[15] (Adapted from Demko and Sharpless, 2001) (1 sentence, 61 words)

When grouping related items to achieve conciseness, pay careful attention to parallelism. The preceding examples illustrate two types of parallelism:

Plural nouns fish, meats, and eggs
reactions, epoxides, and processes
Nominalizations extraction, preconcentration, separation, and detection

Parallelism

See appendix A.

Exercise 6.23

Imagine that you are writing a paper on lead concentrations in particulate matter in Houston, Texas. You want to summarize the results of *Smith and Caine (2007)* in your Introduction section. Your notes are given below. Convert your notes into one or two concise sentences for your Introduction, grouping related items with one of the patterns described above. Check your sentence(s) for proper punctuation and parallelism.

Sara Smith & Tim Caine (2007) studied lead concentrations in particulate matter in Pittsburgh. They wanted to see how much lead there was in the particulate and what the sources of the lead might be. They found about 65% of the lead was from diesel vehicles. About 20% was from emissions from nondiesel vehicles. The remaining 15% was from industrial emissions. The lead concentrations were reported to be somewhere between 2 and 20 parts per billion (ppb).

Exercise 6.24

Imagine you are writing a paper about carbon nanotubes (CNTs). In your Introduction section, you want to summarize some unique properties of CNTs that are reported in the literature. You have listed these in your notes, given below. Convert your notes into one or two concise sentences for your Introduction, grouping related items with one of the patterns described above. Check your sentence(s) for proper punctuation and parallelism.

- CNT have very strong tensile strength with a mean Young's modulus value of 1002 GPa (Yu et al., 2008)
- CNTs have unusually high thermal conductivity (Hone et al., 2007)
- CNTs conduct electricity ballistically (implying a bullet-like trajectory) and as a result do not create resistive heating when they conduct electricity (Frank et al., 1999)
- CNTs can be manipulated to display properties of either semiconductors or metallic conductors (Collins et al., 2006)

Be Fluid

Experienced writers know how to make their words flow, logically linking thoughts and ideas. One way to achieve fluidity in your writing, particularly in the Introduction section, is to use words or short phrases that create obvious linkages between sentences and/or add emphases to your writing. Consider the italicized examples below. Note how many of these words and phrases are followed by a comma.

a. *Furthermore*, large volumes of organic solvents are used and significant amounts of residues are generated. (From Llompart et al., 2001)
b. *Subsequently*, SPME was applied to more complex matrixes, such as solid samples or biological fluids. (From Llompart et al., 2001)
c. *For example*, it has been reported that iron released from airborne fine particles or present on their surfaces plays a role in the generation of ROS, and this is also true for coal fly ash particles (*14–16*). (From Dellinger et al., 2001)
d. *As a result* of these endeavors, we have found that in the presence of zinc salts[9b,c] tetrazole formation proceeds with excellent yields and scope in refluxing water.[16] (From Demko and Sharpless, 2001)
e. *Moreover*, saponification acts as a cleanup step and then improves selectivity and reliability in peak identification. (Adapted from Llompart et al., 2001)
f. *However*, these methods are rather complicated and not highly selective. (From Vesely et al., 2003)
g. *Thus*, PCBs have a long environmental half-life and tend to accumulate in the food chains; the highest concentrations were usually found in human beings and higher animals at the top of the food chain.[3,4] (From Llompart et al., 2001)
h. *In numerous studies*, Cr^{6+} compounds have been shown to be carcinogenic in vivo and mutagenic in vitro. (From Plaper et al., 2002)
i. *In contrast* to Cr^{6+}, trivalent chromium (Cr^{3+}) is actually an essential nutrient, needed for the expression of glucose tolerance (*1*). (From Plaper et al., 2002)

Words and phrases such as these are also used in the middle of sentences or between sentences connected by a semicolon. Consider these passages, paying special attention to the ways in which commas and semicolons are used.

a. Dairy products, and milk *in particular*, have received special interest due to their extensive and elevated consumption by the population.[6] (From Llompart et al., 2001)
b. With these types of samples, direct SPME is not recommended; *nevertheless*, the headspace mode (HSSPME) is an effective alternative to extracting volatile and semivolatile compounds from complex matrixes. (From Llompart et al., 2001)

c. It is *therefore* important that Cr^{3+} toxicity be investigated in further detail, to ensure safer application of this additive. (From Plaper et al., 2002)

d. Purification requires, *for example*, crystallization or chromatography, which may lead to losses. (From Boesten et al., 2001)

e. Although HSSPME has been applied to an enormous variety of matrixes,[22,24] *to date*, the number of publications that apply SPME technology to milk samples is relatively low,[25,26] and most of them are related to the determination of flavors.[27,28] (Adapted from Llompart et al., 2001)

Fluid Writing

See appendix A.

When used properly, such words and phrases contribute to the flow of the written passage. Of course, their use is not confined to the Introduction section of a journal article (although all the examples here come from Introductions). When used appropriately, and in the right places, such words and phrases add cohesiveness to a journal article as a whole. A list of useful phrases, organized by their common functions, is presented in table 6.6. Many, but not all, of these phrases are conventionally followed by commas when they start a sentence.

Exercise 6.25

Read the following passage. Do the italicized terms serve the functions indicated in table 6.6? If not, what functions do they serve to create a cohesive passage?

Initially, SPME was used to analyze pollutants in water[20,21] via direct extraction. *Subsequently*, SPME was applied to more complex matrixes, such as solid samples or biological fluids. With these types of samples, direct SPME is not recommended; *nevertheless*, the headspace mode (HSSPME) is an effective alternative to extracting volatile and semivolatile compounds from complex matrixes. (From Llompart et al., 2001)

Exercise 6.26

Read the following passage. Two words (or phrases) are missing, indicated by the blanks. What words do you think the authors used? Does the inclusion of these words add a sense of fluency to the passage? Do the words add clarity to the passage?

Laser ablation coupled to ion cyclotron resonance Fourier transform mass spectrometry (in both positive and negative ion modes) can be used to distinguish natural and artificial opals. In positive ion mode, species including hafnium and large amounts of zirconium atoms are found to be specific for artificial opal. ____, aluminum, titanium, iron, and rubidium are systematically detected in the study of natural opals. ____, some ions allow us to distinguish between natural opal from Australia and Mexico. Australian gemstones include specifically strontium, cesium, and barium. (Adapted from Erel et al., 2003)

Exercise 6.27

Browse through the Introduction sections of three different journal articles and find at least five examples of sentences that begin with a word or short phrase that serves one or more of the functions listed in table 6.6.

Table 6.6 Common phrases used to create linkages and their functions.[a]

To show contrast	
Conversely,	
However,	
In contrast,	
Nevertheless,	
On the other hand,	
Unfortunately,	
To provide additional information	
Additionally,	
Furthermore,	
In addition,	
Moreover,	
Namely,	
To describe a typical case	
In general,	
Typically,	
Usually,	
To show cause and effect	
Accordingly,	
As a consequence,	
As a result,	
Consequently,	
Hence,	
Therefore,	
Thus,	
To this end,	
To give examples	
For example,	
For instance,	
To add emphasis or clarify	
In particular,	
More specifically,	
Specifically,	
To signal time	
Afterward,	
Initially,	
Previously,	
Simultaneously,	
Subsequently,	
To date,	
Ultimately,	
To refer to something previously stated	
As mentioned/described above,	
In the latter case,	
In this/these/that/those cases(s),	
In this context,	
In this respect,	

a. These phrases were identified through a computer-based analysis of Introductions from 60 published chemistry journal articles.

6F Writing on Your Own: Practice Peer Review

Before you review a peer's Introduction section, practice the peer review process. Imagine that a colleague has asked you for feedback on an Introduction that is currently in draft form. Based on what you have learned in this chapter, read the draft and instructions (included in "Peer Review Practice: Introduction Section" at the end of the chapter) and offer written suggestions for improving the draft.

6G Writing on Your Own: Fine-Tune Your Introduction

By now, you should have a good draft of your Introduction section, having completed the previous Writing on Your Own tasks. Now it is time to revise and edit your Introduction as a whole, using the suggestions provided in the chapter.

Refer to chapter 18 and the questions below to guide you in the revision process. Be sure to focus on each of the following areas:

1. Organization of text: Check your overall organizational structure. Did you follow the move structure outlined in figure 6.1?
2. Audience and conciseness: Will your Introduction draw readers into the paper? Is it written for a broader audience, moving from a general focus to a more specific focus? Have you taken steps to ensure that your writing is concise? Find at least three sentences that can be written more clearly and concisely.
3. Writing conventions: Check to be sure that you have (1) not used direct quotes; (2) paraphrased the literature accurately, giving credit (in the form of citations) where it is due; (3) focused on the science rather than the scientists; (4) used tense and voice purposefully; and (5) created linkages between sentences to enhance the fluidity of your written work.
4. Grammar and mechanics: Check for typos and errors in spelling, subject–verb agreement, parallelism, and punctuation, paying special attention to your more complex sentences.
5. Science content: Have you correctly conveyed the science of others, and your own? If asked, could you define all of the words you have used in this section?

After thoroughly reviewing your own work, it is common practice to have your work reviewed by a peer or colleague. A "new set of eyes" will pick up mistakes that you can no longer see because you are too familiar with your own writing. To facilitate this process, use the Peer Review Memo on the *Write Like a Chemist* Web site. After your paper has been reviewed (and you have reviewed another's paper), make final changes in your Introduction section.

Finalizing Your Written Work

See chapter 18.

Chapter Review

As a self-test of what you've learned in this chapter, define each of the following terms for a friend or colleague who is new to the field:

background information
current work
direct quotes
fill-the-gap statement
gap statement
green chemistry
linking words
parallelism
present perfect
review of the literature

Also explain the following to a friend who hasn't yet given much thought to writing an Introduction section for a journal article:

- Main purposes of an Introduction section
- Moves of an Introduction
- Role of a gap statement in an Introduction
- Types of gaps often identified in an Introduction
- Use of present, past, and present perfect in an Introduction
- Place of direct quotations and paraphrasing in an Introduction
- Common techniques for making writing more concise and fluid

Additional Exercises

Exercise 6.28

Read the following two passages. Which move in the Introduction does each passage correspond to? (Consult figure 6.1.) Rewrite each passage so that it follows the writing conventions of the corresponding move more closely:

a. In the paper that we wrote and have presented below, we make use of fluorescence to accomplish the characterization of the coil-globule transition of isolated PEO chains in toluene. (Adapted from Farinha et al., 2001)

b. A particular type of damage caused by free radicals known as oxidative damage has been associated with vascular disease in people with types 1 and 2 diabetes mellitus (DM) (2). There are several different potential sources of this free radical production in diabetics. For example, one possible source is autoxidation of plasma glucose (4). Another possible source is if the leucocytes get activated (5). A third possible source is if the bioavailability of transition metals is increased (6). (Adapted from Cheng et al., 2004)

Exercise 6.29

Reflect on what you have learned about writing an Introduction section for a journal article. Select one of the reflection tasks below and write a thoughtful and thorough response:

a. Reflect on the unique characteristics of an Introduction to a journal article.
   - In what ways is an Introduction different from the Methods, Results, and Discussion sections of a journal article?
   - In what ways is an Introduction similar to the other sections of a journal article?
   - Which part of an Introduction do you think is most critical?

b. Reflect on the role of a literature review in an Introduction.
   - Why is a literature review so important?
   - What factors should you consider when deciding which articles to include in or exclude from your own literature review?
   - When reading the primary literature, what should you concentrate on for your Introduction?

c. Reflect on the three moves of an Introduction section.
   - As a writer, what challenges do you associate with each move (and corresponding submoves)?
   - Why is the opening sentence of an Introduction so important?
   - Do you think it is a good idea to conclude an Introduction with a final paragraph that summarizes key findings? Or do you think readers should be patient and discover key findings later in the paper, in the Results section? Explain.

Peer Review Practice: Introduction Section

Imagine that a colleague has asked you to review an Introduction that is currently in draft form; your colleague plans to use your feedback to improve the written work.

Using parts 2 and 3 of the Peer Review Memo on the *Write Like a Chemist* Web site, review the Introduction below. Remember to give suggestions that are specific enough to guide your colleague in improving the Introduction. (The Introduction below is adapted from an original source, noted in the Instructor's Answer Key.)

1 Toxicology is the study of harmful effects of chemicals on people, animals and other
2 living organisms. Forensic toxicology involves the analysis of drugs and poisons in
3 biological specimens. It also involves the interpretation of the results to be applied in a
4 court of law. Availability of analytical reference standards becomes a critical factor when
5 a novel target substance is encountered. Reference standards are also important when
6 a comprehensive screening procedure is updated. Commercial drugs can be acquired
7 within a reasonable period of time. Their metabolites generally cannot. The situation is
8 even more complicated in the rapidly changing scene of designer drugs. Identification
9 of low-dose substances in biomatrixes without reference standards is a challenge to any
10 well-equipped research laboratory. The forensic analyst in charge usually has to be satis-
11 fied with comparing sample mass spectra to those published in electronic libraries for
12 electron impact gas chromatography/mass spectrometry (GC/MS).
13 Marquet[1] suggests that the number of liquid chromatography/mass spectrometry
14 (LC/MS) applications in forensic toxicological analysis has increased markedly during
15 the past decade. In comprehensive drug screening, identification has been based on
16 fragment ions or comparison of full mass spectra. This necessitates reference substances
17 for the construction of spectra libraries. Several researchers question the interlaboratory
18 reproducibility of mass spectral libraries obtained by these techniques.[2–4] This lack of
19 reproducibility hinders the creation of universal reference libraries.
20 Burlingame[5] and Lewis et. al[6] use accurate mass in the monitoring of specific
21 compounds in environmental and biological samples with glass capillary gas chroma-
22 tography/high-resolution mass spectrometry. This approach is limited by expensive
23 instrumentation. Orthogonal acceleration time-of-flight mass spectrometry (OATOFMS)
24 allows continuous mass measurement with moderate resolution (5000) and high mass
25 accuracy (5 ppm). Several affordable benchtop liquid chromatography/time-of-flight
26 mass spectrometry (LC/TOFMS) instruments were recently launched onto the market.
27 The accurate mass measurement enables formulation of candidate elemental com-
28 positions for a particular mass. This allows tentative characterization of substances.
29 Predefined exact masses can be searched for identification. A number of studies use
30 OATOFMS in the identification and characterization of: unknown drug metabolites,
31 glucuronide conjugates, pesticides, anabolic steroids, and quantitative drug analysis.[7–15]
32 A preliminary communication from this laboratory introduced the concept of urine
33 drug screening by positive pneumatically assisted electrospray ionization LC/TOFMS
34 with an automated target library search based on elemental formulas.[16] This approach
35 was based on the assumption that tentative identification of drugs in urine is viable
36 without reference standards by use of exact monoisotopic masses and metabolite pat-
37 terns from the literature. The present study evaluated this screening methodology to the
38 full with a series of urine samples taken at autopsy. It showed the scope and limitations
39 of this method in forensic toxicology practice.

7 *Writing the Abstract and Title*

The purpose of the abstract is to inform, to give away the punch line right at the start, and to let your readers decide whether they want to read the full document. Scientific writing is not like mystery writing in which the results are hidden until the end.

—Adapted from Alley (1996)

This chapter addresses how to write abstracts and titles for journal articles. Both the abstract and title provide succinct, informative (not descriptive) summaries of the research. To this end, they are usually written in the final stages of the writing process. After completing this chapter, you should be able to do the following:

- Write a concise and informative abstract
- Write a concise and informative title

As you work through the chapter, you will write an abstract and title for your own paper. The Writing on Your Own tasks throughout the chapter will guide you step by step as you do the following:

7A Read titles and abstracts

7B Prepare to write

7C Write your abstract

7D Write your title

7E Practice peer review

7F Fine-tune your abstract and title

When compared to the Introduction, Methods, Results, and Discussion sections of a journal article, the title and abstract are quite short; the title usually has fewer than 20 words, and many journals limit the abstract to fewer than 200 words. Despite their brevity (and perhaps because of it), the title and abstract are the most widely read sections of the journal article and thus are viewed by many as the most important sections of the journal article.

An Important 200 Words

Titles and abstracts are read by more readers than any other section of the journal article.

Reading and Analyzing Writing

We begin by asking you to read and analyze the title and **abstract** for the aldehydes-in-beer article (excerpt 7A). The questions in exercise 7.1 will guide your analysis of this excerpt.

Abstract

Concise and highly informative, the abstract informs readers about the purpose, the theoretical or experimental approach, principal results, and major conclusions of the work.

Exercise 7.1

As you read the title and abstract in excerpt 7A, consider the following:

a. Read the title. Which of the following are included: research topic, importance, gap statement, procedures, instrumentation, results, interpretations, citations, conclusions?

b. The abstract contains six sentences (107 words). Briefly state the purpose of each sentence. Based on these purposes, propose a move structure for the abstract.

c. Are there any sentences in the abstract that do not include science content? Explain.

d. Based only on the title and abstract, who are the intended audiences for this article (including subdisciplines of chemistry)? Give reasons for your choices.

e. What rules do the authors follow regarding the use of abbreviations in their abstract? What other writing conventions do you notice?

f. What verb tense(s) do the authors use in their abstract?

g. Suggest why the authors include keywords at the end of the abstract. How many of these keywords also appear in the title and abstract?

Abstract Headings and Word Counts

In this chapter, we begin each abstract with a heading (**Abstract**) and conclude each abstract with a word count. We do this for instructional purposes; headings and word counts do not typically appear in published abstracts.

Excerpt 7A (from Vesely et al., 2003)

Abstract

Analysis of Aldehydes in Beer Using Solid-Phase Microextraction with On-Fiber Derivatization and Gas Chromatography/Mass Spectrometry.

A new, fast, sensitive, and solventless extraction technique was developed in order to analyze beer carbonyl compounds. The method was based on solid-phase microextraction with on-fiber derivatization. A derivatization agent, *O*-(2,3,4,5,6-pentafluorobenzyl) hydroxylamine (PFBOA), was absorbed onto a divinyl benzene/poly(dimethylsiloxane) 65-μm fiber and exposed to the headspace of a vial with a beer sample. Carbonyl compounds selectively reacted with PFBOA, and the oximes formed were desorbed into a gas chromatograph injection port and quantified by mass spectrometry. This method provided very high reproducibility and linearity. When it was used for the analysis of aged beers, nine aldehydes were detected: 2-methylpropanal, 2-methylbutanal, 3-methylbutanal, pentanal, hexanal, furfural, methional, phenylacetaldehyde, and (*E*)-2-nonenal. (107 words)

Keywords: Aldehydes; beer analysis; derivatization; SPME; GC/MS

7A Writing on Your Own: Read Titles and Abstracts

Read and review the titles and abstracts of the journal articles that you collected during your literature search (started in chapter 2). How well do they capture the purpose, principal results, and conclusions of the work?

Use these titles and abstracts as models in your own writing.

Abstracts and titles are generally written for expert and scientific audiences; however, parts of each are also typically accessible to a student audience. For example, the abstract in the aldehydes-in-beer article targets professional food chemists and analytical chemists specifically, but a student in organic chemistry could read the abstract and understand which aldehydes are present in aged beer. Moreover, the student could also discern from the title that the article is about aldehydes in beer.

The major purpose of the title is to inform readers about the specific content of the work, ideally identifying both what was studied and how it was studied. The major purpose of an abstract is to summarize, in one clear and concise paragraph, the purpose, experimental approach, principal results, and major conclusions of the work. In most journals, the abstract includes only text; in some journals (e.g., *The Journal of Organic Chemistry* and *Organic Letters*), the abstract also includes a graphic. Importantly, both the abstract and title must be able to stand on their own. This is because these two sections (and only these two sections) are reprinted by abstracting services (e.g., Chemical Abstracts Service, or **CAS**) in separate documents for literature searches. Also, many chemists read titles and abstracts to obtain a quick overview of the journal's contents but do not read the articles in full.

CAS

The Chemical Abstracts Service (CAS) reprints titles and abstracts from refereed journal articles to facilitate searches of the chemical literature.

As you may have noticed in excerpt 7A, some abstracts also include a list of keywords. **Keywords**, required in many journals, help readers locate relevant works when they search the literature. Guidance in selecting keywords is provided in Information for Authors documentation for journals that require them. Even if the journal that you are targeting does not require keywords, it is wise to create a list anyway and incorporate as many of these words as possible into your title and abstract. Doing so will greatly increase the probability that your paper will be found by interested individuals searching the literature.

Keywords

Searchable words that help scientists find relevant works. Keywords are often required in abstracts. Some journals provide a list of keywords for authors to choose from.

Analyzing Organization

We consider the organization for both the abstract and title. Because the abstract is generally written before the title, we begin with the abstract.

The Abstract

The abstract involves three essential moves (figure 7.1). The major objectives of these moves are to state (1) what was done, (2) how it was done, and (3) what was found. Each move is designed to inform, not describe. The first move has one required submove (submove 1.3): state purpose and/or accomplishment(s) of work. In some cases, authors lead up to this submove by first identifying the research area and its importance (submove 1.1) and/or a gap in the field that is addressed by the work (submove 1.2), similar to moves 1 and 2 of the Introduction section (see figure 6.1).

Move 2 summarizes the research methods; procedures and/or instrumental techniques may also be identified. The amount of detail presented varies with the goals of the paper. A paper that describes the development of a novel approach is likely to include more information than one that uses standard

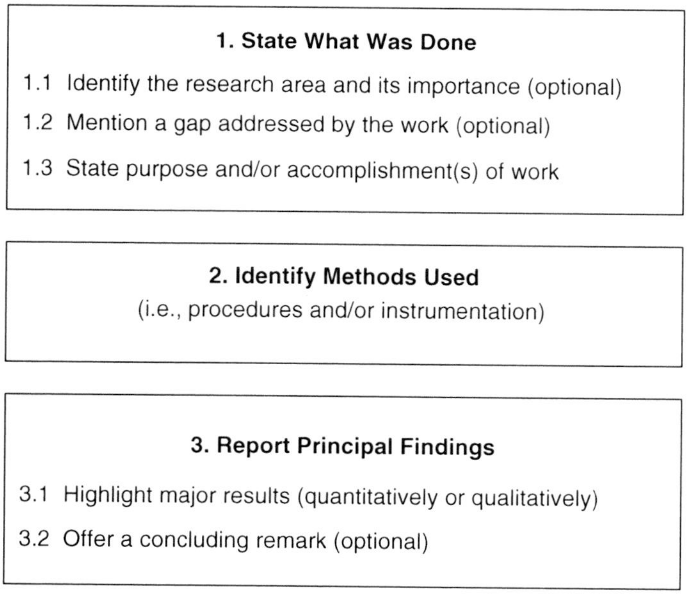

1. State What Was Done

1.1 Identify the research area and its importance (optional)

1.2 Mention a gap addressed by the work (optional)

1.3 State purpose and/or accomplishment(s) of work

2. Identify Methods Used

(i.e., procedures and/or instrumentation)

3. Report Principal Findings

3.1 Highlight major results (quantitatively or qualitatively)

3.2 Offer a concluding remark (optional)

Figure 7.1 A visual representation of the move structure for a typical journal article abstract.

methodologies. In articles that describe chemical syntheses, this move usually includes a graphic.

Move 3 highlights the principal findings of the work and is often the longest segment of the abstract. When practical, numerical values (with their error terms, e.g., standard errors or deviations) are reported; otherwise, only the major trends suggested by the data are summarized. It is not appropriate to include a table in the abstract or to repeat all of the results presented in the paper. In many instances, the abstract ends with a concluding statement that draws attention to the major findings or impacts of the work.

The moves and submoves of the abstract directly parallel moves found in other sections of the journal article. Despite these similarities, it is important that you do not repeat yourself in the abstract. Effective writers resist the temptation to simply copy sentences from other sections of their papers to use in their abstracts.

The Title

The title also has an organizational structure. After analyzing more than 300 titles of chemistry journal articles, we found that titles commonly follow an "X of Y by Z" pattern (table 7.1). In essence, X, Y, and Z are three moves linked together by common words (e.g., *of* or *by*). Y describes what was studied; X and Z modify or extend Y in some way. Y is required; X and Z are optional, but typically at least X or Z is present. Of course, this pattern illustrates only a conventional way to construct a journal article title; countless variations are possible.

Table 7.1 Common examples of the "X of Y by Z" pattern found in journal article titles.

X (optional)		Y (required)		Z (optional)
Basic Pattern				
A nominalization	of	What was studied	on	Target of Y or what was
(e.g., Determination,	in		in	impacted by Y
Investigation, Analysis	for		via	
Measurement)	to		by	Method used (or detail of
	. . .		using	method used) to study Y
A phrase that refers to,			at	
describes, or modifies Y			from	
			. . .	

continued

Table 7.1 (continued)

X (optional)		Y (required)		Z (optional)
Examples				
Preparation	of	5-Substituted 1*H*-Tetrazoles	from	Nitriles in Water[a]
Analysis	of	Aldehydes in Beer	Using	Solid-Phase Microextraction with On-Fiber Derivatization and Gas Chromatography/ Mass Spectrometry[b]
Crystal Structure	of	Native Chicken Fibrinogen	at	2.7 Å Resolution[c]
Heteronuclear Recoupling	in	Solid-State Magic-Angle-Spinning NMR	via	Overtone Irradiation[d]
Cancer-Protective Properties	of	High-Selenium Broccoli[e]		
	—	A Class II Aldolase Mimic[f]		

a. Demko and Sharpless (2001).
b. Vesely et al. (2003).
c. Yang et al. (2001).
d. Wi and Frydman (2001).
e. Finley et al. (2001).
f. Hedin-Dahlström et al. (2006).

Exercise 7.2

Consider the following 10 titles. Do they conform to the pattern presented in table 7.1? When possible, identify X, Y, and Z in each title.

a. Effect of Randomly Methylated β-Cyclodextrin on Physical Properties of Soils (from Jozefaciuk et al., 2003)
b. Determination of Polychlorinated Biphenyls in Milk Samples by Saponification—Solid-Phase Microextraction (from Llompart et al., 2001)
c. Role of Free Radicals in the Toxicity of Airborne Fine Particulate Matter (from Dellinger et al., 2001)
d. Antioxidative Activity of Volatile Chemicals Extracted from Beer (from Wei et al., 2001)
e. Antiadhesive Effect of Green and Roasted Coffee on *Streptococcus mutans*' Adhesive Properties on Saliva-Coated Hydroxyapatite Beads (from Daglia et al., 2002)

f. Biotransformation and Accumulation of Arsenic in Soil Amended with Seaweed (from Castlehouse et al., 2003)

g. In Vitro Effect of Arsenical Compounds on Glutathione-Related Enzymes (from Chouchane and Snow, 2001)

h. Chemical Characterization of Sicilian Prickly Pear (*Opuntia ficus indica*) and Perspectives for the Storage of Its Juice (from Gurrieri et al., 2000)

i. A Method for the Analysis of Low-Mass Molecules by MALDI-TOF Mass Spectrometry (from Guo et al., 2002)

j. Arsenic Contamination of Bangladesh Paddy Field Soils: Implications for Rice Contribution to Arsenic Consumption (from Meharg and Rahman, 2003)

7B Writing on Your Own: Prepare to Write

Before you begin writing your own abstract and title, make a list of keywords that another researcher might use to find your paper in a literature search. (You should incorporate many of these words in your abstract and title.)

Determine the word limit for the abstract that you are writing. Do not surpass this word limit.

Review the move structure for the abstract (figure 7.1). Outline the information that you want to include in each move of your abstract.

Analyzing Abstracts

In this section, we read and analyze abstracts taken from chemistry journal articles. (Later in the chapter, we focus on titles.) We include passages from abstracts (P1–P17) and entire abstracts (excerpts 7B–7H) to illustrate both individual moves and how abstract moves work together as a whole. In part 1, we examine selections move by move. Because abstracts in organic chemistry journals vary slightly from other chemistry journals, we consider them separately at the end of part 1. In part 2, we examine writing practices that span the entire abstract.

Part 1: Analyzing Excerpts

Move 1: State What Was Done

Recall that the first move of the abstract has two optional submoves (1.1 and 1.2) and one required submove (1.3). We begin by examining four passages from abstracts that begin immediately with the required submove. Each example is the first sentence of the abstract. The purpose of this required submove is to state the major purpose or accomplishments of the work. Note that the words used in this submove must be different than the words used in move 3 of the Introduction section, which introduces the reader to the current work.

P1 Antioxidative compounds were isolated from the 50% methanol extract of dried leaves of *Celastrus hindsii*. (From Ly et al., 2006)

P2 In this study, we report the chemical synthesis and functionalization of magnetic and gold-coated magnetic nanoparticles and the immobilization of single-stranded biotinylated oligonucleotides onto these particles. (From Kouassi and Irudayaraj, 2006)

P3 Hydrogen (H_2) concentrations during reductive dechlorination of *cis*-dichloroethene (*c*DCE) and vinyl chloride (VC) were investigated with respect to the influence of parameters entering the Gibbs free energy expression of the reactions. (From Heimann and Jakobsen, 2006)

P4 A method for analyzing ergosterol in a single kernel and ground barley and wheat was developed using gas chromatography-mass spectrometry (GC-MS). (From Dong et al., 2006)

We next examine passages from abstracts that begin with a description of the general research area and/or mention a gap in the field (optional submoves 1.1 and 1.2). In these passages (P5–P7), we include not only the first sentence but several additional sentences to show how the authors lead up to submove 1.3. Note that although others' works are alluded to in these passages, no citations are included. This absence of citations is true not only here but throughout the abstract. If a work must be cited in an abstract (a rare occurrence), the full citation must be included in the abstract; in this way, the abstract can stand alone.

Cite in an Abstract?

In general, most journals prefer that you not cite others' works in the abstract.

P5 Studies have shown that ebselen is an antiinflammatory and antioxidative agent. Its protective effect has been investigated in oxidative stress related diseases such as cerebral ischemia in recent years. However, experimental evidence also shows that ebselen causes cell death in several different cell types. Whether ebselen will have a beneficial or detrimental effect on cells under ischemic condition is not known. Herein, we studied the **effect** of ebselen on C6 glioma cells under oxygen and glucose deprivation (OGD), an in vitro ischemic model. (From Shi et al., 2006)

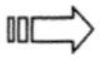

Affect vs. Effect

See appendix A for more information on these easily confused words.

P6 Atrazine (2-chloro-4-[ethylamino]-6-[isopropylamino]-1,3,5-triazine) is one of the most commonly used herbicides in North America and is frequently detected in ground and surface waters. This research investigated possible covalent modifications of hemoglobin following in vivo exposures to atrazine in Sprague Dawley (SD) rats and in vitro incubations with diaminochlorotriazine. (From Dooley et al., 2006)

P7 Over the past decade, electron monochromator-mass spectrometry (EM-MS) has been shown to be a selective and sensitive technique for the analysis of a wide variety of electrophilic compounds in complex matrixes. Here, for the first time, three different dinitroaniline pesticides, flumetralin, pendimethalin, and trifluralin, have been shown to be present in both mainstream and sidestream tobacco smoke using an EM-MS system. (From Dane et al., 2006)

Exercise 7.3

Identify submoves 1.1–1.3 in passages P5–P7. When appropriate, indicate which submoves are absent.

Move 2: Identify Methods Used

The next several passages illustrate how experimental methods are reported in abstracts. We begin with two passages that describe procedures: P8 highlights steps taken to extract pesticides from cigarette smoke; P9 highlights steps taken to separate proteins using novel forms of electrophoresis and chromatography. Note that the authors of P9 define PMMA, the abbreviation for poly(methyl methacrylate), in the title of their paper; hence, it is used without definition in their abstract. All other abbreviations are defined in P9.

Abbreviated Terms in Abstracts and Titles

Acronyms and abbreviations, used only to prevent needless repetition, should be defined in the abstract (or title) so that the abstract, together with the title, can stand alone. These abbreviations should be defined again in the article. Common abbreviations (e.g., NMR) need not be defined.

P8 A number of cigarettes were tested including three pure-tobacco-type cigarettes, an experimental reference cigarette, and 11 commercial cigarettes. Due to the complexity of the smoke particulate matter, the pesticides were identified only after each sample was subjected to a multistep cleanup process that included phenyl solid-phase extraction, an acid wash, aminopropyl solid-phase extraction, and normal phase liquid chromatography fractionation. (Adapted from Dane et al., 2006)

P9 Sodium dodecyl sulfate microcapillary gel electrophoresis (SDS μ-CGE) and micellar electrokinetic chromatography (MEKC) were used as the separation modes for the first and second dimension of the electrophoresis, respectively. The microchip was prepared by hot embossing into PMMA from a brass mold master fabricated via high-precision micromilling. The microchip incorporated a 30-mm SDS μ-CGE and a 10-mm MEKC dimension length. Electrokinetic injection and separation were used with field strengths of up to 400 V/cm. Alexa Fluor 633 conjugated proteins, ranging in size from 38 to 110 kDa, were detected using laser-induced fluorescence with excitation/emission at 633/652 nm. (From Shadpour and Soper, 2006)

P10 and P11 illustrate how authors commonly refer to instrumentation in their abstracts. In most cases, instrument names are written out in full (without acronyms), and no information regarding vendors, model numbers, or operational parameters is included.

P10 Eight phenolic compounds... were... obtained by reversed-phase high-performance liquid chromatography, and their structures were elucidated by NMR spectroscopy and mass spectrometry analyses. (Adapted from Ly et al., 2006)

P11 Particle size and oligonucleotide attachment were confirmed by transmission electron microscopy; oligonucleotide binding was characterized by Fourier transform infrared spectroscopy and hybridization confirmed by fluorescence emission from the fluorophore attached to the target oligonucleotide strand. The rate of hybridization was measured using a spectrofluorometer and a microarray scanner. (From Kouassi and Irudayaraj, 2006)

Move 3: Report Principal Findings

Move 3, the last and most important move of the abstract (and often the longest), highlights the principal findings of the work. Three examples are considered (P12–P14). Only the most essential or representative data are reported, including numerical values, when appropriate. (Note that numerical data should include units and standard errors or deviations. Do not omit units or error terms to conserve space.) Move 3 may also inform readers about the contents of the full article. For example, P12 informs readers that acute toxicity information for three pesticides is presented in the text. Move 3 sometimes ends with a concluding remark. In P12 and P13, the data themselves end the abstract and no summative remarks are made; in P14, the authors add a concluding remark.

Error Terms

Error terms, such as standard deviations and standard errors, should be included in the abstract, when appropriate.

The ACS Style Guide recommends including a space before and after the "±" symbol (e.g., 17 ± 9 mL).

P12 All cigarette types tested showed the presence of the three pesticides in the tobacco smoke, with flumetralin ranging from trace levels up to 37 (±9) ng/cig, pendimethalin ranging from trace levels up to 10.4 (±0.6) ng/cig, and trifluralin ranging from trace levels up to 47 (±17) ng/cig. Acute toxicity information is presented for the three pesticides. (From Dane et al., 2006)

P13 [Note: The authors define ethyl vinyl ether (EVE), propyl vinyl ether (PVE), and butyl vinyl ether (BVE) in the first part of the abstract.] ... Using a relative kinetic method, rate coefficients (in units of cm^3 $molecule^{-1}$ s^{-1}) of $7.79 \pm 1.71 \times 10^{-11}$, $9.73 \pm 1.94 \times 10^{-11}$, and $1.13 \pm 0.31 \times 10^{-10}$ have been obtained for the reaction of OH with EVE, PVE, and BVE, respectively; $1.40 \pm 0.35 \times 10^{-12}$, $1.85 \pm 0.53 \times 10^{-12}$, and $2.10 \pm 0.54 \times 10^{-12}$ for the reaction of NO_3 with EVE, PVE, and BVE, respectively; and $2.06 \pm 0.42 \times 10^{-16}$, $2.34 \pm 0.48 \times 10^{-16}$, and $2.59 \pm 0.52 \times 10^{-16}$ for the ozonolysis of EVE, PVE, and BVE, respectively. Tropospheric lifetimes of EVE, PVE, and BVE with respect to the reactions with reactive tropospheric species (OH, NO_3 and O_3) have been estimated for typical OH and NO_3 radical and ozone concentrations. (Adapted from Zhou et al., 2006)

P14 The recoveries of ergosterol from ground barley were 96.6, 97.1, 97.1, 88.5, and 90.3% at the levels of 0.2, 1, 5, 10, and 20 μg/g (ppm),

respectively. The recoveries from a single kernel were between 93.0 and 95.9%. The precision (coefficient of variance) of the method was in the range 0.8–12.3%. The method detection limit and the method quantification limit were 18.5 and 55.6 ng/g (ppb), respectively. The ergosterol analysis method developed can be used to handle 80 samples daily by one person, making it suitable for screening cereal cultivars for resistance to fungal infection. The ability for detecting low levels of ergosterol in a single kernel provides a tool to investigate early fungal invasion and to study mechanisms of resistance to fungal diseases. (From Dong et al., 2006)

Respectively

See appendix A and chapter 4.

Major findings may also be expressed in more qualitative, and less quantitative, terms by highlighting general trends in the abstract. Three examples are provided below. P16 and P17 offer two additional examples of concluding remarks.

P15 Variations in the temperature between 10 and 30 °C did not affect the H_2 concentration in a fashion that suggested thermodynamic control through a constant energy gain. In another set of experiments, H_2 levels at constant ionic strength were independent of the chloride concentration between 10 and 110 mmol chloride per liter. These findings demonstrate that the partial equilibrium approach is not directly applicable to the interpretation of reductive degradation of chlorinated ethenes. We also present recalculated thermodynamic properties of aqueous chlorinated ethene species that allow for calculation of in situ Gibbs free energy of dechlorination reactions at different temperatures. (From Heimann and Jakobsen, 2006)

P16 The rate of hybridization increased concomitantly with the concentration of the probe and the target in the reaction medium. Furthermore, exposure of probe and target oligonucleotide to a combination of target and noncomplementary DNA strands reduced the rate of hybridization, possibly because of steric crowding in the reaction medium and cross-linking between reacting oligonucleotides and the noncomplementary strands. The study undertaken opens several possibilities in bioconjugate attachment to functionalized iron and iron nanocomposite structures for controlled manipulation and handling using magnetic fields. (From Kouassi and Irudayaraj, 2006)

P17 Oil from the T/V *Exxon Valdez* was found on 14 shorelines, mainly in Herring Bay and Lower Pass, with an estimated 0.43 ha covered by surface oil and 1.52 ha containing subsurface oil. Surface and subsurface oil were most prevalent near the middle of the intertidal and had nearly symmetrical distributions with respect to tide height. Hence, about half the oil is in the biologically rich lower intertidal, where predators may encounter it while disturbing sediments in search of prey. The overall probability of encountering surface or subsurface oil is estimated as 0.0048, which is only slightly greater than our estimated probability of encountering subsurface oil in the lower intertidal of Herring Bay or Lower Pass. These encounter probabilities are sufficient to ensure that sea otters and ducks that routinely excavate sediments while foraging within the intertidal would likely encounter subsurface oil repeatedly during the course of a year. (From Short et al., 2006)

We end this section by examining two complete abstracts from two journals in organic chemistry: *The Journal of Organic Chemistry* and *Organic Letters*. Both journals require authors to include a graphic in their abstracts. This graphic appears in the abstract and in the journal's table of contents, along with the title. The graphic is often a reaction mechanism or scheme described in the article but may also be an illustrative sketch, graph, or spectrum. The graphic has no title or caption but may include text for labeling purposes (i.e., to label compounds, R groups, reaction arrows, etc.). Color is allowed in the abstract graphic.

Excerpt 7B is from *The Journal of Organic Chemistry*. For the most part, this abstract follows the move structure presented in figure 7.1: the opening sentence describes accomplishments, the second sentence identifies methods, the third and fourth sentences summarize key results, and the last sentence offers a conclusion. Notice, however, that no information is given about the synthetic procedure because this information is contained in the graphic.

Excerpt 7B (from Hedin-Dahlström et al., 2006)

Abstract

1a + **2** —MIP→ **3a**

A class II aldolase-mimicking synthetic polymer was prepared by the molecular imprinting of a complex of cobalt (II) ion and either (1*S*,3*S*,4*S*)-3-benzoyl-1,7,7-trimethylbicyclo[2.2.1]heptan-2-one (**4a**) or (1*R*,3*R*,4*R*)-3-benzoyl-1,7,7-trimethylbicyclo[2.2.1]heptan-2-one (**4b**) in a 4-vinylpyridine-styrene-divinylbenzene copolymer. Evidence for the formation of interactions between the functional monomer and the template was obtained from NMR and UV–vis titration studies. The polymers imprinted with the template demonstrated enantioselective recognition of the corresponding template structure, and induced a 55-fold enhancement of the rate of reaction of camphor (**1**) with benzaldehyde (**2**), relative to the solution reactions, and were also compared to reactions with a series of reference polymers. Substrate chirality was observed to influence reaction rate, and the reaction could be competitively inhibited by dibenzoylmethane (**6**). Collectively, the results presented provide the first example of the use of enantioselective molecularly imprinted polymers for the catalysis of carbon–carbon bond formation. (139 words)

Compound Labels

Compound labels are common in abstracts. They should follow the numbering scheme used in the main text. (See chapter 4 and appendix A.)

Excerpt 7C is the abstract from Boesten et al. (2001), one of the six key articles referred to throughout this module. The article is published in *Organic Letters*, which limits abstracts to no more than 75 words. This particular abstract contains 67 words and only three sentences. The first sentence accomplishes move 1, the second sentence (with the graphic) addresses moves 2 and 3, and the last sentence accomplishes move 3. Quite concise, don't you agree?

Excerpt 7C (adapted from Boesten et al., 2001)

Abstract

a) R_1R_2CO (*in situ*)
b) NaCN, AcOH
solvent, time, temp
76–93% yield, dr >99/1
3 steps
73% yield, >98% ee

Diastereoselective Strecker reactions based on (*R*)-phenylglycine amide as chiral auxiliary are reported. The Strecker reaction is accompanied by an in situ crystallization-induced asymmetric transformation, whereby one diastereomer selectively precipitates and can be isolated in 76–93% yield and dr > 99/1. The diastereomerically pure α-amino nitrile obtained from pivaldehyde (R_1 = *t*-Bu, R_2 = H) was converted in three steps to (*S*)-*tert*-leucine in 73% yield and >98% ee. (67 words)

Exercise 7.4

Excerpt 7D, the abstract from Demko and Sharpless (2001) on the synthesis of substituted tetrazoles, is quite concise: two sentences and 57 words! Read excerpt 7D and determine if all of the moves are present. In addition, answer these questions:

a. What move(s) or submove(s) is accomplished in the graphic?
b. What move(s) or submove(s) is accomplished in the text of the abstract?
c. What results do you expect to find in the full text?
d. Do the authors present their results quantitatively or qualitatively?
e. Do the authors include a summative remark?
f. How do the authors use verb tenses?

Excerpt 7D (from Demko and Sharpless, 2001)

Abstract

R—C≡N → (1.1 equiv NaN_3, 1.0 equiv $ZnBr_2$; water, reflux) → 5-R-1*H*-tetrazole (N—NH, N=N), 52–96%

R = Ar, Alk, Vinyl, SR, NR_2

The addition of sodium azide to nitriles to give 1*H*-tetrazoles is shown to proceed readily in water with zinc salts as catalysts. The scope of the reaction is quite broad; a variety of aromatic nitriles, activated and unactivated alkyl nitriles, substituted vinyl nitriles, thiocyanates, and cyanamides have all been shown to be viable substrates for this reaction.

Part 2: Analyzing Writing across the Abstract

In this section, we summarize writing practices that span the entire abstract. Some of these have been mentioned previously but are repeated here for convenience.

- Abbreviations and acronyms: *The ACS Style Guide* suggests that writers use abbreviated terms sparingly in abstracts; they should be used mainly to minimize awkwardness and needless repetition. When abbreviated terms are used in the abstract, they should be defined at first use (unless they are too common to require definition). For those that are defined, they must be defined again at their first mention in the paper.

- Citations: *The ACS Style Guide* states that references should not be cited in the abstract. If a citation cannot be avoided, the full reference must be included so that the abstract may stand on its own.
- Error terms and units: If data are presented quantitatively in the abstract, proper units and errors terms (when appropriate) should be included.
- Formatting: If you glance through several different chemistry journals, you will notice that abstracts are easily identifiable; they are often set apart in some way from the body of the paper by using a different font, margins, or spacing. However, you need not (and should not) attempt to "copy" the formatting that you see in published works as you prepare your manuscript. Unless otherwise stated, the abstract should be double spaced and in the same font as the rest of the manuscript. Typically the first page of the submitted manuscript comprises the title, author list, and abstract; the Introduction section begins on the second page. If your paper is accepted, the journal editors may give you additional formatting instructions to prepare for publication.
- Keywords: The repetition of key terms and concepts in the abstract and title is commonplace. Because both the title and abstract are often used for computer searches, it is important that key terms be included in both.
- Vendors: Instrumentation is identified in the abstract but not vendors or brand names.
- Verb tenses: Verb tenses in the abstract are consistent with conventions used in other sections of the journal article. Past tense is used to refer to work completed in the past and to describe results:

  Past tense

  Antioxidative compounds *were isolated*.
  A number of cigarettes *were tested*.
  All cigarette types tested *showed* the presence of...

  Present tense is used to make statements of fact, to identify information reported in the paper, and to state beliefs or interpretations expected to be true over time:

  Present tense

  Atrazine *is* a commonly used herbicide.
  Diastereoselective Strecker reactions *are* reported.
  The results *provide* the first example of...
  The study undertaken *opens* several possibilities.

Present perfect can be used to summarize the work of others, demonstrate a gap, introduce one's own work, and/or report principal findings:

Present perfect

Studies *have shown* that...
...in vivo toxicity *has not* yet *been proved.*
Here, for the first time, three different dinitroaniline pesticides *have been shown*...

- Voice. Both passive and active voice are used in abstracts, although passive voice is more common:

Passive voice (more common)

Although cyclodextrins *are used* in soil decontamination...
The effects of RAMEB concentrations on clay minerals *were studied*...
A saponification-HSSPME procedure *has been developed*...

Active voice (less common)

The scope of the reaction *is* quite broad.
Variations in temperature *did not affect*...
The results *provide* the first examples...

- We. The word *we* is used only rarely in abstracts; when used, it usually refers to work that the authors present in the paper:

We report the chemical synthesis of...
We present calculated thermodynamic values for...

Present Perfect

See table 6.2.

Tense–Voice Combinations

See tables 4.1, 5.1, and 6.5.

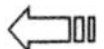

Exercise 7.5

Read excerpts 7E–7H (abstracts from four articles examined in chapters 3–6).

a. Consult figure 7.1. Which moves and submoves are accomplished within the first two sentences of each abstract?

b. Find at least two examples in these abstracts that support and/or refute each general writing practice listed in part 2 (excluding error terms, formatting, and keywords).

c. Excerpts 7E–7H were published in journals that do not require keywords. Using the abstracts and titles as guides, suggest three keywords for each abstract.

d. Which excerpt ends with a concluding remark that goes beyond reporting results?

Excerpt 7E (from Jozefaciuk et al., 2001)

Abstract

Although cyclodextrins are increasingly used in soil decontamination, little is known about their effects on soil physicochemical properties. In this work, the surface and pore properties of randomly methylated β-cyclodextrin (RAMEB) and three typical clay minerals were characterized, and the effects of RAMEB concentrations on clay minerals were studied using water vapor adsorption-desorption and mercury intrusion porosimetry techniques. As compared to clay minerals, for pure RAMEB, very large surface area and volume of nanometer-size pores (micropores) were determined. Energy of interaction with water vapor, volume of micrometer-size pores (mesopores), and fractal dimensions in both pore size ranges of RAMEB were lower than those of the minerals. When increasing amounts of RAMEB were added to the minerals, the surface area and micropore volume decreased and adsorption energy increased. The volume of mesopores decreased after RAMEB treatments for bentonite and kaolin and increased for illite. As deduced from the fractal dimensions increase, the pore structure of the minerals became more complex with RAMEB addition. The observed changes were in general contrary to those expected when RAMEB and minerals coexist as separate, nonreactive phases and suggested strong interaction of RAMEB with clay minerals. (190 words)

Excerpt 7F (from Llompart et al., 2001)

Abstract

A saponification-HSSPME procedure has been developed for the extraction of PCBs from milk samples. Saponification of the samples improves the PCB extraction efficiency and allows attaining lower background. A mixed-level fractional design has been used to optimize the sample preparation process. Five variables have been considered: extraction time, agitation, kind of microextraction fiber, concentration, and volume of NaOH aqueous solution. Also the kinetics of the process has been studied with the two fibers (100-μm PDMS and 65-μm PDMS-DVB) included in this study. Analyses were performed on a gas chromatograph equipped with an electron capture detector and a gas chromatograph coupled to a mass selective detector working in MS-MS mode. The proposed method is simple and rapid, and yields high sensitivity, with detection limits below 1 ng/mL, good linearity, and reproducibility. The method has been applied to liquid milk samples with different fat content covering the whole commercial range, and it has been validated with powdered milk certified reference material. (159 words)

Excerpt 7G (adapted from Dellinger et al., 2001)

Abstract

Exposure to airborne fine particles ($PM_{2.5}$) is implicated in excess of 50 000 yearly deaths in the USA as well as a number of chronic respiratory illnesses. Despite intense interest in the toxicity of $PM_{2.5}$, the mechanisms by which it causes illnesses are poorly understood. Because the principal source of airborne fine particles is combustion and combustion sources generate free radicals, we suspected that $PM_{2.5}$ may contain radicals. Using electron paramagnetic resonance (EPR), we examined samples of $PM_{2.5}$ and found large quantities of radicals with characteristics similar to semiquinone radicals. Semiquinone radicals are known to undergo redox cycling and ultimately produce biologically damaging hydroxyl radicals. Aqueous extracts of $PM_{2.5}$ samples induced damage to DNA in human cells and supercoiled phage DNA. $PM_{2.5}$-mediated DNA damage was abolished by superoxide dismutase, catalase, and deferoxamine, implicating superoxide radical, hydrogen peroxide, and the hydroxyl radical in the reactions inducing DNA damage. (147 words)

Excerpt 7H (from Plaper et al., 2002)

Abstract

Trivalent chromium is a metal required for proper sugar and fat metabolism. However, it has been suggested that it causes DNA damage in in vitro test systems, although in vivo toxicity has not yet been proved. In the present study, the effect of Cr^{3+} on bacterial cells was tested with the Pro-Tox (C) assay, and its cellular uptake was measured with flame atomic absorption spectroscopy. The potential genotoxicity of Cr^{3+} was further examined by the study of its influence on a bacterial type II topoisomerase. Cr^{3+} was shown to cause DNA damage and inhibit topoisomerase DNA relaxation activity, probably by preventing the formation of the covalent link between enzyme and double helix. In addition, Cr^{3+} decreases the viability and/or proliferation rate of eukaryotic cells such as murine B16 melanoma cells and human MCF-10A neoT ras-transformed human epithelial cells. The possible implication for Cr^{3+} intake by humans is discussed. (148 words)

Exercise 7.6

Below is an abstract from the literature (adapted from Cortes et al., 2006). The text has been divided into four passages and their order scrambled. Based on the move structure in figure 7.1, suggest the correct order for passages a–d.

a. Pesticide residues were extracted from samples with a small amount of ethyl acetate and anhydrous sodium sulfate. No additional concentration and cleanup steps were necessary. Analyses were performed by large volume GC injection using the through oven transfer adsorption desorption (TOTAD) interface.

b. A simple, rapid, and sensitive multiresidue method has been developed for the determination in vegetables of organophosphorus pesticides commonly used in crop protection.

c. Results are reported for the analyses of eggplant, lettuce, pepper, cucumber, and tomato.

d. The calculated limits of detection for each pesticide injecting 50 μL of extract and using an NPD were lower than 0.35 μg/kg, which is much lower than the maximum residues levels (MRLs) established by European legislation. Repeatability studies yielded a relative standard deviation lower than 10% in all cases.

Exercise 7.7

The following abstract (adapted from Hageman et al., 2006) has been determined to be too long by its authors (221 words). The authors would like to reduce it to fewer than 200 words and, if possible, to 185 words. Rewrite the abstract to make it more concise.

Abstract

The United States National Park Service has initiated an extensive research campaign on the atmospheric deposition and fate of semi-volatile organic compounds in its alpine, sub-Arctic, and Arctic ecosystems in the Western U.S. Results of the analyses of pesticides in seasonal snowpack samples collected in spring 2003 from seven national parks are presented in the Results section of this paper. From a target analyte list of 47 pesticides and degradation products, the most frequently detected current-use pesticides were dacthal, chlorpyrifos, endosulfan, and α-hexachlorocyclohexane, whereas the most frequently detected historic-use pesticides were dieldrin, α-hexachlorocyclohexane, chlordane, and hexachlorobenzene. The results of several tests are described in this paper to help to explain what sources are responsible for our pesticide results. Correlation analysis with latitude, temperature, elevation, particulate matter, and two indicators of regional pesticide use reveals that regional current and historic agricultural practices are largely responsible for the distribution of pesticides in the national parks in this study. Pesticide deposition in the Alaskan parks is attributed to long-range transport because there are no significant regional pesticide sources. The percentage of total pesticide concentration due to regional transport (%RT) was calculated for the other parks; %RT was highest at parks with higher regional cropland intensity and for pesticides with lower vapor pressures and shorter half-lives in air. These results have many important implications.
(221 words)

7C Writing on Your Own: Write Your Abstract

Using figure 7.1 and the abstracts that you have collected from the literature as guides, write each part of your abstract. Remember to use keywords and to make your text concise and informative rather than descriptive. Include units and error bars when appropriate. Abide by the word limit specified by your targeted journal.

Analyzing Titles

We conclude this chapter and module with a brief look at titles, often the last part of a journal article to be written. The title of a journal article must be as concise, specific, and informative as possible. Also, because the title is written for an expert audience, the title should be formal. Although an informal, catchy title is appropriate in many genres (e.g., in newspaper headlines and popular science articles), it is inappropriate in a journal article.

A common organizational pattern for titles (X of Y by Z) was presented in table 7.1. According to this pattern, titles often begin with a nominalization or phrase (X) that modifies what was studied (Y). Avoid preceding X with empty words (e.g., "The," "A," "An") or redundant phrases (e.g., "A Study of," "Research on"); such redundancy will make your title wordy, without adding relevant information.

Wordy The Preparation of 5-Substituted 1*H*-Tetrazoles...
A Study of the Cancer-Protective Properties of...
An Analysis of Aldehydes in Beer Using...

Better Preparation of 5-Substituted 1*H*-Tetrazoles...
Cancer-Protective Properties of...
Analysis of Aldehydes in Beer Using...

Capitalization in Titles

Capitalize all main words (nouns, pronouns, verbs, adjectives, and adverbs).

Do not capitalize prepositions or the following words (unless they are used as the very first word in a title):

a	and	nor	so
an	but	or	the

See appendix A for more on capitalization.

Two- and three-word modifiers (often requiring the use of hyphens) are common in titles because they can make the title more concise. The hyphenated words may be used in the X, Y, and/or Z parts of the title, as illustrated below. (Note, too, that none of these titles begin with the word "The.")

X Continuous-Flow pI-Based Sorting of Proteins and Peptides in a Microfluidic Chip Using Diffusion Potential (from Song, Y.-A., et al., 2006)

X, Y Synthesis and Self-Assembling Properties of Diacetylene-Containing Glycolipids (from Nie and Wang, 2006)

Y Atmospheric Deposition of Current-Use and Historic-Use Pesticides in Snow at National Parks in the Western United States (from Hageman et al., 2006)

Y Comparison of Odor-Active Volatile Compounds of Fresh and Smoked Salmon (from Varlet et al., 2006)

Y Inhibition of Hemoglobin- and Iron-Promoted Oxidation in Fish Microsomes by Natural Phenolics (from Pazos et al., 2006)

Z Sampling and Determination of Formaldehyde Using Solid-Phase Microextraction with On-Fiber Derivatization (from Martos and Pawliszyn, 1998)

Z Determination of Polychlorinated Biphenyls in Milk Samples by Saponification—Solid-Phase Microextraction (from Llompart et al., 2001)

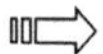

Two-Word Modifiers

See appendix A.

Colons are also common in titles. If you use colons, be sure that the segment of the title that precedes the colon can stand alone as the full title. The passage after the colon may or may not stand alone. It is customary to capitalize the first word after the colon, as if it were the start of a new sentence. Here are a few examples:

A Simple and Rapid Assay for Analyzing Residues of Carbamate Insecticides in Vegetables and Fruits: Hot Water Extraction Followed by Liquid Chromatography-Mass Spectrometry (from Bogialli et al., 2004)

Arsenic Contamination of Bangladesh Paddy Field Soils: Implications for Rice Contribution to Arsenic Consumption (from Meharg and Rahman, 2003)

Effect of Six Decades of Selective Breeding on Soybean Protein Composition and Quality: A Biochemical and Molecular Analysis (from Mahmoud et al., 2006)

Accurate Inertias for Large-Amplitude Motions: Improvements on Prevailing Approximations (from Wong et al., 2006)

To make your title informative, be as specific as possible and avoid words that do not convey the specific content of your study. Also, remember to use keywords in your title. Because keywords promote effective literature retrieval, they should be used liberally in the title. In the following examples, words that were listed as keywords in the abstract are italicized.

Noncovalent Cross-Linking of *Casein* by *Epigallocatechin Gallate* Characterized by *Single Molecule Force Microscopy* (from Jöbstl et al., 2006)

Factors Affecting Transfer of *Polycyclic Aromatic Hydrocarbons* from Made *Tea* to *Tea Infusion* (from Lin et al., 2006)

Herbicidal Effects of *Soil-Incorporated Wheat* (from Mathiassen et al., 2006)

The ACS Style Guide recommends that authors spell out most terms in titles (except for common abbreviations, e.g., NMR, DNA, and UV). Some journals allow a few additional terms to be used in titles without definition (e.g., FTIR, PCBs, GC/MS, and PAHs). In the following examples, NMR, GC/MS, and PCB are not defined; PAH is defined in one title but not in two others, and MDAM is defined.

Monofluorinated Analogues of Polybrominated Diphenyl Ethers as Analytical Standards: Synthesis, NMR, and GC-MS Characterization and Molecular Orbital Studies (from Luthe et al., 2006)

Atmospheric PCB Concentrations at Terra Nova Bay, Antarctica (from Gambaro et al., 2005)

Biodegradation, Bioaccessibility, and Genotoxicity of Diffuse Polycyclic Aromatic Hydrocarbon (PAH) Pollution at a Motorway Site (from Johnsen et al., 2006)

Molecular Simulations of Benzene and PAH Interactions with Soot (from Kubicki, 2006)

Application of Multi-Component Damage Assessment Model (MDAM) for the Toxicity of Metabolized PAH in *Hyalella azteca* (from Lee and Landrum, 2006)

When in Doubt about Abbreviations

If you are unsure about abbreviations in your title or abstract, search for keywords and abbreviations in published abstracts using the ACS Journals Search. In this way, you can determine common practices in your target journal.

 **Exercise 7.8**

Rewrite the following titles so that they better conform to the writing guidelines presented in the chapter.

a. A Study Of How Fermentation and Distillation affect the Oxygen-18/Oxygen-16 Isotope Ratio in Ethanol
b. Are There Chlorogenic Acids and Lactones in Your Caffeinated and Decaffeinated Coffees? HPLC-MS has the Answer
c. Tree Bark: An Analysis of How Brominated Flame Retardants Impact Tree Bark in North America

 **Exercise 7.9**

Using only the information provided, propose titles for the following journal articles. Use the X of Y by Z pattern, whenever possible.

a. A paper that reports the antioxidant activity of the phenolic fraction in extra virgin olive oil (EVOO) using electrochemical methods
b. A paper that describes a novel method used to synthesize quinolines by incorporating allenyl cations in a catalytic intermolecular Friedel–Crafts reaction
c. A paper that describes the results of an inhalation study using Sprague–Dawley rats to investigate the toxicity of low levels of 1,3-butadiene

7D Writing on Your Own: Write Your Title

Identify keywords to include in your title. Then, using table 7.1 as a guide, write a title for your paper, making every attempt to follow the X of Y by Z pattern. Make sure that your title is concise, specific, and informative.

Consider your use of capitalization, colons, and abbreviations. Do they follow recommended guidelines and/or common practices, as seen in others' titles?

Reread your title. Does it capture the essence of your paper, including content and emphases? Have you incorporated keywords in the title?

7E Writing on Your Own: Practice Peer Review

Imagine that a friend has asked you to review an abstract and title that are in draft form (see "Peer Review Practice: Title and Abstract" at the end of the chapter for instructions, background information, and the draft). Based on what you have learned in this chapter, read the draft and offer written feedback.

7F Writing on Your Own: Fine-Tune Your Abstract and Title

After practicing the peer review process with the abstract and title at the end of the chapter, solicit feedback on your own abstract and title from a peer.

Use the Peer Review Memo on the *Write Like a Chemist* Web site to exchange feedback. Use the feedback received to make final changes in your work.

After those changes are made, you will be ready to complete your journal article so that it reads as a single, unified document (with all its essential components, including the title and abstract, Introduction, Methods, Results, and Discussion sections, and references). See chapter 17 for details on formatting references and chapter 18 for hints on finalizing your written work.

Congratulations! You've just about completed your journal article.

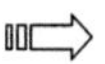

Finalizing Your Written Work

See chapters 17 and 18.

Chapter Review

As a self-test, check what you've learned in this chapter by explaining the purpose and characteristics of each of the following terms to a friend who is new to the field:

abstract	keywords
CAS	title

Also explain the following to a friend in chemistry who has not yet given much thought to the final steps of finishing a journal article, specifically writing an abstract and composing a title:

- Moves of an abstract
- Use of tense in an abstract
- Use of passive and active voice in an abstract
- X of Y by Z pattern commonly used in titles
- Use of capitalization in titles
- Relationship between a title, an abstract, and a computer search

Additional Exercises

Exercise 7.10

The following abstract (adapted from an original source, identified in the Instructor's Answer Key) has 277 words, but it must have no more than 250 words to be submitted for publication.

a. Edit the abstract so that it has no more than 250 words.
b. Propose a title for this paper based on the content of the abstract. Use the standard X of Y by Z pattern, if possible. Follow capitalization rules when finalizing your title.

Abstract. The explosion and collapse of the World Trade Center (WTC) was a catastrophic event that produced an aerosol impacting many workers, residents, and commuters during the first few days after September 11, 2001. During the initial days that followed the collapse, 14 bulk samples of settled dust were collected at locations surrounding the epicenter of the disaster, including a single location that was located inside of a building. Some of the samples collected from these various sites were analyzed for a variety of different potential hazards, such as inorganic and organic constituents, as well as morphology. The results of analyses for a wide variety of persistent organic pollutants are described herein, including polycyclic aromatic hydrocarbons, polychlorinated biphenyls, and select organochlorine pesticides on settled dust samples. The Σ_{86}-PCBs comprised less than 0.001% by mass of the bulk in the three bulk samples analyzed indicating that PCBs were of limited significance in the total settled dust across lower Manhattan. Likewise, organochlorine pesticides, including chlordanes, hexachlorobenzene, heptachlor, 4,4′-DDE, 2,4′-DDT, 4,4′-DDT, and Mirex, were found at low concentrations in the bulk samples. Conversely, the Σ_{37}-PAHs comprised up to nearly 0.04% (<0.005–0.039%) of the bulk in the six bulk samples analyzed. Further size segregation of three initial bulk samples and seven additional samples indicated that Σ_{37}-PAHs were found in higher concentrations on relatively large particles (10–53 μm), representing up to 0.04% of the total dust mass. High concentrations were also found on fine particles (<2.5 μm), often accounting for ~0.005% by mass. Taking all of these many factors into consideration, we estimate that approximately 100–1000 tons of Σ_{37}-PAHs were spread over a localized area immediately after the WTC disaster on September 11. (277 words)

Exercise 7.11

Read and revise the following abstracts so that they are more in line with the abstracts presented in this chapter:

Abstract. Laponite and tempamine were used to compose thin films that were studied by electron paramagnetic resonance and AFM. These studies prove that Laponite films are oriented and that orientation disappears with age. Laponite films create ordered barriers that confine movement of incorporated molecules. Further studies, summarized in the Introduction, involve the investigation of films assembled with additional compounds, such as polymers, that are envisaged to produce ordered films with useful mechanical properties.

Abstract. Grignard reactions are of utmost importance in organic synthesis (Lee, 2005), and finding the prime conditions under which to conduct these reactions is really crucial to their usefulness. This work looks at the effects of time, and temperature on yield in the reaction of isopropyl magnesium bromide with 4-methoxybenzaldehyde to produce 1-(4-methoxyphenyl)-2-methylpropan-1-ol. The reaction was first run for 10 min at 25, 50, 75, and 80 °C. Next it was run at 80 °C for 10, 20, and 30 min (see Methods section for more details). Highest yields (85%) were obtained at 80 °C with 10–20 min reaction times. Utilizing conditions that optimize yields will improve the economic practicality of these reactions, and increase their usefulness as a synthetic tool.

Exercise 7.12

Reflect on what you have learned about writing an abstract, a title, and the journal article as a whole, now that you are just about done writing your own journal article. Select one of the reflection tasks below and write a thoughtful and thorough response.

a. Reflect on the importance of the abstract and title of a journal article.
   - In what ways do the abstract and title prepare readers for the contents of the paper?
   - What are the keys to an effective abstract and title?
   - What makes writing an abstract and title challenging?
   - How might you minimize the challenge(s) to make the task easier?

b. Reflect on your experience writing a journal article, from start to finish.
   - How have your views of scientific writing changed while writing your journal article?
   - What have you learned that has assisted you most in writing your journal article?
   - What have you learned that will make you a better reader of the professional literature?
   - How successful have you been with your journal article? What will you do in the future to write an even more effective journal article?

Peer Review Practice: Title and Abstract

Imagine that a friend has asked you to review the title and abstract of a paper that is being written with other researchers on chromated copper arsenate (CCA), a compound used to preserve wood. The research team examined the chemical structure of arsenic (As) and chromium (Cr) in CCA to determine if the oxidation state of As and Cr changed over time due to weathering.

Using parts 2 and 3 of the Peer Review Memo on the *Write Like a Chemist* Web site, review the title and abstract below. Provide specific suggestions in your memo to help your friend improve the title and abstract. (The title and abstract below are adapted from an original source, noted in the Instructor's Answer Key.)

The Implications of Environmental Weathering: The Chemical Structure of Arsenic and Chromium in Wood Treated with CCA

1 X-ray Absorption Spectroscopy (XAS) is used to evaluate the chemical structure of As
2 and Cr in three samples of CCA-treated materials: newly treated wood, aged wood (5
3 years as decking), and dislodgeable residue from aged (1–4 years as decking) CCA-
4 treated wood. Chromated copper arsenate (CCA) has been used to treat lumber for over
5 60 years to increase the expected lifetime of wood. Since arsenic and chromium are
6 involved in CCA-treated wood, attention has become focused on the potential risks of
7 this practice. In particular, exposure of children to arsenic from CCA-treated wood used
8 in decks and play sets has received considerable attention. We found several important
9 findings in this study. First, Cr and As have the same forms in fresh and aged CCA-
10 treated materials and in dislodged residue. In all cases, the dominant oxidation states
11 are: As(V) and Cr(III). Second, the local chemical environment of the two elements
12 is best represented as a Cr/As cluster consisting of a Cr dimer bridged by an As(V)
13 oxyanion. Long-term stability of the As/Cr cluster is suggested by its persistence from
14 the new wood through the aged wood and the dislodgeable residue.

Module 2

The Scientific Poster

8 *Writing the Conference Abstract and Title*

All too familiar words from research mentor to student: "I'd like you to submit an abstract to the upcoming ACS conference on your research. The submission deadline is in two weeks. Unfortunately, I'll be out of town, but I'm sure you can write it on your own."

This chapter focuses on writing a **conference abstract**, not to be confused with a journal article abstract, which was addressed in chapter 7. A conference abstract is submitted in response to a Call for Abstracts issued by conference organizers, typically 3–6 months before the conference takes place. The abstract is reviewed and, if accepted, allows the authors to present a **contributed paper** (as either an oral or poster presentation) at a conference session. Only accepted papers may be presented at a conference; whether they are oral or poster presentations is a decision usually made by conference organizers. By the end of this chapter, you will be able to do the following:

- Describe the major purposes of the conference abstract
- Recognize a typical organizational structure for the conference abstract
- Identify common writing conventions in the conference abstract
- Write a concise and informative conference abstract with title and author list

The Writing on Your Own tasks throughout the chapter will guide you as you do the following:

8A Prepare to write

8B Decide on an organizational structure

8C Draft your conference abstract and title

8D Fine-tune your conference abstract and title

Conference Abstract

A short text, written in response to a Call for Abstracts, that describes the work to be presented at a conference. Minimally, the abstract includes a title, author list and affiliations, and a short description of the work to be presented.

Contributed Paper

An oral or poster presentation given at a conference. A paper can only be presented after an abstract is submitted by the authors and then accepted by conference organizers.

Conference Session

Because most chemistry conferences address multiple areas of chemistry, abstracts are usually submitted to a division of chemistry (e.g., Division of Organic Chemistry) and to a specific session within that division (e.g., Asymmetric Reactions and Syntheses Session of the Division of Organic Chemistry).

Writing a conference abstract is often the first professional genre that novice writers tackle on their own, with little, if any, guidance from their research mentors. This chapter is intended to prepare you for the task by focusing on the text of the abstract. (If your abstract is accepted and results in a poster presentation, see chapters 9 and 10 for guidance in preparing the poster.)

Like the journal article abstract, most conference abstracts have strict word limitations (typically between 150 and 200 words). Adherence to this limit is expected; your abstract will likely be rejected if the limit is exceeded. Many other formatting specifications must also be followed. For example, most instructions include rules for formatting your title, author list, graphics, and special characters (e.g., μm, α, $\leq$, m^3). The reason for such specificity is that your abstract, if accepted, will be printed (as is) in **conference proceedings**. Adherence to these guidelines by all authors ensures that the proceedings appear uniform and professional.

Conference Proceedings

A written record of conference sessions and events, including a schedule and abstracts of all conference presentations.

Because they both include the word "abstract", you may think that a conference abstract is just like a journal article abstract. In fact, a conference abstract combines features of both the Introduction section and abstract of a journal article. Like the Introduction, a conference abstract introduces work that will be presented (in this case, at a conference). Like the journal article abstract, a conference abstract highlights significant findings.

Reading and Analyzing Writing

We begin by asking you to read and analyze two excerpts. Excerpt 8A includes a partial list of instructions available to authors planning to submit an abstract to the 233rd American Chemical Society (ACS) national meeting, held March 2007. Excerpt 8B is an abstract for a paper that was presented at that meeting.

Exercise 8.1

As you read excerpts 8A and 8B, complete the following tasks:

a. Revise the following title so that it adheres to conference guidelines:

   The Factors That Affect Indoor Air Quality: Exposures to Carbonaceous Aerosols

b. The abstract contains seven sentences (149 words). Briefly state the purpose of each sentence. Based on these purposes, propose a move structure for the conference abstract.

c. Identify the intended audience(s) for the abstract. Give reasons for your answer.

d. Articulate two rules, one on the use of abbreviations and the other on the use of personal pronouns (e.g., *I*, *our*, *we*) in the abstract.

e. Identify the verb tenses used in excerpt 8B. State the function(s) of each tense.

f. Specify two or three keywords from the title and text of excerpt 8B that would facilitate an electronic search, permitting easy retrieval and/or indexing of the abstract.

Excerpt 8A (adapted from Instructions for Authors. On Line Abstract Submission System for ACS Conferences (OASYS). http://oasys.acs.org/acs/233nm/oasys.htm (accessed January 2008))

Abstracts should be 150 words or less. Abstracts may contain one or more graphics. The number of words you can submit may depend on the size of your graphic file (the program will automatically enforce a size restriction), but if you exceed the maximum word count because your graphic is too large, you can scale your graphic in OASYS.

Styles to Observe

The abstract should be entered using the following style:

- The title is in sentence case, with only the first letter of the title capitalized, except words that are proper nouns, acronyms, or words that follow a colon. Here's an example:

  My theories on recombinant DNA: Or, how I spent my summer vacation

- The title is not in bold or italics, except as needed (e.g., foreign words).
- The title does not begin with The or A (these will be removed by our editors).
- The abstract text does not contain the word "abstract" or any author names.
- The abstract text does not contain footnotes.
- The author names do not contain titles or suffixes (suffixes such as "III" can be used).

Excerpt 8B (from Russell and Bahadur, 2007)

Predicting nanoparticle interfaces with molecular dynamics
Lynn Russell and Ranjit Bahadur. Scripps Institution of Oceanography, University of California, San Diego, CA

Nanoparticles exhibit physical properties that reflect the boundary between gaseous and condensed phases, sometimes resulting in behavior that is not predicted by bulk-phase approaches. Recent advances in nanoparticle measurements mean that our theoretical understanding sometimes lags our experimental observations of nanoparticle behavior. To address this gap, molecular dynamics (MD) simulations were designed to investigate the behavior of phase transitions in nanoparticles, in particular their water uptake. Our work addresses the dissolution of nanoparticle salt crystals in condensed water and the surface tension of nanoparticle interfaces that affect the water uptake process. MD simulations have shown that voids play a necessary role in initiating short time scale (1 ns) dissolution. We have also used MD simulations to predict the size dependence of liquid-vapor, solid-vapor, and solid-liquid interfaces for the NaCl-water-air systems. The results indicate Tolman lengths near 0.1 nm for liquid interfaces and below zero for a solid-vapor interface. (149 words)

Analyzing Audience and Purpose

A conference abstract is written for two distinct audiences. Conference organizers represent the first audience; they read the abstract and decide whether to accept it for a conference paper. Several factors influence this decision, not the least of which is the quality of writing in the abstract. A poorly written abstract often leads to a poorly prepared conference presentation; hence, such abstracts are easy for reviewers to dismiss. Other factors include the appropriateness of the abstract topic (does it fit the theme of the conference session?), the quality of the science, and how far the research has progressed. Conferences are forums for presenting new, even late-breaking, results. However, an abstract may be rejected if the work is too new or appears to lack sufficient data to ensure a quality presentation by the time of the conference. (If you find yourself using phrases such as "we will measure" or "we propose to analyze" repeatedly in your abstract, you may want to consider postponing your submission.)

Alternatively, an abstract should not be submitted if the content is too old. For example, an abstract describing work that is already published should not be submitted. Similarly, abstracts describing results that have already been presented at other national conferences should not be submitted, unless significant new progress has been made. Some overlap with previous presentations is allowed (after all, new science builds on past accomplishments), but a good rule of thumb is that at least 75% of the content should be new. (Note: This rule is often relaxed if the conference is local or university-based.)

The second audience comprises conference attendees, who read abstracts to determine whose talk to attend or poster to view. Because national conferences typically have multiple concurrent sessions, attendees rely on abstracts to make their decisions about what sessions to attend, where to go, and when. Individuals who read your abstract are likely to be in a related field of chemistry, but most likely will not be in your specific area of chemistry. Thus, it is important to keep your abstract general enough to be readily understood across different areas of science, thereby targeting primarily a scientific audience. In this regard, the conference abstract is more similar to the Introduction section of a journal article than to the abstract of a journal article, the latter written for an expert audience.

8A Writing on Your Own: Prepare to Write

Locate, and then read, the Call for Abstracts and Instructions for Authors for a conference that you plan to attend. Alternatively, use excerpt 8A for this assignment. How long should the abstract be? What special formatting is required? When is the deadline for submission?

Consider what aspects of your research you want to present. Classify your work into three groups: (1) completed work, (2) nearly completed work, and (3) work that you hope to complete. Base your abstract largely on work in the first two categories. If this is not your first abstract submission, be sure that you have not presented too much of this work at another national conference.

A typical move structure for the conference abstract is shown in figure 8.1. The first move identifies the research topic and suggests why the research is important. These two submoves parallel the first two submoves of the journal article Introduction section (see figure 6.1). The key difference is the absence of a third submove in the conference abstract; unlike the journal article Introduction, the conference abstract devotes little, if any, space to background information or a review of others' works. Recall that the primary purposes of the conference abstract are to tell readers about the work that you will present and to help them decide if they should attend your session. Thus, summarizing others' works is of little use in this regard. Because background information is deemphasized, there are typically no citations in a conference abstract. If citations are included, full references must also be provided so that the abstract can stand on its own.

Moves 2 and 3 of the conference abstract also parallel the moves of the journal article Introduction. Move 2 points out a gap in the field (e.g., a problem that needs to be solved, work that needs to be done) and serves as a transition to move 3, which describes the work to be presented. Depending on the goals of the project, different aspects of the work may be highlighted. In some instances, methods will be emphasized; in others, results will be the focus of attention. When possible and appropriate, numerical data should be included in the conference abstract. If this is not possible, because you are still completing parts of the work, at least state in the abstract what data will be presented at the conference (e.g., "Arsenic levels from two different lakes will be reported."). Results are the

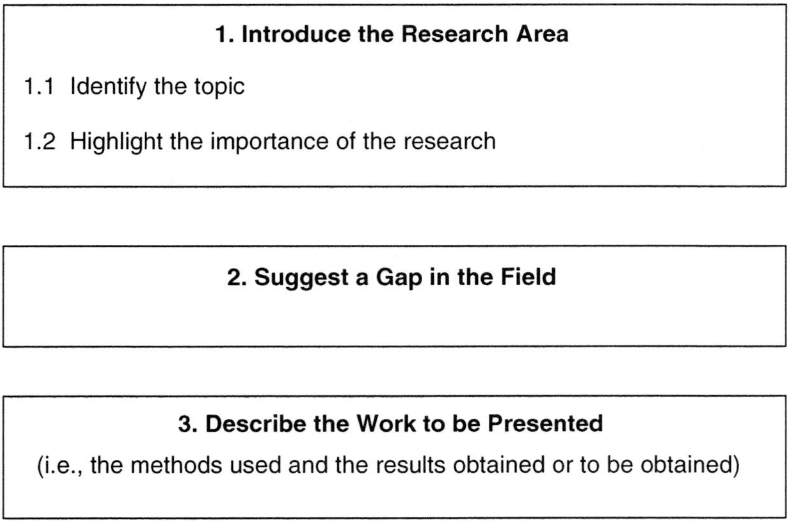

1. Introduce the Research Area

1.1 Identify the topic

1.2 Highlight the importance of the research

2. Suggest a Gap in the Field

3. Describe the Work to be Presented

(i.e., the methods used and the results obtained or to be obtained)

Figure 8.1 A visual representation of the move structure for a typical conference abstract.

punch line of the research story, and an abstract without a punch line is not very satisfying.

Authors who follow the moves depicted in figure 8.1 when writing their conference abstracts will meet most chemists' expectations, but variations are also quite common. For example, authors place varying degrees of emphasis on the different moves. Some authors focus on moves 1 and 2, while others concentrate on move 3. Some even skip moves 1 and 2 entirely. We examine a few such variations in the excerpts that follow.

Analyzing Conference Abstracts

We are now ready to read and analyze abstracts in more depth. We begin by looking at a set of abstracts to identify similarities and differences in organization, emphases, and content. We then summarize writing conventions that run across most conference abstracts.

Analyzing Excerpts

In this section, we analyze abstracts that were published in the proceedings of the 233rd ACS national meeting. Abstracts from three different divisions are examined: Division of Agricultural & Food Chemistry, Division of Environmental Chemistry, and Division of Organic Chemistry.

We begin by analyzing abstracts (printed in their entirety) that follow the move structure in figure 8.1. To facilitate analysis, we identify moves 1, 2, and 3 and include a word count at the end of each abstract. In this way, you can see for yourself how much emphasis the authors placed on each move. (Note that in some cases, 200-word abstracts were permissible, so do not be surprised if abstracts exceed 150 words.)

In excerpts 8C and 8D, moves 1 and 2 are accomplished in a single sentence. (In one case, two statements are linked by a semicolon; in the other case, two statements are linked by "but.") In just a few words, the topic is identified and a gap is suggested. The remainder of both abstracts is devoted to move 3. Methods are mentioned briefly, but results are emphasized and numerical values are included. Both abstracts conclude with a sentence that states the implications of the work—specifically, that litter mercury is accumulating in soils (excerpt 8C) and that stored beer maintains phytonutrients (excerpt 8D). Abstracts do not typically end in this way, in part, because when abstracts are written, implications are usually not known. However, if implications are known, they are often included. (Implications of the work are nearly always addressed in the final oral or poster presentation.)

Exercise 8.2

Read excerpts 8C and 8D and answer the following questions:

a. Do the authors transition from moves 1 and 2 to move 3 in the same way or in a different way? Explain.

b. Identify the numerical values (and units) that are reported in each abstract. What formatting conventions are followed?

c. What verb tense is used in the last sentence of each abstract? Justify this choice.

Excerpt 8C (from Bushey et al., 2007)

[*Move 1*] Plant leaf tissue has been documented to contain significant amounts of mercury (Hg); [*Move 2*] however, the role of leaf tissue Hg relative to atmospheric deposition and soil pools of Hg is not well established. [*Move 3*] A quantitative investigation was conducted to assess the role of plants in the deposition and fate of mercury within an upland forest watershed and the potential implications for soft-water lake ecosystems of the northeastern USA. Plant tissue samples were collected over a two-year period. Leaf mercury content increased approximately 10-fold over the growing season with average uptake rates of 0.21–0.35 ng/g-day. Uptake varied by species and was consistent between the two growing seasons. Leaf total Hg content reached 47–62 ppb within fresh litterfall samples. An annual flux of 180 mg of total Hg per hectare was estimated, representing the largest ecosystem input of Hg. Hydrologic modeling of upland runoff and litter decomposition results suggest that this litter mercury is accumulating within upland soils. (158 words)

Excerpt 8D (from Rohrer and Majoni, 2007)

[*Move 1*] The health benefits of beer phytonutrients, such as reduction in coronary heart disease, have been reported, [*Move 2*] but little information is available on changes in phytonutrients under household refrigerated storage. [*Move 3*] The objective of this study was to determine the phytonutrient concentration in non-alcoholic beer beverages and one alcoholic beer during 60-day storage. Phytonutrient concentration was evaluated as the total polyphenol content and the flavonoid content [(+)-catechin and (–)-epicatechin]. Results found a significant increase ($p < 0.05$) in total polyphenol content in all beer beverages except one non-alcoholic beer. Overall, total polyphenol (505 mg/L) and (+)-catechin content (0.52 mg/L) in O'Doul's non-alcoholic beer were greater than in the other non-alcoholic beverages. This increase during storage indicates that consuming stored beer beverages after two months still allows health benefits to be attained. (129 words)

Next, consider excerpt 8E. Like the abstracts in excerpts 8C and 8D, this abstract accomplishes moves 1 and 2 quickly, in just two sentences. However, because this abstract is about a new method, the authors use most of the abstract to describe that method (Köhler theory analysis), and no results are reported. The authors do indicate, however, that results from a diverse set of aerosol sources will be presented at the conference.

Exercise 8.3

Read excerpt 8E and answer the following questions:

a. What phrase do the authors use to signal the end of the description of their new method?

b. What phrase do the authors use to transition from the description of their method to the results that they intend to present at the conference?

c. The authors use present tense to describe their methods (e.g., "surfactant properties are characterized") rather than past tense ("surfactant properties were characterized"). Find two more examples of present tense in the methods description. Propose a reason for this verb tense choice.

Excerpt 8E (from Nenes et al., 2007)

[*Move 1*] Quantifying the impact of water soluble organic compounds (WSOC) on cloud droplet formation constitutes a substantial source of uncertainty in aerosol-cloud climate interaction studies. [*Move 2*] This uncertainty is a consequence of the plethora and complexity of the compounds that constitute atmospheric organic matter; methods for characterizing its interaction with water vapor are few and require further development. [*Move 3*] This study focuses on a new method, termed "Köhler theory analysis", to characterize the average solubility, molar mass, surfactant characteristics, and droplet growth kinetics of minute amounts of WSOC samples typically collected from an aerosol sampler. Surfactant properties are characterized by surface tension and contact-angle measurements with a pendant drop method tensiometer. Water soluble mass is characterized by a functional group analysis, by separating the samples into hydrophilic, hydrophobic, and deionized components. Finally, the droplet growth kinetics and cloud droplet formation potential of all carbonaceous samples are measured using a Continuous Flow Streamwise Thermal Gradient Cloud Condensation Nucleus counter. We will present results from a diverse set of sources such as rural biomass burning, urban Atlanta aerosol, Mexico City aerosol, and in situ stratocumulus cloudwater samples collected aboard the CIRPAS Twin Otter. (188 words)

Exercise 8.4

Read excerpt 8F. While doing so, complete the following tasks:

a. Identify moves 1 and 2. (Note: move 2 is not a typical gap statement but does imply work that needs to be done.)

b. Consider move 3. Is it more in line with excerpts 8C and 8D (which focus principally on results), excerpt 8E (which focuses mainly on methods), or neither? Justify your answer.

c. Comment on the authors' use of personal pronouns (e.g., *I*, *we*, *our*), verb tense (past, present, and/or future), and voice (active and/or passive).

d. Repeat (c) above for the abstracts in excerpts 8B–8E. What similarities and differences do you notice across all abstracts examined thus far in the chapter?

Excerpt 8F (from Zuo et al., 2007)

Flavonoids are an important natural pigment and are widely distributed in vegetables, berries, and fruits. Interest in the separation and determination of flavonoid and other phenolic compounds in plants has increased in past decades because these compounds have definitive anticarcinogenic and cardioprotective effects on humans. In this work, two flavonol glycosides, quercetin galactoside and quercetin arabinoside, were identified in American cranberry fruit. Analyses included separation, hydrolysis, and structure elucidation of flavonol glycosides. The separation was carried out by solvent extraction, thin-layer chromatography, and high performance liquid chromatography (HPLC). After hydrolysis of the obtained flavonol glycosides, flavonol aglycones and sugars were identified by HPLC and gas chromatography-mass spectrometry (GC-MS), respectively. (109 words)

We now consider the abstract presented in excerpt 8G, which reverses the emphases observed in excerpts 8C–8F. In excerpt 8G, moves 1 and 2 comprise the bulk of the abstract, and move 3 is just a single sentence. Much of the abstract is used to elucidate important gaps in the field: that (1) a size-based standard for regulating atmospheric particulate matter is inadequate because it does not take into account chemical composition, and (2) compositional analyses are difficult, especially for metals, because they often exist in multiple oxidation states. These gaps help to establish the relevance of the authors' work. Only in the last sentence do the authors mention their work specifically (move 3). Here, they identify their method (micro-XANES) and tell readers what results will be presented at the conference.

Excerpt 8G (from Nico et al., 2007)

[*Move 1*] Currently atmospheric particulate matter is regulated based on various size categories because of the apparent association between particle size and adverse health effects. [*Move 2*] However, the current size-based understanding of atmospheric particles is relatively crude because it does not account for differences in the chemical composition of these particles. Presumably a chemically reactive particle has a greater potential for damage than a chemically inert particle of comparable size. Of the metals potentially released as aerosols, Cr is of particular interest because of its potential to exist in at least three oxidation states of very different toxicities: metallic Cr, which is used in a variety of industrial processes, Cr(III), which is an essential, non-toxic micronutrient, and Cr(VI), which is a strong oxidizer, highly toxic and carcinogenic. [*Move 3*] We have employed micro-X-ray absorption near edge spectroscopy (micro-XANES) to determine the chemical form of chromium in ambient $PM_{2.5}$ collected from several locations in northern California. (150 words)

We next consider excerpts 8H and 8I, two abstracts that describe chemical syntheses. In each case, roughly equal space is given to moves 1 and 2 and to move 3. Both abstracts indicate that successful syntheses have been accomplished, although details about the syntheses (e.g., stoichiometry, reaction conditions, and product yields) are not included in the abstract (but will be available at the conference). Note the inclusion of a graphic in excerpt 8I. Graphics are allowed in most abstracts, although their inclusion often limits the number of words that can be used. If you plan to include a graphic in your abstract, be sure to check instructions for how to prepare the graphic (i.e., software programs to use) and save the file (e.g., as HTML, GIF, or JPEG).

Excerpt 8H (from Jiao and Smith, 2007)

[*Move 1*] 2,2′-Bipyrroles are key synthetic precursors for porphycenes, corroles, saphyrins, and other expanded porphyrin analogs. Most of the current bipyrrole syntheses are mainly based on an Ullmann dimerization reaction of a preformed pyrrole or on oxidative coupling. [*Move 2*] However, the Ullmann reaction for bipyrrole synthesis generally requires high temperatures and thus few functional groups can survive the conditions. Moreover, the type of bipyrroles accessible from the Ullmann reaction and from oxidative coupling is very limited and many bipyrroles are still inaccessible or can only be obtained in low yields. Although some improvements have **recently** been made to bipyrrole synthesis, tedious synthetic routes are still involved and yields are low. [*Move 3*] We have developed an efficient novel synthetic route to synthesize bipyrroles and eventually porphycenes based on a Pd(0)-catalyzed reaction. In this novel synthetic route, the reaction can be performed at room temperature under very mild reaction conditions and provide good yields of bipyrroles in most cases. The synthesis of a series of bipyrroles using this novel synthetic route will be presented and the scope and limitations of the reaction will be discussed. Further usage of the bipyrroles in synthesis of porphycenes will also be reported. (192 words)

Recently

The ACS Style Guide recommends avoiding the word *recently* in articles and books. This recommendation is relaxed in conference abstracts. However, *recently* is a good word to cut if you are near the abstract word limit.

Excerpt 8I (from Arimitsu et al., 2007)

[*Move 1*] Furans and hydrofurans are well known compounds found in natural products and pharmaceuticals; currently, the synthesis of these compounds using catalytic methods constitutes an important focus of research. [*Move 2*] Although fluorine substituted furans and hydrofurans are attractive targets from a biological standpoint, there are few reports on practical methods for preparing them, and no method is catalytic. [*Move 3*] Recently, a cost-effective synthesis of gem-difluoro homopropargyl alcohols **1** was reported by our group. Using the triple bond of gem-difluoro homopropargyl alcohols **1** as a synthetic handle, we have uncovered novel transformations toward fluorinated furans and hydrofurans. (94 words)

1) ICl
2) Pd(0), R″-I

1) $AgNO_3$
2) Silica-gel

1) $AgNO_3$
2) Pd/H_2

The abstracts in excerpts 8B–8I all follow the moves in figure 8.1, although not in exactly the same manner. The difference is in how much emphasis the authors place on each move. We conclude this section by examining two abstracts that do not follow the move structure. In excerpts 8J and 8K, moves 1 and 2 are skipped entirely, and only move 3 is addressed. The advantage of this approach, which is quite common, is that authors can focus solely on the work that will be presented, without allocating precious space to less essential information.

Excerpt 8J describes a chemical synthesis. It is quite short (55 words) and includes a graphic. "This work" (i.e., the work to be presented) is addressed in the first sentence. Compound labels (bolded numbers **1**, **2**, **3**, and **4**) are used to link the compounds named in the text to their respective structures in the graphic. Excerpt 8K is longer (150 words). Again, the authors mention their own work in the first sentence (the fractionation and analysis of cranberry fruit for flavonoids). However, flavonoids are not defined, their importance is not emphasized, and no gap is suggested.

Bolded Numbers

See "Abbreviations, Acronyms, and Compound Labels" in appendix A for more on bolded numbers.

Excerpt 8J (from Lee et al., 2007)

[*Move 3*] This work surveys complementary routes for the synthesis of pyrazolo[1,5-α]pyrimidine-7-ones **1** and pyrazolo[1,5-α]pyrimidin-5-ones **2**. The use of 1,3-dimethyluracil **3** as an electrophile for pyrimidine ring construction affords pyrazolo[1,5-α]pyrimidin-5-ones **2**, contrary to literature reports. Novel use of *trans*-3-ethoxyacrylate **4** as an electrophile also afforded **2**, and the isolated intermediates from this reaction support our proposed mechanism. (55 words)

1 **2** **3** **4**

Excerpt 8K (from Liberty et al., 2007)

[*Move 3*] Cranberry fruit of Early Black cultivar was fractionated chromatographically and fractions were analyzed for flavonoid content. The effects of the flavonoid fractions and ursolic acid, an abundant triterpenoid in cranberry peel, were assessed in two models of colon cancer and one model of breast cancer. Clonogenic soft agar assays were used to determine the effect of these compounds on tumor colony formation in HCT-116, HT-29 and MCF-7 cells. MTT and trypan blue assays were performed to assess their ability to inhibit tumor cell proliferation. TUNEL assays were performed to assess apoptotic response to the cranberry compounds. The proanthocyanidins inhibited tumor colony formation in HCT-116 and HT-29 cells in a dose-dependent manner, with greater effect on the HCT-116 cell line. Ursolic acid strongly inhibited tumor colony formation in both colon cell lines. These compounds also decreased proliferation in all three tumor cell lines with the HCT-116 cell line most strongly affected. (150 words)

Exercise 8.5

Compare the two abstracts about flavonoids in excerpts 8F and 8K. Excerpt 8F includes moves 1–3; excerpt 8K includes only move 3. Which approach do you prefer? Explain your choice.

Exercise 8.6

Sentences a–f below are all from conference abstracts. Read each sentence and indicate which move the sentence accomplishes (move 1, 2, or 3). Refer to figure 8.1 as needed.

a. Single-walled carbon nanotubes (SWNTs) are filamentous manifestations of a repeating aromatic carbon structure formed into an open cylinder. (From Ferguson and DeMarco, 2007)

b. We have developed a novel method of estimating the fraction of open-ended carbon nanotubes in samples with porosity accessible for adsorption and gas storage. (From Agnihotri et al., 2007)

c. The discovery that negatively charged aggregates of C60 are stable in aqueous environments has elicited concerns regarding the potential environmental and health effects of these aggregates. (From Duncan and Vikesland, 2007)

d. However, the specific electronic properties of individual nanotubes remain untapped on the industrial scale. (From Jackson and Scott, 2007)

e. The synthesis, properties, and dynamic reactions of four heteroaromatic systems will be presented. (From Philp, 2007)

f. The purpose of this investigation was to better understand the behavior of multiwalled carbon nanotubes (CNTs) during a simulated drinking water treatment process. (Adapted from Mansfeldt et al., 2007)

8B Writing on Your Own: Decide on an Organizational Structure

Review the move structure in figure 8.1 and the various ways in which authors have adapted these moves in excerpts 8B–8K. In addition, consider the data that you want to present and whether you will include a graphic in your abstract.

With these considerations in mind, sketch out a move structure for your abstract. Estimate how much space you will devote to each move and which move(s) will be assigned the most and least importance. For guidance, consider browsing through the Technical Program Archive of past national meetings on the ACS Web site.

Writing an Abstract Title

A title must be submitted with the conference abstract. Because only the title, author list, and abstract are printed in the conference proceedings, the title is one of the few ways that interested individuals will be able to find your presentation. Hence, the title should be highly informative and use keywords that others in your field will recognize. If you plan to present a poster, keep in mind that your title should be short enough to fit on a single line across the top of your poster. A few examples of abstract titles included in the *Proceedings of the 231st American Chemical Society National Meeting* in Atlanta, Georgia (March 26–30, 2006) are listed below. Note that titles follow the same structure as journal article titles (see table 7.1).

1. Analysis of Bluntnose Minnow Growth in Differently Treated Mine Water
2. Direct Electron Transfer at the Anode of an Ethanol/Air Biofuel Cell

3. Enantioselective Biodegradation of Metalaxyl by Sewage Sludge and Screening Bacteria
4. On the Cooling Time of an Orange in the Refrigerator
5. Use of Pumpkin Pectin for Concentration of Ions of Heavy Metals

Abstract Titles

Abstract titles often follow an X of Y by Z pattern (see table 7.1).

Exercise 8.7

Examine abstract titles 1–5 above and answer the following:

a. Which titles follow the X of Y by Z pattern presented in table 7.1? Explain.
b. Make a list of keywords that others in the field might recognize.

Adding an Author List

A complete list of authors (not just the presenting author) and their affiliations should be submitted with the conference abstract (e.g., see the author list in excerpt 8B). Review abstract guidelines for formatting instructions. Typically, first and last names of all authors (e.g., Wilhelmus H. J. Boesten) are included. The sequencing of authors varies with each area of chemistry. The author giving the presentation is often listed first or distinguished with an asterisk or underlining.

Analyzing Writing across the Conference Abstract

Many of the writing conventions used in conference abstracts are those used in other chemistry genres. Important conventions are summarized here.

- Abbreviations and acronyms: As with the journal article abstract, most abbreviations and acronyms are defined at first use in the conference abstract (e.g., MD for molecular dynamics in excerpt 8A, WSOC for water soluble organic compounds in excerpt 8E, micro-XANES for micro-X-ray absorption near edge spectroscopy in excerpt 8G). Abbreviations and acronyms need not be defined in the conference abstract if they are not considered essential for understanding the work (e.g., CIRPAS Twin Otter, the name of a ship, in excerpt 8E).

- Citations: It is rare to see citations in conference abstracts. If a citation cannot be avoided, a full reference must be included in the abstract (see chapter 17).
- Keywords: A list of keywords is not included in a conference abstract (as it is in a journal article abstract), but it is wise to use keywords in the abstract text and title. The inclusion of keywords facilitates electronic searches, making the work more accessible to others interested in your field.
- Verb tense: Verb tenses in conference abstracts are consistent with conventions described elsewhere in the textbook (e.g., see tables 4.1, 5.1, 6.5). For example, past tense is used to describe work done in the past (e.g., "The rates were measured"), and present tense is used to make statements of fact or state information that is expected to be true over time (e.g., "Furans are found in natural products"). Some authors use future tense (in active or passive voice) in conference abstracts to refer to the work that will be presented (e.g., "We will present these findings" and "The synthesis will be presented"). Others prefer to state such intentions in present tense–passive voice (e.g., "The findings are presented").
- Voice: Both active and passive voice are used in conference abstracts. Active voice is especially common in move 1 (e.g., "2,2′-Bipyrroles are key synthetic precursors") but is also used in the rest of the abstract. Passive voice is often used to refer to work done in the past (e.g., "Molecular dynamics simulations were designed to investigate . . . nanoparticles").
- *We*: The word *we* often appears in conference abstracts, particularly in move 3 (e.g., "We have used" and "We will present"). In cases where there is a single author, *I* is used.

Exercise 8.8

Sentences a–f are all from conference abstracts. Select the correct verb for each sentence based on the move or submove indicated.

a. Sonication and heat treatment *are/were* common steps employed during purification of as-produced nanotubes. [Submove 1.1] (From Agnihotri et al., 2007)

b. Thermal-optical analysis (TOA) *is/was* widely used to classify carbonaceous aerosol into organic and elemental carbon. [Submove 1.1] (From Subramanian et al., 2007)

c. However, the Ullman reaction . . . generally *requires/required* high temperature and thus few functional groups can survive. [Move 2] (From Jiao and Smith, 2007)

d. The rates of superoxide and singlet oxygen production *are/were* measured for three distinct varieties of fullerene suspension. [Move 3] (From Hotze et al., 2007)

e. The performance of the simulated process *is/was* monitored by nephelometric turbidity, pH, and UV–vis absorbance. [Move 3] (Adapted from Mansfeldt et al., 2007)

f. Further usage of bipyrroles in the synthesis of porphycenes *was also reported/are also reported/will also be reported*. [Move 3] (From Jiao and Smith, 2007)

8C Writing on Your Own: Draft Your Conference Abstract and Title

Using the organizational structure developed in Writing on Your Own task 8B, write the first full draft of your conference abstract, title, and author list. Remember to define abbreviations and acronyms that are critical to understanding your work, incorporate keywords into your abstract and title, and use tense and voice in conventional ways. Avoid the use of citations. Make sure that you write for the appropriate audience.

8D Writing on Your Own: Fine-Tune Your Conference Abstract and Title

Find several people (including all co-authors) to read over the draft of your abstract and give you feedback. You will benefit the most by finding readers who can evaluate your abstract both for its scientific merit and writing quality. Another pair of eyes will catch mistakes that you have missed.

Do not submit your abstract until it has been peer reviewed and you have had the chance to improve it. Some online abstract submission programs allow you to edit your work up until the abstract deadline. After that date, no revisions are allowed. Before the deadline arrives, double-check your word count, capitalization, spelling, punctuation, units, and other writing conventions. See chapter 18 for additional hints on finalizing your work.

Chapter Review

Check your understanding of what you've learned in this chapter by defining each of the following terms for a friend or colleague who is new to the field:

abstract title	conference proceedings
author list	conference session
conference abstract	contributed paper

Also, answer the following questions for a friend who has just been encouraged to submit a conference abstract:

- What is the purpose of the conference abstract?
- What audience should be addressed in the conference abstract?
- How are conference abstracts and journal article Introduction sections similar? How are they different? What are the similarities and differences between conference and journal article abstracts?
- What are the typical moves of a conference abstract? What are some common variations of these moves?
- What are common writing conventions regarding the use of verb tense, voice, and personal pronouns in a conference abstract?
- What information should be included in the title and author list?

Additional Exercises

Exercise 8.9

Conference abstracts need not follow the move structure in figure 8.1 to accomplish their purpose. Consider, for example, the abstract in excerpt 8L below.

a. How do the moves in excerpt 8L compare to the moves presented in figure 8.1? Identify the moves in the abstract and their order.
b. Rewrite the abstract so that it (1) begins by introducing the topic, (2) suggests a gap, and (3) describes the work to be presented.
c. Suggest a title.

Excerpt 8L (from Phares, 2007)

The performance of an inlet for the size-resolved collection of aerosols is presented. The device resembles a cylindrical differential mobility analyzer (DMA) in that a sample flow is introduced around the periphery of the annulus between two concentric cylinders, and charged particles migrate inward towards the inner cylinder in the presence of a radial electric field. Instead of being transmitted to an outlet flow, the sample is collected onto a Nichrome filament located on the inner cylinder. The primary benefit of this mode of size-resolved sampling, as opposed to aerodynamic separation into a vacuum, is that chemical ionization of the vapor molecules is feasible. Because there is no outlet aerosol flow, the collection efficiency is determined by desorption of the particles from the filament, chemical ionization of the vapor, separation in a mobility drift cell, and continuous measurement of the current produced when the ions impinge on a Faraday plate.

Exercise 8.10

Imagine that you are a conference organizer with instructions to accept only five abstracts. The abstracts in excerpts 8B–8L have been submitted. Which five abstracts would you select? Explain your answer. Give reasons both for accepting and rejecting the abstracts.

Exercise 8.11

Reflect on what you have learned about writing a conference abstract and title. Select one of the reflection tasks below and write a thoughtful and thorough response:

a. Reflect on the different ways in which conference abstracts are organized.
   - Why does so much variation in the move structure exist? Would there be any advantages to a more rigid organizational template? Explain.
   - Does the flexibility that exists make it easier or more difficult to write an abstract? Explain.
   - Do any of the variations in excerpts 8B–8L have a greater appeal to you? Explain.

b. Reflect on the different purposes for a conference abstract and their impact on the writing process.
   - Does the fact that the abstract is evaluated by external reviewers make it easier or more difficult to write? Explain.
   - Does the fact that the abstract will endure years beyond the conference, in conference proceedings, add a level of seriousness to the task? Explain.
   - Should the fact that the abstract will draw interested conference attendees to your presentation influence the way you write your abstract? Explain.

c. Reflect on your experience writing your own conference abstract and title.
   - What challenges did you encounter? How did you resolve those challenges?
   - What parts of this chapter have helped you the most in writing your abstract and title? How did you use chapter information to assist you in your writing?
   - What have you learned from the experience of writing your conference abstract and title that will help you the most in the future?

9 Writing the Poster Text

At chemistry conferences, poster sessions are nearly always more engaging than talks. As proof, consider how few people fall asleep while viewing a poster! This is because the poster, by its very nature, is interactive, inviting people in to talk with you about your science.

Congratulations! If you are reading this chapter, you are likely preparing a poster for a scientific conference. This means that your conference abstract (chapter 8) was accepted and that you have been invited to give a poster presentation. In this chapter, we focus on the various sections of the poster and how to write them. In chapter 10, we highlight the visual attributes of the poster (layout, font size, color schemes, etc.). By the end of this chapter, you will be able to do the following:

- Address the correct audience in your poster
- Write the major sections of your poster
- Use bulleted lists and graphics appropriately
- Add title, references, and acknowledgments to your poster

As you work through the chapter, you will compose the text and graphics for your own poster. The Writing on Your Own tasks throughout this chapter guide you step by step as you do the following:

9A Prepare to write

9B Draft your poster Methods section

9C Draft your poster Results section

9D Draft your poster Discussion section

9E Draft your poster Introduction section

9F Add your poster title, author list, acknowledgments, and references

Reading and Analyzing Writing

We begin by asking you to read and analyze a poster that we created based on the journal article by Vesely et al. (2003) regarding aldehydes in beer (figure 9.1). Journal articles usually include far too much information for a single poster; hence, in the poster in figure 9.1, we include only a fraction of the information presented in the full journal article. (For the full article, see excerpts 3A, 4A, 6A, and 7A.) The hypothetical poster focuses on what Vesely's group might have presented early in their research project, specifically, the methods that they developed to analyze their samples. A black-and-white version of the poster is presented in figure 9.1; a full-color version of the poster is available on the *Write Like a Chemist* Web site. Exercise 9.1 guides you in the analysis of the poster and lays the groundwork for the rest of the chapter.

Exercise 9.1

Browse through the poster in figure 9.1 and answer the following questions:

a. How much content is included in the poster? To answer this question, read the poster in two minutes and then (without looking at the poster again) jot down two or three ideas that you have learned. Use this task to generalize how much information you should include in a poster.

b. Identify the major sections of the poster and the headings and subheadings that are used to signal these sections. What types of information are included (and excluded) in each section?

c. What writing conventions predominate in the poster? Examine, for example, the formatting of headings and lists, capitalization, units, abbreviations, references, and citations.

d. What audience is targeted in the poster? (There may be more than one.)

Analyzing Audience and Purpose

To benefit most from a poster presentation, you should consider it primarily an opportunity for an exchange of ideas and dialogue, rather than merely a forum for data presentation.

—Anholt (1994)

There is no one right audience to target in a poster; rather, the correct audience will vary with the conference. Some conferences are primarily for experts; others, such as the national meeting of the American Chemical Society, attract a wide range of individuals, from high school teachers and undergraduate students to

On-Fiber Derivatization to Improve Aldehyde Extraction

Petr Vesely[†,‡], Lance Lusk,[†] Gabriela Basarova,[‡] John Seabrooks,[†] and David Ryder[†]
Miller Brewing Company[†] and Institute of Chemical Technology,[‡] Prague, Czech Republic

Introduction

OVERVIEW. Aldehydes are believed to contribute to the loss of flavor in beer during aging. However, aldehyde levels in beer are usually very low and have been difficult to extract. Methods such as liquid-liquid extraction[1] and sorbent extraction[2] have been used, but these methods are rather complicated and not highly selective.

A promising way to improve aldehyde extraction is through on-fiber derivatization. The derivatizing agent *O*-(2,3,4,5,6-pentafluorobenzyl)-hydroxylamine (PFBOA) has been used with some success,[3] and on-fiber derivatization with PFBOA has been used successfully to extract formaldehyde.[4]

THIS WORK. Building on these techniques, we have optimized a method for extracting methional from a 5% ethanol solution using solid-phase microextraction (SPME) with PFBOA on-fiber derivatization. Methional was selected as the test aldehyde because it is one of the more difficult aldehydes to extract from beer.

Methods

Chemicals

- The PFBOA solution was prepared in water (6 g/L).
- The SPME fiber was coated with 65 μm poly(dimethyl-siloxane)/divinyl benzene (PDMS/DVB SPME).
- The optimization solution contained 5% EtOH (pH 4.5) spiked with 5 ppb methional ($CH_3SCH_2CH_2CHO$).

On-Fiber Derivatization Procedure

- 100 μL PFBOA was added to 10 mL deionized water in 20 mL vial with crimp cap.
- The SPME fiber was placed in the headspace of the PFBOA solution for 10 min (50 °C), then the PFBOA-loaded fiber was placed in the headspace of the methional solution (10 mL in a 20 mL vial). Temperature and time were optimized.
- The derivatized methional was detected in single-ion mode using GC/MS (HP6890 GC with DB-5 column and mass-selective Agilent 5972A detector).

Results and Discussion

(1) Detection and calibration of derivatized methional

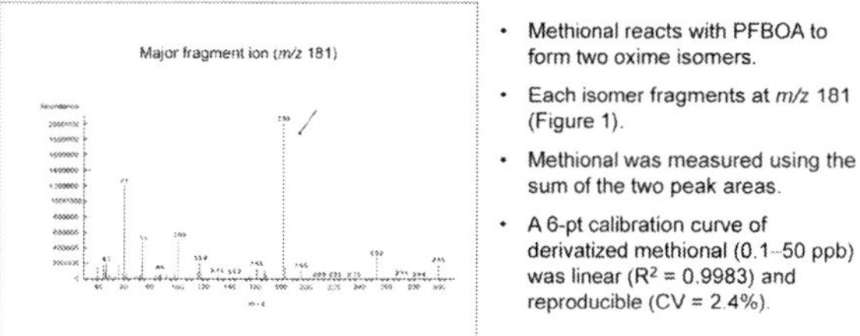

Figure 1. Spectrum of derivatized methional.

- Methional reacts with PFBOA to form two oxime isomers.
- Each isomer fragments at *m/z* 181 (Figure 1).
- Methional was measured using the sum of the two peak areas.
- A 6-pt calibration curve of derivatized methional (0.1–50 ppb) was linear ($R^2 = 0.9983$) and reproducible (CV = 2.4%).

(2) Optimization of derivatization temperature

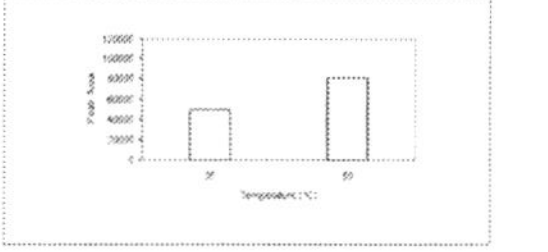

Figure 2. Detector response with temperature.

- A temperature increase favored partitioning of methional to the headspace (Figure 2).

Optimized temperature: 50 °C

(3) Optimization of derivatization time

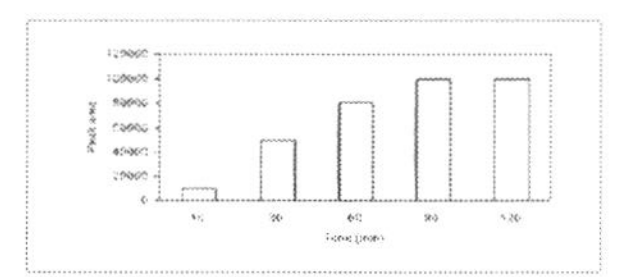

Figure 3. Detector response with time.

- Time to reach equilibrium between fiber and headspace was 90 min (50 °C), but 60 min was chosen as a compromise (Figure 3).

Optimized time: 60 min

(4) Measurement of methional recovery rates

The methional recovery rate was determined using the method of standard addition: a 5 ppb methional solution was spiked with 100 ppb methional.

$$\text{Recovery} = \frac{\text{measured conc.}}{105 \text{ ppb}} = 90\%$$

Conclusions

A new, fast (60 min), sensitive (0.1-50 ppb), reproducible, solventless extraction technique was developed for methional in 5% ethanol.

This technique can be used to measure low aldehyde levels in beer and to assess the role aldehydes play in the loss of beer flavor with aging.

References

1. Tressl et al. *J. Agric. Food Chem.* **1978,** *26*, 422.
2. Hawthorne et al. *J. Am. Soc. Brew. Chem.* **1987**, *45*, 23.
3. Koshy et al. *J. Chromatogr. Sci.* **1975,** *13*, 97.
4. Marton & Pawliszyn. *Anal. Chem.* **1998,** *70*, 2311.

Acknowledgments

This work received financial support from the Miller Brewing Company. The authors also gratefully acknowledge P. R. Armstrong for technical assistance.

Figure 9.1 A poster created from the content in Vesely et al., 2003.

research professors and Nobel Laureates. Sometimes the title of the conference suggests the predominant audience. For example, a conference titled "Colloidal, Macromolecular, and Polyelectrolyte Solutions" is likely for experts, whereas a conference titled "The National Conference on Undergraduate Research" is clearly for college students. Most conferences send presenters a list of poster guidelines, and information about audience is often addressed in those guidelines.

Poster Audience

The audience for a poster depends on the conference. Some conferences are more specialized than others.

In general, however, it is best to assume that your audience knows little about your topic.

If, after reviewing the conference title and poster guidelines, you are still unsure about the correct audience for your poster, it is best to target a scientific audience (e.g., graduate students, faculty, or professionals who are in a different discipline or subfield of chemistry). Don't assume that most viewers will be knowledgeable about your research area; more than likely, only a few will have such specialized knowledge. Thus, prepare your poster for your most probable audience: chemists and chemistry students with a research emphasis different from yours. If viewers want more information, they can ask you questions. Your thoughtful answers to these questions will satisfy their need for more information. Indeed, the best part of poster sessions is the exchange of ideas that takes place when passersby stop and ask you questions about your work.

Exercise 9.2

Based on the titles of the following conferences, indicate which audience(s) should be targeted: student, scientific, or expert.

a. Great Lakes College Chemistry Conference (GLCCC)
b. Annual Conference on Analytical Chemistry and Spectroscopy
c. The 18th International Symposium on Supercritical Fluid Chromatography and Extraction
d. National Undergraduate Chemistry Conference
e. The 26th International Conference on Molten Salt Chemistry & Technology
f. The 21st International Symposium on the Organic Chemistry of Sulfur
g. The 16th Annual Chemistry and Biochemistry Graduate Research Symposium

Exercise 9.3

Search the Internet for three conferences that would be appropriate for your research area (or area of interest). Try to find conferences that have different target audiences (student, scientific, or expert).

9A Writing on Your Own: Prepare To Write

To make sure that you are prepared to create your poster, think about the research that you have completed so far. What do you want your audience to know after they have viewed your poster?

Analyzing Organization

Many people believe scientists do not need to learn writing skills. I would argue that scientists need these skills more than most professions. Concise, informative, flowing, and intelligent verse can make the difference between a funded and rejected grant, a published and rejected scientific publication, or a well-received and botched poster. Writing skills are not only encouraged, they are mandatory for success in all scientific fields.

—Bradley F. Schwartz, Southern Illinois University School of Medicine

The poster text is divided into the same general IMRD sections as the journal article: Introduction, Methods, Results, and Discussion. Similarly, most posters include an Acknowledgments section, some have an abbreviated References section, and all have a title and author list. Most posters do not include an abstract, in part because of space limitations and in part because an abstract already appears in the conference proceedings. Like the journal article, the IMRD structure of the poster follows an hourglass shape. The top (Introduction) and bottom (Discussion) sections have a broader focus, while the middle sections (Methods and Results) have a narrower focus. Each section of the poster can be divided into individual "moves" or steps that guide viewers in a conventional way through the content of each section. These moves are analyzed in the next part of the chapter.

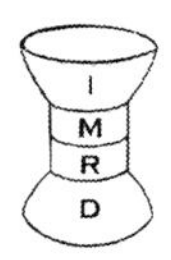

Analyzing Excerpts

We now read and analyze excerpts from each major section of the poster, beginning with Methods (the section that you are most likely to prepare first) and continuing through the Results, Discussion, and Introduction. The title and author list, also major sections of the poster, are addressed in chapter 8. Toward the end of the chapter, we address two less prominent sections: Acknowledgments and References. Because posters are not published, we cannot include excerpts from the literature; instead, we present excerpts that we have created based on three journal articles from module 1: Llompart et al. (2001), Vesely et al. (2003), and Boesten et al. (2001). For each poster section, you will examine its move structure, read and analyze the hypothetical excerpts, and review general writing practices.

The Poster Methods Section

A poster cannot communicate as much information as a journal article can. When designing a poster, the presenter should accept this constraint and limit the poster to the essential information. If too much information is included, the poster will overwhelm the audience and in many cases cause passersby to give up on that poster and move on to the next one.
—Alley (2003)

The poster Methods section offers a brief snapshot of the methods used in the presented work. (Of course, the Methods section will be more involved if the poster focuses on the development of a new method or procedure.) The essential moves of the Methods section are presented in figure 9.2. First materials and then methods are presented. The term "materials" is used loosely and refers to chemicals, solvents, standards, samples, and so forth. Similarly, the term "methods" refers to instrumentation, experimental methods, and/or numerical procedures. Because materials and methods are rather specialized, this section targets a relatively narrow audience.

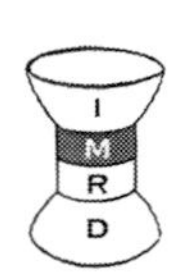

Note that the moves in figure 9.2 are characterized by the words *highlight*, *summarize*, and *identify*. These words emphasize the importance of brevity in

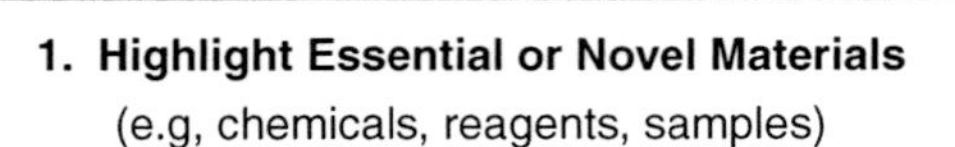

Figure 9.2 A visual representation of the move structure for a typical poster Methods section.

the poster Methods section. The average person will likely view your poster for only a few minutes. Complex experimental procedures are too difficult to grasp in such a short time. Instead, focus only on the key steps used to obtain the information presented in the poster. If more details are needed, viewers can ask for them.

With these two moves in mind, we examine the Methods sections of three hypothetical posters. The first poster concerns the detection of PCBs in full-fat milk. We include both the poster Methods section (excerpt 9A) and the journal article Methods section (excerpt 9B). In this way, you can see for yourself the differences and similarities between the two genres. A key difference is that the poster includes far less information than the journal article. For example, the poster addresses only full-fat and skim milk, whereas the journal article also includes half-fat milk. Half-fat milk is an intermediate case and serves only to confirm the two extreme cases. Thus, the intermediate case can be omitted without changing the essential message of the poster.

Exercise 9.4

Compare excerpts 9A and 9B. What similarities and differences do you notice? Consider science content, organization, and writing conventions (e.g., subheadings, capitalization, abbreviations, and parentheses).

Excerpt 9A (a poster Methods section based on Llompart et al., 2001)

METHODS

Reagents and Materials

- Full-fat (3.61%) and skim milk (0.34%) samples were obtained from local supermarkets and spiked with one of eight PCB congeners (1–100 ng/mL).

PCB-28 PCB-52 PCB-101 PCB-105

PCB-118 PCB-138 PCB-153 PCB-180

- Spiked samples were homogenized, held at 4 °C for 24 h, frozen, then thawed 1 h before analysis.

- SPME holders with one of two fiber assemblies were used:

  100 μm PDMS (poly(dimethylsiloxane))

  65 μm PDMS-DVB (poly(dimethylsiloxane)divinylbenzene)

Saponification and HSSPME Procedures

- PCB-spiked milk samples were placed in headspace vials. When saponification was performed, samples were treated with NaOH, sealed, immersed in water (100 °C), and equilibrated for 6 min.
- During HSSPME, the fiber was exposed to the headspace over the sample for 5–240 min as the sample was stirred.

Sample Analysis

- PDMS-fibers were analyzed by GC-electron capture detection using an HP 5890 series II GC.
- PDMS-DVB fibers were analyzed by GC/MS-MS using a Varian 3800 GC with ion trap (Varian Saturn 2000).

Abbreviations in Posters

Most abbreviations are defined at first use in posters. In excerpt 9A:

PDMS is first used and defined in the Methods section.

PCB is first used and defined in the Introduction, thus does not need to be defined again.

Excerpt 9B (adapted from Llompart et al., 2001)

Experimental Section

Reagents and Materials. The PCB congeners, 2,4,4′-trichlorobiphenyl (PCB-28), 2,2′,5,5′-tetrachlorobiphenyl (PCB-52), 2,2′,4,5,5′-pentachlorobiphenyl (PCB-101), 2,3,3′,4,4′-pentachlorobiphenyl (PCB-105), 2,3′,4,4,′5-pentachloro-biphenyl (PCB-118), 2,2′,3,4,4′,5′-hexachlorobiphenyl (PCB-138), 2,2′,3,4,4′,5′-hexachlorobiphenyl (PCB-153), and 2,2′,3,4,4′,5,5′-heptachlorobiphenyl (PCB-180) (PCB numbering according to IUPAC) were supplied by Ultra Scientific (North Kingstown, RI). Isooctane, acetone, and sodium hydroxide were obtained from Merck (Mollet del Valles, Barcelona, Spain). All the solvents and reagents were analytical grade.

Isooctane, acetone, and sodium hydroxide were obtained from Merck (Mollet del Valles, Barcelona, Spain). All the solvents and reagents were analytical grade.

The full-fat milk (3.61% fat), half-fat milk (1.55% fat), and skimmed milk (0.34% fat) were purchased from local supermarkets. Spiked milk samples were prepared by addition of a small volume of acetone solutions containing the target analytes in the concentration range of 1–100 ng/mL. The spiked samples were homogenized in an ultrasonic bath for 5 min; later, the samples were kept at 4 °C for 24 h, to allow analyte–matrix interactions. Afterward, they were frozen at −20 °C until 1 h before the analysis.

The certified reference material (CRM 450), used for the validation of the method, is real contaminated powdered milk with a certified content in PCB-52, PCB-101, PCB-118, PCB-156, and PCB-180. This material contains approximately 3.9% water and 25% fat. It is used after reconstituting and was supplied by the EC Community Bureau of Reference (BCR).

HSSPME Extraction Procedure. Manual SPME holders were used with a 100-μm poly(dimethylsiloxane) (PDMS) and 65-μm poly(dimethylsiloxane)-divinylbenzene (PDMS-DVB) fiber assembly (Supelco, Bellefonte, PA). The fibers were conditioned as recommended by the manufacturer.

The samples were placed in headspace vials. When saponification was performed, a few milliliters of NaOH solution were added to the sample. The vial was sealed with a headspace aluminum cap furnished with a Teflon-faced septum, immersed in a water bath maintained at 100 °C, and let equilibrate for 6 min before HSSPME. Afterward, the fiber was exposed to the headspace over the sample for 5–240 min, depending on the experiment. The sample was magnetically agitated during sampling. Once the exposition period was finished, the fiber was immediately inserted into the GC injector and the chromatographic analysis was carried out. Desorption time was set at 5 min.

Chromatographic Conditions. GC-ECD analyses were performed in an HP 5890 series II GC equipped with an electron capture detector and a split/splitless injector, operated by an HP Chemstation software. PCBs were separated on a 25 m length × 0.32 mm i.d., HP-1 column coated with a 0.17 μm film. The GC oven temperature program was as follows: 90 °C hold 2 min, rate 20 °C/min to 170 °C, hold for 7.5 min, rate 3 °C/min, to final temperature 280 °C, and hold for 5 min. N_2 was employed as carrier and makeup gas, with a column flow of 1.2 mL/min at 90 °C. Split flow was set at 50 mL/min. Injector and ECD temperatures were 260 and 280 °C, **respectively**. Injector valve time was set at 2 min.

The GC/MS-MS analyses were performed on a Varian 3800 gas chromatograph (Varian Chromatography Systems, Walnut Creek, CA) equipped with a 1079 split/splitless injector and a ion trap spectrometer (Varian Saturn 2000, Varian Chromatography Systems) with a waveboard for MS-MS analysis. The system was operated by Saturn GC/MS WorkStation v5.4 software. The MS-MS detection method was adapted from elsewhere.[29] PCBs were separated on a 25 m length × 0.32 mm i.d., CPSil-8 column coated with a 0.25 μm film. The GC oven temperature program was as follows: 90 °C hold 2 min, rate 30 °C/min to 170 °C, hold for 10 min, rate 3 °C/min to 250 °C, rate 20 °C/min to a final temperature of 280 °C, and hold for 5 min. Helium was employed as a carrier gas, with a constant column flow of 1.0 mL/min. Injector was programmed to return to the split mode after 2 min from the beginning of a run. Split flow was set at 50 mL/min. Injector temperature was held constant at 270 °C. Trap temperatures, manifold temperatures, and transfer line temperatures were 250, 50, and 280 °C, respectively.

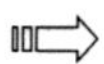

Respectively

See appendix A.

Even at first glance, several differences and similarities are apparent between the poster and journal article excerpts. Consider the first move. Both the poster and journal article begin by presenting materials (chemicals, samples, reagents, etc.), and both use subheadings to signal this move. In the poster, however, the PCBs are identified by chemical structure rather than by chemical name. Structures, although not required, are easier to grasp at a glance, enhancing the audience's ability to understand the compounds quickly. Note, too, that product information (vendor, purity) is not included in the poster (although this information may be included if deemed essential by the authors).

Differences are apparent in the second move as well. Both the poster and journal article signal the second move with subheadings, but in the poster, the procedures are presented in bulleted lists of sentences or phrases. Such lists, which could also be numbered, are common in posters. Essential details are presented, specifically those that are needed to understand the rest of the poster. In this case, we chose to highlight a few details about the saponification and extraction procedures. Note, however, that we did not include operating parameters for the GC-electron capture or GC/MS-MS instruments (other than the vendors and model numbers). In a poster, such parameters are optional; they may be included, but they are not required. Also, in a poster (but not a journal article), the parameters can be listed in phrases rather than in complete sentences. An illustrative sketch, diagram, or photograph of the SPME holder, fiber, or extraction setup would also be appropriate to include at this point. Finally, if any numerical methods have been used, they should be mentioned in an additional bullet.

As a second example, we present another poster Methods section (excerpt 9C) based on the same research presented in figure 9.1. Recall that the hypothetical poster in figure 9.1 focuses on methods development; the poster in excerpt 9C focuses on methods application. If you examine the Methods sections in the two posters, you will see that figure 9.1 describes how the on-fiber derivatization procedure was optimized; excerpt 9C describes how the optimized procedure was used to analyze beer samples.

Learning how to focus on specific aspects of your work is an important skill in poster preparation.

Exercise 9.5

Compare the Methods section of the poster in figure 9.1 with the Methods section in excerpt 9C. What similarities and differences do you notice? Consider science content, organization, and writing conventions (e.g., subheadings, capitalization, abbreviations, and parentheses).

Excerpt 9C (a poster Methods section based on Vesely et al., 2003)

METHODS

Samples

- The beer samples (American lagers) were stored at 30 °C for 4, 8, or 12 weeks.
- The controls were stored at 0 °C for 12 weeks.

Methods Validation

- Calibration curves were prepared for 9 aldehydes (linearity >0.96).
- Each beer sample was measured 10 times (variance <5.5%).
- Recovery was determined by spiking beer samples with 10 ppb of the standard aldehyde (recoveries 89–110%).

On-Fiber Derivatization and Detection of Aldehydes

- The derivatizing agent *O*-(2,3,4,5,6-pentafluorobenzyl)-hydroxylamine (PFBOA) was absorbed onto a 65 μm PDMS/DVB SPME fiber (10 min, 50 °C).
- The fiber was placed in the sample headspace (50 °C, 60 min).
- The beer aldehydes selectively reacted with the PFBOA in the fiber.
- The oximes that formed were desorbed and detected by GC/MS (DB-5 30 m × 0.25 mm × 0.50 μm; He carrier gas at 1.1 mL/min; splitless, injector 250 °C).

Bulleted Lists

- Start full sentences with a capital letter and end them with a period.
- Start sentence fragments with either a capital or lowercase letter; a period is not needed.
- If possible, use one list style consistently within a poster.

As the last example, we present a poster Methods section based on Boesten et al. (2001) concerning the asymmetric Strecker synthesis of an α-amino acid (excerpt 9D and at the end of chapter 2). The poster Methods section presents only part of the content from the article because (once again) the full content would be too much for a poster. The poster focuses on only the first step of the synthesis: the asymmetric Strecker reaction of (*R*)-phenylglycine amide **1** and pivaldehyde **2** to preferentially crystallize the (*R*,*S*) amino nitrile product **3**. The poster Methods section includes relevant general information and the optimized steps used in the synthesis of (*R*,*S*)-**3**. (In the Results section of this poster, the authors explain how these steps were optimized.) An equation is used to illustrate the general reaction, introduce the compounds and their compound labels, and identify the reaction variables (solvent, time, temperature). The steps of the synthesis are summarized in a bulleted list following the equation.

Excerpt 9D (a poster Methods section based on Boesten et al., 2001)

EXPERIMENTAL SECTION

General. (*R*)-Phenylglycine amide **1** was purchased from DSM (Netherlands). The dr of (*R*,*S*)-**3** and (*R*,*R*)-**3** was determined by ^{1}H NMR using the relative integration between the **t**-Bu signals at 1.05 ppm for (*R*,*S*)-**3** and 1.15 ppm for (*R*,*R*)-**3**. NMR spectra were recorded using a Varian VXR-300 spectrometer (300 mHz) and referenced to residual solvent.

The synthesis of amino nitrile (R,S)-3

Ph, H_2N, $CONH_2$ + O, H → NaCN, HOAc / solvent, time, temp → Ph, HN, $CONH_2$, CN, H + Ph, HN, $CONH_2$, CN, H

(*R*)-**1**   **2**   (*R*,*S*)-**3**   (*R*,*R*)-**3**

- A stirred suspension of **1** (400 mmol) in H_2O (400 mL) was added to pivaldehyde **2** (419 mmol) and stirred for 30 min.
- NaCN/HOAc was added at 23–28 °C. The mixture was heated to 70 °C and stirred for 24 h.
- After cooling to 30 °C, the product was filtered, washed (500 mL H_2O), dried, and analyzed by ^{1}H NMR.
- (*R*,*S*)-**3** formed as a colorless solid (92.4%, dr > 99/1).

Exercise 9.6

Review excerpts 9A–9D. Based on these excerpts, answer the following questions:

a. What information is generally included in the Methods sections of posters? What information is typically excluded?

b. What conclusions can you draw about the audience that is targeted in the Methods section of a poster?

c. How well does each poster excerpt follow the move structure in figure 9.2?

Writing conventions typical of a poster Methods section are summarized below. Many of these conventions apply to other sections of the poster, as well.

- Abbreviations: Abbreviations are usually defined in parentheses at their first use, for example, "pentafluorobenzene (PFB)". Readily understood symbols or abbreviations may be used without definition (e.g., GC/MS, MeOH, R-OH, P_{gas}, cat.). A few unconventional abbreviations may be used in posters if space is tight (e.g., "temp" for "temperature," "rt" for "room temperature," "wk" for "week," "exp" for "experiment," "sat'd" for "saturated," "&" for "and"). Additionally, some abbreviations, such as those for long chemical names, need not be defined if an exact identity is not essential for understanding the poster.

  Be careful, however, not to become too casual with abbreviations. For example, popular shorthand is not appropriate (e.g., "B4" for "before" or "FYI" for "for your information"). Also, use abbreviations consistently throughout the poster. For example, "eq" should not be used for "equation" and "equivalent" in the same poster.
- Bulleted or numerical lists: Lists are common in posters. It is customary to capitalize the first letter of the first word in a bulleted list if the list contains full sentences ending with periods; lowercase is preferred for lists of phrases or fragments (not ending with periods). When possible, use the same types of lists (bulleted or numbered) throughout your poster. (See comments on parallelism below for related issues.)
- Capitalization: Molecular formulas (NaOH), abbreviations (PCBs), instrument vendors (Varian 3800 GC), and chemical compounds (e.g., "2-methylpropanal") are capitalized as recommended in *The ACS Style Guide*.
- Numbers and units: Conventional scientific units should be used in poster Methods sections (mL, μm, mol, M, etc.); however, some formatting conventions may be relaxed. For example, the conventional space between a number and its unit may be omitted in a poster, and the numerical form of a number may be used instead of its word form (even at the start of a sentence). If space allows, however, follow conventional practices. One convention that should never be relaxed is the use of leading zeros for numbers <1 (e.g., use 0.35 not .35).

Conventional formatting	2 μL, nine aldehydes, 30 m × 0.25 mm
Relaxed formatting	2μL, 9 aldehydes, 30m×0.25mm

- Parallelism: Parallel language should be used in bulleted lists in poster Methods sections and elsewhere. Not only is parallelism grammatically correct, but its repetitive pattern also helps viewers comprehend poster information more quickly. Pay particular attention to parallel subheadings and lists:

Nonparallel list	1. American lager samples were stored... 2. Load PFBOA on fiber... 3. Concentration of aldehydes on...
Parallel list	1. American lager samples were stored... 2. PFBOA was loaded on fiber... 3. Aldehydes were concentrated on...
Nonparallel subheadings	**Chemicals** **Extracting the Filters** **Analyze using GC**
Parallel subheadings	**Chemicals** **Extraction Procedure** **GC Analysis**

- Verb tense: Poster Methods sections are written predominantly in the past tense (e.g., used, exposed, inserted, were stored) because they describe work done in the past.
- Voice and *we*: Poster Methods sections are written largely in passive voice; active voice is used less often. *We* is commonly avoided.

Passive voice	Aldehydes were concentrated and derivatized... (*R*)-Phenylglycine amide **1** was purchased from DSM.
Active voice	PFBOA was the derivatizing agent.

Unconventional Abbreviations

Some unconventional abbreviations are allowed in posters (e.g., "temp" for "temperature"). However, shorthand (e.g., "B4" for "before") should not be used.

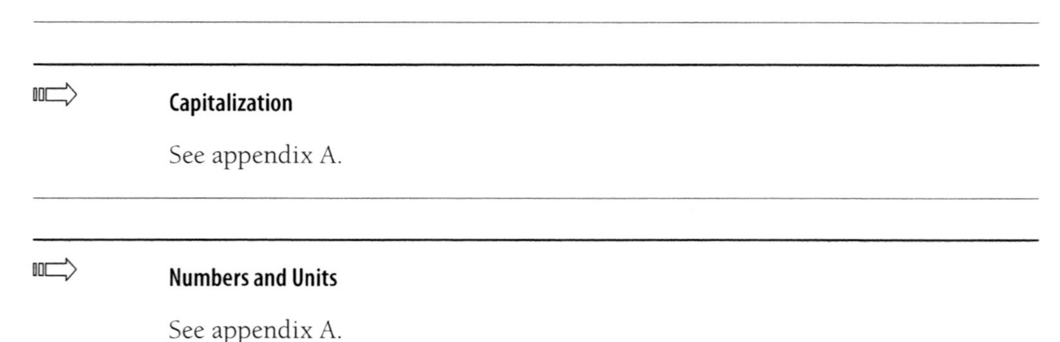

Capitalization

See appendix A.

Numbers and Units

See appendix A.

Parallelism and Fragments

Bulleted fragments often replace complete sentences in posters. No period is needed after a fragment. When possible, use parallel language in lists and headings. (See appendix A.)

Exercise 9.7

Reconsider poster excerpts 9A, 9C, and 9D and complete the following tasks:

a. Make a list of the abbreviations used in each poster. Which abbreviations do you think were defined earlier in these posters (e.g., in the Introduction section)? Which will most likely not be defined at all?

b. Examine the lists contained in each poster. Are all items in a single list parallel to each other? Are punctuation and capitalization used consistently?

c. Is past or present tense preferred? Is passive or active voice preferred? Support your answer with at least three examples.

Exercise 9.8

When poster excerpts 9A, 9C, and 9D were prepared, it was assumed that we had plenty of space. However, assume now that space is tight. Conserve space in the *first* list of each poster by doing one or more of the following:

a. Convert the bulleted lists (which currently are written in full sentences) into shorter fragments. (Be sure to keep the fragments parallel and use proper punctuation.)

b. Abbreviate words that will be easily understood in their abbreviated forms (e.g., "temp" for temperature).

c. Reformat numbers and units to make them more condensed (even though such formatting would be incorrect in a journal article).

d. Omit some information or make the writing more concise.

Exercise 9.9

Consider the following excerpts taken from experimental sections of journal articles. Convert each procedure into a single item or a bulleted list of items that would be appropriate for a poster Methods section. Omit information that would not be included in a poster.

a. FTIR spectra were recorded using a Nicolet model 870 spectrometer (Madison, WI) equipped with a deuterated tryiglycine sulfate (DTGS) detector. (From Kizil et al., 2002)

b. The liquid chromatograph (LC) was an Agilent (Waldbronn, Germany) 1100 series system consisting of vacuum degasser, autosampler, binary pump, column oven, and diode array detector. (From Pelander et al., 2003)

c. **2-(*p*-Toluenesulfonyl)-4′-methoxyacetophenone (2a)**. A mixture of 2-bromo-4′-methoxyacetophenone (45.8 g, 200 mmol) and *p*-toluenesulfinic acid sodium hydrate (35.6 g, 200 mmol) in ethanol (1 L) was heated at reflux for 1.5 h. The mixture was stirred and cooled to room temperature, and the resulting solid was collected, washed with ethanol (2 × 50 mL), and dried to give 54.6 g (90%) of pure **2a**. (Adapted from Swenson et al., 2002)

d. *Delphinium* × *cultorum* cv. Magic Fountains dark blue/white bee seeds were donated by Bodger Seed, Ltd., South El Monte, CA. The seeds were germinated in high porosity, peat-based growing mix in Styrofoam trays (4 in. × 6 in. × 2 in.) and grown until the first leaf appeared. The seedlings were then transplanted into pots (4 in. × 4 in. × 6 in.) and then into large clay pots (10 in. × 6 in. × 10 in.) and were grown until maturity. The plants were raised in the Bioactive Natural Products Laboratory Greenhouses, Michigan State University. The plants were subjected to a 12 h photoperiod, watered once daily, and maintained at 75 °F. Plant parts were harvested when 75–80% of the florets were fully expanded on each raceme and stored at –20 °C. (From Miles et al., 2000)

9B Writing on Your Own: Draft Your Poster Methods Section

Create a list of the steps that you took in your research. Which of these steps are necessary to include in your poster? Which can be excluded?

Consult figure 9.2 to review the moves that make up the Methods section of a poster. After you have decided what to include (and what to exclude), write the Methods section of your poster. Remember that past tense is preferred and that bulleted lists (with parallel language) are common. Use standard formatting, number, and abbreviation conventions at first; convert to abbreviated formats if space is tight. Consider adding a photograph or illustration to your Methods section.

The Poster Results Section

Like any scientific presentation, the poster should tell a story. . . . Include only materials relevant to the story line.

—Anholt (1994)

The Results section is undoubtedly the most important section of the poster. The chance to share your results with others is the predominant reason for preparing your poster in the first place. However, you must be careful not to overwhelm

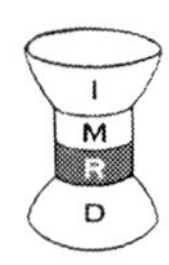

your readers by presenting too many results. Focus on one or two key findings, targeting a reasonably narrow audience (hence, the narrow part of the hourglass). Avoid the temptation to crowd too much information into a single poster. Limit yourself to results that can be digested in just a few minutes. Deciding which results to include (and which to exclude) is one of the most challenging steps in preparing a poster.

Note that "raw" data (e.g., spreadsheet calculations, uncalibrated measurements, original spectra before they have been baseline corrected) should not be included in a poster. It is important that you present results only after they have been carefully analyzed and prepared for public view.

Most results in posters are presented in graphical form (e.g., figures, graphs, photographs, schemes, illustrations, spectra). Graphics are preferred over text because they allow the reader to comprehend the data more quickly (provided the authors do not overcrowd their graphics). The text accompanying the graphics serves largely to guide viewers through the data. In general, figures are preferred over tables because they are easier to read and understand at a glance; however, tables may be included if they provide useful information and/or illustrate important trends.

As you prepare your Results section, imagine "walking" someone through your poster. Your results should be organized so that they guide viewers through your story of scientific discovery. (Like the journal article, a poster should also tell a story.) The moves for the poster Results section (figure 9.3) are designed to facilitate this process. In move 1, you set the stage and prepare viewers for moves 2 and

1. Share Preliminary Results

Prepare viewers for principal findings by doing one or more of the following:

Share results that build confidence in your approach	Share results that motivated your study	Share results that lay groundwork for your principal findings

2. Share Principal Results

(i.e., share key findings; identify and summarize key trends)

3. Share Related Results (optional)

(i.e., share results that support, extend, or strengthen your principal findings)

Figure 9.3 A visual representation of the move structure for a typical poster Results section.

3 by presenting preliminary results, that is, results that lead toward your principal findings (move 2). Preliminary results can introduce or motivate your claims (e.g., a spectrum that shows the presence of a pollutant), build confidence in your measurements (e.g., a calibration curve), or lay groundwork needed to understand your principal findings (e.g., a chromatogram that illustrates how a complex mixture was separated). As you prepare move 1, imagine yourself pointing to your poster and saying to your viewers "First, we wanted to be sure we could do ____", "In previous work, we successfully prepared ____", or "This spectrum confirmed that ____ is present in our sample." By preparing your viewers for what lies ahead, you strengthen their confidence in themselves (to understand your results) and in you (to present data that they can trust).

Walking Viewers through a Poster

At most conferences, there is a 2 h block of time when you are expected to stand by your poster. That time allows you to "walk" viewers through your poster and engage in dialog.

After the foundations are laid (move 1), you are ready to share principal findings (move 2). Using two to four graphics (and accompanying text), focus your viewers' attention on only the most important results. In posters, these results can report what has already succeeded or what has yet to succeed. Use graphics to display the trends visually, and use text to highlight and summarize those trends. Label relevant peaks and features in your figures to draw your viewers' attention to the most important areas. (This is like using a pointer in a slide show presentation.) These labels will make your key findings more conspicuous. As in journal articles, poster figures include captions, placed either below or to the side of the figures. Unlike figures in journal articles, poster figures may also include titles above them.

Poster Figures

In journal articles, figures have captions but no titles. In posters, figures often use captions, titles, and other descriptive labels to highlight important points.

Some poster Results sections conclude with move 2 (which is fine), but in many posters, authors share additional results. Move 3 allows you to do this without overwhelming your viewers. The key is to relate these results to your principal findings. Viewers will more easily grasp a second set of data if it supports, extends, or strengthens what they have just learned.

To summarize, as you prepare your Results section, divide your results into three groups: (1) preliminary results, (2) principal results, and (3) related results. Share preliminary results first (move 1), principal results second (move 2), and related results third (move 3). How you categorize your results is your decision; what is important is the logical thread connecting all the data that you share. Viewers will be able to absorb more information if the data build on themselves in a logical manner.

Let's examine these moves in three hypothetical poster Results sections. We begin with excerpt 9E, the poster about PCBs in milk. The authors present their results in four steps, walking the reader through a story that might be titled "Detecting PCBs: Before and After Saponification." The steps are numbered, rather than bulleted, to make the sequencing clear. First, preliminary results are presented, highlighting what happens before saponification. The data are presented in two chromatograms (Figure 1). The first chromatogram shows PCBs detected in skim milk; the second, in full-fat milk. (Both samples were spiked with equal amounts of PCBs.) Side by side, these two chromatograms make the problem clear: PCBs are more difficult to detect in full-fat milk because of matrix effects (i.e., the extra fat in full-fat milk retains the PCBs). Such results develop the viewer's confidence in the data (e.g., without the skim milk chromatogram, viewers might think that the problem is a poorly operating GC). The data also prepare viewers for similar graphs later on in the poster. The accompanying text helps focus the viewer's attention on the relevant information.

Principal results (move 2) are presented in steps 2 and 3. In step 2, the authors describe the saponification process, which involves the optimization of five reaction conditions. These conditions are listed in a table for easy viewing (Table 1). In step 3, the success of the technique is showcased with an "after saponification" chromatogram of PCBs in full-fat milk (Figure 2). Because the groundwork was laid in Figure 1 ("before saponification"), the viewer is prepared to understand and appreciate the importance of this chromatogram. The essential result is highlighted in the text, that saponification increased the PCB sensitivity in full-fat milk by a factor of 4 to 8 (depending on the PCB congener).

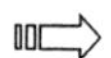

Poster Tables

See chapter 16 for table formatting.

The results are extended (move 3) in step 4, "Procedure Validation." The extended information informs the viewer that (1) results were linear and reproducible and (2) the technique also works with a certified reference standard (Table 2). These results build logically on the principal finding of the work, that with saponification PCBs can be detected in full-fat milk.

Excerpt 9E (a poster Results section based on Llompart et al., 2001)

RESULTS

1. **Before saponification.** PCBs were detected in skim milk (Fig. 1A) but not full-fat milk (Fig. 1B) because of matrix effects in full-fat milk.

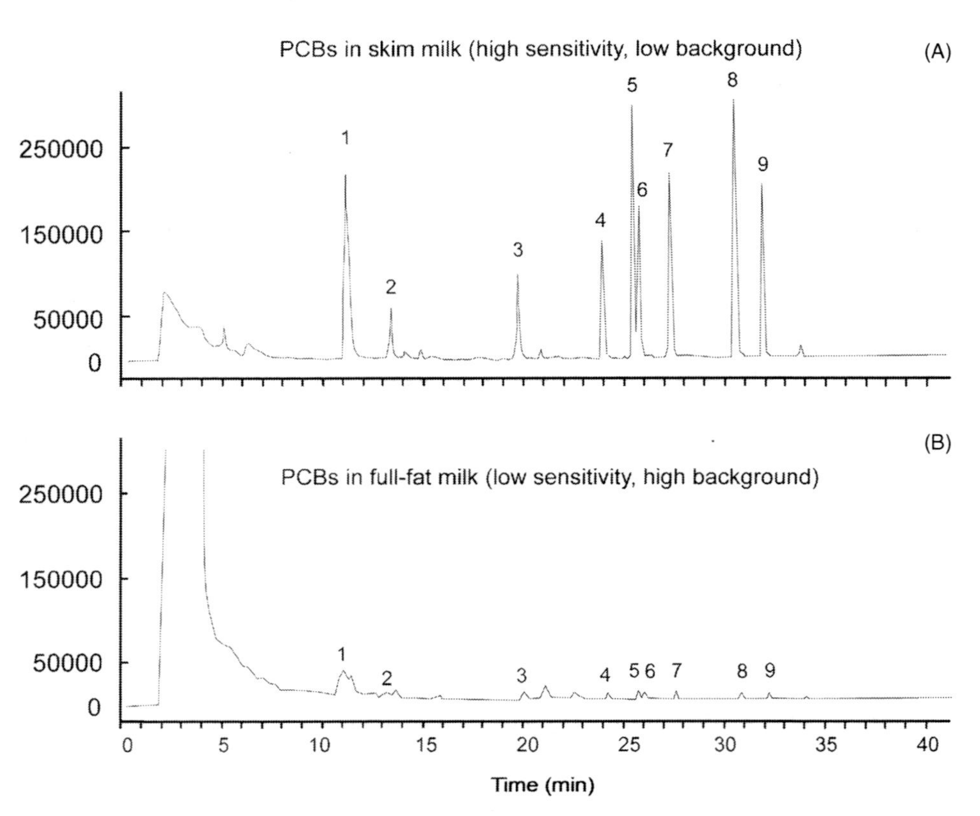

Figure 1. PCBs in spiked skim milk (A) and spiked full-fat milk (B).

2. **Saponification.** Extraction of PCBs from full-fat milk matrix was improved by saponification (Table 1).

Table 1. Optimization of saponification process.

Factor	Options Tried	Best Option
NaOH vol (mL)	0.5, 2.0, 3.5	2.0–3.5
NaOH conc. (%)	20, 30	20
Time (min)	30, 60	60[a]
Fiber	PDMS, PDMS-DVB	PDMS-PVB
Stirring	yes, no	yes

[a] Longer times were better but too slow for desired approach.

3. **After Saponification.** Skim milk response increased 5 to 10 times; full-fat milk response increased 4 to 8 times, depending on PCB congener (Fig. 2).

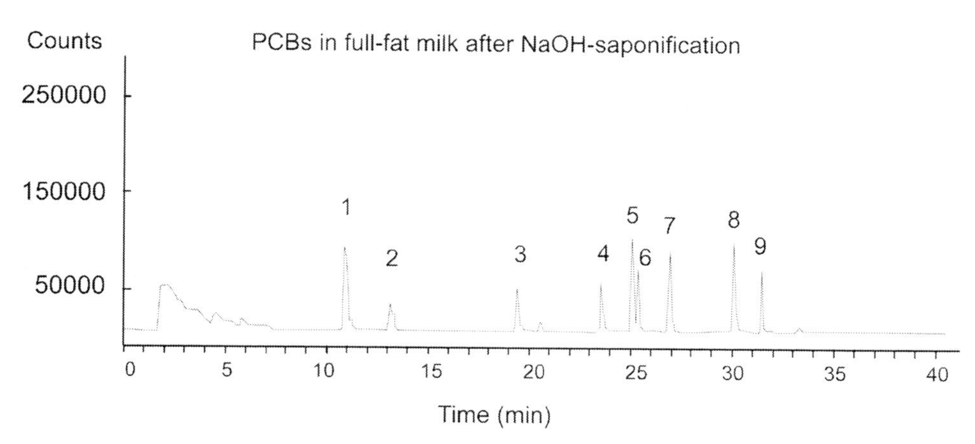

Figure 2. PCBs in full-fat milk with saponification.

4. **Procedure Validation.** Calibration results were linear ($R^2 > 0.994$) and reproducible (RSD < 11.0 %) over a wide range of concentrations. Recoveries ≥75% were achieved with a certified milk sample (Table 2).

Table 2. Procedure validation with certified milk (25% fat).

PCB	Certified Value (ng/g)	Recovery (%)
52	1.16 ± 0.17	75
118	3.19 ± 0.24	97
153	8.60 ± 1.10	98
156	1.64 ± 0.11	102
180	9.29 ± 0.26	88

We next consider the two posters concerning aldehydes in beer, the first on methods development and the second on methods application. In particular, note what information is included in each Results section. In the poster in figure 9.1, the optimization of temperature and time comprises the bulk of the Results section. In the second poster, these "results" have been moved to a single bulleted item in the Methods section (excerpt 9C), which states simply that "The fiber was placed in the sample headspace (60 min, 50 °C)." The new Results section (excerpt 9F) focuses on the aldehydes. The authors use two figures to convince the viewer that the aldehydes can be detected: (1) a chromatogram showing the retention times of the nine derivatized aldehydes and (2) a mass spectrum of methional, showing that *m/z* 181 is a fragment ion that can be used to detect the aldehydes.

The principal findings are shared in Figure 3. A bar graph summarizes the concentrations of the nine aldehydes in beer after 4, 8, and 12 weeks of storage. The graph groups the aldehydes by their most likely source. This grouping will be useful in the Discussion section of the poster, where the individual aldehydes are discussed. A graph is preferred over a table for these data because a graph makes the trends easier to see. The text accompanying the graph reinforces these trends.

Bar Graphs

In posters, bar graphs are used to illustrate even simple relationships. In journal articles, such relationships are usually expressed in words.

The authors expand these results in Figure 4; in this case, they elucidate the relationships between aldehyde concentrations and beer flavor thresholds. Based only on Figure 3, a viewer might assume that furfural affects beer flavor because it increased the most during beer storage. However, as shown in Figure 4, even at high concentrations, furfural is only a small fraction of its flavor threshold (FT). Instead, (*E*)-2-nonenal has the greatest impact, although it is still <30% FT. Figure 4 also makes clear the important message that none of the aldehydes in stored beer exceeded their respective flavor thresholds.

Excerpt 9F (a poster Results section based on Vesely et al., 2003)

RESULTS

- A beer sample was spiked with 9 aldehydes and derivatized. The chromatogram is shown in Fig. 1. Aldehydes were detected by *m/z* 181 (Fig. 2).
- All aldehydes increased during storage (Fig. 3), but none exceeded their flavor threshold (Fig. 4).

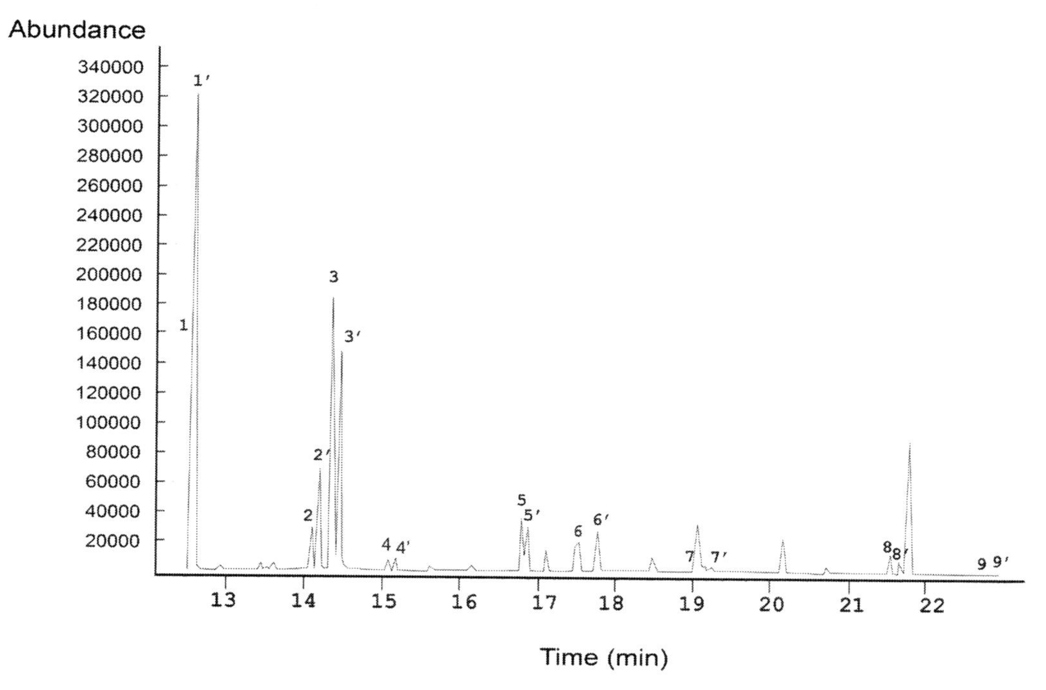

Fig. 1. Chromatogram of 9 aldehydes in spiked beer. Each aldehyde has an isomer (e.g., 1, 1′).

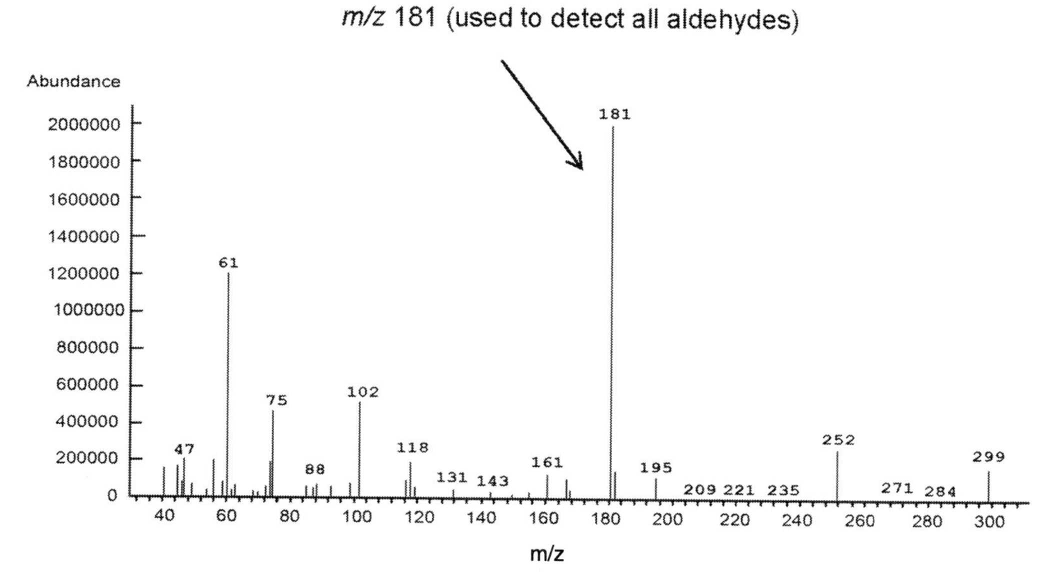

Fig. 2. Aldehydes were identified by *m/z* 181 (methional shown).

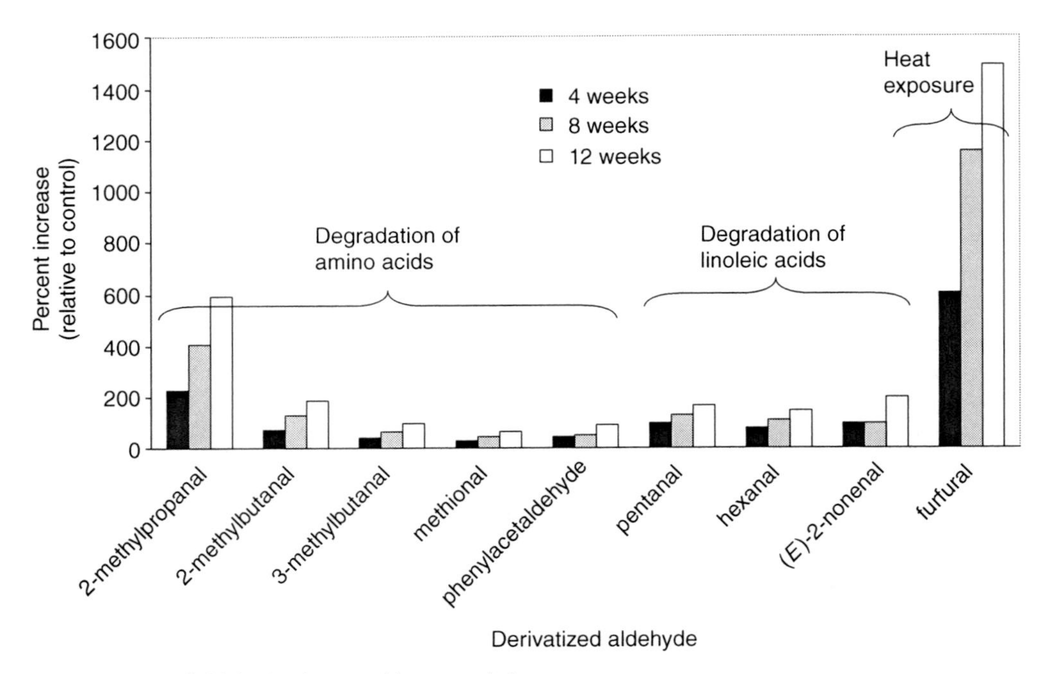

Fig 3. Increase of aldehydes (grouped by source) during storage.

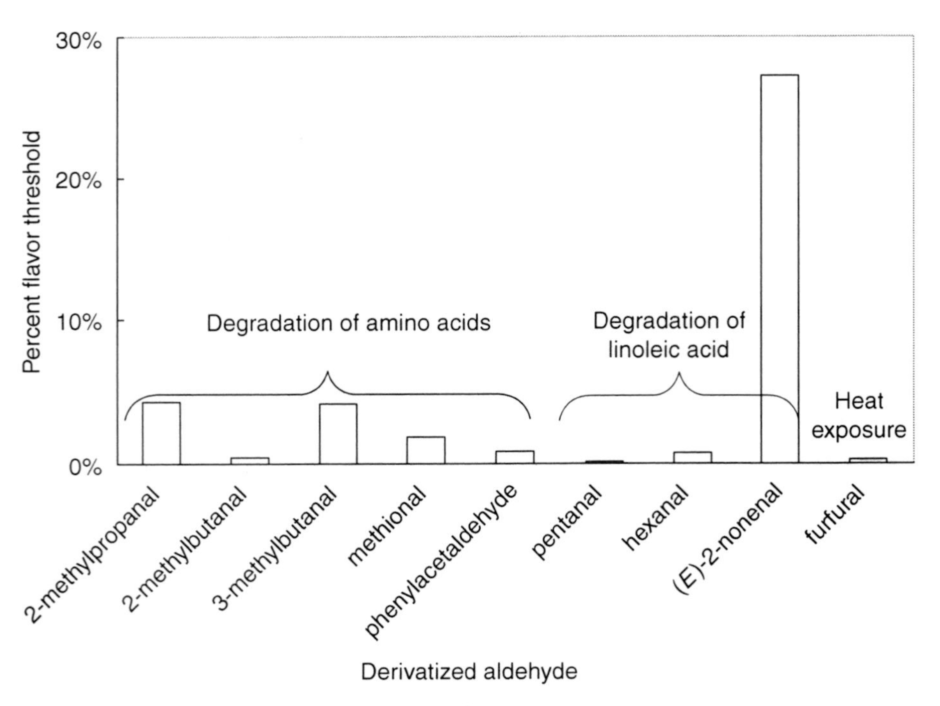

Fig. 4. Percent of flavor threshold at 12 weeks for 9 aldehydes.

Exercise 9.10

How is the attention of viewers drawn to important data and trends in excerpt 9F?

Exercise 9.11

In excerpts 9E and 9F, text is limited; the graphics tend to speak for themselves. Review the excerpts and answer the following questions:

a. What purpose(s) do the graphics and text serve?
b. What general rule of thumb may guide you in finding the right balance between text and graphics in posters?

Lastly, we consider the Results section for the poster that describes the asymmetric Strecker synthesis (excerpt 9G). The results comprise three bulleted items. The first item presents evidence that the targeted diastereomer (*R*,*S*)-**3** can be verified by X-ray analysis. The second item identifies the reaction conditions that were optimized (solvent, time, and temperature) and graphs one of them (time). The principal findings are presented in the third item. The authors share as their principal findings the reaction conditions that were tried and ultimately produced (*R*,*S*)-**3**. The results are presented in a table for easy viewing (Table 1). The optimal conditions are listed (logically) in the last row of the table (in bolded font) and reiterated in the text. The Results section ends here; no supporting or extending evidence is included.

Excerpt 9G (a poster Results section based on Boesten et al., 2001)

RESULTS

- The absolute configuration of (*R*,*S*)-**3** was confirmed by X-ray analysis (Fig. 1).

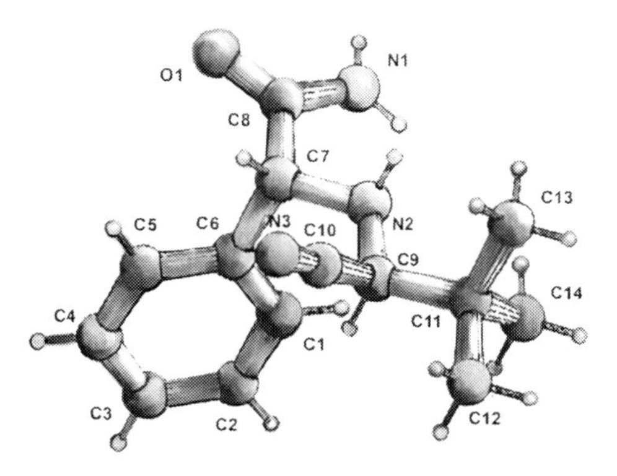

Fig. 1. X-ray structure of (*R*,*S*)-**3**.

- Solvent, time, and temperature were optimized. Time was optimized at 30 min with dr > 99/1 (Fig. 2).

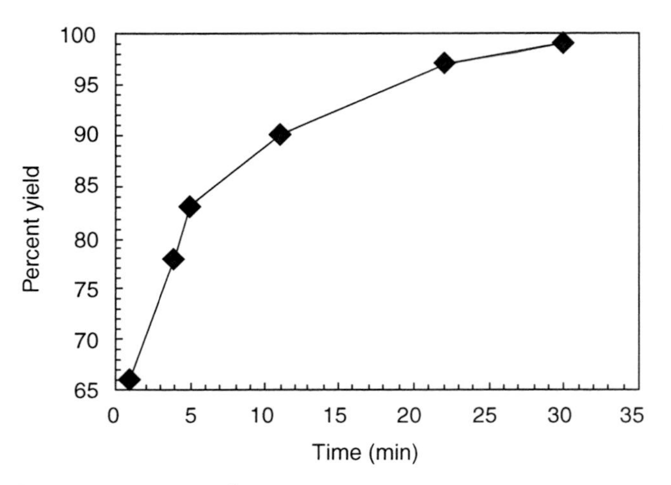

Fig. 2. Optimization of reaction time.

- Optimization trials (Table 1) showed that water at 70 °C for 24 h produced (*R*,*S*)-**3** in 93% yield (dr > 99/1).

Table 1. Optimization trials for the synthesis of (*R*,*S*)-**3**.

solvent(s)	temp (°C)	time (h)	yield (%)	dr (*R*,*S*)-**3**/(*R*,*R*)-**3**
MeOH	rt	20	80	65/35
MeOH/2-PrOH	rt	22	51	99/1
2-PrOH	rt	22	84	88/12
2 PrOH/*t*-BuOH	rt	20	65	96/4
MeOH/H_2O	rt	20	69	81/18
H_2O	55	24	81	85/15
H_2O	60	24	84	96/4
H_2O	65	24	84	98/2
H_2O	**70**	**24**	**93**	**>99/1**

Compound Labels

Standard conventions for using compound labels such as (*R*,*S*)-**3** should be followed in a poster. (See appendix A.)

The writing conventions listed below are commonly used in poster Results sections.

- Abbreviations: Due to limited space, the abbreviation "Fig." may be used to refer to figures. "Table" is not generally abbreviated. Other readily understood abbreviations may be used when space is tight (e.g., temp, conc, MeOH).
- Bulleted or numerical lists: Lists are frequently used when highlighting or summarizing results. Follow the formatting guidelines presented above in the discussion of poster Methods sections. When possible, use the same type of formatting throughout your poster.
- Graphics and graphics labels: In general, poster graphics follow the same formatting guidelines as journal articles (see chapter 16). Two differences are that (1) titles may be used in poster figures (along with captions, which are required), and (2) color may be used freely in poster graphics (first to illustrate the science but also to add visual appeal). As in journal articles, poster graphics should be labeled. Table titles are placed above the table. Figure captions are placed either below or beside the figure. **Dangling graphics**, that is, graphics that are never referenced in the text, should be avoided. Exceptions to this rule are graphics added primarily for visual appeal (e.g., a logo, photograph, or clipart). Photographs used to convey science (e.g., a photograph of an instrument or a map of a sampling site), however, should be labeled.
- Verb tense: Poster Results sections use past tense to describe results obtained in the past and present tense to describe "facts" expected to be true over time:

Past tense	High sensitivity to PCBs *was observed* in skim milk. The best yield *was obtained* with H_2O.
Present tense	Each aldehyde *produces* two peaks in the chromatogram. Each derivatized aldehyde *fragments* to give a major peak at *m/z* 181.

- Voice and *we*: Poster Results sections use both passive and active voice, and the word *we* is typically avoided:

Passive voice	High sensitivity was achieved (not "we achieved high sensitivity"). Five factors were optimized (not "we optimized five factors").
Active voice	Calibration results were linear. Furfural had the highest concentration.

Dangling Graphics

Graphics that are included in a poster, but are never referenced in the text.

Graphics should not dangle in posters, unless they are included for aesthetic purposes only.

Tense–Voice Combinations in Results Sections

See table 4.1.

Exercise 9.12

Revisit the graphics in excerpts 9E–9G and answer the following questions:

a. Are the figure captions and table titles placed in conventional places?

b. Do the authors add titles to their figures? If so, where are those titles placed?

c. Are there any dangling graphics in these excerpts? Explain.

Exercise 9.13

Consider the following sets of bulleted items (adapted from Jozefaciuk et al., 2003; Dellinger et al., 2001; Jozefaciuk et al., 2003; respectively). Which item, in each set, is most appropriate for a poster Results section? Consider standard tense, voice, and conciseness practices to make your decision.

Set 1

- We present the results of water uptake measurements in Figure 1.
- Results of water uptake measurements are presented in Fig. 1.

Set 2

- Exposure of the cells to H_2O_2 (100 μM) produced 100% damaged DNA.
- Exposure of the cells to 100 μM hydrogen peroxide produced 100% damaged DNA.

Set 3

- In our work, we found that RAMEB alone absorbed a high amount of water.
- RAMEB alone absorbed a high amount of water.

Exercise 9.14

Create two bulleted items to accompany the following table (adapted from Plaper et al., 2002) that would be appropriate for a poster Results section:

Table 1. Uptake of chromium by *E. coli* cells treated with different Cr^{3+} compounds.

Cr^{3+} compound	concentration added (mM)	intracellular uptake of Cr (mg of Cr/g of dry weight)
$CrCl_3 \times 6H_2O$	0	<0.021
	0.63	1.8
	1.25	6.7
$Cr(NO_3)_3 \times 9H_2O$	0	<0.021
	0.63	2.4
	1.25	5.7
$KCr(C_2O_4)_2 \times 3H_2O$	0	<0.025
	0.63	<0.022
	1.25	<0.02

9C Writing on Your Own: Draft Your Poster Results Section

Decide what results you are going to present in your poster. Be careful not to include too many results or to present results that are still in "raw" form. Categorize your results into three groups according to the move structure in figure 9.3: preliminary, principal, and related. These groupings will help you organize your results so that they weave a consistent thread throughout your Results section.

Prepare the graphics for your Results section first. Remember that, in general, figures are easier for viewers to read than tables. Add titles, labels, and captions to your figures to focus your viewers' attention on the important points.

Next, write the text that will accompany the graphics. Consider using a bulleted list of short sentences. Remember to refer readers to appropriate graphics and to adhere to conventions for the use of present and past tense.

The Poster Discussion Section

The Discussion section of a poster has two major purposes: (1) to interpret or explain the results presented and (2) to summarize the most important findings of the work. These two purposes form the move structure for the Discussion section (figure 9.4).

1. Interpret or Explain Results

2. Conclude with a Take-Home Message

Figure 9.4 Visual representation of the move structure for a typical poster Discussion section.

Move 1 (Interpret or Explain Results) is often integrated into the poster Results section, thereby becoming a combined Results and Discussion (R&D) section. An example of a combined R&D section is shown in figure 9.1. In such posters, interpretative remarks (Discussion) are included right along with the graphics (Results). In this way, space is conserved, and viewers can read and interpret the data simultaneously (usually easier than looking back and forth between the two sections). For instructional purposes, however, we have placed move 1 in the Discussion section, and we use a stand-alone Discussion section in the three hypothetical posters presented below. We follow this approach, in part, to maintain a clear distinction between results (just the facts) and discussion (interpretation of the facts).

The poster Discussion section ends with a take-home message (move 2). The take-home message sums up the essential conclusions of the presented work (i.e., those points you most want your viewers to remember). Because of its importance, move 2 is often given a separate heading (e.g., "Conclusions" or "Key Points"). Most take-home messages are only a few sentences (or bulleted items) in length. Some authors also include brief remarks about work in progress or future work at the end of move 2. Move 2 is generally written for a broad audience and thereby completes the bottom of the hourglass.

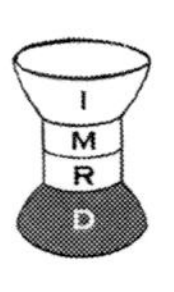

With these issues in mind, let's examine the Discussion sections of our three hypothetical posters. Each section is divided into two parts: Discussion and Conclusions. Excerpt 9I uses a bulleted list to present the discussion points; excerpts 9H and 9J use a paragraph format. Both styles are common and illustrate that there is no one right way to present information. Excerpt 9J also includes a scheme, which is common in posters that present a synthesis. In all three excerpts, the Conclusions sections comprise one or two sentences and are written without bullets.

As you read through these Discussion sections, you will notice that they are quite short. Moreover, the discussion points within these sections are often used to highlight, explain, or reiterate key findings rather than to truly interpret data. In addition, unlike a journal article, few references are made to the literature. These common practices illustrate that the main emphasis in posters is to present, rather than interpret, results. This emphasis reflects both the newness of the data presented (late-breaking results may not yet be fully understood) and the interactive role of the poster (to promote an exchange of ideas and dialog).

Nevertheless, a few interpretative remarks are encouraged; viewers want to know what you think about your data.

Exercise 9.15

Consider the poster Discussion sections in excerpts 9H–9J as you answer the following questions:

a. What writing features do you notice? Consider such features as the use of fragments versus complete sentences and bulleted lists versus paragraphs.
b. How well do the excerpts adhere to the move structure in figure 9.4?
c. Which excerpts use concluding remarks to highlight work in progress?

Excerpt 9H (a poster Discussion section based on Llompart et al., 2001)

DISCUSSION

PCBs are more difficult to extract from full-fat milk than skim milk, suggesting that PCBs are more strongly retained in the full-fat matrix. Even at 100 °C, SPME without saponification was not efficient. Saponification of milk fats to their corresponding glycerols and carboxylates appears to facilitate the release of PCBs.

CONCLUSIONS

Saponification enables a more efficient SPME process for the detection of PCBs in full-fat milk. An increase in response of 4 to 9 times can be achieved depending on the PCB congener.

Excerpt 9I (a poster Discussion section based on Vesely et al., 2003)

DISCUSSION

During long-term storage (4–12 wk) at elevated temperatures (30 °C), American-style beers develop a stale flavor. Nine aldehydes were analyzed as possible contributors to this flavor loss:

- **5 Strecker aldehydes** (degradation products of amino acids). All 5 aldehydes increased during storage, but none exceeded more than 4.3% of their respective flavor thresholds.
- **Pentanal and hexanal** (degradation products of linoleic acid). Both aldehydes were less than 1% of their flavor thresholds.
- **(*E*)-2-nonenal** (a degradation product of linoleic acid). This aldehyde increased to 27% of its flavor threshold but is not expected to contribute to stale beer flavor (Schieberle & Komarek, ACS, Chicago, Fall 2001).
- **Furfural** (a heat exposure indicator). This aldehyde increased more than 2-fold but reached only 1% of its flavor threshold.

CONCLUSIONS

The increase after 12 weeks for most aldehydes was significant (16-fold for furfural, 7-fold for 2-methylpropanal); however, no aldehyde approached its respective flavor threshold.

We are currently exploring additive or synergistic effects **among** these aldehydes to explain the stale flavor of aged beer.

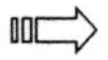

Between vs. Among

See appendix A for more information on these commonly confused words.

Excerpt 9J (a poster Discussion section based on Boesten et al., 2001)

DISCUSSION

A proposed mechanism is shown in Scheme 1. The *re*-facial attack of CN^- to the intermediate imine **4** appears to be preferred, forming (*R*,*S*)-**3**. (*R*,*S*)-**3** is less soluble and precipitates out of solution; (*R*,*R*)-**3** is more soluble and epimerizes in solution via the imine **4**.

Scheme 1

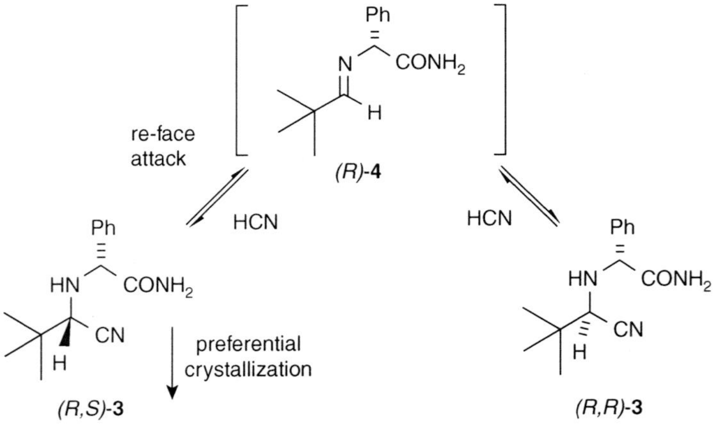

Ph
N CONH$_2$
H
(*R*)-**4**
re-face attack
HCN
HCN
Ph
HN CONH$_2$
CN
H
(*R*,*S*)-**3**
preferential crystallization
Ph
HN CONH$_2$
CN
H
(*R*,*R*)-**3**

CONCLUSIONS

(*R*)-Phenylglycine amide **1** is an excellent chiral auxiliary in the asymmetric Strecker reaction of pivaldehyde **2**. In water at 70 °C, the (*R*,*S*)-**3** product was isolated in 93% yield and dr > 99/1. Work is underway to convert (*R*,*S*)-**3** to (*S*)-*tert*-leucine and thereby complete the asymmetric Strecker reaction.

Exercise 9.16

Write a combined Results and Discussion section:

a. Either combine excerpts 9E (Results) and 9H (Discussion) or combine excerpts 9F (Results) and 9I (Discussion). Approach the task by rewriting the information in 9H or 9I (stand-alone Discussion sections) so that they accompany the graphics in 9E or 9F (Results).

b. Which approach do you prefer: stand-alone Results and Discussion sections or merged Results and Discussion sections? Why?

We conclude this part of the chapter by identifying some writing conventions that are commonly used in the poster Discussion section:

- Bulleted or numerical lists: Discussion items and conclusions may be presented in paragraph or list form. In Discussion section lists, complete sentences are more common than fragments. (See the guidelines presented in the poster Methods section for formatting these lists.) Conclusions are usually set apart from the Discussion section with a separate heading.
- Citations: Citations may be included in a poster Discussion section. (See the discussion of the poster Introduction section, below, for information on citations and references.)
- Hedging: Hedging words should be used to soften interpretive remarks in posters. A few examples are presented below; hedging words are italicized:

  Saponification of milk fats to their corresponding glycerols and carboxylates *appears* to facilitate the release of PCBs from the sample matrix.

  These results, however, do not rule out the *possibility* that additive or synergistic effects among these aldehydes *may* contribute to the stale flavor of aged beer.

  The latter process is *attributed to* the reversible reactions of the amino nitriles.
- Verb tense: Poster Discussion sections use both past tense (to summarize findings observed in the past) and present tense (to state findings and conclusions that are expected to be true over time). Because the Discussion section focuses on conclusions, present tense is used more often.

Past tense	With saponification, the SPME process *was* 4 to 9 times more effective.
	None *reached* more than 4.3% of their flavor threshold.
Present tense	During long-term storage, American-style beers *develop* a stale flavor.
	PCBs *are* more difficult to extract from full-fat milk than skim milk.
	In methanol at room temperature, the *re*-face attack of CN^- to the intermediate imine **4** *is* preferred.
	(*R*)-Phenylglycine amide **1** *is* an excellent chiral auxiliary in the asymmetric Strecker reaction of pivaldehyde **2**.

- Voice and *we*: Both active and passive voice may be used in the poster Discussion section. The word *we* should be used sparingly (and is not used in excerpts 9H–9J), but it can be used to signal a decision, interpretation, or conclusion made by the authors.

Active voice	Furfural *increased* more than 2-fold. SPME (without saponification) *is* not an efficient technique. *We* next *plan* to convert (*R*,*S*)-**3** to (*S*)-*tert*-leucine. Saponification *enables* a more efficient SPME process.
Passive voice	Nine aldehydes *were analyzed* as possible contributors to this flavor loss. This aldehyde . . . *is* not *believed* to be a key contributor to flavor loss. The latter process *is attributed* to the reversible reactions of amino nitriles.

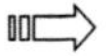

Hedging

See appendix A.

Active and Passive Voice

See appendix A.

Exercise 9.17

Examine the use of past and present tense in excerpts 9H–9J. What conclusions can you draw?

Exercise 9.18

Find instances of hedging in excerpts 9H–9J. Propose generalizations about when hedging is appropriate and when it is not necessary in a poster Discussion section.

9D Writing on Your Own: Draft Your Poster Discussion Section

Consider the key results of your work. How can you use text to highlight, explain, and interpret these results? Decide if you want a combined Results and Discussion section, with explanatory remarks included along with the graphics, or a stand-alone Discussion section.

Think about the take-home message that you want your viewers to get from your poster. How can you best summarize your work? Try to limit your take-home message to one or two sentences.

The Poster Introduction Section

The poster Introduction is typically not written until you know the full and final content of your poster. The first move of the Introduction focuses on the importance of your research area (figure 9.5). It should be clear to your viewers, after reading only the first few sentences of your Introduction, what your general research area is and why the area is important (submove 1.1). Background information is shared largely for the purpose of establishing this importance and should be kept to a minimum (submove 1.2). Importance is also commonly established by pointing out gaps in the field (i.e., work that has yet to be done or problems that need to be solved). In-text citations to essential works are often included (although they are not required); an abbreviated citation format may be used to conserve space (see the end of this part of the chapter for more information on citations). If citations are not included, you should still know what major works influenced your work, in case someone asks.

The second move of the poster Introduction previews the specific accomplishments of the work and is often given its own subheading (e.g., **Research Objectives** or **Goals**). The focus should be on research goals that have been achieved and are presented in the poster. Move 2 has a narrower focus than move 1; hence, the poster Introduction follows the broad-to-narrow hourglass structure. As a test to see if your Introduction addresses moves 1 and 2 sufficiently, ask yourself if a viewer, after reading only your Introduction, could answer the following questions: (1) What research area is addressed? (2) Why is this area important? (3) What specific accomplishments will the authors present in their poster?

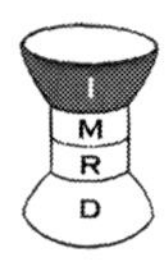

1. Establish Importance of Research Area

1.1 Introduce the research area

1.2 Emphasize importance (through background information, gap statements)

Cite essential works (optional)

2. Preview Accomplishments
(Identify major goals that have been achieved)

Figure 9.5 A visual representation of the move structure for a typical poster Introduction section.

With the organizational structure of the poster Introduction in mind, let's examine the Introduction sections of our three hypothetical posters. As you glance through these excerpts, notice that the subheadings (e.g., **Overview and Importance** and **Research Objectives**) direct the reader's attention to essential information in the most efficient manner. Also note that no bulleted lists are used in these Introductions. Although bulleted lists are common in other sections of the poster, they are less common in the Introduction. Finally, we include a word count in each Introduction, to give you a better idea of how long an Introduction should be. Most poster Introductions are under 150 words.

Conciseness

For the poster Introduction, a good rule of thumb is to stay under 150 words.

Exercise 9.19

Read excerpts 9K–9M. Based only on the information presented in these Introduction sections, answer the three questions that viewers should be able to answer after reading a poster Introduction:

a. What research area is addressed?

b. Why is this area important?

c. What specific accomplishments will the authors present in their poster?

Excerpt 9K (a poster Introduction based on Llompart et al., 2001)

INTRODUCTION

Overview and Importance. The pollutants known as polychlorinated biphenyls (PCBs) are widely distributed in the environment due to their extensive use in the past, lipophilic (fat-loving) character, and general chemical stability. Moreover, PCBs tend to accumulate in the food chain. Thus, several countries have established recommended maximum limits for PCBs in food products. Among these products, milk is especially important because of its extensive and widespread consumption by humans.

Objective. To date, there is no simple or rapid procedure for testing PCBs in milk. Headspace solid-phase microextraction (HSSPME) is a promising approach, but only a few works have applied this technique to milk (*1*, *2*). Here, we present a simple and rapid saponification-HSSPME procedure for extracting PCBs from milk. (119 words)

Excerpt 9L (a poster Introduction based on Vesely et al., 2003)

INTRODUCTION

Overview and Importance. American packaged beers lose their flavor and become stale during storage. Carbonyl compounds, particularly aldehydes, may be involved in this process. Aldehydes are formed in beer through one of several possible mechanisms including Strecker degradation of amino acids or linoleic acid (Pollock. In *Brewing Science*, Acad. Press, 1981, 371). In most cases, their concentrations are very low. As a result, it has been difficult for brewing chemists to monitor their levels in beer and thereby deduce their role in affecting beer flavor.

Goal. Using a newly optimized derivatization and solid-phase microextraction (SPME) process, low-level aldehyde concentrations in beer stored for 4, 8, and 12 weeks at 30 °C were monitored. Although all aldehydes increased in concentration, no aldehyde exceeded its flavor threshold. (126 words)

Excerpt 9M (a poster Introduction based on Boesten et al., 2001)

INTRODUCTION

The synthesis of α-amino acids is important because they are used extensively in pharmaceuticals, agrochemicals, and as chiral ligands. The Strecker reaction is historically one of the most versatile ways to synthesize α-amino acids, but this method yields only 50% of a single enantiomer. Higher yields can be achieved by using chiral auxiliaries, but auxiliaries are often high in cost and low in availability.

Overview. To solve these problems, we present the first example of a crystallization-induced asymmetric transformation using optically pure (*R*)-phenylglycine amide **1** as a chiral auxiliary. The (*R*,*S*)-**3** diastereomer precipitates out of solution in 76–93% yield with a diastereomeric ratio (dr) > 99/1. (106 words)

Exercise 9.20

Glance again at excerpts 9K–9M. How well do the Introduction sections adhere to the move structure in figure 9.5? What generalizations can you make about sentence format, citation format, verb tense, and voice?

We conclude this part of the chapter by identifying some writing conventions that are characteristic of poster Introduction sections:

- Bulleted or numbered lists: Lists are uncommon in poster Introductions. More often, the Introduction is written in paragraph form with the text left-only or right-and-left justified (see chapter 10).
- Citations: In-text citations (typically fewer than four) are often included in posters, most commonly in the Introduction. Either a numerical or

author–date format may be used. Numerical formats are often preferred because they conserve space. When citations are used, a References section must also be included (see below). In some instances, authors insert the reference information (not merely a citation) into the text (e.g., "(Boesten et al. *Org. Lett.* **2001**, 3, 1121)"). In this case, a References section is not needed.

- Verb tense: The poster Introduction is written primarily in the present tense, though instances of the present perfect (to signal that knowledge gained from work completed in the past is still believed to be true in the present) are also possible.

Present	In most cases, their concentrations *are* very low.
	We *present* the first example of a crystallization-induced asymmetric synthesis.
Present perfect	Thus, countries *have established* recommended maximum limits for PCBs...
	As a result, it *has been* difficult for brewing chemists to monitor their levels...

- Voice and *we*: Both active and passive voice are used in the poster Introduction, although active voice is more common. The word *we* can be used to signal the current work, that is, the work that the authors will present in the poster.

Passive voice	Low-level aldehyde concentrations *were monitored.*
Active voice	*We present* an asymmetric Strecker reaction where...

Citations and References

Posters generally include a few citations and a References section. (See chapter 17.)

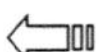

Present Perfect

See table 6.2.

Exercise 9.21

Rewrite the following passages (adapted from Webber et al., 2004, and Wei et al., 2001, respectively) so that they are more suitable for a poster Introduction section. Look for problems with lists, citations, verb tenses, voice, and conciseness.

a. This poster is about asbestos fibers that are naturally occurring hydrated silicate mineral fibers that found myriad uses in the twentieth century. However, airborne asbestos fibers became the well-recognized cause of asbestosis, bronchogenic carcinoma, and mesothelioma during the latter half of that century (Hansen et al., 1998; Timbrell, 1982). In this poster, we demonstrate the following:

- how we reconstructed airborne asbestos concentrations from the last century through a combination of
- paleolimnological methods
- particle-separation techniques
- analytical transmission electron microscopy

b. As everyone knows, plants have been used for centuries in herbalism, homeopathy, and aromatherapy because of their medicinal qualities. The long-term use of plants has led to recent observations about their antioxidant properties (*1*, *2*). Many scientists have observed antioxidant activities in compounds derived from the volatile constituents (*3*, *4*) and essential oil extracts[5,6] of plants. They have reported that ingestion of these volatile chemicals can prevent lipid peroxidation, which is associated with diseases such as cancer, leukemia, and arthritis. In the present study, analysis and antioxidative tests on the volatile extract isolated from a commercial beer were performed. Why did we choose beer? We chose beer because

   1. it is one of the most popular beverages in the world
   2. its popularity as a beverage is second only to soft drinks

---

9E Writing on Your Own: Draft Your Poster Introduction Section

Consider the importance of your research area, more generally, and your own research, more specifically. What background information will your audience need to grasp the importance of this project? Identify a few key references that you could include in your Introduction. List the key objectives of your research that are addressed in the poster. Then write the Introduction to your poster.

Considering Additional Poster Sections

We conclude the chapter with a brief examination of two additional poster sections: Acknowledgments and References.

The Poster Acknowledgments Section

Funding agencies and any individuals who helped you in the work and who are not named as authors of your poster should be acknowledged. If space permits, consider including a funding agency logo. (See figure 9.1 for an example of a poster Acknowledgments section with a logo.)

The Poster References Section

A References section is needed if you cite others' works in your poster, unless you inserted an abbreviated reference directly into the text. If you include citations in only one section of your poster (e.g., the Introduction), the References section can be placed at the end of that section; otherwise, include the references at the end or bottom of your poster. Format the references with the citation format used in the poster: number them in citation order (if you used numerical citations) or arrange them alphabetically (if you used author–date citations). Because of space limitations, references may be abbreviated. Consider the following examples for a poster with numerical citations:

REFERENCES

1. Smedes & Boer. *Trends Anal. Chem.* **1997**, *16*, 503.
2. Vesely et al. *J. Agric. Food Chem.* **2003**, *51*, 6941.
3. Erickson. In *Anal. Chem. of PCBs*, Lewis Publ. 1997.

Exercise 9.22

Rewrite these references using an abbreviated format:

a. Fredriksson, S.-Å.; Hulst, A. G.; Artursson, E.; de Jong, A. L.; Nilsson, C.; van Baar, B. L. M. Forensic Identification of Neat Ricin and of Ricin from Crude Castor Bean Extracts by Mass Spectrometry. *Anal. Chem.* **2005,** *77,* 1545–1555.

b. Hansen, N.; Klippenstein, S. J.; Miller, J. A.; Wang, J.; Cool, T. A.; Law, M. E.; Westmoreland, P. R.; Kasper, T.; Kohse-Hoinghaus, K. Identification of C_5H_x Isomers in Fuel-Rich Flames by Photoionization Mass Spectrometry and Electronic Structure Calculations. *Phys. Chem. A* **2006,** *110,* 4376–4388.

c. Heimann, A. C.; Jakobsen, R. Experimental Evidence for a Lack of Thermodynamic Control on Hydrogen Concentrations during Anaerobic Degradation of Chlorinated Ethenes. *Environ. Sci. Technol.* **2006,** *40,* 3501–3507.

d. Lockshin, R. A., Tilly, J. L., Zakeri, Z., Eds. *When Cells Die: A Comprehensive Evaluation of Apoptosis and Programmed Cell Death;* Wiley-Interscience: New York, 1998.

9F Writing on Your Own: Add Your Poster Title, Author List, Acknowledgments, and References

Add the title and author list to your poster (see chapter 8). If needed, write References and Acknowledgments sections following recommended formats. Proceed to chapter 10 to work on the layout of your poster.

Chapter Review

Check your understanding of what you've learned from this chapter by defining each of the following terms, in the context of the chapter, for a friend or colleague new to the field:

conference proceedings
dangling graphics
parallelism
poster sessions
sentence fragments

As a review, explain the answers to the following questions to a friend or colleague who has not yet written text for a scientific poster:

- What is the purpose of a poster?
- Who is the audience of a poster?
- What is the broad organizational structure of a poster?
- What are the moves of each section of a poster?
- What are common conventions regarding abbreviations, capitalization, and parentheses in a poster?
- In what sections of a poster are incomplete sentences appropriate? When should they not be used? In what sections are bulleted lists appropriate?
- How and where should hedging be used in a poster?
- In which section are citations most common? What is the average number of citations in a poster? Where are references listed?

Additional Exercises

Exercise 9.23

Sentences a–c are taken from journal articles. For each, (1) indicate which poster section the information would belong in (Introduction, Methods, Results, Discussion) and (2) rewrite the information so that it is appropriate for a poster.

a. Filters containing $PM_{2.5}$ collected at the five sites were extracted with PBS to give a solution/suspension of $PM_{2.5}$. (From Dellinger et al., 2001)

b. The isotherms for RAMEB-enhanced minerals show lower adsorption as compared to the pure minerals for all but KA + 9% RAMEB samples. (From Jozefaciuk et al., 2001)

c. Our results suggest that Cr^{3+} has an impact on DNA, DNA topology, and consequently processes leading to cell growth and proliferation. This could ultimately lead to the mutagenic and carcinogenic potential of Cr^{3+}. Because it is known that humans exposed to different Cr^{3+} species accumulate high levels of Cr^{3+} intracellularly (*17*), presented results may have an impact on human intake of Cr^{3+} as a nutrition additive. (Adapted from Plaper et al., 2002)

Exercise 9.24

Reflect on what you have learned about writing text for a poster. Select one of the reflection tasks below and write a thoughtful and thorough response:

a. Posters must be written and formatted to communicate important ideas with a minimal amount of effort on the part of viewers.
   - What are the keys to an effective poster?
   - In what ways do bulleted and numbered lists help?
   - How does the inclusion of graphics assist viewers?

b. Reflect on the challenges associated with deciding what to include in and what to exclude from a poster.
   - What types of information seem to be obligatory for inclusion?
   - What types of information seem to be optional?
   - What types of information should be avoided?
   - What questions will you pose to yourself to guide you in making these decisions?

c. Reflect on the differences between a poster and a journal article.
   - In what ways are posters and journal articles similar?
   - In what ways are posters and journal articles different?
   - What particular features contribute to poster effectiveness?

10 *Designing the Poster*

The visual appeal of a poster is important, but clever graphics and design are no substitute for good science.
—Frances Blanco-Yu, Seton Hill University

The purpose of this chapter is to help you design a poster that is visually appealing. Specific attention is paid to poster layout, font, and color. These design elements are illustrated with posters that we have created using the text introduced in chapter 9. Of course, what makes a poster attractive is (at least in part) a matter of taste, and many new design features will likely gain (and lose) popularity in the next decade. We cannot anticipate these changes; hence, we focus on a few basic principles of poster design that are likely to hold true over time. The guiding principle is to present your science in a way that is clear, crisp, and uncluttered. By the end of this chapter, you will be able to do the following:

- Select the most appropriate layout for your poster
- Select the font and font attributes for your poster
- Select the color scheme for your poster

The Designing on Your Own activities throughout the chapter will guide you in preparing your poster as you do the following:

10A Select a poster layout

10B Choose a font and font size

10C Add color and artwork

10D Finalize your poster

Although the focus of this chapter is on visual appeal, a good-looking poster is not a substitute for good science. Viewers visit your poster to learn about your science, not the latest trends in graphic design. Therefore, conservative, but effective, use of design elements is preferred over flashy, distracting design. Commonly used graphic design elements, such as photos, backgrounds, shadowing of text, and "artsy" fonts can dramatically enhance the appeal and clarity of a poster, but if used carelessly, they can turn the poster into a scattered and confusing mess. The goal is to use your sense of aesthetics for color and your creative energies to communicate your science and to make the poster inviting, accessible, and memorable for your audience.

Not long ago, a "poster" consisted of 8–12 sheets of paper cut and pasted onto individual pieces of colored construction paper. Today, most institutions have plotters, which can print multicolored single-page posters (e.g., 3 ft × 5 ft or 0.91 m × 1.52 m) from software files (e.g., Microsoft PowerPoint or Adobe Illustrator). We assume that you have access to such plotters and software, and we present only single-page posters in this textbook. Such advances make it possible for even the first-time presenter to create professional quality posters.

Reading and Analyzing Writing

We begin by asking you to read guidelines (excerpt 10A) for authors presenting a poster at an American Chemistry Society (ACS) conference. Most conferences offer a similar set of guidelines to poster session authors. Much of the advice will be the same, though some details will vary conference to conference (e.g., the permissible dimensions of the poster). Hence, we strongly encourage you to check the guidelines for your own specific conference.

Exercise 10.1

As you read excerpt 10A, look at the poster in figure 9.1. If possible, view the color version on the *Write Like a Chemist* Web site. Comment on how well the poster adheres to ACS guidelines.

Excerpt 10A (adapted from Instructions, Regulations for Speakers, Authors. American Chemical Society Web site. http://portal.acs.org/portal/Navigate?nodeid=907 (accessed January 3, 2008))

Instructions for Poster Session Authors

ACS Policies for Poster Sessions

- Each horizontal poster board measures 4 ft high × 6 ft wide (including frame). All presentations must be confined to the poster board itself.
- Each author is responsible for mounting his/her material prior to the opening of the poster session and for removing it IMMEDIATELY after the close of the session.

- Authors must remain with their posters for the duration of the session or as long as they are scheduled by their division, as indicated in the technical program. You are expected to display and discuss your results and answer questions from other attendees.

Design Suggestions for Scientific Posters

- Allow ample time to prepare your poster. All poster materials (illustrations, charts, and text) must be prepared in advance.
- All posters should feature a title, your name, the name of the institution(s) where the research was performed, and should credit other contributors, as appropriate.
- Use a crisp, clean design. All lettering should be legible from about 5 ft (1.5 m) away. Title lettering should be about 2 to 3 in. (5 to 7.5 cm). Subheading lettering should be 1/2 to 1 in. high (1.25 to 2.5 cm). Text lettering should be approximately 24 points (1/4 in. or 0.625 cm).
- Make illustrations simple and bold, with captions at least 3/8 in. high. Enlarge photos, tables, and charts to show pertinent details clearly.
- Do not tell the entire research history. Present only enough data to support your conclusions and show the originality of the work. The best posters display a succinct statement of major conclusions at the beginning, followed by supporting text and a brief summary at the end.
- Displayed materials should be self-explanatory, freeing you for discussion.
- Enhance your effectiveness by using a solid, colored background.
- Utilize other techniques to improve the graphic impact. Use color to add emphasis and clarity. Simplicity, ease of reading, etc. are more important than artistic flair. Keep in mind that lighting may be dim inside large poster sessions, so make sure your contrast and color combinations are easy to read.
- You may want to bring handouts of your abstract or copies of your data, poster, or conclusions to share with interested viewers.

Analyzing Poster Size and Layout

When well designed, posters are not simply journal papers pasted onto boards. Nor are they mounted sets of PowerPoint slides. Rather, posters... are a medium distinct in typography, layout, and style.
—Alley (2003)

When confronted with a single, very large "page" that comprises a poster, it is sometimes difficult to know where to begin. The first step is to determine the size of your poster because most conferences restrict poster size. The dimensions are limited by the physical size of the poster board available; hence, conference specifications must be followed. Unless stated otherwise, it is best to assume that the poster board includes a frame, so your poster must be smaller than the specifications given (e.g., for an ACS conference, the poster must be smaller than 4 ft × 6 ft). Remember, too, that your final poster will have margins; the dimensions of these margins will be determined by the size of the plotter paper. Hence, unless you want to trim the margins (this requires a poster cutter), you will need to adjust for this space. Common poster dimensions (without margins) are 3 × 4, 3 × 5, or 4 × 6 (height × length in feet) or 0.91 × 1.22, 0.91 × 1.52, or 1.22 × 1.83 (height × length in meters).

After the poster size is selected, you are ready to lay out the poster. (We mean this in a virtual sense, the layout on the "page" that you have created in your poster-making software program.) A well-designed layout makes the poster **flow**, that is, where the poster starts, how it moves (turns), and where it ends should be clear to the reader. Viewers read from left to right and from top to bottom; thus, the upper left corner is the "start" and the lower right corner is the "end." Guiding the reader through the "turns" is a bit more challenging. Two basic layout schemes are presented (with their respective turning schemes): the column layout and the row layout. Both are described in more detail below.

Flow

Flow refers both to the writing in a poster and to a poster's layout.

"Good flow" transitions the viewer logically and smoothly from one idea or section to the next.

Whichever layout you choose, be sure to include explicit headings in your poster. Headings serve as navigational signposts for the reader. Because viewers will spend only a few minutes at each poster, they need headings to guide them through the poster and help them locate the information that they seek. Most poster headings follow IMRD sequencing; for example, viewers expect introductory information (indicated by such headings as **Introduction**, **Background**, or **Objectives**) to precede Methods. Furthermore, they expect Methods to precede Results, and Results to precede the Discussion and Conclusions. In chapter 9, we noted that a variety of terms can be used for section headings; for example, **Overview and Importance** may be substituted for **Introduction** or used as a subheading within the Introduction section. There are no fixed conventions about the exact words to be used in headings and subheadings; what is anticipated, however, is that the introductory information comes first and the conclusions come last.

Poster Headings

Explicit headings serve as navigational signposts; clear headings contribute to the flow of an effective poster.

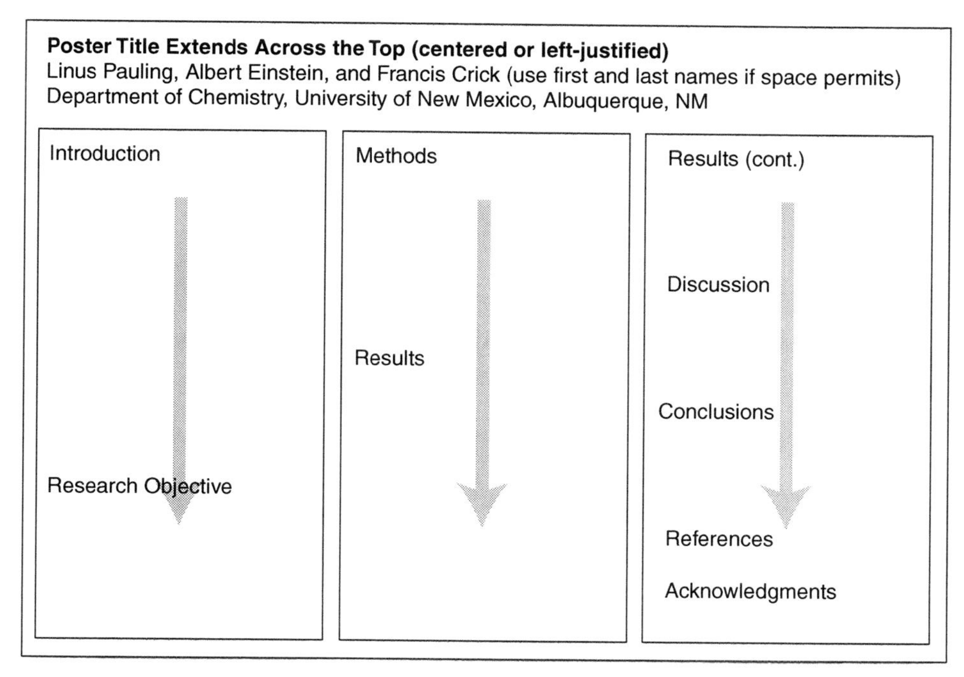

Figure 10.1 A typical column (or newspaper) layout for a scientific poster.

The column layout, illustrated in figure 10.1, is modeled after a newspaper; thus, it should be familiar to most readers. The column format is used by the vast majority of authors. Columns make it easier for multiple viewers, standing side by side, to read your poster at once. The title, authors, and affiliations go across the top of the poster and can be left-justified (as shown in figure 10.1) or centered. The sections of the poster begin in the upper-left corner. The viewer reads one column, reaches the bottom, and moves over to the top of the next column (to the right). The text itself moves the reader's attention down the column, making the flow easy to follow. The number of columns varies with the size of the poster, but three or four columns are typical.

We illustrate the column layout in an actual poster in figure 10.2. The poster (using the written text, tables, and figures from chapter 9) includes three columns. In this case, the three columns are approximately equal in width, but unequal column widths are also allowed (and common). Vertical lines may be used to separate the columns, although we did not do this in figure 10.2. Note, too, that Table 1 is not split between two columns; you must never start a table (or figure) at the bottom of one column and then continue it in the next. Another example of a three-column layout is illustrated in figure 9.1, where numbered subheadings (1, 2, 3, 4) are used to guide readers through a combined Results and Discussion section.

The row layout is illustrated in figure 10.3. This design inevitably includes two rows; typically, there is not enough space for more than two. The advantage

PCBs in Milk Samples by Saponification Solid-Phase Microextraction

María Llompart, Manuel Pazos, Pedro Landín, and Rafael Cela
Departamento de Química Analítica, Nutrición y Bromatología, Facultad de Química, Universidad de Santiago de Compostela, Spain

INTRODUCTION

Overview and Importance. The pollutants known as polychlorinated biphenyls (PCBs) are widely distributed in the environment due to their extensive use in the past, lipophilic (fat-loving) character, and general chemical stability. Moreover, PCBs tend to accumulate in the food chain. Thus, several countries have established recommended maximum limits for PCBs in food products. Among these products, milk is especially important because of its extensive and widespread consumption by humans.

Objective. To date, there is no simple and rapid procedure for testing PCBs in milk. Headspace solid-phase microextraction (HSSPME) is a promising approach, but only a few have applied this technique to milk (*1, 2*). Here, we present a simple and rapid saponification-HSSPME procedure for extracting PCBs from milk.

References

1. DeBruin et al. *J. Anal Chem.* **1998,** 70, 1986-1992.
2. Röhring & Meische. *J. Anal. Chem.* **2000,** 366, 106-111.

METHODS

Reagents and Materials

- Full-fat (3.61%) and skim milk (0.34%) samples were obtained from local supermarkets and spiked with one of eight PCB congeners (1-100 ng/mL).

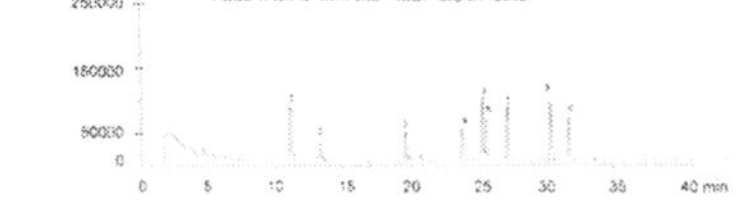

- Spiked samples were homogenized, held at 4 °C for 24 h, frozen, then thawed 1 h before analysis.
- SPME holders with one of two fiber assemblies were used:
  100 μm PDMS (poly(dimethylsiloxane))
  65 μm PDMS-DVB (poly(dimethylsiloxane)divinylbenzene)

METHODS (CONT.)

Saponification and HSSPME Procedures

- PCB-spiked milk samples were placed in headspace vials. When saponification was performed, samples were treated with NaOH, sealed, immersed in water (100 °C), and equilibrated for 6 min.
- During HSSPME, the fiber was exposed to the headspace over the sample for 5-240 min as the sample was stirred.

Sample Analysis

- PDMS-fibers were analyzed by GC -electron capture detection using a HP 5890 series II GC. PDMS-DVB fibers were analyzed by GC/MS-MS using a Varian 3800 GC with ion trap (Varian Saturn 2000).

RESULTS

(1) Before saponification. High sensitivity to PCBs in skim milk (Fig. 1A) but not full-fat milk (Fig. 1B) was achieved.

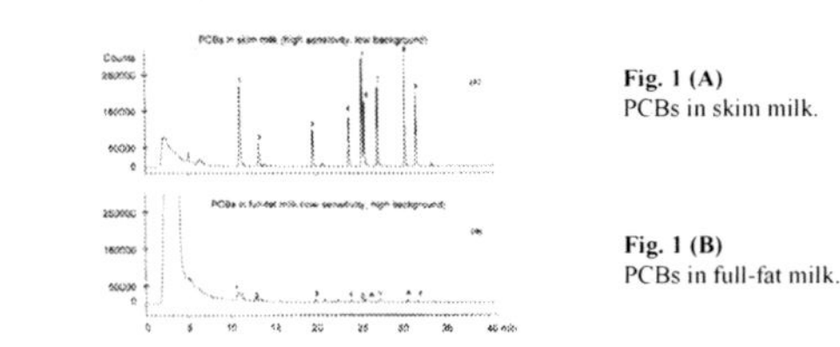

Fig. 1 (A) PCBs in skim milk.

Fig. 1 (B) PCBs in full-fat milk.

(2) Saponification Trials. Factors varied to improve extraction (Table 1).

Table 1. Factors varied in optimization process.

Factor	Values tried	Optimized value
1. NaOH vol (mL)	0.5, 2.0, 3.5	2.0-3.5
2. NaOH conc. (%)	20, 30	20%
3. Time (min)	30, 60	60*
4. Fiber	PDMS, PDMS-DVB	PDMS-DVB
5. Stirring	yes, no	yes

*longer times were better but too slow for desired approach

RESULTS (CONT.)

(3) After Saponification. Skim milk response increased 5 to 10 times; full-fat milk response increased 4 to 8 times, depending on PCB congener (Fig. 2).

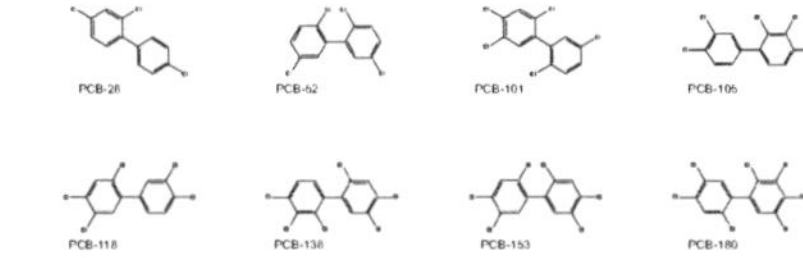

Fig. 2. PCBs in full-fat milk with saponification.

(4) Procedure Validation. Calibration results were linear ($R^2 > 0.994$) and reproducible (RSD < 11.0 %) over a wide range of concentrations. Good results were also achieved with a certified milk sample (Table 2).

Table 2. Procedure validation with certified milk (25% fat).

PCB	certified value (ng/g)	recovery (%)
PCB-52	1.16 ± 0.17	75
PCB-118	3.19 ± 0.24	97
PCB-153	18.6 ± 1.1	98
PCB-156	1.64 ± 0.11	102
PCB-180	9.29 ± 0.26	88

DISCUSSION

PCBs are more difficult to extract from full-fat milk than skim milk, suggesting that PCBs are more strongly retained in the full-fat matrix. Even at 100 °C, SPME without saponification was not efficient. Saponification of milk fats to their corresponding glycerols and carboxylates appears to facilitate the release of PCBs.

CONCLUSIONS

Saponification enables a more efficient SPME process for the detection of PCBs in full-fat milk. An increase in response of 4 to 9 times can be achieved depending on the PCB congener.

Acknowledgment. We gratefully acknowledge financial support from the Xunta de Galicia (Conselleria de Medio Ambiente), project PGIDT99MA23701.

Figure 10.2 The column layout, illustrated in a poster based on Llompart et al. (2001).

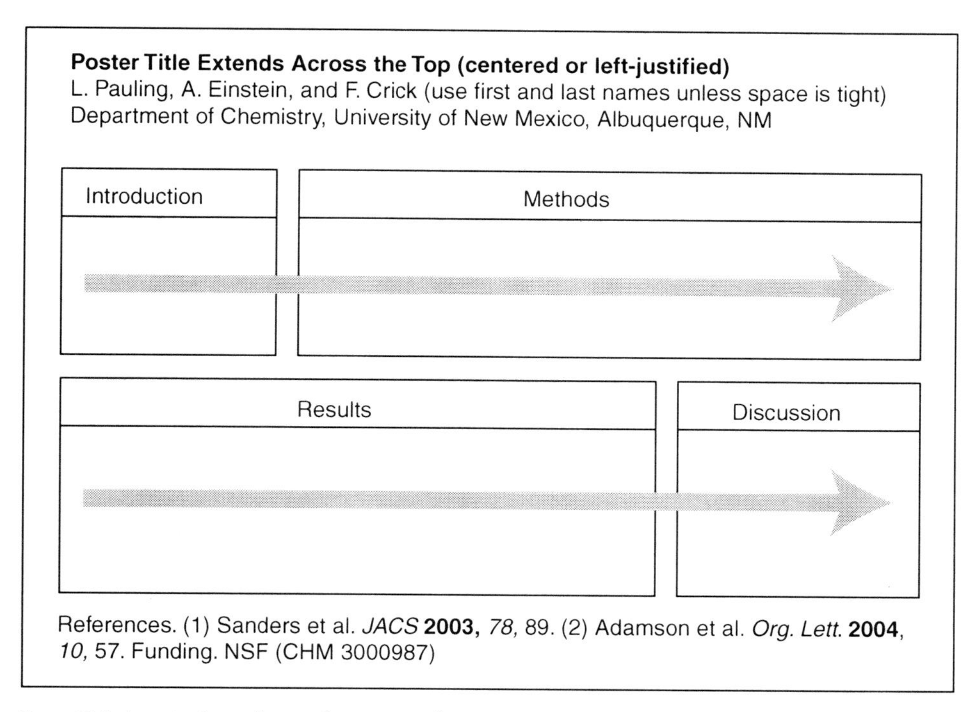

Figure 10.3 A typical row layout for a scientific poster. Pure row layouts are uncommon in posters.

of this approach is that it allows two or more figures to be arranged side by side for easy comparison. However, this layout is difficult to follow visually, and it requires viewers to move from left to right twice (for two rows) as they read the poster, which is an inconvenience if the conference is crowded. For these reasons, most authors avoid the pure row layout and instead used a combined column-and-row approach, illustrated in figure 10.4. The left-hand column reads from top to bottom (column layout), but the Results section (with two sets of figures placed side by side) and the Discussion and Conclusions sections are in row layout.

Figures 10.1–10.4 represent just a few examples of how to lay out a poster. Each poster is different; hence, each will require variations on these themes. The only hard-and- fast rule is that the flow should be logical and clear to your viewers.

Our discussion thus far might suggest that one writes the text of the poster first and then simply "cuts and pastes" it into the appropriate layout. In fact, the text evolves in an iterative way. One drafts the text first, capturing the key ideas that will be shared, but then the text is altered (numerous times) until it fits into the allotted space. Typically, you have less space than you first imagined, requiring that you reduce the number of lines and words in the text. The final texts presented in chapter 9 went through just such an iterative process. We wrote the text first, realized (all too often) that the text would not fit, and then revised it to make it more concise.

Detection of Low-Level Aldehydes in Aged Beer

Petr Vesely†,‡, Lance Lusk,† Gabriela Basarova,‡ John Seabrooks,† and David Ryder†
Miller Brewing Company† and Institute of Chemical Technology,‡ Prague, Czech Republic

Introduction

Overview and Importance
American packaged beers lose their flavor and become stale during storage. Carbonyl compounds, particularly aldehydes, may be involved in this process. Aldehydes are formed in beer through one of several possible mechanisms including Strecker degradation of amino acids or linoleic acid (Pollock. In *Brewing Science*, Acad. Press, 1981, 371). In most cases, their concentrations are very low. As a result, it has been difficult for brewing chemists to monitor their levels in beer and thereby deduce their role in affecting beer flavor.

Goal. Using a newly optimized derivatization and solid-phase microextraction (SPME) process, low-level aldehyde concentrations in beer stored for 4, 8, and 12 weeks at 30 °C were monitored. Although all aldehydes increased in concentration, no aldehyde exceeded its flavor threshold.

Methods

Samples. The beer samples (American lagers) were stored at 30 °C for 4, 8, or 12 wk. The controls were stored at 0 °C for 12 wk.

Methods Validation
- Calibration curves were prepared for 9 aldehydes (linearity >0.96).
- Each beer sample was measured 10 times (variance <5.5%).
- Recovery was determined by spiking beer samples with 10 ppb of the standard aldehyde (recoveries 89-110%).

On-Fiber Derivatization and Detection of Aldehydes
- The derivatizing agent *O*-(2,3,4,5,6-pentafluorobenzyl)-hydroxylamine (PFBOA) was absorbed onto a 65-_m PDMS/DVB SPME fiber (10 min, 50 °C).
- The fiber was placed in the sample headspace (50 °C, 60 min).
- The beer aldehydes selectively reacted with the PFBOA in the fiber.
- The oximes that formed were desorbed and detected by GC/MS (DB-5 30 m _ 0.25 mm _ 0.50 _m; He carrier gas at 1.1 mL/min; splitless, injector 250 °C).

Acknowledgment. The authors gratefully acknowledge support from the Miller Brewing Co.

Results

Spiked Beer Samples

A beer sample was spiked with 9 aldehydes and derivatized. The chromatogram is shown in Fig. 1. Aldehydes were detected by *m/z* 181 (Fig. 2).

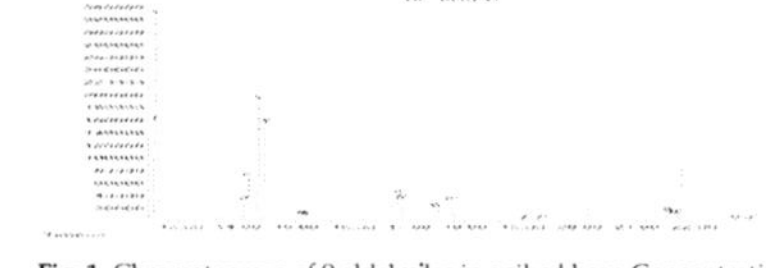

Fig. 1. Chromatogram of 9 aldehydes in spiked beer. Concentrations were determined by adding together both isomers (e.g., 1, 1′).

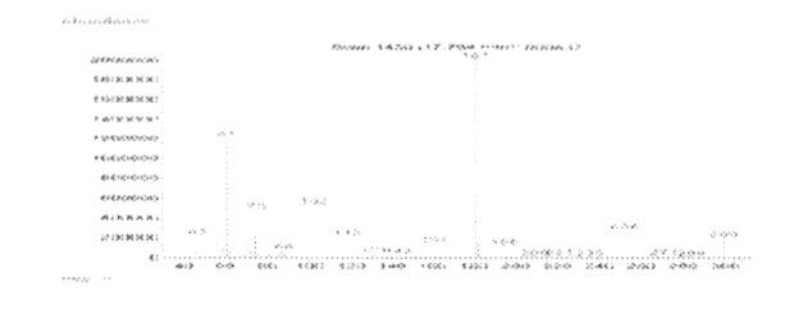

Fig. 2. Aldehydes were identified by *m/z* 181 (methional is shown).

Aged Beer Samples

All aldehydes increased during storage (Fig. 3), but none exceeded their flavor threshold (Fig. 4).

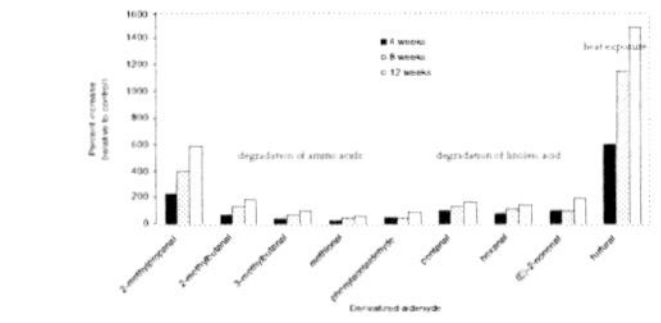

Fig. 3. Increase of aldehydes (grouped by source) during storage.

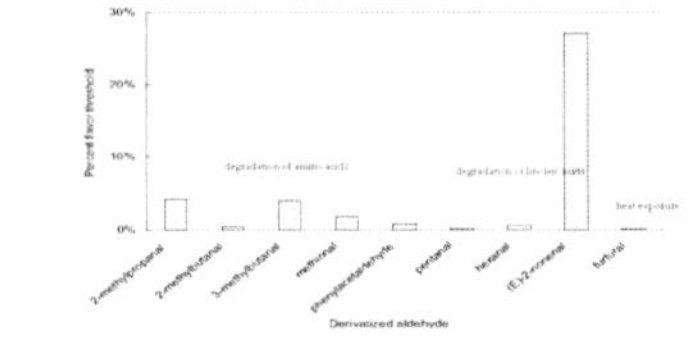

Fig. 4. Percent of flavor threshold at 12 weeks for 9 aldehydes.

Discussion

During long-term storage (4–12 wk) at elevated temperatures (30 °C), American-style beers develop a stale flavor. Nine aldehydes were analyzed as possible contributors to this flavor loss:
- 5 Strecker aldehydes (Figs. 3 and 4). All 5 increased during storage, but none exceeded >4.3% of their respective flavor threshold.
- Pentanal and hexanal (degradation products of linoleic acid). Both aldehydes were <1% of their flavor thresholds.
- (*E*)-2-nonenal (a degradation product of linoleic acid). This aldehyde increased to 27% of its flavor threshold but is not expected to contribute to stale beer flavor (Schieberle & Komarek, ACS, Chicago, Fall 2001).
- Furfural (a heat exposure indicator). This aldehyde increased more than 2-fold but reached only 1% of its flavor threshold.

Conclusions

The increase after 12 wks for most aldehydes was significant (16-fold for furfural, 7-fold for 2-methylpropanal); however, no aldehyde approached its respective flavor threshold.

We are currently exploring additive or synergistic effects among these aldehydes to explain the stale flavor of aged beer.

Figure 10.4 A poster based on Vesely et al. (2003), which illustrates a combined column-and-row layout. This approach allows figures to be placed side by side and viewed from left to right.

Exercise 10.2

Below are early drafts of the text written for two posters introduced in chapter 9. The drafts were ultimately shortened because of space limitations in the final posters. Without referring to the final posters, rewrite each text so that it is more concise; aim for the final word counts noted for each draft.

a. Here is the initial draft of the Discussion section for the poster based on Llompart et al. (2001). This draft is 83 words. The final version is 50 words (with no bullets).

   - PCBs are more difficult to extract from full-fat milk than skim milk, suggesting that PCBs are more strongly retained in the full-fat sample matrix. Even at 100 °C, SPME (without saponification) is not an efficient technique.
   - Saponification of milk fats to their corresponding glycerols and carboxylates appears to facilitate the release of PCBs from the sample matrix. With saponification, the SPME process was 4 to 9 times more effective, depending on the fat content of the milk and the PCB congener.

b. Here is the initial draft of the first paragraph in the Introduction section for the poster based on Boesten et al. (2001). This draft is 98 words, and the final version is 64 words.

   The asymmetric synthesis of α-amino acids is an important topic due to their extensive use in pharmaceuticals and agrochemicals and as chiral ligands. The Strecker reaction is historically one of the most versatile ways to produce α-amino acids, but this method has a maximum yield of only 50% for a single enantiomer. Higher yields can be achieved by using chiral auxiliaries, but auxiliaries have other drawbacks, such as high cost, low availability, the need for purification, and high loss rates. A possible solution to these problems would be to use a chiral auxiliary in a crystallization-induced asymmetric transformation.

10A Designing on Your Own: Select a Poster Layout

Look over the text that you have written for your poster (chapter 9). Next, decide what section headings you will use, what graphics you will include in each section, and how much space each section will require. When these pieces are in place, experiment with layout and select a format (i.e., the column, row, or column-and-row format) that is best for your poster. Begin revising your text and graphics so that they fit the allotted spaces.

Analyzing Fonts and Font Size

Fonts can enhance or detract from an effective poster. There are two categories of fonts: **serif** and **sans-serif**. Serif fonts have letters with little "tags" at the end of each straight line. The lines in sans-serif fonts end abruptly without tags. Serif fonts are considered more traditional, while sans-serif fonts have a more modern look. The two most commonly used fonts among chemists are Times New Roman (a serif font) and Arial (a sans-serif font). These and other fonts are illustrated in table 10.1.

Serif and Sans-Serif Fonts

Serif font: A font with short, light lines or curves (called "serifs") that project from the top or bottom of a main stroke of a letter.

Sans-serif font: A clean modern font that has letters without serifs.

Many authors combine two different fonts in a poster. One approach is to use a sans-serif font for the title and headings (e.g., Arial, Arial Black, Franklin Gothic Heavy, or Tahoma) and a highly legible serif font for the text (e.g., Times New Roman or Bookman Old Style). Arial is also a popular font for the poster text. Three different two-font combinations are illustrated in figure 10.5. We caution against using more than two fonts in your poster, or choosing overly decorative or fancy fonts (there are hundreds to choose from). Select a crisp, clean combination that is easy to read, and then continue with the rest of your poster design.

Table 10.1 Some common serif and sans-serif fonts (all shown in 12 point size).

Font type	Font example
Serif	This is an example of Century.
	This is an example of Book Antiqua.
	This is an example of Bookman Old Style.
	This is an example of Times New Roman.
Sans-Serif	This is an example of Arial.
	This is an example of Century Gothic.
	This is an example of Comic Sans MS.
	This is an example of Franklin Gothic Medium.
	This is an example of Tahoma.

Introduction	**Introduction**	**Introduction**
The synthesis of ∝-amino acids is important because they are used extensively in pharmaceuticals, agrochemicals, and as chiral ligands. The Strecker reaction is historically one of the most versatile ways to synthesize ∝-amino acids, but this method yields only 50% of a single enantiomer. Higher yields can be achieved by using chiral auxiliaries, but auxiliaries are often high in cost and low in availability. **Overview**. To solve these problems, we present the first example of a crystallization-induced asymmetric transformation using optically pure (*R*)-phenylglycine amide **1** as a chiral auxiliary. The (*R*,*S*)-**3** diastereomer precipitates out of solution in 76–93% yield with a diastereomeric ratio (dr) > 99/1.	The synthesis of ∝-amino acids is important because they are used extensively in pharmaceuticals, agrochemicals, and as chiral ligands. The Strecker reaction is historically one of the most versatile ways to synthesize ∝-amino acids, but this method yields only 50% of a single enantiomer. Higher yields can be achieved by using chiral auxiliaries, but auxiliaries are often high in cost and low in availability. **Overview**. To solve these problems, we present the first example of a crystallization-induced asymmetric transformation using optically pure (*R*)-phenylglycine amide **1** as a chiral auxiliary. The (*R*,*S*)-**3** diastereomer precipitates out of solution in 76–93% yield with a diastereomeric ratio (dr) > 99/1.	The synthesis of ∝-amino acids is important because they are used extensively in pharmaceuticals, agrochemicals, and as chiral ligands. The Strecker reaction is historically one of the most versatile ways to synthesize ∝-amino acids, but this method yields only 50% of a single enantiomer. Higher yields can be achieved by using chiral auxiliaries, but auxiliaries are often high in cost and low in availability. **Overview**. To solve these problems, we present the first example of a crystallization-induced asymmetric transformation using optically pure (*R*)-phenylglycine amide **1** as a chiral auxiliary. The (*R*,*S*)-**3** diastereomer precipitates out of solution in 76–93% yield with a diastereomeric ratio (dr) > 99/1.
Heading font: Century Gothic Text font: Arial	Heading font: Comic Sans MS Text font: Times New Roman	Heading font: Tahoma Text font: Bookman Old Style

Figure 10.5 Three examples of two-font combinations; one font is used for the word "Introduction" (bolded), and a different font is used for the text. All font sizes are 9 points.

In addition to font style, authors must also pay careful attention to font size. The rule of thumb recommended by the ACS (excerpt 10A) is that the text should be easy to read from a distance of at least 5 feet (1.5 m). One way to check this on your computer screen is to set your poster to 100% scale and literally step back 5 feet. This is often difficult, however, because the text is so large that only a small portion of the poster can be viewed at any one time on the screen. Alternatively, you can refer to table 10.2, which gives a range of font sizes that are common in different sections of a poster using Arial font. You will need to adjust the font size accordingly if you use a font larger or smaller than Arial. Most agree that font sizes less than 24 points (1/4 in.) are too small and should not be used.

A poster should be easily read from a distance of 5 feet (1.5 m). Your smallest font size should be at least 24 points (1/4 in.). (Point, abbreviated as pt., refers to font size.)

Text style and special effects (e.g., bold, italics, shadows) must also be considered when making decisions about fonts. Bolding can be used effectively, primarily for the title, authors' names and affiliations, section headings, and graphics labels (e.g., **Table 1**, **Fig. 3**) contained in titles or captions (not in the main text). Too much bolding, however, can distract the viewer (or, worse, divert the viewer's attention from important content) and make your poster difficult to read.

Table 10.2 Common font sizes for different sections of a poster in Arial font.

Poster Section	Height (in.)	Point size (Arial)	Example (points in Arial font)
Titles	3	288	96
	2	192	
	1	96	
Headings, subheadings, names, affiliations	1	96	48
	1/2	48	
Text, captions, references	1/2	48	36
	3/8	36	
	1/4	24	24

Right and
left
justificatio
n can
cause odd
spacing
between
words.

Right and
left
justification
can cause
odd spacing
between
words.

Right and left justification can
cause odd spacing between
words.

Figure 10.6 Odd spacing created in justified text can sometimes be improved by changing the width of the text box.

The poster title and section headings are the only appropriate place for most other text effects (e.g., *italics*, shadow, outline, ALL CAPS), because they generally make the text more difficult to read.

Text **justification** (i.e., alignment) must also be considered. For many years, scientists preferred to leave the right edge unjustified or jagged. With word processing, however, justified text (like a newspaper column that is aligned on both the left- and right-hand margins) has become more common. In a poster, it is common to see large blocks of text (e.g., the Introduction section) justified, and short sections of text (e.g., bulleted lists) left-justified (jagged on the right). If you use full justification (i.e., left and right edges), look out for odd spacing between words. Awkward spacing can usually be fixed by changing the column width, as shown in figure 10.6, where the problem is solved by widening the column. If changing the width does not work, consider using left-justified text.

Justification

Justification refers to the alignment of the left and right edges of the text.

Whether to align both the right- and left-hand margins ("justified") or leave the right margin jagged ("left-justified") is a matter of taste, although justified text is becoming more common.

10B Designing on Your Own: Choose a Font and Font Size

Experiment with the font style and font sizes for your poster. Choose a style that is easy to read and large enough to see from 5 feet (1.5 m) away. Continue to revise your text as needed to accommodate the layout, font(s), and font sizes selected.

One hint for working with text: Use text boxes rather than placing your text directly into the poster. Text boxes make the text easier to move around and to resize, if needed.

Analyzing Color and Artwork

Color is potentially the most useful tool for creating an effective and aesthetically pleasing poster, but it is also the most subjective. As with layout and font, colors should be chosen to help communicate your scientific message. Follow these few simple guidelines:

- Use contrast: Make sure that your text contrasts sharply with the background. You may use the traditional approach, a dark text on a light background, or an **inverted** color scheme, light text on a dark background. If you use an inverted color scheme, consider using larger font sizes; the dark background tends to encroach on light letters, making them appear smaller.
- Avoid red and green: Most colorblind individuals cannot distinguish red and green. Avoid using red text on a green background (or vice versa) and red and green in graphs.
- Include text boxes: If you want a dark background, you can still achieve good contrast with dark text by placing text in "boxes" filled with a light color. The borders of the boxes can be rounded or square, depending on the effect that you are trying to achieve.
- Limit yourself to three or four colors: Use colors consistently throughout your poster (i.e., use one color for text, one color for backgrounds, one color for borders, etc.).
- Use conservative colors for the majority of your poster; save bright colors for accents and highlights.
- Use color to enhance your science: Colors should facilitate understanding, be clearly distinguished, and, when possible, convey meaning (e.g., blue for water, red for heat). Create a "color code" for your own consultation and use it consistently throughout the poster. For example, if you use blue and gray to represent compounds **1** and **2** in your first graph, use blue and gray for these compounds in additional graphs. Readers will grow accustomed to what each color represents and will be able to interpret your poster more quickly.
- Minimize background distraction: The background color or graphic should not overpower the text or distract viewers from the scientific content. For a brief period of time, blurred and faded images were popular for poster backgrounds, but too often they interfered with the text. Now consensus favors a one-color, solid background, although shading is sometimes used.
- Preprint your poster to check colors: Remember that the colors that you view on your computer screen often look different when printed; print a small-scale color draft of your poster before you print the full-scale version.

Inverted Color Scheme

An inverted color scheme uses light letters on a dark background. Because dark backgrounds appear to "shrink" light letters, use larger font sizes for inverted color schemes.

Posters may also contain different types of illustrative artwork, such as photographs, maps, and logos. When used appropriately, artwork can greatly enhance the visual appeal of the poster and add to the viewer's understanding of the research. Photographs of equipment, a field-sampling site, or even scientists performing a particular technique are all appropriate and commonplace in posters. Maps are also common, illustrating the location of a research study. Finally, small color logos can be placed near the title or Acknowledgments section of posters to indicate the researchers' institutions and funding sources, respectively. Logos should not be used in place of text; rather, they should be used to reinforce what is stated in the text.

Logos

Logos are commonly included in titles and Acknowledgments sections of posters.

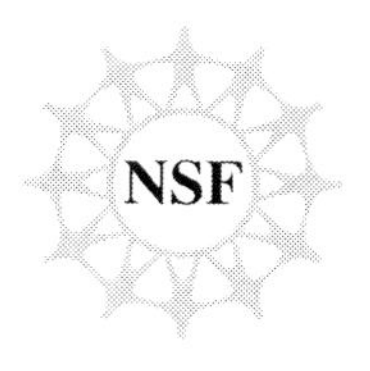

It is important to distinguish between scientific graphics and artwork. Scientific graphics (figures, tables, charts) should be mentioned in the text (e.g., "Figure 1") and include a title or caption, just as they would in a journal article or research proposal. Illustrative artwork, however, can be left "dangling" (i.e., included without being mentioned in the text). Photographs may fit into either category, depending on their content. If an instrument, site, or technique is featured in the photograph, it should have a caption and should be referenced in the text. A photograph of a research group, research building, or familiar campus landmark can be left dangling.

To conclude this section and chapter, we present three sample posters (figures 10.7–10.9) created from text developed in chapter 9 and design elements discussed in this chapter. (See also figure 9.1 in chapter 9 for a fourth sample poster.) Table 10.3 summarizes some of the design elements of these posters. Full-color versions of these posters are available at the *Write Like a Chemist* Web site. Together, we hope that these four posters stimulate ideas for preparing your own poster.

Detection of Low-Level Aldehydes in Aged Beer

Petr Vesely†,‡, Lance Lusk,† Gabriela Basarova,‡ John Seabrooks,† and David Ryder†
Miller Brewing Company† and Institute of Chemical Technology,‡ Prague, Czech Republic

Introduction

Overview and Importance. American packaged beers lose their flavor and become stale during storage. Carbonyl compounds, particularly aldehydes, may be involved in this process. Aldehydes are formed in beer through one of several possible mechanisms including Strecker degradation of amino acids or linoleic acid (Pollock. In *Brewing Science*, Acad. Press, 1981, 371). In most cases, their concentrations are very low. As a result, it has been difficult for brewing chemists to monitor their levels in beer and thereby deduce their role in affecting beer flavor.

Goal. Using a newly optimized derivatization and solid-phase microextraction (SPME) process, low-level aldehyde concentrations in beer stored for 4, 8, and 12 weeks at 30 °C were monitored. Although all aldehydes increased in concentration, no aldehyde exceeded its flavor threshold.

Methods

Samples

- The beer samples (American lagers) were stored at 30 °C for 4, 8, or 12 weeks.
- The controls were stored at 0 °C for 12 weeks.

Methods Validation

- Calibration curves were prepared for 9 aldehydes (linearity > 0.96).
- Each beer sample was measured 10 times (variance < 5.5%).
- Recovery was determined by spiking beer samples with 10 ppb of the standard aldehyde (recoveries 89–110%).

On-Fiber Derivatization and Detection of Aldehydes

- The derivatizing agent *O*-(2,3,4,5,6-pentafluorobenzyl)-hydroxylamine (PFBOA) was absorbed onto a 65 µm PDMS/DVB SPME fiber (10 min, 50 °C).
- The fiber was placed in the sample headspace (50 °C, 60 min).
- The beer aldehydes selectively reacted with the PFBOA in the fiber.
- The oximes that formed were desorbed and detected by GC/MS (DB-5 30 m × 0.25 mm × 0.50 µm; He carrier gas at 1.1 mL/min; splitless, injector 250 °C).

Acknowledgment. The authors gratefully acknowledge support from the Miller Brewing Co.

Results

Spiked Beer Samples

A beer sample was spiked with 9 aldehydes and derivatized. The chromatogram is shown in Fig. 1. Aldehydes were detected by *m/z* 181 (Fig. 2).

Fig. 1. Chromatogram of 9 aldehydes in spiked beer. Each aldehyde has an isomer (e.g., 1, 1′).

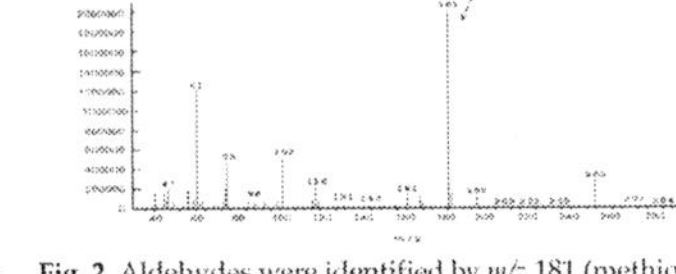

Fig. 2. Aldehydes were identified by *m/z* 181 (methional shown).

Aged Beer Samples

All aldehydes increased during storage (Fig. 3), but none exceeded their flavor threshold (Fig. 4).

Fig. 3. Increase of aldehydes (grouped by source) during storage.

Fig. 4. Percent of flavor threshold at 12 weeks for 9 aldehydes.

Discussion

During long-term storage (4–12 weeks) at elevated temperatures (30 °C), American-style beers develop a stale flavor. Nine aldehydes were analyzed as possible contributors to this flavor loss:

- **5 Strecker aldehydes** (Figs. 3 and 4). All 5 aldehydes increased during storage, but none exceeded more than 4.3% of their respective flavor thresholds.
- **Pentanal and hexanal** (degradation products of linoleic acid). Both aldehydes were less than 1% of their flavor thresholds.
- **(*E*)-2-nonenal** (a degradation product of linoleic acid). This aldehyde increased to 27% of its flavor threshold but is not expected to contribute to stale beer flavor (Schieberle & Komarek, ACS, Chicago, Fall 2001).
- **Furfural** (a heat exposure indicator). This aldehyde increased more than 2-fold but reached only 1% of its flavor threshold.

Conclusions

The increase after 12 weeks for most aldehydes was significant (16-fold for furfural, 7-fold for 2-methyl-propanal); however, no aldehyde approached its respective flavor threshold.

We are currently exploring additive or synergistic effects among these aldehydes to explain the stale flavor of aged beer.

Figure 10.7 The final design of the poster based on Llompart et al. (2001).

PCBs in Milk Samples by Saponification–Solid-Phase Microextraction

María Llompart, Manuel Pazos, Pedro Landín, and Rafael Cela

Departamento de Química Analítica, Nutrición y Bromatología, Facultad de Química, Universidad de Santiago de Compostela, Spain

INTRODUCTION

Overview and Importance. The pollutants known as polychlorinated biphenyls (PCBs) are widely distributed in the environment due to their extensive use in the past, lipophilic (fat-loving) character, and general chemical stability. Moreover, PCBs tend to accumulate in the food chain. Thus, several countries have established recommended maximum limits for PCBs in food products. Among these products, milk is especially important because of its extensive and widespread consumption by humans.

Objective. To date, there is no simple and rapid procedure for testing PCBs in milk. Headspace solid-phase microextraction (HSSPME) is a promising approach, but only a few have applied this technique to milk (*1, 2*). Here, we present a simple and rapid saponification-HSSPME procedure for extracting PCBs from milk.

References

1. DeBruin et al. *J. Anal Chem.* **1998,** 70, 1986–1992.
2. Röhring & Meische. *J. Anal. Chem.* **2000,** 366, 106–111.

METHODS

Reagents and Materials

- Full-fat (3.61%) and skim milk (0.34%) samples were obtained from local supermarkets and spiked with one of eight PCB congeners (1–100 ng/mL).

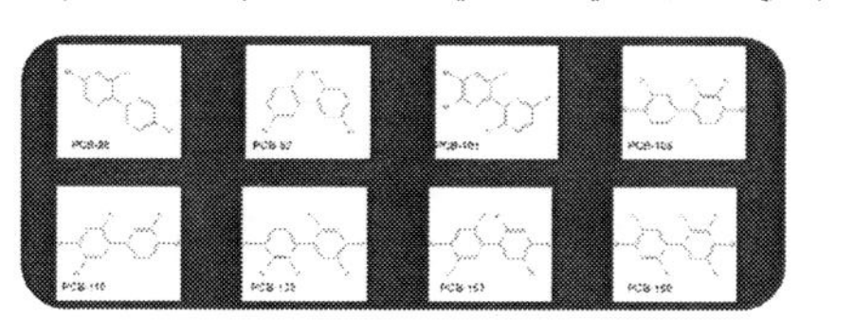

- Spiked samples were homogenized, held at 4 °C for 24 h, frozen, then thawed 1 h before analysis.
- SPME holders with one of two fiber assemblies were used:
  100 µm PDMS (poly(dimethylsiloxane))
  65 µm PDMS-DVB (poly(dimethylsiloxane)divinylbenzene)

METHODS (CONT.)

Saponification and HSSPME Procedures

- PCB-spiked milk samples were placed in headspace vials. When saponification was performed, samples were treated with NaOH, sealed, immersed in water (100 °C), and equilibrated for 6 min.
- During HSSPME, the fiber was exposed to the headspace over the sample for 5-240 min as the sample was stirred.

Sample Analysis

- PDMS-fibers were analyzed by GC-electron capture detection using a HP 5890 series II GC.
- PDMS-DVB fibers were analyzed by GC/MS–MS using a Varian 3800 GC with ion trap (Varian Saturn 2000).

RESULTS

(1) Before saponification. PCBs were detected in skim milk (Fig. 1A) but not full-fat milk (Fig. 1B) because of matrix effects in full-fat milk.

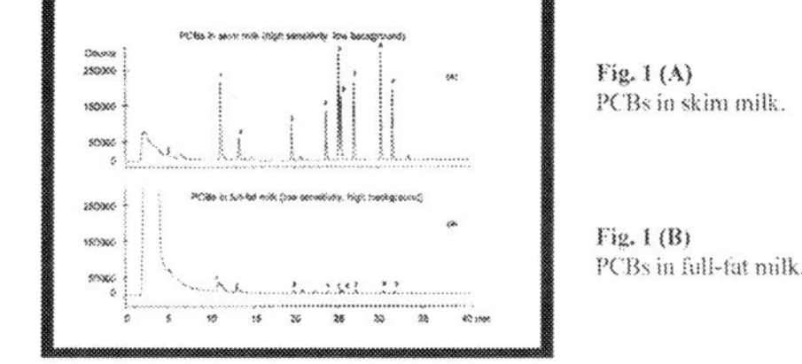

Fig. 1 (A) PCBs in skim milk.

Fig. 1 (B) PCBs in full-fat milk.

(2) Saponification. Extraction of PCBs from full-fat milk matrix was improved by saponification (Table 1).

Table 1. Optimization of saponification process.

Factor	Options Tried	Best Option
NaOH vol (mL)	0.5, 2.0, 3.5	2.0-3.5
NaOH conc. (%)	20, 30	20%
Time (min)	30, 60	60*
Fiber	PDMS, PDMS-DVB	PDMS-DVB
Stirring	yes, no	yes

*Longer times were better but too slow for desired approach

RESULTS (CONT.)

(3) After Saponification. Skim milk response increased 5 to 10 times; full-fat milk response increased 4 to 8 times, depending on PCB congener (Fig. 2).

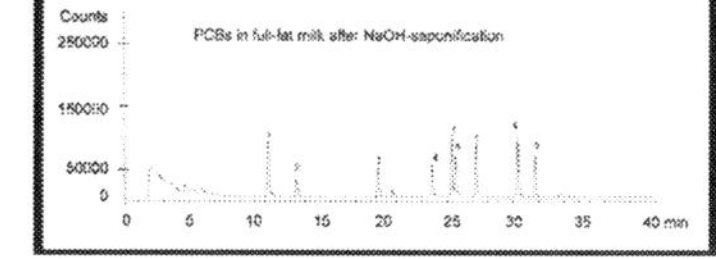

Fig. 2. PCBs in full-fat milk with saponification.

(4) Procedure Validation. Calibration results were linear ($R^2 > 0.994$) and reproducible (RSD < 11.0 %) over a wide range of concentrations. Recoveries ≥75% were achieved with a certified milk sample (Table 2).

Table 2. Procedure validation with certified milk (25% fat).

PCB	Certified Value (ng/g)	Recovery (%)
52	1.16 ± 0.17	75
118	3.19 ± 0.24	97
153	18.60 ± 1.10	98
156	1.64 ± 0.11	102
180	9.29 ± 0.26	88

DISCUSSION

PCBs are more difficult to extract from full-fat milk than skim milk, suggesting that PCBs are more strongly retained in the full-fat matrix. Even at 100 °C, SPME without saponification was not efficient. Saponification of milk fats to their corresponding glycerols and carboxylates appears to facilitate the release of PCBs.

CONCLUSIONS

Saponification enables a more efficient SPME process for the detection of PCBs in full-fat milk. An increase in response of 4 to 9 times can be achieved depending on the PCB congener.

Acknowledgment. We gratefully acknowledge financial support from the Xunta de Galicia (Consellería de Medio Ambiente), project PGIDT99MA23701.

Figure 10.8 The final design of the poster based on Vesely et al. (2003).

Asymmetric Strecker Synthesis via a Crystallization-Induced Asymmetric Transformation using a Chiral Auxiliary

Wilhelmus H. J. Boesten, Jean-Paul G. Seerden, and Ben de Lange

DSM Research Life Sciences-Organic Chemistry and Biocatalysis, The Netherlands

Introduction

The synthesis of α-amino acids is important because they are used extensively in pharmaceuticals, agrochemicals, and as chiral ligands. The Strecker reaction is historically one of the most versatile ways to synthesize α-amino acids, but this method yields only 50% of a single enantiomer. Higher yields can be achieved by using chiral auxiliaries, but auxiliaries are often high in cost and low in availability.

Overview. To solve these problems, we present the first example of a crystallization-induced asymmetric transformation using optically pure (*R*)-phenylglycine amide **1** as a chiral auxiliary. The (*R,S*)-3 diastereomer precipitates out of solution in 76-93% yield with a diastereomeric ratio (dr) > 99/1.

Experimental Section

General. (*R*)-Phenylglycine amide **1** was purchased from DSM (Netherlands). The dr of (*R,S*)-**3** and (*R,R*)-**3** was determined by ^{1}H NMR using the relative integration between the *t*-Bu signals at 1.05 ppm for (*R,S*)-**3** and 1.15 ppm for (*R,R*)-**3**. NMR spectra were recorded using a Varian VXR-300 spectrometer (300 mHz) and referenced to residual solvent.

DSM

Experimental Section (cont.)

The synthesis of amino nitrile (*R,S*)-3.

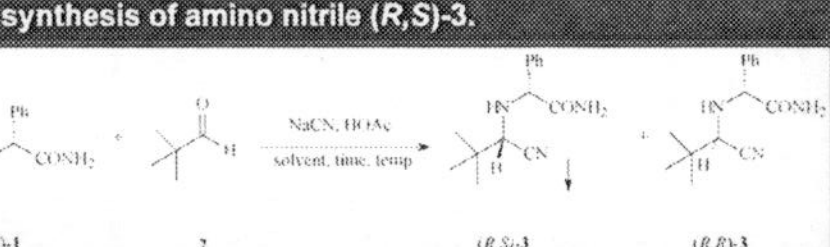

- A stirred suspension of **1** (400 mmol) in H_2O (400 mL) was added to pivaldehyde **2** (419 mmol) and stirred for 30 min.
- NaCN/HOAc was added at 23–28 °C. The mixture was heated to 70 °C and stirred for 24 h.
- After cooling to 30 °C, the product was filtered, washed (500 mL H_2O), dried, and analyzed by ^{1}H NMR.
- (*R,S*)-3 formed as a colorless solid (92.4%, dr > 99/1).

Results

The absolute configuration of (*R,S*)-**3** was confirmed by X-ray analysis (Fig. 1).

Fig. 1. X-ray structure of (*R,S*)-3.

Solvent, time, and temperature were optimized. Time was optimized at 30 min with dr > 99/1 (Fig. 2).

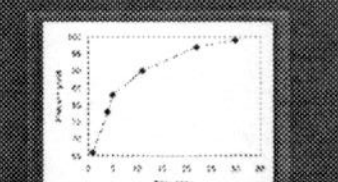

Fig. 2. Optimization of reaction time.

Results (cont.)

Optimization trials (Table 1) showed that water at 70 °C for 24 h produced (*R,S*)-3 in 93% yield (dr > 99/1).

Table 1. Optimization trials for the synthesis of (*R,S*)-3.

solvent(s)	temp (°C)	time (h)	yield (%)	dr (*R,S*)-3/(*R,R*)-3
MeOH	rt	20	80	65/35
MeOH/2-PrOH	rt	22	51	99/1
2-PrOH	rt	22	84	88/12
2 PrOH/*t*-BuOH	rt	20	65	96/4
MeOH/H_2O	rt	20	69	81/18
H_2O	55	24	81	85/15
H_2O	60	24	84	96/4
H_2O	65	24	84	98/2

Discussion

A proposed mechanism is shown in Scheme 1. The *re*-facial attack of CN^- to the intermediate imine **4** appears to be preferred, forming (*R,S*)-**3**. (*R,S*)-**3** is less soluble and precipitates out of solution; (*R,R*)-**3** is more soluble and epimerizes in solution via the imine **4**.

Scheme 1

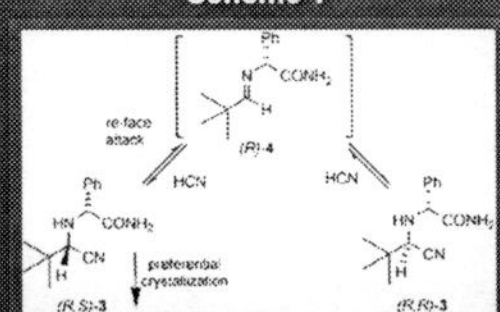
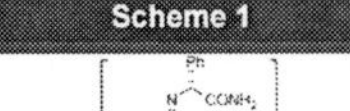

Conclusions

(*R*)-phenylglycine amide **1** is an excellent chiral auxiliary in the asymmetric Strecker reaction of pivaldehyde **2**. In water at 70 °C, the (*R,S*)-**3** product was isolated in 93% yield and dr > 99/1. Work is underway to convert (*R,S*)-**3** to (*S*)-*tert*-leucine and thereby complete the asymmetric Strecker reaction.

Figure 10.9 The final design of the poster based on Boesten et al. (2001).

Table 10.3 Selected design elements in sample posters. Fonts include Arial (Ar), Book Antiqua (BA), Comic Sans MS (CS), Tahoma (Th), and Times New Roman (TNR). Posters are 36 in. tall × 56 in. wide (~91 cm × 142 cm).

Design Element	Figure 9.1	Figure 10.7	Figure 10.8	Figure 10.9
Title	Ar 136 bold	CS 100 bold	Ar 112 bold	Th 100 bold
Author list	Ar 60 bold	CS 50 bold	Ar 72 bold	Th 60 bold
Affiliation	Ar 50 bold	CS 50 bold	Ar 60	Th 60
Headings	CS 72 bold	BA 72 bold	Ar 72 bold	Th 72 bold
Text	Ar 36	BA 36	TNR 36	Ar 39
Layout	column	column/row	column	column
Color, background	orange (border)	dark red	white	dark blue
Color, text boxes	white	cream	none	none
Color, text	black	black	dark blue	white/yellow
Alignment, Introduction	left-justified	left-justified	justified	justified

Exercise 10.3

Analyze the black-and-white posters in figures 9.1 and 10.7–10.9. Evaluate layout, font, and font size. Then view the color versions of the posters on the *Write Like a Chemist* Web site. How does color further enhance poster impact?

10C Designing on Your Own: Add Color and Artwork

Now that your poster design is almost complete, you can experiment with a color scheme. Try several different combinations of colors until you find one that is pleasing and easy to read. Make sure that the colors you select enhance your poster rather than distract the viewer from the scientific content. Choose poster and text backgrounds that enhance the readability of your poster. Decide whether there are any additional graphics, such as logos and photos, that you would like to add to your poster.

Exercise 10.4

The *Write Like a Chemist* Web site includes six sets of "initial" and "revised" student posters. Glance at these posters. Based on what you have learned in this chapter, identify at least three improvements that have been made in each revised poster. Suggest at least three additional improvements that could still be made.

10D Designing on Your Own: Finalize Your Poster

Before you print a final version of your poster, revise and edit your poster as a whole. We recommend that you reread and edit your poster, focusing on each of the following areas:

1. Organization: Check your overall organizational structure. Did you follow the move structures that were suggested in chapter 9? Is each section of the poster easy to identify? Does each section contain appropriate headings and information?
2. Audience, conciseness, and fluency: Verify that your poster addresses the correct audience. Is your poster crisp and uncluttered, leaving out unnecessary details? Could some of the content be presented in a bulleted list of phrases rather than in complete sentences?
3. Science content: Are you prepared to discuss your poster with an expert audience? Have you correctly conveyed the science in your work? Have you used words and units correctly? If asked, could you define all of the words that you have used in your poster?
4. Writing conventions: Check to be sure that you have followed poster writing conventions, including the use of abbreviations, bulleted fragments, lists, and citations.
5. Grammar and mechanics: Check for typos and errors in spelling, subject–verb agreement, and punctuation. Be sure that you have used troublesome words such as *effect*, *affect*, and *data* correctly.
6. Visual design: Consider the overall appearance of your poster. Is it easy to read? Will a reader be able to naturally follow the flow of your poster? Is it free of distracting, unnecessary graphics? Have you used fonts and colors in a consistent manner?

Finalizing Your Poster

See chapter 18.

Chapter Review

As a self-test of what you've learned in this chapter, define each of the following terms, in the context of the chapter, for a friend or colleague who is new to designing posters:

column layout	inverted color scheme	row layout
dangling graphics	justification	sans-serif font
flow	logos	serif font

As a follow-up self test, explain the basic principles governing these aspects of poster design:

- Layout
- Fonts and font sizes
- Colors

Additional Exercises

Exercise 10.5

Reflect on what you have learned about designing a scientific poster. Select one of the reflection tasks below and write a thoughtful and thorough response:

a. Reflect on the poster as a visual display of information.
   - Which visual features (e.g., bullets, colors, font sizes, graphics, layout, photographs) seem to be most important for flow?
   - Which visual features seem to be most important for the clear, unambiguous presentation of information?
   - Which visual features are most likely going to draw viewers in?
   - Which visual features might distract readers, thus the need for care in using (or avoiding) them?

b. It was stated early in the chapter that the goal of a poster is to present your science in a way that is clear, crisp, and uncluttered.
   - What layout principles allow for a clear presentation?
   - What layout principles allow for a crisp presentation?
   - What layout principles allow for an uncluttered presentation?

c. Reflect on the similarities and differences between a journal article and poster.
   - Consider the two genres in terms of audience, organization, and writing conventions.
   - Consider the two genres in terms of scientific content.
   - Consider the two genres in terms of visual presentation.

Module 3 *The Research Proposal*

11 *Overview of the Research Proposal*

The primary purpose of reports and articles is to inform; the primary purpose of proposals is to persuade. Your goal in proposal writing is not just to inform your audience about a solution to a problem, but to convince your audience to give you funds so that you can solve a problem.

—Adapted from Alley (1996)

In this module, we focus on writing a **research proposal**, a document written to request financial support for an ongoing or newly conceived research project. Like the journal article (module 1), the proposal is one of the most important and most utilized writing genres in chemistry. Chemists employed in a wide range of disciplines including teaching (high school through university), research and technology, the health professions, and industry all face the challenge of writing proposals to support and sustain their scholarly activities.

Research Proposal

A document written to solicit financial support for an ongoing or new research project.

Research proposals, sometimes referred to by the more generic term *grant proposals*, are written to solicit a research grant.

Before we begin, we remind you that there are many different ways to write a successful proposal—far too many to include in this textbook. Our goal is not to illustrate all the various approaches, but rather to focus on a few basic writing skills that are common to many successful proposals. These basics will get you started, and with practice, you can adapt them to suit your individual needs. After reading this chapter, you should be able to do the following:

- Describe different types of funding and funding agencies
- Explain the purpose of a Request for Proposals (RFP)
- Understand the importance of addressing need, intellectual merit, and broader impacts in a research proposal
- Identify the major sections of a research proposal
- Identify the main sections of the Project Description

Toward the end of the chapter, as part of the Writing on Your Own task, you will identify a topic for the research proposal that you will write as you work through this module.

Reading and Analyzing Writing

Consistent with the read-analyze-write approach to writing used throughout this textbook, this chapter begins with an excerpt from a research proposal for you to read and analyze. Excerpt 11A is taken from a proposal that competed successfully for a graduate fellowship offered by the Division of Analytical Chemistry of the American Chemical Society (ACS). As is true for nearly all successful proposals, the **principal investigator** (PI) wrote this proposal in response to a set of instructions. We have included the instructions with the excerpt so that you can see for yourself how closely she followed the proposal guidelines.

Principal Investigator

The principal investigator (PI) is the main author of a proposal. Additional authors are termed co-principal investigators (co-PIs).

Exercise 11.1

Read excerpt 11A. First read the Program Description and Proposal Instructions, and then read the proposal excerpt. How well does the author follow the instructions? Comment briefly on the excerpt in terms of its audience and purpose, organization, writing conventions, grammar and mechanics, and science content.

Excerpt 11A

Graduate Fellowship Announcement (adapted from the ACS Division of Analytical Chemistry Graduate Fellowship announcement. http://www.wabash.edu/acsgraduatefellowship/ (accessed May 31, 2005))

Program Description

- The purposes of these Fellowships are to encourage basic research in the field of analytical chemistry, to promote the growth of analytical chemistry in academic institutions and industry, and to provide recognition of future leaders in the field of analytical chemistry.
- The Graduate Fellowship Committee of the ACS Division of Analytical Chemistry, which evaluates the applications and makes the fellowship awards, comprises representatives from the sponsoring companies, analytical faculty from undergraduate institutions, and scientists from national laboratories.
- The student applicant must be a full-time student working toward a Ph.D. in analytical chemistry and must have completed the second year of graduate study by the time the fellowship period begins.

Proposal Instructions

IN NO MORE THAN TWO PAGES attached to this application, (1) provide your thesis title, (2) summarize the objectives of your thesis, (3) summarize work already accomplished, (4) summarize work planned for the term of the fellowship, and (5) explain the relevance of your work to analytical chemistry.

Research Proposal Excerpt (from a successful proposal submitted for this award: Haes, 2003)

Thesis Title and Objectives. The current title of my thesis is "The Characterization and Development of the Localized Surface Plasmon Resonance Nanosensor." The overall goal of my thesis work is to elucidate the sensing mechanism of the localized surface plasmon resonance (LSPR) of triangular Ag nanoparticle biosensors and to apply that knowledge to optimize their use as a novel analytical tool. The combination of UV–vis spectroscopy and surface-enhanced Raman scattering (SERS) will allow for high throughput general analyte screening and specific analyte identification, respectively. The individual objectives leading to my overall goal include the following: (1) systematically study the short-range behavior of alkanethiol self-assembled monolayer adsorbates on nanosphere lithography (NSL) derived Ag nanoparticles in order to maximize the LSPR response for a single adsorbate, (2) systematically study the long-range behavior of multilayer adsorbates on various sized and shaped NSL derived nanoparticles, (3) investigate the model system of streptavidin binding onto biotinylated Ag nanoparticle sensors, (4) determine the level of nonspecific binding interactions of the LSPR nanosensor response, (5) perform an immunoassay using anti-biotin on biotinylated Ag nanoparticles, (6) systematically study how the LSPR nanosensor responds to resonant molecules by varying the extinction maximum (that is, produce on and off resonance conditions) of Ag nanoparticles, (7) attach cytochrome P450 (cyt P450) to the surface of Ag nanoparticles to study their interactions with drug molecules, (8) engineer a microfluidic chip to decrease analyte solution volumes, and (9) combine the high throughput screening of the LSPR nanosensor with SERS for chemical identification.

Background and Summary of Accomplished Work. Before I entered graduate school, the Van Duyne group had already demonstrated that NSL is a simple, inexpensive, and versatile technique for the fabrication of nanoparticles[1] and that NSL-derived Ag nanoparticles could be used as biological and chemical optical nanosensors by monitoring the maximum LSPR peak extinction wavelength.[2] Since joining the group, additional progress has been made on Objectives 1–6. For example, Objective 1, which measures the short range (viz., 0–2 nm) distance dependence of the electromagnetic fields that surround resonantly excited Ag nanoparticles, has been tested. Results indicate that the response is linear and can be systematically tuned by changing the structure and composition of the Ag nanoparticles.[2,5] . . . The long range distance dependence of the LSPR of noble metal nanoparticles (Objective 2) has also been elucidated.[5] Measurement of the LSPR extinction peak shift versus number of layers and adsorbate thickness is non-linear and has a sensing range that is dependent on the composition, shape, in-plane width, and out-of-plane height of the nanoparticles. This remarkable set of experiments confirms that the sensing capabilities of noble metal nanoparticles can be tuned to match the size of biological and chemical analytes by adjusting the aforementioned properties. The optimization of the LSPR nanosensor for a specific analyte will improve an already sensitive nanoparticle-based sensor.

. . . [*Section omitted that describes work toward Objectives 3–5*]

To date, all LSPR nanosensor experiments have been performed using non-resonant molecules. Because the **effect** that molecular resonances have on nanoparticle sensing is unknown, my current studies are aimed at answering this question (Objective 6). Preliminary results indicate that the resonant molecule, $Fe(bpy)_3^{2+}$ (bpy = 4,4′-bipyridine), dramatically enhances the sensitivity of the LSPR nanosensor when the extinction maximum of the nanoparticles is slightly red-shifted from the molecular resonance.

Planned Research for the Duration of the ACS Analytical Fellowship. During the remainder of my graduate school career, I will concentrate on completing the aforementioned resonant molecule studies and on fulfilling my final three thesis objectives: attaching cyt P450 (a resonant molecule) to Ag nanoparticles in order to study their interactions with drug molecules, engineering a microfluidic chip to decrease analyte solution volumes, and combining the high throughput screening of the LSPR nanosensor with SERS for chemical identification. These accomplishments will allow the objectives of my thesis to be fulfilled and will aid in the development of a more powerful biosensor, a useful analytical tool.

The motivating factor behind my current studies with $Fe(bpy)_3^{2+}$ is to best design a sensor surface for cyt P450 modification. Like $Fe(bpy)_3^{2+}$, cyt P450 has a molecular resonance in the visible region of the electromagnetic spectrum. The cyt P450 family of enzymes participates in the metabolism of a large fraction of all drugs in medicine. Interactions with cyt P450 enzymes are often the major limitation of a drug's usefulness. To optimize the LSPR nanosensor for drug screening using cyt P450 functionalized nanoparticles, the role molecular resonances have on the LSPR sensing mechanism must be better understood.

One of the characteristics of the LSPR biosensor is its generality. This is a powerful attribute for fast, high throughput screening of adsorbates, but, at times, specific analyte identification is needed. SERS, an extremely sensitive analytical tool, yields detailed chemical and conformational information of adsorbates near roughened noble metal surfaces.[8] SERS signals are generally 10^6 times larger than normal Raman signals. Because NSL-derived Ag nanoparticles exhibit large SERS enhancements, the combination of UV–vis and SERS will produce a biosensor capable of both high throughput screening and exact chemical identification.

Relevance to Analytical Chemistry. The relevance of my work to analytical chemistry is apparent on two levels. First, the understanding of metal particles in the nanoregime is not fully understood. Additionally, many chemical processes occur in this size scale. Clearly, this research advances that knowledge. Second, the analytical technique of biosensing is the practical application of this work. Our results suggest that in the near future Ag nanotriangle biosensors could be used for the detection of a wide variety of biomolecules. Binding of DNA, proteins, and possibly eukaryotic cells (by using protein ligand intermediates) to noble metal nanoparticles opens a window of opportunity in medical diagnostics and could greatly simplify often tedious immunohistochemical detection tasks performed regularly in biomedical research. Future work on miniaturization of the sensor, linkage of the sensor to drug delivery chips, and biocompatibility could make this laboratory-based device into a portable analytic and diagnostic tool. Finally, by combining the powerful techniques of LSPR screening and SERS identification on a microscale system, new limits to chemical and biological sensing processes will be reached.

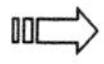

Affect vs. Effect

See appendix A for more information on these commonly confused words.

Several features are worth noting in excerpt 11A. First, consider the Proposal Instructions. In addition to a title, the instructions ask for (in no more than two pages) a statement of project objectives, a summary of previous work, a description of the proposed work, and an explanation of why the work is relevant. These four items typify the content requested in most proposals. Second, consider the proposal itself. What writing components did you notice as you read the excerpt? Below we highlight some of the components that are addressed in this module. Additional features are highlighted in exercises 11.2–11.8.

- Audience and purpose: Haes writes her proposal for an expert audience. The proposed work is in a specialized subfield of analytical chemistry (nanosensors), and she assumes that her readers are knowledgeable in this area. For example, terms such as *localized surface plasmon resonance* and *nanosphere lithography* are used without definition. The Program Description states that evaluators are "representatives from the sponsoring companies, analytical faculty from undergraduate institutions, and scientists from national laboratories"; hence, the author has targeted the right audience. We will see, however, that not all proposals are written for an expert audience and, in many proposals, more than one audience is targeted.

- Organization: Haes closely follows ACS guidelines and uses subheadings that directly correspond to points 1–5 in the Proposal Instructions (a practice that we highly recommend). She begins by describing the broad goal of the research project and then enumerates the individual objectives that lead to that goal. She delineates work that has already been accomplished, as well as work that remains to be done. The proposal concludes by emphasizing why this work is relevant to the field of analytical chemistry. Her proposal mirrors the general organizational structure of many proposals.
- Writing Conventions: Haes uses predominantly present tense, present perfect, and future tense. Present tense is used for project goals and objectives and for knowledge believed to be true over time; present perfect is used for work conducted in the past; future tense is used for work to be done in the future. Haes also uses personal pronouns in her proposal (e.g., "*my* thesis work", "*our* results", and "before *I* entered graduate school"). Her use of personal pronouns and verb tenses follows writing conventions common in research proposals. Haes also makes use of enumerated lists in her proposal. Most chemists agree that this is an effective and space-saving formatting technique; hence, lists are commonplace in proposals.
- Grammar and mechanics: Haes pays careful attention to parallelism and punctuation, most notably in her enumerated lists. Haes's lists are grammatically parallel, and she uses punctuation (i.e., colons, commas, and semicolons) appropriately.

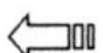

Present Perfect

See table 6.2.

Exercise 11.2

Look for the following formulas, abbreviations, and acronyms in the proposal in excerpt 11A: Ag, SERS, UV–vis, cyt, and bpy. Which are defined at first use, and which are not? Are these choices appropriate? Explain.

Exercise 11.3

Research proposals often make use of two related words: *goals* and *objectives*. Describe how Haes uses these two words in her proposal. Are the terms used consistently throughout?

Exercise 11.4

Consider passages 1–4 below, taken from the Haes proposal:

a. Two of the four passages depict moves commonly encountered in journal articles. Match the two passages with their correct move(s), choosing from the following: (1) introduce topic/purpose, (2) present background information, (3) identify a gap, (4) fill a gap, and (5) describe experimental methods.

b. The remaining two passages depict new moves (i.e., moves not listed in 11.4a). Which passages are they? Suggest the purpose(s) of these passages, thereby identifying new moves.

1. Because the effect that molecular resonances have on nanoparticle sensing is unknown, my current studies are aimed at answering this question (objective 6).
2. Preliminary results indicate that the resonant molecule, $Fe(bpy)_3^{2+}$ (bpy = 4,4′-bipyridine), dramatically enhances the sensitivity of the LSPR nanosensor when the extinction maximum of the nanoparticles is slightly red-shifted from the molecular resonance.
3. Like $Fe(bpy)_3^{2+}$, cyt P450 has a molecular resonance in the visible region of the electromagnetic spectrum. The cyt P450 family of enzymes participates in the metabolism of a large fraction of all drugs in medicine.
4. The overall goal of my thesis work is to elucidate the sensing mechanism of the localized surface plasmon resonance (LSPR) of triangular Ag nanoparticle biosensors and to apply that knowledge to optimize their use as a novel analytical tool.

Exercise 11.5

The following two sentences use colons, commas, and/or semicolons incorrectly; correct the mistakes:

a. The techniques include: (1) GC/MS; (2) UV–vis spectroscopy; and (3) NMR spectroscopy.

b. Several techniques were used: (1) GC/MS, (2) IR, UV, or UV–vis spectroscopy, and (3) NMR spectroscopy.

Colons, Commas, and Semicolons

See appendix. A.

Exercise 11.6

Consider the following passage from a section in Haes's proposal, omitted from excerpt 11A. We have removed all colons, semicolons, and commas from the passage. Decide which form of punctuation belongs in each empty space or if the space should remain blank.

The hexadecanethiol (HDT) induced LSPR peak shift for Ag nanotriangles as a function of structure has been determined ___ (1) HDT sensitivity decreased by 0.11 and 0.52 nm per nm increase in in-plane width at fixed out-of-plane heights of 50.0 and 30.0 nm ___ respectively ___ (2) HDT sensitivity increased by 0.33 nm per nm increase in out-of-plane height at fixed in-plane widths of 100 nm ___ (3) HDT sensitivity was 1.5 times larger for chopped tetrahedra than for hemispheres with equal volumes ___ and (4) HDT sensitivity of Ag nanotriangles was 3.5 times larger than Au nanotriangles with identical in-plane widths and out-of-plane heights.

Exercise 11.7

Find one sentence in the proposal in excerpt 11A that illustrates each of the following:

a. The correct use of present tense to describe the goal(s) and objective(s) of the proposed work.
b. The correct use of present tense to describe knowledge (or findings) expected to be true over time.
c. The correct use of present perfect for work done in the past.
d. The correct use of future (with *will*) to describe future work.

Exercise 11.8

Each of the following sentences includes a list or series that is not parallel. Revise the sentences so that the lists or series within them are parallel.

a. The method we now describe for deducing the individual *J* values from any first-order multiplet requires two principal operations: (1) to assign each of the individual 2″ components (cf. Figure 1) and (2) systematic identification of the individual J values.[5] (Adapted from Hoye and Zhao, 2002)

b. Specific particulate-associated diseases implicated by these studies include the exacerbation of asthmatic episodes, the induction of chronic bronchitis, and causing the induction of interstitial fibrosis. (Adapted from Kristovich et al., 2004)

c. However, during incineration, the following major problems were encountered: (1) excessive bed temperature, (2) frequent clinker formation, (3) flue gases reaching high temperatures, and (4) excessive pressure drop. (Adapted from Shie et al., 2004)

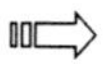

Parallelism in Lists

Lists should be parallel in language, numbering, punctuation, and formatting. (See appendix A.)

Selecting a Funding Agency and RFP

The most important criteria for all proposal writing are the explicit instructions in the RFP.

—Paradis and Zimmerman (1997)

The first step in writing a successful proposal is to select a funding agency. This is no easy task. The number of funding sources is mind-boggling and includes federal, state, and local government organizations, as well as public and private corporations, foundations, and individuals. To begin to appreciate the range of possibilities, browse through the funding opportunities listed on the Web pages of the ACS, USDA, EPA, NIH, or NSF. Or, if you have a specific topic, create a list of keywords and search for them using searchable databases such as www.grants.gov (a free portal site to more than 1,000 federal grant-making agencies) or the Illinois Researcher Information System, IRIS (a fee-based site to more than 8,600 federal and private funding agencies in the sciences, social sciences, arts, and humanities). You can also search online using Google.com or other Internet search engines. (If you use Google, try placing quotation marks around the searchable terms and add (without quotation marks) "+grant" following the terms. E.g., if you are searching for grants on water pollution, you might try the following: "water pollution" +grant.)

Abbreviations for Funding Agencies

Many funding agencies have abbreviated names. Here are five used in this textbook:

ACS: American Chemical Society

USDA: U.S. Department of Agriculture

EPA: Environmental Protection Agency

NIH: National Institute of Health

NSF: National Science Foundation

To navigate this complex world of funding successfully, it helps to seek advice from others. Whenever possible, talk with individuals at your institution (e.g., peers, colleagues, or research mentors) who have received grants that match your particular research interests or funding needs or who are familiar with the funding process (e.g., grant officials). Find out what worked for them. Here are a few additional guidelines to help you in the agency-selection process:

- Look for **internal funding** first (funding from within your institution), as opposed to **external funding**. Many institutions offer competitive internal awards for undergraduate and graduate students. Such awards often provide a research stipend and may also offset expenses for tuition, books, travel (to conferences or field sites), and/or laboratory supplies. For faculty, internal funding may serve as a stepping stone to external funding (and is sometimes called "seed" money).
- Look for funding opportunities that target researchers like you (e.g., undergraduate- or graduate-level students, beginning faculty members, non-U.S. citizens, members of an underrepresented population).
- Look for funding opportunities that target your specific need (e.g., a summer research fellowship, a travel grant, a new piece of equipment).
- Look for funding opportunities that match your project goals. Some grants support **fundamental research**, research that targets new insights and knowledge (e.g., the ACS Graduate Fellowship described in excerpt 11A). Other funding agencies target more **applied research**, research designed to solve a specific problem.

Internal and External Funding

Internal funding originates in your home institution; proposals are evaluated by anonymous internal reviewers.

External funding originates outside your institution (e.g., the NSF); proposals are judged by anonymous external reviewers.

Fundamental and Applied Research

Fundamental research attempts to uncover scientific principles and the laws of nature.

Applied research uses acquired knowledge to solve problems.

After you have narrowed your search to a specific agency and grant, the next step is to read in detail the grantor's instructions for the application process. Such instructions are typically included in a **Request for Proposals (RFPs)** or **Request for Applications (RFAs)**. RFPs will vary in length from a few paragraphs to more than 10 pages; regardless of length, these documents are your most valuable resource for writing a proposal. The RFP will help you organize your proposal, format it correctly, and address the specific needs of your funders. Because of their importance, we examine several RFPs in this chapter and use them to guide our discussion of how to write a proposal (here and in chapters 12–15).

RFPs and RFAs

RFP: Request for Proposals

RFA: Request for Applications (another term for RFPs)

Exercise 11.9

Browse through the grant and funding opportunities listed on the Web site of a large institution, such as the NSF, NIH, ACS, or EPA, and complete the following tasks:

a. Determine if there are funding opportunities that target specific researcher populations (e.g., undergraduate or graduate students, new faculty, underrepresented populations).
b. Determine if there are funding opportunities that target a specific need (e.g., summer research fellowships, travel grants, equipment grants).
c. Identify five areas of research that the agency will fund. If possible, classify each area as fundamental or applied research.
d. Locate the RFP for one funding opportunity. Browse through the RFP. Compare this RFP to the Proposal Instructions in excerpt 11A for the ACS Graduate Fellowship in Analytical Chemistry.

Writing a Persuasive Proposal

After you have selected a funding agency and read the RFP, the task of writing the proposal is at hand. Before you begin, keep in mind that your proposal must be persuasive. Think of yourself as an entrepreneur asking others to invest in your project; your proposal must convince them to give you money. It is not enough to limit your persuasive remarks to a single paragraph or section of the proposal; your proposal should be compelling throughout. This can be accomplished by using different persuasive tactics throughout your proposal. Collectively, the tactics involve persuading your funders that your proposed work

- meets their needs
- has intellectual merit
- has broader impacts for society

We look briefly at each tactic below (abbreviated as need, intellectual merit, and broader impacts) and consider them again as we work through the various sections of the proposal.

Need

Your research idea must overlap with a need that the funding agency has identified. Funding will be denied—or worse, the proposal will be returned unread—if the needs of the funding agency are not addressed. The need is often suggested in the RFP title and spelled out in the RFP text. The RFP may also refer you to a Web site with the agency's mission statement, strategic plan, or rationale for the current funding initiative. Read these documents! Knowledge of the agency's mission allows you to align your research interests and goals with theirs.

Exercise 11.10

Consider the following RFP titles for EPA grants offered in 2006. What need is addressed in each title? How does each need fit into the overarching mission of the EPA, which is to protect human health and the environment?

- Allergenicity of Genetically Modified Foods
- Ecological Impacts from the Interaction of Climate Change, Land Use Change, and Invasive Species
- Sources, Composition, and Health Effects of Coarse Particulate Matter
- Fate and Effects of Hormones in Wastes from Concentrated Animal Feeding Operations

Intellectual Merit

When I was Program Officer at the National Science Foundation, I witnessed without a doubt that simply and clearly written proposals sailed through more easily than ones that were not so. Surely there is a lesson there.

—Robert Damrauer, University of Colorado–Denver

It is not enough that your proposal targets the needs of a funding agency; your proposed idea(s) must also be intellectually sound. Typically, reviewers judge the intellectual merit of a proposal according to three criteria, each of which is described in more detail below:

- the creativity of your ideas
- the credibility of your work plan
- your competence as a researcher

Three C's

Creativity

Credibility

Competence

Demonstrate your intellectual merit in a research proposal by emphasizing the three C's.

Proposals score high points if they present a creative approach or an innovative idea. A good idea that has already been explored, by you or by others, is seldom funded. A winning proposal must demonstrate that the proposed work will accomplish something new. **Gap statements** and gap-fillers, as in journal articles (see chapter 6), are used in research proposals to call attention to the new steps that your proposed work will take. By addressing problems that were left unsolved in the past, and by showing how your new approach will solve them, you underscore the creativity of your work.

Gap Statements

See table 6.1.

Intellectual merit is also based on the credibility of your work plan, including both the proposed research methods and projected timeline. To be credible, your proposed methods must be sound. Knowledge of standard techniques must be evident (in the text and/or through citations to the literature), even if you are proposing a new or innovative approach. Credibility is also enhanced by describing the inherent limitations of your methods (e.g., by including error bars, detections limits, or the range over which a calibration is accurate). Timelines should be ambitious but realistic. If you are an undergraduate student, you are likely to underestimate how long your project will take. Consult your research mentor for ideas about realistic timelines.

Writing and Science

Writing skills cannot replace good science, but writing skills are needed to convey good science. Your competence as a scientist will be judged, in large part, by your writing.

Your competence to do the proposed work is also evaluated. Competence is judged largely by your past achievements (thus, the importance of doing excellent work if you receive a research grant). Early in your career (e.g., as an undergraduate or graduate student), transcripts, test scores, and letters of recommendation are used to judge your competence. As you progress in your career, your competence is judged by your level of education, work history, publications, grants received, conferences presentations, and other scholarly contributions.

Intellectual merit is also judged by your writing ability (hence the reason for including this module in our textbook). The value of a well-written proposal cannot be overstated. An articulate and clearly organized proposal promotes the original qualities of your work and convinces readers that you can carry out the work described. It is not an exaggeration to say that your competence as a scientist will be judged by your writing. A competent scientist pays attention to detail; a carelessly written proposal that includes typos, poor organization, and confusing sentence structure leaves a negative impression and undermines any confidence readers might have had with the author. Thus, although writing skills cannot replace good science, writing skills are needed to convey good science.

Broader Impacts

A creative but excessively narrow proposal has limited chances of being funded. Proposals that demonstrate the potential for contributions to a greater good and the advancement of future scientists' knowledge are likely to be supported. Minimally, you should explain how your work will benefit other scientists (both within your discipline and in general); optimally, you should explain how your work will benefit society at large. Importantly, **broader impacts** must be spelled out in language that a general audience can understand. Granting officials often use your words to share your ideas with potential donors and the general public. Hence, the burden is on you to express your science in a way that is easily understood.

Broader Impacts

In this textbook, the term *broader impacts* refers to how your proposed work will benefit other scientists, society at large, and the training of future scientists.

Another broader impact of research is the training and education of future scientists. Trained scientists are a national (and international) resource; research activities create a "pipeline" of individuals—undergraduate, graduate, and postdoctoral researchers—to renew this resource. Ideally, this pipeline reflects the ever-changing composition of society. To this end, federally funded programs (e.g., the NSF) now require that investigators describe how their research activities will add to this pipeline by training new scientists and encouraging participation by underrepresented groups. In this way, funding initiatives advocate the development of not only new science but also new scientists.

Exercise 11.11

Glance at an RFP for a research grant. Use an RFP provided by your instructor or search for a funding agency on the Web such as the EPA, NIH, NSF, Research Corporation, or the ACS Petroleum Research Fund.

a. Describe the activities that the grant will fund.
b. Identify whether the grant targets fundamental or applied research.
c. Describe how the RFP addresses issues related to need, intellectual merit, and broader impacts.

Addressing need, intellectual merit, and broader impacts are essential for a successful proposal. The converse is also true; proposals that lack these qualities are unlikely to be funded. Bowman and Branchaw (1992) corroborated this assertion by compiling a list of the most common reasons why proposals are rejected (table 11.1). As you glance through the list, note how many reasons link back to need, intellectual merit, and broader impacts. Note, too, that proposals are rejected because they are not submitted on time and do not follow RFP guidelines. Paying careful attention to "picky" requirements such as page length and formatting is essential. An improperly formatted proposal immediately suggests

Table 11.1 Common reasons for the rejection of research proposals (adapted from Bowman and Branchaw, 1992).

(a) The author did not demonstrate a clear understanding of the problem.

(b) The proposal did not arrive by the submission deadline.

(c) The information requested in the RFP was not provided.

(d) The objectives were not well defined.

(e) The wrong audience was addressed.

(f) The procedures and methodology were not specific.

(g) The overall design was questionable.

(h) Cost estimates were not realistic: either too high or too low.

(i) Resumes of key personnel were inadequate.

(j) Personnel lacked experience or the required qualifications.

(k) The proposal was poorly written and not well organized.

(l) The proposal did not follow the organizational pattern specified in the RFP.

(m) The proposal did not provide adequate assurance that completion deadlines would be met.

(n) Essential data were not included in the proposal.

(o) The proposed time schedule was unrealistic.

(p) The project objectives were not clearly linked to an agency-based need.

(q) The author did not adequately address why this project is important or how it will contribute to society at large.

that you are incapable of carrying out a well-organized, disciplined research project.

Exercise 11.12

Review the list in table 11.1. Identify reasons for the rejection of research proposals that link directly to the following:

a. the authors' failure to address need (identified by the funding agency)

b. the authors' failure to establish intellectual merit (creativity, credibility, and competence)

c. the authors' failure to address broader impacts

d. the authors' writing skills

e. other

Selecting an Audience

A successful proposal must also target the proper audience. For proposals, the audience typically comprises a panel of preselected reviewers. The task of the reviewers is to read and score the submitted proposals. Because the background and training of panel members will vary from proposal to proposal, you must consult the RFP for information about who your reviewers will be. For example, in excerpt 11A, the Program Description states that the reviewers will be "representatives from the sponsoring companies, analytical faculty from undergraduate institutions, and scientists from national laboratories." As a result, the author wrote her proposal largely for an expert audience. If, however, you are an undergraduate writing a proposal to support your undergraduate research, your audience will be much different. Undergraduate research awards are typically reviewed by a broad cross section of faculty, including faculty from both the physical and social sciences. In this case, you should introduce your ideas at a scientific or even an "advanced" general audience level. If sociology professors learn some chemistry while reading your proposal, all the better. They will be inclined to regard your proposal more highly, and they will trust that you understand the proposed work, an essential prerequisite for funding.

If the audience is not clearly defined in the RFP, we recommend a two-tiered approach. When you introduce your ideas and describe the importance of your work, target a general to scientific audience:

- Explain important concepts using analogies that less expert readers will understand.
- Define acronyms and terms that some readers will be unfamiliar with.
- Use graphics to illustrate ideas or share preliminary results.

In the Experimental Approach section, shift to a more expert audience (e.g., PhD-level chemists):

- Demonstrate your expertise and ability to do the proposed work.
- Use acronyms as you would in a journal article.

By following this two-tiered approach, a less expert reviewer will be able to follow the main ideas of your work, whereas an expert reviewer will be able to judge the intellectual merit of your proposal fully.

Another aspect of audience is readability. Although your science must be top-notch (complete with sophisticated terminology), your proposal must also be highly readable. This apparent conflict can be resolved by using an accessible writing style: keep sentences short, use bulleted lists and other easy-to-read formats, and include graphics to break up the text. Reiterating themes and "telling a story" throughout your proposal (as we emphasize in chapters 12–14) will also increase your proposal's readability. As a rule of thumb, a 15-page proposal should take about an hour to read (less time than it typically takes to read a five-page journal article). If your proposal is too difficult to read in this time frame, it may be written for the wrong audience.

Exercise 11.13

Refer back to the RFPs that you found for exercise 11.11, or select another RFP. Read the RFP for information about audience. What hints are offered about audience? In your own words, describe the target audience to authors who will be writing this proposal.

Organizing the Proposal

In this section, we describe the organizational structure modeled in this textbook for the research proposal. In truth, there is no one "right" structure for a proposal. Unlike journal articles, which conventionally follow the IMRD format (see chapter 2), no consensus structure exists for proposals. They are all a bit different, each requiring a slightly different organizational approach. Thus, we have developed a generic organizational structure, one that captures the essence of many proposals but is not specific to any single proposal. We leave it up to you to adapt this structure to meet the needs of your specific proposal.

If you are lucky, your RFP will offer organizational guidelines. If it does, by all means, follow them, to the point of mirroring the suggested headings and subheadings in your proposal. This will help reviewers locate pertinent information in your proposal and will improve your proposal's readability (and rating). If, however, your RFP does not provide such guidance, or you are writing a fictitious proposal to improve your writing skills, our generic structure will get you started. Below we outline the generic proposal structure used in this textbook, including (1) proposal headings, (2) major divisions of the proposal, and (3) main sections of the Project Description (the primary major division of the proposal). This information can serve as a roadmap to guide you through chapters 12–14.

Proposal Headings

Because the organizational structure of most research proposals is complex, headings of different hierarchical levels (level 1, level 2, level 3, etc.) are needed to help clarify the structure. Two commonly used styles are illustrated in generic

Table 11.2 Generic forms of two heading styles (levels 1–3) commonly used in research proposals.

Style 1	Style 2
MAJOR DIVISION TITLE	**MAJOR DIVISION TITLE**
LEVEL 1 HEADING	**I. LEVEL 1 HEADING**
Level 2 Heading	**(A) Level 2 Heading**
Level 3 Heading. Text . . .	**(1) Level 3 Heading.** Text . . .

form in table 11.2. (Style 1 is illustrated again in table 11.3, with actual headings.) The only difference between the two styles is that style 2 includes numbers and/or letters to signal level changes, and style 1 does not. In both styles, major section titles are bolded, centered, and written in all capital letters. Level 1 and level 2 headings differ in their use of capitalization, but both styles are bolded and left-justified and are given a line of their own. Level 3 headings are bolded and indented one tab position and conclude with a bolded period. For level 3 headings, the text begins after the period, on the same line as the heading. Authors are free to choose their own style, as long as they use it consistently throughout the proposal.

Major Divisions of the Research Proposal

Research proposals typically comprise three major divisions. Each division is demarked with its own title (table 11.2). These titles differ from other headings in the proposal because they are centered rather than left-justified. The three division titles used in this textbook (and in most proposals) are

PROJECT SUMMARY
PROJECT DESCRIPTION
REFERENCES CITED

The Project Summary is a one-page document, suitable for publication, that offers a self-contained description of the proposed research activities (see chapter 15). The References Cited section includes a complete listing of all sources cited in the proposal and is typically not page restricted (see chapter 17 for directions on formatting citations and references). The Project Description is the heart of the research proposal. It is page restricted (usually 4–30 pages, depending on the proposal) and is divided into sections and subsections, each with its own heading. The Project Description is the focus of chapters 12–14.

Table 11.3 The three main sections of the Project Description addressed in this textbook, with suggested headings for the moves in each section (level 1–3 headings in style 1).

Section	Moves	Suggested Headings	Level
Goals and Importance (chapter 12)	1. State goals and objectives	GOALS AND OBJECTIVES	1
	2. Establish importance	PROJECT SIGNIFICANCE	1
	3. Introduce proposed work		
		EXPERIMENTAL APPROACH	1
Experimental Approach (chapter 13)	1. Share prior accomplishments	**Prior Accomplishments**	2
	2. Share preliminary results	**Preliminary Results**	2
	3. Describe proposed methodology	**Proposed Methods**	2
		Objective 1.	3
		Objective 2.	3
Outcomes and Impacts (chapter 14)	1. Present a project timeline	PROJECT TIMELINE	1
	2. List expected outcomes	EXPECTED OUTCOMES	1
	3. Conclude the proposed work	CONCLUSIONS	1

Main Sections of the Project Description

The Project Description is typically divided into three main sections (table 11.3). The first main section introduces project goals and importance (chapter 12). The second section describes the experimental approach (chapter 13). The third section summarizes project outcomes and impacts (chapter 14). Each main section (and corresponding chapter) is organized by **moves**. The major moves are listed in table 11.3, along with headings that authors commonly use in their proposals to signal these moves. (Note: For instructional purposes, we have reformatted the headings in proposal excerpts included in this module to conform to style 1, as depicted in table 11.3.)

Examining Sample RFPs

We conclude this chapter with two sample RFPs. Excerpt 11B is a generic RFP created for instructional purposes, to aid readers who have not yet found a suitable RFP and funding agency. The generic RFP, although highly abbreviated, captures the essence of many RFPs and can be used to guide your writing as you prepare or revise a research proposal.

Excerpt 11C is from the RFP for the NSF Faculty Early Career Development Award in Chemistry (CAREER), a prestigious award offered by the NSF to exceptional junior faculty. In the rest of this module, we use excerpts from successful CAREER proposals to illustrate effective proposal-writing techniques and to apply the read-analyze-write approach to writing; hence, it is important that you review the RFP that motivated these proposals. (Note: CAREER proposals require authors to describe, in what the NSF calls a Career-Development Plan, both research and educational activities. In this textbook, we focus almost exclusively on the research activities proposed by the CAREER authors.)

Exercise 11.14

Read through the generic RFP in excerpt 11B and answer the following questions:

a. What are the three major divisions of the generic RFP?
b. Consider the information requested in the Project Description. What headings might you use to signal the requested information? Suggest a list of headings for the Project Description using style 1 or 2. (Consult tables 11.2 and 11.3 as needed.)
c. Give examples of instructions that are quite specific.
d. What audience is identified?
e. What, if anything, appears to be missing from this generic RFP?

Excerpt 11B (an abbreviated, generic RFP that models essential features of scientific research RFPs)

Project Summary. In one page or less, write a summary, suitable for publication, of proposed activities. It should not be an abstract of the proposal but rather a self-contained description of the proposed work. The summary should be written in the third person and include descriptions of (1) the goals and objectives of the project, (2) the importance of the project, (3) the proposed methods, and (4) the broader impacts of the work. It should be written for a scientifically literate reader, but not necessarily an individual in your field or discipline.

Project Description. The Project Description **may not exceed 5 pages** (Times New Roman 12-pt font, 1.5 line spacing, numbered from page 1 to 5). Tables, figures, and all graphics must be included within the 5-page limit. The Project Description should include a title and address the following:

- a clear statement of the project goals and objectives
- the importance and significance of the proposed work
- essential background information, relating the proposed work to current knowledge in the field

- a clear description of the experimental methods
- preliminary results, when appropriate
- a timeline or work plan
- projected outcomes
- broader impacts of the work

References Cited. Begin References Cited on a new page (Times New Roman 12-pt font, 1.5 line spacing). There is no page limit for this section, but we encourage you to limit references to those that have most greatly influenced your work. Format your references appropriately (numerical or alphabetical), paralleling the format used for in-line citations. Include full journal article titles and all authors listed in the order in which they appear in the journal article.

Exercise 11.15

Read excerpt 11C and answer the following questions:

a. What are three major divisions of the CAREER proposal?
b. The Project Description is divided into two sections. What is each section called? Suggest a purpose for each section.
c. Based on the information presented in the CAREER RFP, suggest headings that could be used to organize the Career-Development Plan.
d. Assign each proposed heading from (c) to one of the following sections of the Project Description used in this textbook: goals and importance, experimental approach, or outcomes and impacts. (Consult table 11.3.)

Excerpt 11C (adapted from NSF Program Solicitation: Faculty Early Career Development (CAREER) Program. Proposals for Fiscal Years 2003, 2004, and 2005. http://www.nsf.gov/pubs/2002/nsf02111/nsf02111.htm (accessed January 3, 2008))

Project Summary: See GPG Section II.C.1.

Summarize the **integrated research and education activities** of the proposed career-development plan.

Project Description: *See GPG Section II.C.3.*

Note: the project description may not exceed 15 pages.

a. **Results from Prior NSF Support**, if applicable.
b. **Career-Development Plan**. Provide a well-argued and specific proposal for activities that will, over a 5-year period, build a firm foundation for a lifetime of integrated contributions to research and education. (For examples of possible activities, refer to CAREER Program Description in Section II.A. and the document "NSF Merit Review Broader Impacts Criterion: Representative Activities," located on the NSF Web site at http://www.nsf.gov/pubs/2002/nsf022/bicexamples.pdf.)

The plan should be developed in consultation with the department head or equivalent organizational official and include

- The objectives and significance of the proposed integrated research and education activities;
- The relationship of the research to the current state of knowledge in the field, and of the education activities to the current state of knowledge on effective teaching and learning in one's field of study;
- An outline of the plan of work, describing the methods and procedures to be used, including evaluation of the education activities. Both research and education activities should be included in the plan for each year, but the relative amount of effort devoted to each may vary from year to year;
- The relation of the plan to the PI's career goals and job responsibilities, and to the goals of his/her department/organization; and
- A summary of prior research and educational accomplishments.

***References Cited**: See GPG Section II.C.4.*

11A Writing on Your Own: Identify a Proposal Topic and RFP

Identify a topic for your research proposal. Find a funding agency that might support your topic. (Hint: Look at the Acknowledgments sections of journal articles in your research area. What funding agencies have been acknowledged by the authors?) If you cannot find a suitable funding agency, assume that you are seeking an internal award through your home institution.

Find an RFP that you can follow to write your proposal (ideally, one from the funding agency that you selected). If you cannot find a suitable RFP, use the generic RFP in excerpt 11B.

Jot down at least five reasons why your topic is important. How might society benefit from your proposed work?

Chapter Review

To check what you've learned in this chapter, define each of the following terms and explain its importance, in the context of this chapter, to a chemistry colleague who is new to the field:

applied research	external funding	principal investigator
broader impacts	fundamental research	RFA
co-principal investigator (co-PI)	internal funding	RFP

Also explain the following to a friend or colleague who hasn't yet given much thought to writing a research proposal:

- Main purpose(s) of a research proposal
- Difference(s) between a research proposal and a Project Description
- Three tactics used to write a persuasive proposal
- Three C's of intellectual merit
- Common reasons for the rejection of a research proposal
- Three major sections of a research proposal
- Three main sections of a Project Description
- Challenges associated with defining an appropriate audience for a research proposal

Additional Exercises

Exercise 11.16

Excerpt 11D, from the RFP for the NSF Collaborative Research in Chemistry (CRC) Grant, begins with a description of the program, which identifies the purpose of the grant and eligibility requirements. That description is followed by Full Proposal Instructions, which provide step-by-step guidance for writing three major divisions of the proposal: Project Summary, Project Description, and References Cited. Read excerpt 11D and answer questions a–h.

(Note that the original RFP required six additional sections, not included in excerpt 11D: Bibliographical Sketches, Budget, Current and Pending Support, Facilities, Supplementary Documentation, and Suggested Reviewers. These sections are beyond the scope of this textbook.)

a. Describe how the RFP addresses the three essential features of a proposal: need, intellectual merit, and broader impacts.
b. What three sections are required in the Project Description? How many pages should each section contain?
c. Create a list of headings that you might use if you were writing this proposal.
d. The RFP lists eight items (labeled a–h) to be addressed. Which of these items are common to many proposals? Which items seem to be specific to this proposal?
e. The RFP makes reference to the Grant Proposal Guide (GPG). Locate the guide on the NSF Web site (www.nsf.gov) by searching for GPG. Browse through the guide and comment on the types of information that it contains.

f. Like most proposals, this RFP requests a Project Summary. What should be included in the summary? How long should it be?
g. What format should the author(s) use in preparing references?
h. What audience should the authors address in the proposal?

Excerpt 11D (sample RFP adapted from NSF Collaborative Research in Chemistry (CRC): Program solicitation, NSF 03–583. http://www.nsf.gov/pubs/2003/nsf03583/nsf03583.htm (accessed January 3, 2008))

Program Description

The CRC Program enables researchers from diverse scientific and engineering backgrounds to respond to recognized scientific needs, to take advantage of current scientific opportunities, or to prepare the groundwork for anticipated significant scientific developments in chemistry. CRC proposals will involve three or more investigators with complementary expertise. The members of the collaborative team can come from more than one institution and can include non-academic and international scientists. The principal investigator will most likely be a chemist; however, there is no restriction on the scope of disciplines represented by the co-investigators. Investigators may include, in addition to chemists, researchers from other science and engineering disciplines appropriate to the proposed research.

The use of **cyber-infrastructure** to enable and enhance collaborations is encouraged.

Full Proposal Instructions

Proposers are strongly encouraged to consult the proposal submission checklist included in the Grant Proposal Guide as they prepare their proposal. Proposals not compliant with the proposal preparation guidelines, as supplemented by the following instructions, may be returned without review.

Project Summary. One-page limit, including the names and affiliations of all senior personnel. The project summary must address both the intellectual merit and broader impacts of the proposed CRC project.

Project Description. A total of twenty (20) pages, including Results from Prior Support, Modes of Collaboration and Education, and Management Plan. CRC proposals are likely to be read by non-specialists at some stage of the review process. It is, therefore, particularly important that they be written to emphasize the impact of the projects on the chemical sciences in a broad context.

1. Research Plan. Narrative, not to exceed eighteen (18) pages, consisting of the following items:
   a. An explanation of the scientific context, intellectual merit, relevance to chemistry and timeliness of the proposed project.
   b. A description of the proposed research.
   c. A justification for why a collaborative effort involving at least three investigators is necessary to carry out the proposed project.

d. A description of the contribution to be made by each senior investigator.
e. A discussion of the broader impacts of the proposed work.
f. A timeline for the planned work and a justification for the duration.
g. Plans for disseminating the results.
h. Results from prior NSF support.

2. Modes of Collaboration and Education [*section omitted from excerpt*]

3. Management Plan [*section omitted from excerpt*]

References Cited. References should include full titles of articles and book chapters cited. This section includes bibliographic citations only and must not be used to provide parenthetical information outside of the Project Description. Please indicate with an asterisk (*) references co-authored by two or more proposal investigators.

Cyber-infrastructure

Computers, technologies, wireless connections, and other information technology services that enable and enhance collaboration within and between institutions.

Exercise 11.17

Excerpt 11E is from the National Cancer Institute's Quick Guide for Grant Applications. It is not an RFP but rather a tool for helping investigators prepare a proposal. We include (1) parts of a section that outline the typical components of a proposal and (2) parts of another section that describe the contents of a Project Description (referred to as a "Research Plan"). Read over the excerpt and answer these questions:

a. In what ways do these sections reinforce what has already been stated in this chapter?
b. What differences do you notice?
c. Consider the four mandated sections of the Research Plan (A–D, itemized under "General Proposal Outline") and the five bulleted questions listed under "Content". In which sections of the Research Plan should each of the questions be answered?
d. Look over the list of suggestions. Which heading style (style 1 or style 2) is inferred? (Refer to tables 11.2 and 11.3, if needed.)

Excerpt 11E (adapted from the National Cancer Institute's Quick Guide for Grant Applications. http://deainfo.nci.nih.gov/extra/extdocs/gntapp.htm (accessed January 3, 2008))

General Proposal Outline

Abstract
Research Plan

A. Specific Aims
B. Background and Significance
C. Preliminary Results/Progress Report
D. Research Design and Methods

Research Plan (Overview)

Purpose: The purpose of the research plan is to describe the what, why, and how of the proposal. This is the core of the proposal and will be reviewed with particular care. . . . The assessment of this research plan will largely determine whether or not the proposal is favorably recommended for funding.

Recommended Length: The maximum length of the research plan is 25 pages.

Content: The research plan should answer the following questions:

- What do you intend to do?
- Why is this worth doing? How is it innovative?
- What has already been done in general, and what have other researchers done in this field? Use appropriate references. What will this new work add to the field of knowledge?
- What have you (and your collaborators) done to establish the feasibility of what you are proposing to do?
- How will the research be accomplished? Who? What? When? Where? Why?

Suggestions

1. Make sure that all sections (A, B, C, and D—the what, why, and how of the proposal) are internally consistent and that they dovetail with each other. Use a numbering system, and make sections easy to find. Lead the reviewers through your research plan. One person should revise and edit the final draft.
2. Show knowledge of recent literature and explain how the proposed research will further what is already known.
3. Emphasize how some combination of a novel hypothesis, important preliminary data, a new experimental system, and/or a new experimental approach will enable important progress to be made.
4. Establish credibility of the proposed principal investigator and the collaborating researchers.

Exercise 11.18

Reflect on what you have learned about writing a research proposal thus far by writing a thoughtful response to one of the following:

a. Reflect on the three persuasive tactics essential for funding (i.e., establishing a need, demonstrating intellectual merit, and explaining broader impacts) in relation to a research topic of interest to you.

- How might you establish a need for your area of interest?
- How will you demonstrate intellectual merit?
- What are the broader impacts of the research area?

b. Reflect on the vital importance of research proposals in the lives of chemists.

- What skills do you want to master to be able to write an effective research proposal?
- How will you go about mastering those skills?
- What strengths and weaknesses do you bring to the process of writing a research proposal at this point in your professional development?

c. Reflect on the interrelationships among creativity, credibility, and competence (the three C's) when establishing intellectual merit in a research proposal.

- What strategies will you use to demonstrate your creativity in a research proposal?
- What have you done that will assist you in establishing your credibility in a research proposal?
- How will you emphasize your competence?

d. Reflect on the similarities and differences between a research proposal, journal article, and/or scientific poster.

- What are the major differences between/among these genres?
- What are the similarities between/among these genres?
- How do the similarities and differences influence the ways in which each genre is written?

12 Writing the Goals and Importance Section

Good proposals stimulate my curiosity. As I read a proposal, I begin to think up new questions about the project and its science. The best proposals answer those questions before I'm done reading. I have a lot of respect for writers who can anticipate the questions that a reader will have and provide answers.

—Alexander Grushow, Rider University

This chapter focuses on writing the first section of the Project Description. The central purposes of this section are to identify project goals and objectives, highlight the importance of the research, provide relevant background information, and introduce the proposed research. By the end of this chapter, you should be able to do the following:

- Distinguish between broad goals and specific objectives
- Format a list of objectives correctly
- Emphasize the importance of your research
- Affirm your intellectual merit
- Know when and how to introduce your proposed work
- Select appropriate headings

Staying on Track

This chapter covers what normally is presented in the first major section of the Project Description.

Common level 1 headings in this section are

GOALS AND OBJECTIVES

PROJECT SIGNIFICANCE

As you work through the chapter, you will write the opening section of your Project Description. The Writing on Your Own tasks throughout the chapter guide you step by step as you do the following:

12A Prepare to write

12B Create a list of project goals and objectives

12C Introduce and develop the research story

12D Introduce your proposed work

12E Complete the opening section

Reading and Analyzing Writing

We begin with excerpt 12A for you to read and analyze on your own. The excerpt contains only parts of the author's original Goals and Importance section. Her full section starts with a statement of goals and significance, which is followed by individual descriptions of three separate but related studies. In excerpt 12A, we include only (1) the statement of goals and significance and (2) the description of the second study.

Exercise 12.1

Read and analyze excerpt 12A and answer the following questions:

a. In the first paragraph, the author introduces the research area and its significance. Restate the area and its significance in your own words.

b. In the second paragraph, the author describes the principal goal of the research. Restate this goal in your own words.

c. In Study 2, the author describes specific objectives of her project. State two of them in your own words.

d. How does Study 2 link back to the goals and significance stressed in the opening section of the proposal?

e. The author cites five works in Study 2. How do the citations strengthen her proposal?

Excerpt 12A (from Aga, 2002)

PROJECT DESCRIPTION

GOALS AND SIGNIFICANCE OF PROPOSED WORK

The increased use of agricultural chemicals such as herbicides and animal antibiotics has been mirrored by an increased public concern regarding the impact of xenobiotic compounds on the environment and human health. The pressure to provide new information on the fate of pesticides in the environment has become a monumental task for government and industry. In addition to pesticides, antibiotics are now being recognized as emerging contaminants because of their high potential to enter surface and ground waters from animal confinement operations and manure-treated agricultural fields. The widespread use of antibiotics in animal production is an important issue that has currently attracted attention due to the increased incidence of antimicrobial resistance in food-borne microorganisms. Knowledge of the mechanism for degradation and transport of environmental contaminants is essential in developing guidelines for effective water-quality management actions.

The **principal** goal of the PI's planned research activities is to develop innovative analytical methods, such as immunochemical techniques, and apply these methods in environmental investigations. Emphasis is placed on the development of immunochemical methods because of their cost effectiveness and general applicability in conducting environmental investigations. In addition, there is an increased interest from the regulatory agencies to develop analytical methods that are fast and field-portable for on-site monitoring; these are characteristics of immunochemical methods. There are three closely related studies described in this proposal, all of which involve analytical methods development and have applications to studies that examine the fate and behavior of important agricultural contaminants in the environment.

STUDY 1: [not included]

STUDY 2: STEREOSELECTIVITY OF METOLACHLOR DEGRADATION IN SOIL

Objectives and Significance

This study will provide fundamental information on the effect of stereoisomerism on the environmental fate of a widely used chloroacetanilide herbicide, **metolachlor**. **Metolachlor** is classified as a potential carcinogen and is the second most extensively used herbicide in the United States (*7*). Biological dechlorination of metolachlor leads to the formation of more polar metabolites (*8*), metolachlor oxanilic acid (OXA), and metolachlor ethanesulfonic acid (ESA) (Figure 3). Metolachlor OXA and metolachlor ESA are found at higher concentrations and are more frequently detected in surface and ground water than their parent compound (*9*).

Metolachlor is applied as a mixture of eight different stereoisomers, only four of which have herbicidal activity (*10*). This implies that the other four isomers are applied as contaminants, with no additional benefit to crop production. Whether the chirality of metolachlor influences its degradation rate is unknown. This lack of information is mainly due to the difficulty of separation and analysis of its isomers. Because the stereochemistry of compounds plays an important role in their biological activity and degradation pathways, it is valuable to investigate the influence of stereochemistry on the rate of metolachlor degradation in soil.

Analytical methods to separate and quantify individual isomers of metolachlor and its metabolites are imperative to accomplish this investigation. Chromatographic separation of metolachlor and its metabolites is complicated due to the existence of several rotational isomers for each compound. Thus, these compounds will be fractionated by solid-phase extraction (SPE) into metolachlor (neutral fraction), metolachlor OXA (carboxylic fraction), and metolachlor ESA (sulfonic acid fraction). Newly developed stereoselective immunoassays (specific only to the herbicidally active *S*-isomers) and chiral chromatographic techniques will be employed to achieve quantification of individual isomers. These methods will be used to measure the enantiomeric ratios (ER) of metolachlor and its metabolites in pesticide-treated soil samples. Also, the rate of disappearance of metolachlor versus the rate of formation of metabolites will be followed under controlled conditions.

Pesticide enantiomers are useful as tracers of the soil and water-air exchange process for the following reasons. Although a few chiral pesticides are manufactured as single-enantiomer products, most are racemic mixtures having a 1:1 enantiomeric ratio. Enantiomers have the same physical and chemical properties. As a result, transport processes (leaching, volatilization, and atmospheric deposition) and abiotic reactions (hydrolysis and photolysis) do not discriminate between the enantiomers, leaving ERs unaffected. In contrast, metabolism of pesticides by microorganisms in water and soil and by enzymes in higher organisms often precedes stereoselectively, leading to non-racemic residues and alteration of the original ER (*11*). By examining ERs, it is possible to differentiate the relative importance of biological and physical processes in pesticide degradation. This information will have important consequences on water treatment and pesticide application management.

Information on the possible stereoselectivity of metolachlor degradation will prove useful in understanding degradation pathways and in identifying components in natural environments that shorten or enhance herbicide half-life. In addition, should the degradation of metolachlor be found stereoselective, it will be important for the manufacturers to consider production of the pure isomeric forms of this herbicide and the other chloroacetanilides (e.g., alachlor and acetochlor) to minimize pollution. Lastly, regulatory agencies may also need to re-examine the water-quality standards set for chloroacetanilide herbicides taking into consideration the differences in degradation rates of the stereoisomers, if there are any.

STUDY 3: [not included]

Principle vs. Principal

Principle (noun): a basic truth

Principal (adj.): main; chief

(See appendix A.)

Capitalization

Chemical names are not capitalized unless they are at the start of a sentence. (See appendix A and also table 3.1.)

 **Exercise 12.2**

In anticipation of the moves for this section (described below), find one sentence in excerpt 12A that accomplishes each of the following tasks:

a. Identifies a specific project objective
b. Emphasizes the importance of the research area
c. States a fundamental concept
d. Provides relevant background information
e. Identifies a gap in the field

12A Writing on Your Own: Prepare To Write

After selecting a topic for your research proposal (refer back to the Writing on Your Own task in chapter 11), you are ready to begin your review of the literature. Find and read at least four papers in your research area. As you identify papers that you are likely to cite in your proposal, begin compiling your References Cited section. (See chapter 17.)

Take notes as you read, jotting down key ideas from each paper. State why each work is important. Organize what you have learned from these papers in your own way. Think of approaches for sharing this knowledge and its importance with your readers.

Jot down ideas about why your research area is important. Think of at least three ideas on your own, and then augment your list with ideas suggested in the literature. Is it a growing area of interest? What benefit(s) or new knowledge will your research area provide? (Search for these ideas in the Introduction and Conclusions sections of journal articles.)

Analyzing Organization

A typical move structure for the opening section of the Project Description is presented in figure 12.1. (Of course, if your Request for Proposals recommends a different organizational structure, use that one instead.) Collectively, these moves and submoves depict conventional steps taken by writers to share the goals and objectives of the research, establish its importance, provide relevant background information, and introduce the proposed work. Citations to others' works are common in these moves, particularly in the second move, which has the purpose of placing the proposed work within the context of current knowledge in the field. Because move 1 states specific goals and objectives, it has a more narrow focus than moves 2 and 3.

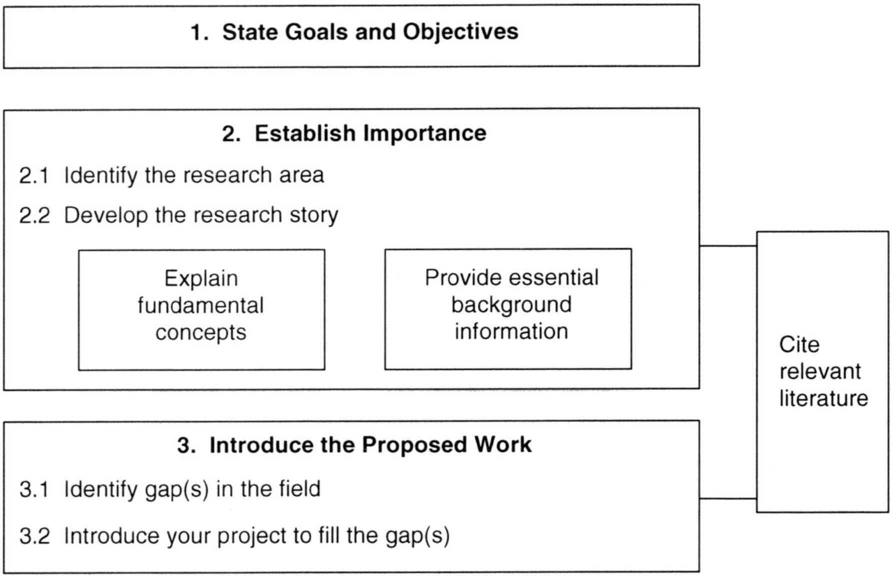

1. State Goals and Objectives

2. Establish Importance

2.1 Identify the research area

2.2 Develop the research story

Explain fundamental concepts

Provide essential background information

Cite relevant literature

3. Introduce the Proposed Work

3.1 Identify gap(s) in the field

3.2 Introduce your project to fill the gap(s)

Figure 12.1 A visual representation of the suggested move structure for a Goals and Importance section of the Project Description.

With figure 12.1 in mind, we are ready to analyze how authors accomplish these moves in authentic proposals. Moves 1–3 are examined below. Headings for each move are suggested, and an analysis of writing features for each move is provided.

Exercise 12.3

Reread the first two paragraphs of excerpt 12A and answer the following questions:

a. What moves are present?

b. Does the author follow the move structure presented in figure 12.1? If not, why might there be differences?

Move 1: State Goals and Objectives

Many, but not all, Project Descriptions begin with a short overview of project goals and objectives. Some Requests for Proposals (RFPs) require this explicitly, but many Project Descriptions start this way even if not required (e.g., 16 of the 22 CAREER proposals cited in this textbook begin by stating the goals and objectives of the proposed work). The purpose of this move is to highlight expected

Table 12.1 Common level 1 headings for move 1 of the Goals and Importance section.

GOALS AND OBJECTIVES	SPECIFIC AIMS
GOALS AND SIGNIFICANCE	RESEARCH OBJECTIVES

accomplishments of the proposed work, allowing reviewers to quickly judge if the proposal is in line with funding objectives. The move stands apart from the rest of the proposal and is typically not viewed as the true start of the Project Description. The move is commonly signaled with a level 1 heading (see tables 11.2 and 11.3). We caution against using a colon after the heading:

Not recommended	**GOALS AND OBJECTIVES:**
Recommended	**GOALS AND OBJECTIVES**

Whatever heading style you choose, be sure to use the same style for equivalent headings throughout your proposal. A few common headings for move 1 are shown in table 12.1.

Goals vs. Objectives

In this textbook, we use the term *goals* to refer to broader purposes and the term *objectives* to refer to the specific steps taken to reach a goal.

Occasionally, you will find the definitions for the two terms reversed.

What is most important is consistency. Avoid using the terms interchangeably within a single document.

Analyzing Excerpts

One's research funding is dependent on clear thinking and the expression of ideas in a coherent manner. In other words, you have to convince readers that you know what you are talking about and that you will be able to meet your established goals.

—Richard Malkin, University of California–Berkeley

Let's examine this move in authentic proposals. In most proposals, the first sentence, immediately after the major division title (e.g., Project Description), states the overall goal of the proposed research (e.g., "The goal of this proposed research is to . . ."). Words like *long-term*, *overarching*, *ultimate*, or *overall* often precede the word *goal*, to denote that the goal is broad in scope and may take several years (and several grants) to accomplish. Most proposals include only one or two goals, in large part because more than two may be viewed as unrealistic for a single proposal. A few examples are shown in P1–P4; each is the opening sentence of the author's Project Description. Note that these goals, although broad in scope, are also rich in scientific content and often target an expert audience:

P1 The goal of the proposed research is the development of new stereoselective [3 + 2] cycloaddition reactions based on Lewis-acid promoted ylide formation for the preparation of complex organic heterocycles (eq 1). (From Johnson, 2003)

P2 The long-term goal of the proposed research is to create functional model complexes of metalloprotein active sites. (From Houser, 2001)

P3 The goal of the proposed research is to investigate the structural and dynamic properties of integral membrane proteins through the use of EPR spin-label spectroscopy and solid-state NMR spectroscopy. (From Lorigan, 2002)

P4 The ultimate goal of the proposed research is the identification and development of fluorescent chemosensors for heavy metal ions and small organic molecules in water. (From Finney, 1999)

The goal statement is followed by specific objectives, which spell out the steps needed to accomplish the larger goal. Ideally, these objectives should pass the SAM test: they should be *specific* (specifying what you will do and how you will do it), *achievable* (within a realistic time frame), and *measurable* (implying ways that you will measure success). Because of their specificity, objectives are usually written for an expert audience. (Paul, A. The Grant Institute's Grants 101, personal communication, 2006)

The SAM Test

Effective objectives should pass the SAM test by being

Specific

Achievable

Measurable

Examples of goal statements and specific objectives are shown in excerpts 12B–12E. The first two examples (excerpts 12B and 12C) enumerate objectives; excerpt 12B lists the objectives in continuous text, while excerpt 12C presents objectives in a displayed list. The latter approach is easier to read and preferred if space permits. (For another example of a list, refer to the list of nine objectives in excerpt 11A.) Note that each excerpt includes a heading, a general goal, and a list of objectives.

Lists in Research Proposals

Lists are common in research proposals. They are used to delineate project objectives (chapter 12) and expected outcomes (chapter 14).

Lists are not common in journal articles; hence, this is an important style difference between the two genres.

Excerpt 12B (from Lyon, 2000)

SPECIFIC AIMS

The goal of this research is to develop a new class of bioresponsive materials that undergo rapid, large-magnitude, volume-phase transitions in response to specific biological stimuli. Our approach to these materials is based on two fundamental aspects of hydrogels: (1) hydrogel solvation/desolvation thermodynamics can be perturbed . . . and (2) hydrogel chemistry can be reduced. . . . Accordingly, this research will involve the following specific objectives: (1) the synthesis of polymer-protein bioconjugates, (2) the fabrication of nano-sized hydrogel thin films, (3) the development of synthetic models for particle size **control, and** (4) the design and synthesis of new core-shell hydrogel particles. By completely characterizing the resultant materials with respect to their structure, morphology, thermodynamics, and kinetics, we can obtain a fundamental understanding of the hydrogel structure-function relationship. In turn, this will allow us to develop rational designs for truly bioresponsive devices.

Commas

Commas are used to separate items in a list of three or more items. Include a comma before the "and" preceding the last item. (See appendix A.)

Excerpt 12C (from Patrick, 2000)

RESEARCH OBJECTIVES

The overall goal of this research is to explore the use of thermotropic **liquid crystal (LC)** solvents as new and versatile media for the engineered growth of molecular thin films. This goal will be pursued in two parts. First, a set of experiments is planned to investigate fundamental aspects of thin film growth in **LC** media. These experiments will help elucidate important physical phenomena and deepen our understanding of the role played by the LC solvent. A second group of experiments will build on this foundation to prepare model thin film materials from several different molecular building blocks. The specific research objectives are

1. to determine the influence of LC surface anchoring and bulk fluid curvature elasticity on orientational order in thin films deposited from LC solvents. This will be

accomplished through two sets of experiments separately measuring the contribution of each effect.

2. to develop a model that explicitly articulates the roles of these factors.
3. to apply this knowledge toward the preparation of two thin film systems:
   a. bi-component composite films in which the alignment of one component is unidirectional, and the alignment of the second component is random,
   b. oriented molecular films for use as re-writable anchoring layers.

Abbreviations

Most abbreviations (or acronyms) are defined at first use and then used without definition. Abbreviations often omit periods (e.g., LC not L.C.). (See appendix A.)

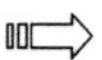

Parallelism in Lists

See appendix A.

Exercise 12.4

Reread passages P1–P4 and excerpts 12A–12C, and answer the following questions:

a. Who is the intended audience for move 1?
b. How are the words "goal" and "objectives" used in move 1?
c. What verb tense(s) is used in move 1 (past, present, future)?
d. When should you use a colon (:) in an enumerated list?
e. When numbers are used to enumerate objectives, are they Roman (i, ii, iii,. . . .) or Arabic (1, 2, 3, . . .)? How are the numbers formatted?
f. Why do you think Aga (12A) and Lyon (12B) include their lists in continuous text rather than in actual list form, as in excerpt 12C?
g. How well do the objectives in 12B and 12C meet the SAM test? Explain.

Excerpt 12D illustrates a slight variation of move 1. Rather than state project goals and objectives explicitly, the author frames them as research questions, again in list form. (Such an approach underscores the inquisitive nature of science.)

Excerpt 12D (from Rose-Petruck, 2000)

SCIENTIFIC GOALS OF THE PROPOSED STUDIES

The following questions will be addressed in the proposed **studies:**

1. How exactly does the solvent modify the ligand motions during the first few hundreds of femtoseconds?
2. Does the ultrafast dissociation process of $Fe(CO)_5$ in solution, and that of related compounds in general, occur in a concerted or a rapid sequential way?
3. Which vibrational modes of the iron pentacarbonyl interact most strongly with the solvation shells, and which vibrational modes lead to the eventual escape of only a single carbonyl ligand?
4. If the dissociation process occurs in a concerted way causing instantaneous loss of four CO-ligands, which ligand motions does the caging solvation shell inhibit?
5. The ultrafast laser pulse excites many vibrational modes of the metal carbonyl simultaneously (i.e., the vibrational modes are phase-coupled). Are the subsequent ligand motions coherent or do ultrafast dephasing processes hinder coherent motions?

Colons (:)

A colon may be used to introduce a list of items, provided that the colon follows a complete sentence. (See appendix A.)

Exercise 12.5

Rephrase the questions in excerpt 12D as statements of objectives. Which approach do you prefer? Why?

If you have only a few objectives, your proposal may read more smoothly if you use ordinal language (e.g., *first* and *second*), as illustrated in excerpt 12E, rather than a numbered list (as in excerpt 12D). Alternatively, you may leave out enumeration entirely. For example, in excerpt 12A, Aga describes her principal goal (to develop innovative analytical methods for environmental applications) and objectives (to develop immunochemical methods that are cost-effective, fast, and field-portable and use them to monitor the fate of agricultural contaminants) without any enumeration.

Excerpt 12E (from Lee, 2001)

SIGNIFICANCE AND OBJECTIVES

The overall goal of the proposed work is to broaden our understanding of nucleotide chemistry through gas-phase mass spectrometric and theoretical quantum mechanical studies. The objectives are twofold. **First**, the gas-phase acidity and nucleophilicity of different sites on nucleobases will be explored to understand their fundamental electronic reactivity. **Second**, the chemistry of orotic acid will be examined to elucidate the mechanism of conversion of this important nucleic biosynthetic precursor to uracil. The significance of each facet of this program is outlined separately below.

First,

Ordinal language may be used in proposals. Use a comma after ordinal terms and other introductory linking words and phrases, e.g.,

First,...

Second,...

...; and third,...

However,

In contrast,...

(See appendix A and table 6.6.)

Exercise 12.6

Let's assume that Lee (excerpt 12E) had said "Two research questions are addressed" instead of "The objectives are twofold." Rewrite the objectives as two research questions.

Exercise 12.7

What headings are used in excerpts 12A–12E to identify move 1? Add new ones to the list of common headings in table 12.1.

Beginning writers often choose words for their goals and objectives that either are too nonspecific or overstate their case. For example, a novice writer might say "We will *see* if this is the case" or "We will *discover* important new compounds." Although both sentences are grammatically correct, the first is too nonspecific (what exactly does the writer mean by the word *see*?) and the second is overstated (what if new compounds are not discovered?). Another common mistake is to overuse the word *experiment*. The word *experiment* usually connotes a single event (e.g., "In today's experiment, we will synthesize cyclohexane."). Research involves countless experiments, which can span months or years; hence, scientists prefer words such as *work*, *project*, or *study*. These and other words and

Table 12.2 Words and phrases to avoid and their more appropriate alternatives.

Avoid words and phrases that	Instead, use words and phrases that
are informal and nonscientific, e.g., "We will (*see, find out, look into, try to figure out if*)..."	are more formal and scientific, e.g., "We will (*monitor, measure, determine, explore, examine, investigate, analyze*)..."
overstate what may be achievable, e.g., "We will (*discover, prove, create, cure, find, put an end to*)..."	lead to measurable results, e.g., "We will (*develop, apply, measure, quantify, explore, compile, examine, characterize, synthesize, modify, adapt*)..."
promise success prematurely (before the work has been done), e.g., "We will (*show, demonstrate, find, prove, discover, create*) *that*..."	acknowledge assumptions or predictions, e.g., "We (*hypothesize, expect, predict*) that..." "*Preliminary* results *suggest*..."
refer to a long-term project as an "experiment" or "lab", e.g., "In this *experiment*, we propose to..."	convey the long-term nature of the work, e.g., "In this *project*, we propose to..." "In this *work*, we propose to..."

phrases to avoid are presented in table 12.2, along with their more appropriate alternatives.

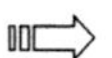

Formal Vocabulary

See appendix A.

Exercise 12.8

Consider the following sentences. Identify words that should be avoided and suggest more appropriate alternatives (see table 12.2).

a. In our experiment, we will show that polyurethane films can be successfully attached to biosensors and used to detect nerve agents.

b. We will discover new types of sandwich compounds, composed of polycyclic aromatic hydrocarbons surrounding a single metal atom.

c. The goal of our research is to see if we can find novel biomolecules that will bind to hairpin and i-Motif DNA.

d. We will look at how injecting swine with artificial porcine somatotrophin affects carcass mass and composition.

Before we conclude our discussion of move 1, we call your attention to the pointers accompanying excerpts 12A–12E. These pointers, together with their corresponding excerpts, highlight correct usages of the following words, punctuation, or other writing features:

- principle or principal (excerpt 12A)
- capitalization (excerpt 12A)
- commas in a list of three or more items (excerpt 12B)
- abbreviations and acronyms (excerpt 12C)
- parallelism in enumerated lists (excerpt 12C)
- colons before a list (excerpt 12D)
- commas after introductory linking words or phrases (excerpt 12E)

To test your knowledge of some of these features, complete exercises 12.9–12.12. Refer back to the excerpts and pointers as needed.

Exercise 12.9

We have intentionally removed commas from the following passage adapted from Fairbrother (2000). Where are additional commas needed?

For example a polymer's interfacial characteristics determine chemical and physical properties such as permeability, wettability, adhesion, friction, wear and biocompatibility.[1–3] However polymers frequently lack the optimum surface properties for these applications.[1,4] Consequently surface modification techniques have become increasingly desirable in technological applications of polymers.[5,6]

Exercise 12.10

Select the appropriate words and correct the punctuation in the following paragraph.

Our *principal/principle* objective is to investigate the long-term effects of these pollutants. Our methods will include: (1) extraction, (2) concentration and (3) purification of liquid and solid samples. In *principal/principle*, the liquid samples should be easier to extract.

Exercise 12.11

Which of the following sentences are punctuated correctly?

a. Our approach is based on: hydrogel thermodynamics and hydrogel chemistry.

b. Our approach is based on two properties: hydrogel thermodynamics and hydrogel chemistry.

c. Our approach is based on hydrogel thermodynamics and hydrogel chemistry.

Exercise 12.12

Revise the last three items in this list so that they are parallel with the first two items (adapted from Lyon, 2000).

This research will involve the following specific objectives:

a. investigation of hydrogel solvation/desolvation thermodynamics
b. synthesis of polymer-protein bioconjugates
c. we will fabricate hydrogel thin films
d. synthetic models will be developed
e. hydrogel particles will be designed

Lastly, we briefly consider verb tense and voice used in move 1. Goals and objectives are commonly stated in present and/or future tense. Both active and passive voice may be used. Common tense and voice combinations are summarized in table 12.3. Personal pronouns (e.g., *we, my, our*) should be avoided in the opening goal statement but may be used elsewhere in move 1:

Opening goal statement

Inappropriate My research goal is to...
Appropriate The goal of the research is to...

Elsewhere

Appropriate Our approach is based on...
Appropriate We will gain a fundamental understanding...

Table 12.3 Common functions of different verb tense–voice combinations in move 1 of the Goals and Importance section.

Function	Tense–Voice Combination	Example
State project goals and objectives	Present–active	The goal of the research *is to identify* ...
	Present–passive	A set of experiments *is planned* to ...
	Future–active	The research *will address* the following...
	Future–passive	This goal *will be pursued* in two parts.

12B Writing on Your Own: Create a List of Project Goals and Objectives

After a Research Proposal topic is selected (refer back to the Writing on Your Own task in chapter 11), it is time to specify project goals and objectives. Consider questions such as these to begin the brainstorming process:

1. What is the overarching goal of your proposed work?
2. What steps will you take to achieve that goal?

Prepare a list of goals and objectives for yourself initially, to guide you in organizing the rest of your proposal. Depending on the RFP that you have chosen, you may want to include the list of goals and objectives in your final proposal, too. Make sure that you use the words *goal* and *objective* as specified in the RFP. Usually *goals* refer to broader purposes and *objectives* refer to specific steps taken to reach a goal. If the RFP uses the terms in reverse, you should do so, too.

Move 2: Establish Importance

When I began writing grant proposals, I had very little success but did not understand what I was doing wrong. At that time, I thought that luck was a big factor and that I just hadn't been lucky yet. So to increase my odds, I sent out more proposals. Basically, this just led to an increased number of rejections. After several years of seemingly fruitless effort, I finally really asked myself, what am I doing wrong? Why are reviewers not recommending my proposals for funding? That is when my thinking began to change.

I started thinking about the proposal as a communication tool to market my ideas. I no longer expect the reviewer to do a lot of work to understand my ideas or recognize the significance of my proposal. I assume it is my job to convince the reader that the problem I am working on and the solution I have proposed are compelling and have high value. I work to grab the reader's attention immediately and then serve up a well-organized, persuasive, and easily readable presentation of my ideas. I provide convincing answers to any obvious questions about the proposed approach so that doubt does not have a chance to take root in the reader's mind. In short, in proposal writing, I think of the reader as the customer, and I work to provide the best service possible.

—Joan Curry, University of Arizona

Table 12.4 Common level 1 headings for move 2 of the Goals and Importance section.

PROJECT SIGNIFICANCE	PROJECT RATIONALE
SIGNIFICANCE OF PROPOSED RESEARCH	PROJECT BACKGROUND

Move 2 represents the true beginning of the Project Description. It is here where authors establish importance, the unifying theme of the move. Each submove in move 2, in some way, contributes to this theme. Establishing importance is crucial to a successful proposal. Projects that are trivial or inconsequential are unlikely to be funded. It is not enough to say, "This project is important." Rather, you must offer compelling reasons why it is important.

Move 2 is also where the research story begins; fundamental concepts are explained, and background information is shared. The research story begins by highlighting what others have accomplished, thereby laying the groundwork for the proposed work in move 3. Common headings for move 2 (table 12.4) incorporate such words as *significance* and *background*. The heading for move 2 is typically a level 1 heading (parallel to the heading for move 1).

Analyzing Excerpts

To illustrate how authors progress through move 2 in their proposals, let's examine some excerpts. The first submove of move 2 identifies the research area and, whenever possible, also hints at why the research area is important. For example, in the following passages (all first sentences of move 2), the broad research area is identified and its importance is suggested. The proposed work is not mentioned. This does not occur until move 3.

P5 Maximizing the world's agricultural efficiency depends on controlling unwanted pests—especially weeds. (From Vyvyan, 2001)

P6 **Preparation** of new materials by directed synthesis at molecular-length scales **is** the object of much current research in materials science, surface chemistry, and the emerging field of crystal engineering. (From Patrick, 2000)

P7 The increased **use** of agricultural chemicals such as herbicides and animal antibiotics **has been mirrored** by an increased public concern regarding the impact of xenobiotic compounds on the environment and human health. (From Aga, 2002)

P8 Thermal conrotatory 4π **ring opening** of aziridines and epoxides **is** a well-recognized method of generating azomethine and carbonyl ylides.[1] (From Johnson, 2003)

Subject–Verb Agreement

See appendix A.

Exercise 12.13

Identify each research area addressed in P5–P8. Based only on the authors' words, suggest why each research area is important.

Exercise 12.14

Consider the following sets of sentences adapted from two research proposals. In each set, determine which sentence is the opening sentence of move 2. Use the move structure in figure 12.1 to assign a move to the other sentence. (Note: Citations have been removed to make this exercise more challenging.)

Set 1 (adapted from Kohen, 2002)

a. The overall goal of this project is to seek a better understanding of how enzymes activate covalent bonds.

b. The role of protein fluctuations in enzyme rearrangement, reactant binding, and product release is well established.

Set 2 (adapted from Hergenrother, 2002)

a. Research in my laboratory will utilize combinatorial chemistry and modern high-throughput screening techniques in an effort to make fundamental biological discoveries.

b. In the past decade, combinatorial chemistry has exploded onto the scene of modern science.

After introducing the topic, the research story is developed (submove 2.2). Submove 2.2 involves two steps (in any order): (1) explain fundamental concepts and (2) provide essential background information. In many proposals, explanatory remarks are included to help reviewers more fully comprehend the proposed science (recall that not all reviewers will be experts in the author's field). Furthermore, explanatory remarks demonstrate the authors' understanding of the proposed science. These remarks may occur anywhere in a proposal, but they are particularly common in move 2. Let's consider a few examples. In the simplest case, an explanatory remark may be embedded within a single sentence, as illustrated in P9–P11:

P9 Allelopathy, the chemical interaction between plants and microorganisms, has been known for thousands of years. (From Vyvyan, 2001)

P10 Gas hydrates are nonstoichiometric compounds consisting of hydrogen-bonded water molecules in a cagelike structure, which traps small-diameter gas molecules. (From Tuckerman, 1999)

P11 Nitrite reductases (NiRs)—enzymes found in several strains of denitrifying bacteria—catalyze the one-electron reduction of nitrite anion to nitric oxide. (From Houser, 2001)

Alternatively, some concepts may be described in more detail, approaching a near textbook-like description. Colored illustrations may be used and textbooks may be cited, if they were used as resources for this information. For example, Hergenrother includes an introduction to apoptosis in his proposal, using both text and graphics. A short passage from his text is reproduced in excerpt 12F, where he describes how either underactive or overactive apoptotic processes can lead to dire cellular consequences. Understanding apoptotic processes is important to his proposed work, which involves apoptotic proteins.

Citing Textbooks?

Authors may cite textbooks (or other authoritative sources) that are used to prepare explanatory remarks in a research proposal.

Textbooks are not usually cited in journal articles (see chapter 6).

Excerpt 12F (from Hergenrother, 2002)

Apoptosis, or programmed cell death, is an essential process that results in the methodical destruction of unwanted or potentially harmful cells.[7] The proper functioning of this pathway is critical for cellular growth and maintenance, and modulation of apoptosis has dire cellular consequences.[8] For example, a down regulation of the apoptotic cascade can cause the uninhibited growth of tumor cells. In such cases, small molecules that restore proper apoptosis have the potential as powerful chemotherapeutic agents.[9] Conversely, in certain degenerative disorders, the apoptosis pathway is overactive, resulting in premature cellular death. In these cases, inhibitors of apoptosis are desired.

Another example is presented in excerpt 12G, where the author describes, in near textbook-like detail, the phenomenon of quantum-mechanical tunneling. A full-color illustration is included in the proposal (although the figure here is in black and white), depicting the different tunneling behaviors of light (e.g., hydrogen) and heavy isotopes (e.g., deuterium and tritium). Later on in the proposal (not included here), the author goes on to describe how the kinetic isotope effect (KIE) can be used as a probe for tunneling. The KIE is defined ("the ratio of the reaction rates of two isotopes of the same element"), and a second full-color illustration is included to depict both a semiclassical model of the KIE (without tunneling) and a quantum-mechanical model (with tunneling). The author includes this information because hydrogen tunneling is integral to his proposed research.

Full-Color Illustrations

Proposals often include full-color graphics to elucidate important concepts. Color graphics should also make sense in gray scale, in case reviewers print out a hard copy.

Excerpt 12G (from Kohen, 2002)

Quantum mechanical tunneling. Tunneling is the phenomenon by which a particle transfers through a reaction barrier due to its wave-like property.[30,37] Figure 1 graphically illustrates this for a carbon-hydrogen-carbon double-well system.... Hydrogen is so light that there is a significant uncertainty in its location at a given energy (the Heisenberg uncertainty principle). When it is close to the reaction's barrier, there is some probability of finding it on the product side, if the barrier is narrow enough and the reactant and product energy levels are degenerate. As illustrated in Figure 1, the lighter isotope has a higher tunneling probability than the heavier one, suggesting that kinetic isotope effects (KIEs) should be effective tools in studying tunneling.

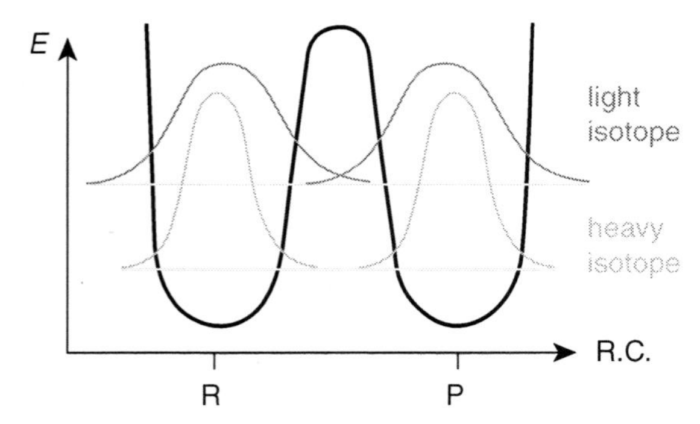

Figure 1. An example of ground state nuclear tunneling along the reaction coordinate (R.C.). The reactant well (R) is on the left side and the product well (P) is on the right. The blue and red lines describe a light and a heavy isotope probability function, respectively. The greater the overlap of the R and P probability functions, the higher the tunneling probability.

Formatting Figures, Tables, and Schemes

See chapter 16.

Exercise 12.15

Select a topic that you are currently studying in research or coursework. First, write a single sentence that includes an explanatory remark about the topic. Next, write a paragraph that describes the topic in more detail, with an illustration if possible. Include citations, as appropriate, in the paragraph.

Another step in developing the research story (submove 2.2) is to provide essential background information. In providing background information, authors summarize (and cite) works that influenced their proposed ideas. Because new ideas nearly always build on earlier contributions, effective writers learn to integrate background information into their own research stories, highlighting the works that laid the groundwork for their proposed work. Providing background information also helps authors establish intellectual merit by demonstrating knowledge of pivotal works in their field.

Effective writers also learn to present background information in ways that further establish the importance of the proposed work (move 2). To underscore this point, we organize the excerpts that follow around three approaches for sharing background information. Each approach provides background information and, at the same time, stresses the importance of the proposed work.

As each approach is presented, we use italics to highlight words and phrases that are particularly effective in conveying importance. These words are commonly used in proposal writing, and we encourage you to use them in your own writing as well.

Three Approaches to Providing Background Information and Emphasizing Importance

Document interest

Emphasize benefits

Establish need for new knowledge

Important Words

Effective proposal writers use words that stress the importance of their research as they provide background information.

Document Interest

Your work acquires credibility when you review the literature and show that your contribution extends from a solid foundation of respected research.

—Paradis and Zimmerman (1997)

One way authors stress importance as they share background information is to document widespread interest in the research area (i.e., if others studied it, it must be important). Slightly different tactics are used depending on how new the research area is. If the field is already well established, phrases such as *long-term interest*, *renewed interest*, *extensively studied*, or *for decades, chemists have studied* are commonly employed. If the field is relatively new, such phrases as *emerging interest*, *increased attention*, *previously unstudied*, *promising new approach*, *unexplored area*, or *an area of mounting concern* are more appropriate.

The following excerpts illustrate this approach. Vyvyan (excerpt 12H) establishes the importance of herbicides by (1) acknowledging the *historic role* of synthetic herbicides *over 50 years*, (2) pointing out *shifting attention* to alternative approaches, and (3) emphasizing that the area has *recently opened up* to significant research. Gudmundsdottir (excerpt 12I) demonstrates that aryl nitrenes are important because they have been *studied extensively over the last decades*. Moreover, *renewed interest* has been *sparked* because aryl nitrenes are *ideal candidates* for organic magnetic materials. Note, too, in excerpts 12H and 12I, how the authors incorporate citations to others' works as they develop their research stories (the history of weed control in excerpt 12H and the history of aryl nitrenes in excerpt 12I).

Excerpt 12H (from Vyvyan, 2001)

BACKGROUND AND SIGNIFICANCE

Maximizing the world's agricultural efficiency depends on controlling unwanted pests—especially weeds. Synthetic herbicides have met the weed control needs of industrialized nations for over 50 years, but mounting ecological and human health concerns are shifting attention to alternative weed control technology based on cues from nature.[1,2] Allelopathy, the chemical interaction between plants and microorganisms, has likely been known for thousands of years. This area, however, has only recently opened up to significant, specific research aimed at determining which natural products are responsible for these interactions.[3]

Excerpt 12I (from Gudmundsdottir, 2001)

INTRODUCTION

Aryl nitrenes have been studied extensively over the last decades and are used in several industrial processes such as microlithography.[1] Aryl nitrenes have also been used in photoaffinity labeling bioorganic molecules. The pursuit for organic magnetic material has sparked renewed interested in nitrene intermediates, which are ideal candidates for magnetic material because of their high spin properties.[2]

Spain (excerpt 12J) also builds her research story as she demonstrates interest in her field. Sixteen studies are cited to document interest, stating that *more and more examples* of kinetic-energy-dependent surface dynamics *are appearing in the literature*. Key findings from these works are summarized (citations 6–21), providing readers with necessary background information and attesting to the authors' intellectual merit. Collectively, the cited works *point to the importance of the emerging discipline* addressed in her proposal, hot atom chemistry.

Excerpt 12J (from Spain, 1997)

More and more examples of kinetic-energy-dependent surface dynamics of reagents on low temperature substrates are appearing in the literature. A few examples follow. Although Cl_2 does not spontaneously etch Si at room temperature, kinetic-energy-enhanced Cl_2 does,[6–11] as observed by mass spectrometric detection of products. Similar studies involving F and F_2 have appeared in the literature.[12–13] The growth of thin films with pulsed supersonic jets prompted the study of CO_2 chemisorption on the Si(100) surface. The initial sticking coefficient of CO_2 on Si(100)2 × 1 was found to be negligible with thermal CO_2 but significant with hyperthermal CO_2.[14,15]

The first direct evidence for the Eley–Rideal mechanism was determined in 1991.[16] In that study, it was demonstrated that kinetic-energy-enhanced $N(C_2H_4)_3N$ reacts with H on a Pt(111) surface to form an ion. The ion leaves the surface with a translational energy that depends on the energy of the incident species. In a study of H atoms incident on D/Cu(111),[17,18] direct evidence that HD is formed by the Eley–Rideal mechanism was found, in part, by varying the translational energy of the H reagent.

Other groups have investigated how incident kinetic energy affects film growth processes. For example, kinetic-energy-dependent surface trapping probabilities were measured with hyperthermal beams of alkali ions directed at a Cu surface.[19] In general, trapping efficiency decreases as incident ion kinetic energy increases. However, at lower incident energies, a minimum trapping probability was observed. Co adsorption to Si(100) has also been shown to depend on Co translational energy.[20] Collectively, these studies point to the importance of the emerging discipline called "hot atom chemistry."[17,21]

Superscript Numbers

Using superscript numbers is one of three common ways to cite others' works. (See chapter 17.)

Similarly, Tuckerman (excerpt 12K) cites works that emphasize *widespread interest* in the research area, highlighting, for example, that crystal hydrates *have attracted the attention* of crystallographers and spectroscopists *over several decades* (*46–28, 123–125*). Specific benefits of crystal hydrates are touted, including their *possible use* as proton conductors (*126*) and as important media for the study of proton motion. The latter is *currently of interest* in the field of low temperature spectroscopy (*127–129*).

Excerpt 12K (from Tuckerman, 1999)

Crystal hydrates of strong acids and bases have attracted the attention of crystallographers and spectroscopists over several decades (*46–48, 123–125*). The interest in these crystal systems lies in their possible use as proton conductors (*126*). Moreover, as proton transfer events often occur in complexes such as $H_5O_2^+$ and $H_3O_2^-$, these systems are important media for the study of proton motion through hydrogen bonds in a crystalline environment, a process currently of interest in the field of low temperature spectroscopy (*127–129*).

Italic Numbers in Parentheses

Using italic numbers in parentheses is one of three common ways to cite others' works. (See chapter 17.)

Exercise 12.16

Reread excerpts 12H–12K and answer the following questions:

a. What research topic is identified in the opening sentence of each excerpt (confirming that excerpts follow submove 2.1)?
b. How many works are cited in these excerpts? Comment on how the cited works help to stress importance, develop the research story, and establish intellectual merit.
c. Are these excerpts written for a scientific or expert audience? Justify your answer.
d. Of the key words and phrases used in these excerpts to document interest in the research area, which three are your favorites (i.e., ones that you would likely use in your own writing)?

Exercise 12.17

Read excerpt 12L and answer the following questions:

a. Briefly describe how the excerpt establishes importance by demonstrating interest in the research area.
b. In addition to interest, what other features are stressed to emphasize the importance of combinatorial chemistry?
c. What words and phrases are used to emphasize the importance of the area?
d. Is this researcher proposing work in an established or emerging field? Explain.

Excerpt 12L (from Hergenrother, 2002)

In the past decade, combinatorial chemistry has exploded onto the scene of modern science. The pharmaceutical industry quickly adopted this burgeoning technology as a method to rapidly derivatize and evaluate "**lead** compounds"—compounds that have already shown promise in some biological assay. As such, combinatorial chemistry has often been relegated to a late role in the drug discovery process and is largely thought of as an "industrial" type of science. However, it is clear that the potential for combinatorial chemistry is much greater than has been realized thus far. Indeed, there exists a real opportunity for the creative uses of combinatorial chemistry and for new paradigms in the development and application of combinatorial technologies.

Lead vs. Led

Lead (n.): Pb (an element)

Lead (v.): to show the way

Lead (adj.): serving as a model or leader

Led (v.): past tense and past participle form of *lead*

The element *lead* (noun) and the verb *led* (past tense/past participle) have the same pronunciation. Be careful not to confuse them in your writing.

Emphasize Benefits

A second way authors stress importance as they share background information is to describe how the research will benefit others. For example, Houser (excerpt 12M) points to environmental benefits; he makes the case that understanding the process of dentrification can lead to *improvements* in *air pollution*, *greenhouse gas production*, and *eutrophication*. Finney (excerpt 12N) points to technological benefits; he argues that advances in fluorescent chemosensors can both *simplify measurements* and *reduce cost*, by using *much less expensive* materials. Similarly, Lyon (excerpt 12O) emphasizes medical benefits; he investigates bio-responsive gels and points out (as he describes previous accomplishments in the field) the already *remarkable benefits* of these gels in drug-delivery systems. Once again, the literature is cited in each excerpt, serving both to develop the research story and to establish the intellectual merit of the investigator.

Excerpt 12M (from Houser, 2001)

Copper-Nitrite Reductase (Cu-NiR)

Nitrite reductases (NiRs)—enzymes found in several strains of denitrifying bacteria—catalyze the one-electron reduction of nitrite anion to nitric oxide (Equation 1).[1,2] In addition to the importance of this process in the global nitrogen cycle (Figure 1), further incentive for the study of the denitrification process is provided by its environmental impact, ranging from the production of NO as a pollutant and N_2O as a potent green-house gas, to lake eutrophication due to farm runoff that contains high concentrations of nitrates and nitrites.

Excerpt 12N (from Finney, 1999)

There are two primary reasons why a fluorescent chemosensor with visible emission would be desirable, in addition to its inherent aesthetic appeal. First, it simplifies qualitative experimental measurement: the human eye is sensitive enough to detect extremely small changes in intensity (the limit of visual detection by a dark-adjusted eye has been estimated as $\approx$ 1 nW),[22] which would allow a simple visual assessment of whether fluorescence enhancement was occurring in the presence of an analyte. Second, the materials associated with the quantitative measurement of visible light are much less expensive than those for measurement of **UV emission; visible emission** would allow the use of disposable plastic curvettes for fluorescence measurements in the lab and would be compatible with the least expensive forms of fiber optic technology.

Semicolons (;)

Use semicolons to join two related statements. (See appendix A.)

Excerpt 12O (from Lyon, 2000)

Among the most remarkable responsive gels have been those involving drug delivery.[14,35–37] An artificial pancreas has been developed,[13,14,38] as has a secretory granule mimic.[39] In these applications, the gel acts as a drug entrapment matrix in its collapsed state and then expands in response to temperature or salinity changes thereby releasing the drug to the environment. Others have modified the active sites of a variety of proteins with short thermoresponsive oligomers[40,41] or have attempted to construct a hydrogel-based artificial muscle **composed of** bundled pH-responsive fibers.[42] Still others have created enzyme-polymer conjugates which swell in response to enzyme activity.[43,33] In these cases, the ionic strength or pH of the environment changes as the enzyme turns over a substrate.

Comprise vs. Compose

See appendix A for more information on these commonly confused words.

Exercise 12.18

Reread excerpts 12M–12O and complete the following tasks:

a. List at least four benefits that the authors have identified in their writing. Suggest two additional benefits that can be addressed, not identified in these excerpts.

b. Comment on the number of citations in these excerpts. Give an example of how the cited works further the research stories and attest to the authors' intellectual merit.

c. Are these excerpts written for a scientific or expert audience? Justify your answer.

Exercise 12.19

Read excerpts 12P and 12Q. One excerpt promotes importance by describing a research area that will have a positive impact on society. The other does so by highlighting a potentially negative consequence to society. Which is which? What benefits will research in each area provide to society?

Excerpt 12P (from Fairbrother, 2000)

Organic surfaces are encountered in a wide range of situations where interfacial properties impact a material's performance characteristics.[1] For example, a polymer's interfacial characteristics determine chemical and physical properties such as permeability, wettability, adhesion, friction, wear, and biocompatibility.[1–3] However, polymers frequently lack the optimum surface properties for these applications.[1,4] Consequently, surface modification techniques have become increasingly desirable in technological applications of polymers.[5,6] These processes are capable of tuning the properties of organic surfaces without affecting the bulk composition, thereby transforming inexpensive raw materials into highly valuable finished products.[7] Within the field of surface modification treatments, vacuum based strategies, including plasma processing,[8] have become a preeminent means for tailoring the properties of organic surface.[9] Compared to wet-chemical treatments for surface modification, these technologies are dry, fast, and environmentally benign.[10]

Excerpt 12Q (from Aga, 2002)

Approximately 19 million pounds of antibiotics are used each year in U.S. cattle, hogs, poultry, and other food animals; this is over 40% of the antibiotics sold in the U.S. (*17*). The routine use of antibiotics on farms to accelerate growth and prevent diseases has been speculated to have created strains of disease-causing bacteria that are resistant to antibiotics, which in turn infect more human beings every year (*18*). Because of the widespread use of antibiotics in livestock production, it would not be surprising to see antibiotic contamination of the aquatic environments situated near animal feedlots and confinements. The presence of antibiotics in water resources is suspected to contribute to the proliferation of resistant microorganisms. It is not clear how much of these chemicals finds its way into the surface water near livestock operations.

Exercise 12.20

Reread excerpt 12P and answer the following questions:

a. Examine the ways in which the author creates linkages, adds emphasis, and contributes to the fluency of his writing with such words and phrases as *for example*, *however*, and *consequently*. Consult table 6.6 and determine the functions of these words.
b. The word *thereby* is also used in excerpt 12P; it is not listed in table 6.6. Nonetheless, consult the table to assist you in determining its function in the excerpt.

Establish Need for New Knowledge

A third way authors establish importance as they share background information is to highlight the need for new knowledge. Few scientists can resist the appeal of new knowledge, and a project will likely be successful if it has the potential to offer new insights, depth, or detail. For example, Fairbrother (excerpt 12R) uses this approach when he states that the surface science associated with polymer modification treatments *remains virtually unexplored*. He goes on to cite a panel report that attests both to this *lack of mechanistic understanding* and to the need for *concerted experimental research* in this area.

Excerpt 12R (from Fairbrother, 2000)

Despite their technological importance, the surface science associated with polymer modification treatments remains virtually unexplored. This lack of a mechanistic understanding can be ascribed to the complexity of typical reactive media present in modification treatments, as well as to the heterogeneous/polyfunctional nature of the organic surface. In the absence of a molecular level understanding, modification strategies have developed along empirical lines. The lack of a fundamental mechanistic understanding of surface modification processes has been highlighted in a recent panel report.[11] The panel notes that to develop an understanding of the physical and chemical processes accompanying surface modification treatments, concerted experimental research is required.[11]

Similarly, Rose-Petruck (excerpt 12S) asserts that current knowledge of ultrafast molecular motions is *not sufficient*, and more *detailed knowledge* of these motions is of *fundamental importance* for understanding reaction mechanisms. Warren (excerpt 12T) suggests *a better mechanistic understanding* of metal-catalyzed processes is needed and may be possible with *new insights* into the patterns of selectivity of metal-catalyzed processes. Walker (excerpt 12U) cites several different phenomena (chemical transport, surface reactivity, electron transfer, and self-assembly) that could benefit from *deeper insights* into interfacial solvation interactions. Such insights could be gained through predictive models of solvent-solute interactions.

Excerpt 12S (from Rose-Petruck, 2000)

A fundamental goal of chemical research has always been to understand the reaction mechanisms leading to specific reaction products. Reaction mechanisms, in turn, are a consequence of the structural dynamics of molecules participating in the chemical process, with atomic motions occurring on the ultrafast timescale of femtoseconds (10^{-15} s) and picoseconds (10^{-12} s). Although kinetic studies often allow reaction mechanisms as well as the kind and properties of reaction intermediates to be determined, the obtained information is not sufficient to deduce the ultrafast molecular dynamics. Because these ultrafast motions are the essence of every chemical process, detailed knowledge about their nature is of fundamental importance.

Excerpt 12T (from Warren, 2002)

Carbene, nitrene, and oxo complexes of Mn–Cu represent challenging synthetic targets that would enable a better mechanistic understanding of metal-catalyzed cyclopropanation, aziridination, and oxidation reactions promoted by these complexes. By offering insights into the patterns of selectivity of these complexes, these studies may pave the way toward a rational design of a future generation of more efficient catalysts.

Excerpt 12U (from Walker, 2001)

Given that interfacial solvation affects chemical transport,[16] surface reactivity and electron transfer,[17–20] and macromolecular self-assembly,[21] predictive models of solvent-solute interactions near surfaces will afford researchers deeper insights into a host of phenomena in biology, physics, and engineering. Research in this area should aid efforts to develop a general, experimentally tested, and quantitative understanding of solution-phase surface chemistry.

Exercise 12.21

Read excerpt 12V. What unique opportunity to gain new knowledge is described? Can you think of other research projects that may be unique because of their location in space or time?

Excerpt 12V (from Harpp, 1998)

OBJECTIVES AND SIGNIFICANCE

The Galápagos Islands provide a unique opportunity to study detailed lithospheric structure and processes, primarily because of their location adjacent to a mid-ocean ridge system. They lie on the Nazca plate, just south of the Galápagos Spreading Center (GSC), with an east-west trending ridge serving as the boundary **between** the Nazca and Cocos plates (Figure 1). Hotspots and the magmas they produce serve as effective probes of the earth's deep interior, particularly when examined in conjunction with geophysical exploration.

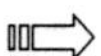

Between vs. Among

See appendix A for more information on these commonly confused words.

Exercise 12.22

Read excerpt 12W. Tuckerman uses all three approaches for establishing importance; he demonstrates interest, emphasizes a benefit, and establishes the need for new knowledge. Find a sentence that illustrates each approach.

Excerpt 12W (from Tuckerman, 1999)

Several hundred to several thousand feet beneath the ocean floor in permafrost and continental edge regions lies a potentially vast source of natural gas: in excess of 10^{16} cubic meters of gas hydrates, consisting largely of methane clathrate (53–55). Gas hydrates are nonstoichiometric compounds consisting of hydrogen-bonded water molecules in a cagelike structure, which traps small-diameter gas molecules. The methane clathrate, in particular, is stable in the pressure range of roughly 20–40 MPa and temperature range of 175–250 K (55). Interest in these vast beds of crystallized gas hydrate is rapidly increasing for several reasons. They constitute perhaps the largest untapped source of hydrocarbon energy available. They also pose a possible threat as an enormous source of greenhouse gas. As global warming continues to cause oceanic temperatures to rise and permafrost regions to recede, the risk of destabilization of the beds increases and with it the risk of releasing large quantities of trapped greenhouse gas into the atmosphere. Given the potential importance of gas hydrates, it is vital that we expand our knowledge of their physical properties, which as yet, are only poorly understood.

Analyzing Writing

We conclude our analysis of move 2 with a brief summary of writing features common in this move. We placed pointers near selected excerpts in this move to call your attention to the following:

- subject–verb agreement (passages P6 and P7)
- formatting figures (excerpt 12G)
- in-line citations (excerpts 12J and 12K)
- lead vs. led (excerpt 12L)
- semicolons connecting two sentences (excerpt 12N)
- comprise vs. compose (excerpt 12O)
- between vs. among (excerpt 12V)

To test your knowledge of some of these features, complete exercises 12.23–12.25.

Exercise 12.23

In each sentence, what is the subject of the italicized verb? Does the verb form agree with the subject? If not, correct the verb form.

a. The preparation of new materials at molecular-length scales and in different molecular orientations *is* the subject of much current research.
b. The increased use of chemicals in agricultural products and antibiotics *have raised* public concerns.
c. Thermal conrotatory 4π ring opening of aziridines and epoxides *is* a well-recognized method of generating azomethine and carbonyl ylides.[1] (From Johnson, 2003)

d. How much of these chemicals finds its way into the surface water near livestock operations is not clear. (Adapted from Aga, 2002)

Exercise 12.24

Select the correct word in each of the following sentences:

a. Combinatorial chemistry has *lead/led* the chemical industry in developing *lead/led* antibiotic compounds.

b. Artificial muscles are *composed of/comprised of* pH-responsive fibers.

c. *Among/Between* the three possibilities, only the two involving active sites will be considered.

Exercise 12.25

Consider the following passage adapted from Aga (2002) that contains five sentences and no semicolons. Rewrite the passage so that two pairs of adjacent sentences are connected by semicolons to emphasize their close relationships.

After elution from the SPE cartridges, the eluents will be evaporated slowly under nitrogen gas. The concentrated samples containing metolachlor will be analyzed by GC/MS while the fractions containing the polar metabolites will be analyzed by HPLC. Both the GC and HPLC will be equipped with chiral columns. For GC/MS, a fused silica column coated with *tert*-butyldimethylsilyl-β-cyclodextrin will be used. This column has been shown to partially separate metolachlor isomers (*13*).

Lastly, we briefly consider verb tense and voice used in move 2. As shown in table 12.5, past and present tenses and present perfect may all be used. Past tense and present perfect are used to provide background information (work done in the past), typically in passive voice. Passive voice allows the writer to focus on the science rather than the scientist:

Present perfect–passive (more common) — Aryl nitrenes have been studied extensively over the last decade.

Present perfect–active (less common) — Scientists have studied aryl nitrenes extensively over the last decade.

Present tense and active voice are commonly used for statements of importance, knowledge, or fact. *We* is not used in move 2 because the proposed work is not mentioned.

Table 12.5 Common functions of different verb tense–voice combinations in move 2 of the Goals and Importance section.

Function	Tense–Voice Combination	Example
Present background information (work done in past)	Past–passive	The first direct evidence for the Eley-Rideal mechanism *was determined* in 1991.[16] (From Spain, 1997)
	Present perfect–passive	An artificial pancreas *has been developed*.[13,14,38] (From Lyon, 2000)
Stress importance	Present–active	The catalytic transformation of organic molecules *is* of tremendous importance.
Report existing knowledge/facts	Present–active	Nitrite reductases *catalyze* the one-electron reduction of nitrite. (From Houser, 2001)
		The Galápagos Islands *sit* on a large volcanic platform in the Pacific Ocean. (From Harpp, 1998)

Exercise 12.26

The following sentences from excerpt 12J use past tense, present tense, and/or present perfect correctly. Identify which is used in each italicized segment. What general purpose is conveyed? (Consult table 12.5, if needed.)

a. Although Cl_2 *does* not spontaneously *etch* Si at room temperature, kinetic-energy-enhanced Cl_2 *does*,[6–11] as observed by mass spectrometric detection of products.

b. In general, trapping efficiency *decreases* as incident ion kinetic energy *increases*. However, at lower incident energies, a minimum trapping probability *was observed*.

c. Other groups *have investigated* how incident kinetic energy affects film growth processes.

d. Co adsorption to Si(100) *has been shown* to depend on Co translational energy.[20]

e. Collectively, these studies *point* to the importance of the emerging discipline called "hot atom chemistry." [17,21]

Exercise 12.27

Rewrite the following sentences so that they are in passive voice and emphasize the science rather than the scientist:

a. We describe two practical tools: the Swain–Schaad Exponential relationship and the temperature dependency of KIEs.

b. People have recognized allelopathy, the chemical interaction between plants and microorganisms, for hundreds of years.

c. Chemists have studied aryl nitrenes extensively over the last decades and have used them in several industrial processes such as microlithography.[1]

12C Writing on Your Own: Introduce and Develop the Research Story

Draft move 2 of your Goals and Importance section. In the first sentence, identify the research area in general terms. (Do not introduce your specific research project.) Develop the research story in the next few paragraphs. Explain fundamental concepts and provide essential background information, using the notes that you prepared in Writing on Your Own task 12A. Share what you learned from your literature review, and its importance, with your readers.

Explain at least one general concept at a scientific or general audience level (keeping in mind that reviewers may not be familiar with your research area). Include an illustration, if applicable. Provide relevant background information at a more advanced level. Be sure that you paraphrase and incorporate citations appropriately. As you share background information, stress the importance of the research using at least one of the three suggested approaches: document interest, emphasize benefits, and establish a need for more knowledge.

Remember that this move gives you the opportunity to demonstrate your grasp of the research area and establish your intellectual merit as a researcher, two keys to an effective proposal.

Move 3: Introduce the Proposed Work

Move 3 shifts the focus from the general research topic and others' works (moves 1 and 2) to the specific work proposed by the author(s). The transition is typically accomplished in two submoves (figure 12.1). Normally, move 3 is not given its own heading; rather, move 3 and its submoves continue under the heading for move 2. (Excerpt 12X is an exception to this practice.)

In the first submove, the need for the proposed work is established in one or two gap statements. Gap statements, as the name implies, point out gaps in the field, in the form of questions that need to be answered, techniques that need to be developed, areas that need further exploration, and so forth. In the second submove, the gap is filled (at least in part) by the proposed work. In these two submoves, the focus of the proposal shifts from the general research area to the more specific research project, setting the stage for the next section of the proposal, Experimental Approach (see chapter 13).

Gap Statements

The gap statement can suggest

- questions that need to be answered
- techniques that need to be improved or developed
- areas that need to be explored

and much more. (See table 6.1.)

Analyzing Excerpts

We are now ready to examine gap and fill-the-gap statements in authentic proposals. Because these two statements work as a pair, we examine them together. Four excerpts are included; for clarity, we have italicized the words that signal the beginning of the gap and fill-the-gap statements in each excerpt.

Excerpt 12X uses lists to delineate both the gap (*the experimental challenge*) and fill-the-gap statements (*facing the challenge*); excerpts 12Y–12AA use narrative form. In each case, the depiction of the work following the fill-the-gap statement is brief, laying the groundwork for a more detailed description of the proposed work.

In these excerpts, the authors assist the reader by signaling gap and fill-the-gap statements. To signal a gap statement, the authors forewarn the reader of *obstacles* and *difficulties* or use such terms as *unfortunately*. To indicate that a fill-the-gap statement is coming, the authors use different forms of *propose* or other words indicating intent.

The four excerpts also illustrate the use of personal pronouns (specifically, *I* and *we*) in the fill-the-gap statement. One author (Spain) uses *I* and *we* ("*I* propose" and "*we* will employ"). All others use only *we* ("*We* propose to address", "*We* will synthesize", "*we* intend to pursue"). Using both *I* and *we* in a single proposal is not uncommon, particularly in proposals that are written by one investigator (e.g., the CAREER proposal). A sole author (the PI) is proposing the work (hence, *I*), but a group of individuals (the PI, students, postdocs, and collaborators) will conduct the work (hence, *we*).

Exercise 12.28

Read excerpts 12X–12AA and answer the following questions:

a. What words do the authors use to signal gaps and gap fillers?

b. Consider the different ways in which the excerpts are formatted (i.e., the lists in 12X and continuous text in 12Y–12AA). Which form of presentation do you prefer? Why?

Excerpt 12X (from Kohen, 2002)

The Experimental Challenge

There are three main obstacles to any attempt to experimentally evaluate the effect of enzyme dynamics on covalent bond activation. Namely, it is difficult to

1. extract effects on the chemical step (e.g., C–H bond cleavage) and its transition state within the multi-step kinetic cascade of enzyme catalysis.
2. measure the motion of a protein on a wide range of time scales, with high regional resolution.
3. interpret information on protein dynamics as vibrational modes coupled to the reaction coordinate. NMR relaxation experiments, for example, yield little information on the direction of internal motion. Experimental correlation between the rigidity of proteins (as measured by H/D amide exchange)[28] and the degree of H-tunneling in the reactions also offers little information about which protein motions are involved.

Facing the Challenge

We propose to address these obstacles by

1. studying the chemical step within a complex kinetic cascade. Hydrogen tunneling and the coupling of primary and secondary hydrogens at the same carbon center will be used as probes of the C–H bond activation.... Theoretical studies predict that dynamic processes will markedly affect these phenomena in enzyme-catalyzed reactions.[26,29]
2. measuring the motion of a protein. The protein dynamics will be measured by various techniques, primarily focusing on the ps-ns time scale. DHFR is small enough to allow NMR relaxation experiments.
3. interpreting the data. DHFR is a small enzyme that catalyzes C–H bond cleavage. Not only experimental data acquisition but also several theoretical approaches benefit from the small size of the enzyme. Molecular dynamics simulation (MD), for example, can assist in the assessment of DHFR dynamics,[35] although it is currently limited to a few ns.[36] We propose to test several theoretical disciplines while attempting to correlate H-transfer data to dynamic fluctuation data. These attempts should yield predictions that we will examine experimentally.

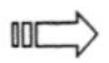

Parallelism in Enumerated Lists

See appendix A.

Excerpt 12Y (from Spain, 1997)

The ever decreasing dimensions of semiconductor devices impose paradoxical requirements on the plasma etching and deposition processes involved in their manufacture.[4]

Low substrate temperatures are required to prevent thermal damage to layers already fabricated, while, at the same time, low energy plasma conditions are required to limit damage by high energy ion implantation. *Unfortunately, many films, especially metal chalcogenides, require high temperature annealing (typically 300 to 500 °C) after deposition in order to obtain crystalline films, violating the low temperature requirement.* The use of kinetic-energy-selected neutral beams, however, may open the door to metal chalcogenides deposition, allowing deposition at low substrate temperatures without damage from high energy ion exposures.

To explore this potential, I propose to study the fundamental chemical dynamics of materials deposition by varying the kinetic energy of reagent atoms impinging on a surface. Specifically, we wish to understand the growth and nucleation of transition metal chalcogenide thin films on chemically modified and solid surfaces as a function of reagent kinetic energy. We will employ laser-induced vaporization by back illumination (LIVBI) to produce translationally hot and neutral metal atom beams. In this method, which has successfully grown many types of metal-containing films including superconductors, the vaporized material is generated as a plasma containing electrons, neutral and ionized atoms, molecules, and perhaps some clusters. Therefore, LIVBI is ideal to study the fundamental aspects of film growth because the beams are directed, neutral, and kinetic-energy resolved from near thermal to hyperthermal.

Excerpt 12Z (from Vyvyan, 2001)

Another obstacle to advancing this field is the minute quantities of the active compounds that can be isolated from the natural sources.. . . For example, the heliannuols (**1–11**, Figure 1) are a promising group of phenolic allelochemicals that exhibit activity against dicotyledon plant species (*13*). Heliannuols A and D (**1** and **4**, respectively) are the most active members of the family with effective concentrations of 10^{-4} to 10^{-9} M. . . . The recent isolation of heliannuols F–K (**6–11**), in which over 6 kg of sunflower leaves was required to obtain just 1–2 mg of each pure compound, clearly illustrates their scarcity (*16*).

Thus, a synthetic source of promising allelochemicals is essential if we are to comprehensively study the agent's mode of activity and establish its basic structure-activity profile. *The proposed work addresses this need. We will synthesize* alleopathic natural products isolated from the sunflower (the heliannuols), and structurally related compounds, in optically pure form based on biomimetic phenol-epoxide cyclizations. The bioactivity of the targets and intermediates will be evaluated through laboratory tests on plant germination and growth. Bioassays will be performed on the synthetic intermediates to allow for the development of a preliminary structure-activity profile for these novel natural herbicides.

Compound Labels

To conserve space, chemical names are often represented by bolded numbers (e.g., **4**) or numbers and letters (e.g., **4a**). Like other abbreviations, they are defined at first use, often in a figure or table. (See appendix A.)

Excerpt 12AA (from Finney, 1999)

Despite the remarkable progress in the field of molecular recognition, *it is still extremely difficult to predict a priori* the structure of a selective, high-affinity ligand for a metal ion such as Hg^{2+}. *The approach we intend to pursue is to take advantage of combinatorial chemistry methods*—a collection of technical advances that allow one to seriously consider undertaking the synthesis of hundreds or even thousands of compounds simultaneously[29]—to carry out the parallel synthesis of a set of 100 potentially selective high-affinity fluorescent chemosensors for mercury.

Exercise 12.29

Read excerpt 12BB and answer the following questions:

a. Identify the gap and fill-the-gap statements.
b. What word(s) does the author use to signal the gap and gap filler?
c. Does the author use personal pronouns as you might expect? Explain.

Excerpt 12BB (from Rose-Petruck, 2000)

However, atom motions cannot be unambiguously "imaged" by time-resolved optical spectroscopic methods as they do not directly measure the structural dynamics but instead characterize energetic properties. Consequently, novel methods that enable the direct measurement of molecular motions during chemical processes are needed. Furthermore, chemical reactions often occur in solution and, consequently, it is desirable that such methods are applicable to chemical processes in the liquid phase.

I propose to develop and apply such methods, based on ultrafast X-ray absorption spectroscopy, to study the ultrafast molecular motions of organometallics in solutions. In particular, initial studies will focus on photo-induced ligand dissociation and substitution reactions of transition metal carbonyls and related compounds in various solvent systems.

Analyzing Writing

We conclude our analysis of move 3 by summarizing writing conventions commonly observed in this move. We used pointers near selected excerpts to point out the following:

- parallelism in enumerated lists (excerpt 12X)
- labels for chemical names (excerpt 12Z)

Test your knowledge of these conventions by completing exercises 12.30 and 12.31.

Exercise 12.30

Reread the lists in excerpt 12X. Specify examples that illustrate parallelism in language, numbers, and punctuation.

Exercise 12.31

Glance ahead to excerpts 13L and 13O. Both excerpts include bolded numbers. What purpose do these bolded numbers serve?

We close our discussion of move 3 by analyzing common verb tense and voice combinations used in this move (table 12.6). Gaps are often stated in present tense (in active and passive voice), sometimes in combination with a present perfect–passive statement; fill-the-gap statements are usually in present or future tense and active voice. Personal pronouns (*I* or *we*) are common in fill-the-gap statements.

Table 12.6 Common functions of different verb tense–voice combinations in move 3 of the Goals and Importance section.

Function	Tense–Voice Combination	Example
Identify a gap in the field	Present–active	Another obstacle *is* the . . .
	Present–passive	Consequently, new methods *are needed* that . . .
	Present perfect–passive Present–passive	Although . . . *has been studied* in depth, little *is known* about . . .
Fill the gap by introducing proposed work	Present–active	I *propose* to overcome these difficulties by . . .
	Future–active	We *will overcome* these difficulties by . . .

Exercise 12.32

Read passages 1–3 below, in which words in the gap statements have been deleted:

a. Suggest the words that have been deleted so that the gap statements read well.
b. Identify the verb tense and voice combinations used in the gap statements.

1. Aside from their applications to polymer systems (see below), the ____ of prior published research involving thermotropic liquid-crystal (LC) solvents for controlled crystallization and materials synthesis is striking. A detailed literature search for applications of thermotropic LCs to film growth revealed a single citation (a patent) describing a method for controlling molecular alignment in an organic film. (From Patrick, 2000)
2. To date, effective, positive-signaling fluorescent chemosensors have been developed for cations such as Ca^{2+} and Zn^{2+}.[4] ____, the development of fluorescent chemosensors for heavy metals . . . or small organic molecules . . . generally does not exist. (Adapted from Finney, 1999)
3. The bent plume model does not take into consideration any plume-ridge interaction, except at the most vague of levels (Harpp, 1995). This is primarily due to the ____ data from this area; the observed patterns are ____ complex to draw any solid conclusions with such sparse sampling. (From Harpp, 1998)

Exercise 12.33

Integrate each set of information (set A and set B) into a concise and effective gap statement:

Set A (adapted from Gunes et al., 2002)

- Controlled-atmosphere (CA) storage can extend the storage life of fruits and vegetables by decreasing metabolism and suppressing postharvest decay.
- Initial CA treatments with N_2 gas have been shown to improve cranberry storage life by reducing fungal decay (*12*).
- Researchers have never examined the effect of CA conditions on the level of antioxidants found in cranberries.

Set B (adapted from Yamashita et al., 2004)

- The unique physicochemical properties of polyfluorinated compounds (PFCs) render several challenges to analytical chemists endeavoring to measure PFCs at trace levels.
- One of the major problems associated with trace level analysis of perfluorinated acids is background contamination in the analytical blanks. Because of the contamination in blanks, the limits of detection (LOD) of perfluorochemicals in water samples are high, in the range of several tens to hundreds of ng/L to a few μg/L (*11–15*).

- Contamination sources of PFCs in laboratories have not been well characterized. Two distinct sources of contamination, instrumental and procedural, are expected in PFC analysis.
- The current study improves PFC methods by reducing the above-mentioned sources of blank contamination.

12D Writing on Your Own: Introduce Your Proposed Work

You are nearly ready to complete the first draft of the Goals and Importance section of your proposal, but first you must introduce your proposed work. Think about the gaps that others' works have left. Focus on those gaps, and then briefly introduce your work by showing how it will fill those gaps.

12E Writing on Your Own: Complete the Opening Section

Combine the three moves of your Goals and Importance section (from the previous Writing on Your Own tasks).

Revise and edit the entire section, paying careful attention to audience (e.g., level of detail, word choice), organization, writing conventions (e.g., verb tenses, voice, personal pronouns, formatting of lists, and citations), and grammar and mechanics (e.g., parallelism, punctuation). Revise the Goals and Importance section so that the individual parts work together as one document.

You may want to consult chapter 18 for proofreading tips. If possible, have a colleague peer review (formally or informally) the section before turning it in to your instructor. Make time to incorporate any feedback that you receive.

Chapter Review

Check your understanding of what you've learned in this chapter by defining each of the following terms, in the context of this chapter, for a friend or colleague who is new to the field:

fill-the-gap statement	goals	parallels lists
formal vocabulary	intellectual merit	research story
gap statement	objectives	SAM test

As a review, try explaining the following to a friend or colleague who has not yet given much thought to writing the Goals and Importance section of a research proposal:

- Main purpose of the Goals and Importance section
- Differences between goals and objectives
- Role of enumerated lists
- Three tangible ways to emphasize importance
- Three approaches to defining gaps in the field
- Ways to establish intellectual merit
- Value of word choice

Additional Exercises

Exercise 12.34

Consider the following list adapted from Vyvyan's (2001) research proposal. We have intentionally added errors in parallelism to his list. Revise the list to correct these errors.

The specific objectives of the proposed work are as follows:

1. to synthesize the heliannuols, allelopathic natural products isolated from the sunflower, in optically pure form.
2. We also plan to perform bioassays on synthetic intermediates and target compounds to allow for development of a preliminary structure-activity profile for these novel herbicides.
3. Important natural products structurally related to the heliannuols will also be synthesized in optically pure form (e.g., helinorbisabone, heliespirone, and helianane);
4. And finally, to use the synthetic methods developed to prepare simplified structural analogues of the heliannuols that may retain allelopathic activity.

Exercise 12.35

Correct errors in grammar, punctuation, and wording so that the following sentences are more in line with conventional writing practices. To help you with this exercise, we have indicated the submove for each sentence.

a. [Submove 2.2] Scientists have studied extensively the mechanism of cyclization.

b. [Submove 2.2] Researchers have explored a variety of synthetic routes for the preparation of carbene and nitrene complexes.

c. [Submove 3.1] However the development of fluorescent chemosensors for heavy metals remain an important yet elusive goal. (Adapted from Finney, 1999)

d. [Submove 3.1] Despite their technological importance the surface science associated with polymer modification treatments remained virtually unexplored.

Exercise 12.36

Read over the combined Results and Discussion section in excerpt 4A, written for a journal article. In one or two sentences, summarize the key findings of that study as they might appear in submove 2.2 (provide essential background information) of a research proposal. Refer to Writing on Your Own task 6A for tips on paraphrasing.

Exercise 12.37

Consider the following sentences taken from excerpt 12J. Using table 12.5, identify and justify the tense used for the italicized verbs:

a. It *was demonstrated* that kinetic-energy-enhanced $N(C_2H_4)_3N$ *reacts* with H on a Pt(111) surface to form an ion.

b. In a study of H atoms incident on D/Cu(111),[17,18] direct evidence that HD *is formed* by the Eley–Rideal mechanism *was found*, in part, by varying the translational energy of the H reagent.

c. However, at lower incident energies, a minimum trapping probability *was observed*.

Exercise 12.38

Check the following sentences for subject–verb agreement. Correct those sentences with faulty subject–verb agreement; write "OK" for sentences that are correct. Refer to appendix A for help, when needed.

a. In these cases, the ionic strength or pH of the environment change as the enzyme turns over a substrate. (From Lyon, 2000)

b. Most of the theoretical work regarding the utility of time-resolved photoelectron angular distributions to probe electronic, vibrational, and rotational dynamics has also concerned neutral photoionization. (From Sanov, 2002)

c. Crystal hydrates of strong acids and bases have attracted the attention of crystallographers and spectroscopists over several decades (*46–48*, *123–125*). (From Tuckerman, 1999)

d. The pressure to provide new information on the fate of pesticides in the environment have become a monumental task for government and industry. (Adapted from Aga, 2002)

e. The catalytic transformation of organic molecules are of tremendous importance on laboratory and industrial scales. (Adapted from Warren, 2002)

Exercise 12.39

Rewrite the following excerpt (adapted from Houser, 2001) to make it more concise:

This type of copper cluster has not previously been observed in any metalloprotein. Despite the fact that the structure of this important crystal has been solved, much remains for scientists to learn about the physical, spectroscopic, and chemical properties of the Cu_Z catalytic center. For example, we could ask ourselves questions such as these: First of all, are there copper-copper bonds present in the cluster? Second, why are four copper ions needed for a two-electron reduction? (78 words; aim for 62 words)

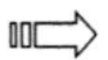

Conciseness

See appendix A and chapter 2.

Exercise 12.40

Consider the following passage adapted from Dyer's (2001) research proposal. We have intentionally added errors to the passage. Revise the passage so that it is more concise and has no punctuation errors.

We are currently initiating three research projects that include: (1) the synthesis of reflective liquid crystal/polymer composite films, (2) a study of microphase separation in hyperbranched block copolymers, and (3) the design and synthesis of polar organic thin films, which is the subject of this proposal. (47 words; aim for 41 words)

Exercise 12.41

Read the following research proposal passages. Assign each passage to the appropriate submove in move 2 (establish importance) or move 3 (introduce the proposed work). Refer to figure 12.1, if needed. Justify your answer.

a. Two separate combinatorial approaches will be taken to find ligands for apoptotic proteins: (1) A focused library of compounds will be synthesized on the solid phase that targets the cysteine protease activity of the caspases. (2) A primary library of compounds will be synthesized and screened for selective binding to the various regulators of apoptosis. The size of the libraries will initially be limited to <10,000 compounds such that the structure of any hits can be elucidated by simple mass spectrometric analysis. (From Hergenrother, 2002)

b. The catalytic transformation of organic molecules is of tremendous importance on laboratory and industrial scales. Not only do catalysts accelerate chemical reactions of interest and enhance product selectivity, they allow chemical transformations to be performed with increased efficiency, minimized waste, and lower energy consumption.[1,2] A vast majority of products of the chemical industry, from bulk to fine chemicals, involve catalysts at some stage of their manufacture. (From Warren, 2002)

c. To date, effective, positive-signaling fluorescent chemosensors have been developed for cations such as Ca^{2+} and Zn^{2+}.[4] However, the development of fluorescent chemosensors for heavy metals, which typically quench fluorescence, or small organic molecules for which simple binding motifs generally do not exist, remains an important yet elusive goal. (From Finney, 1999)

d. Fourier-transform ion cyclotron resonance mass spectrometry (FTICR) is a high-resolution, high-sensitivity technique that allows the entrapment and detection of gas phase species.[17,18] Gas phase ions are trapped in a magnetic field, much like a reactant sits in a flask in solution. (From Lee, 2001)

Exercise 12.42

Reflect on what you have learned from this chapter. Select one of the reflection tasks below and write a thoughtful and thorough response:

a. Reflect on the moves generally included in the Goals and Importance section of a research proposal. (Refer to figure 12.1.)
   - Which of the moves do you think has the greatest potential to sway the opinion of a funding agency?
   - Which move do you think requires the greatest effort on the part of the writer? Explain.
   - Which move might be the most challenging to write? Why?

b. Reflect on the power of language in the Goals and Importance section of a research proposal.

- Based on your reading of the excerpts in this chapter, what specific types of vocabulary do you intend to incorporate into your own research proposal?
- What particular phrases caught your attention in the chapter? Which phrases do you want to try to include in your own written work?
- What kind of language can you use to emphasize your intellectual merit as a researcher?

c. In most research proposals, investigators communicate the importance of the topic by demonstrating interest, emphasizing benefits, and establishing a need for more knowledge in the area.

- Of the three typical ways of communicating importance, which do you think has the potential to be most persuasive? Why?
- How did you (or do you plan to) emphasize the importance of your topic?
- If you had to rate your importance statement(s) on a scale of 1–10 (1 least persuasive, 10 most persuasive), what would your rating be? Explain.

13 *Writing the Experimental Approach Section*

A research proposal must present an effective argument that will persuade those who read and evaluate proposals . . . that the research proposed is significant, that the methods of pursuing this research are well designed, and that the researcher is in fact capable of carrying out the research.
—Beal and Trimbur (2001)

This chapter focuses on writing the experimental section of the Project Description, the section in which you tell readers how you will conduct the proposed work. Unlike the experimental section of a journal article, which is written largely after a work has been completed, here you describe *proposed* methodology. Reviewers, after reading this section, should be convinced that the work is plausible and that the investigator has the background and expertise necessary to carry out the proposed work. By the end of this chapter, you should be able to do the following:

- Know ways to establish your expertise as a researcher
- Be able to share and build on preliminary results
- Present your proposed research in a logical order
- Identify obstacles in your proposed research plan

Staying on Track

This chapter covers what is normally presented in the second main section of the Project Description.

Common headings in this section are

EXPERIMENTAL APPROACH (level 1)

Prior Accomplishments (level 2)

Preliminary Results (level 2)

Proposed Methodology (level 2)

Objective 1. (level 3)

Objective 2. (level 3)

As you work through the chapter, you will write the Experimental Approach section of your own proposal. The Writing on Your Own tasks throughout the chapter guide you step by step as you do the following:

13A Share prior accomplishments

13B Share preliminary results

13C Describe your proposed methods

13D Complete the Experimental Approach section

Reading and Analyzing Writing

We begin with excerpt 13A, part of an Experimental Approach section that you can read and analyze on your own. Excerpt 13A is a continuation of excerpt 12A, where Aga makes the case that we need to better understand the stereochemistry of metolachlor, a widely used herbicide and potential carcinogen, as it degrades in soil. In excerpt 13A, she describes two specific research objectives needed to accomplish this task.

Exercise 13.1

As you read and analyze excerpt 13A, complete the following tasks:

a. Explain why Aga organized this section of her proposal in two parts.

b. Identify the moves in this excerpt, making use of the suggested headings in Staying on Track above. Note that this excerpt is only part of an Experimental Approach section.

c. Determine what verb tense–voice combination predominates in this excerpt.

Excerpt 13A (Aga, 2002)

Proposed Methodology

Analytical Methods Development. A solid-phase extraction (SPE) method that will isolate and fractionate metolachlor and its oxanilic acid (OXA) and ethanesulfonic acid (ESA) metabolites from soil and water will be developed. The SPE method will be used to separate the compounds into sub-classes to simplify the separation of each compound into its individual isomers by high-performance liquid chromatography (HPLC) or by GC/MS. SPE will be used to isolate metolachlor and its degradation product from soil and aqueous samples by adsorbing these compounds into a non-polar material, such as a C-18 resin, followed by elution with organic solvents. With properly designed SPE procedures, closely related compounds may be separated into fractions by using different eluting solvents of varying degrees of polarity. Because SPE is also a pre-concentration technique, the method detection limit for the quantification of individual isomers could be improved by several orders of magnitude.

Several commercially available SPE packing materials, such as silica-based C-18, polymer-based reversed-phase hydrophilic-lipophilic balance (HLB™), and graphitized carbon cartridges will be tested for their potential use in the isolation of the target compounds. Different combinations of organic solvents and buffers of varying pHs will be used to determine the optimum eluting solvent that gives maximum recovery of the analytes.

Our previous work showed that alachlor, a compound that is structurally similar to metolachlor, could be separated from its acidic metabolites using a C-18 SPE cartridge (*12*). However, this procedure does not allow the separation of the OXA and the ESA. Thus, different adsorption mechanisms and solvent systems will be explored to separate the analytes into three fractions: metolachlor, OXA, and ESA. This procedure is depicted in Figure 4 [*not included in excerpt*]. The SPE procedure will be necessary to avoid overlapping of the eight isomers of each compound in the chiral chromatographic analysis.

After elution from the SPE cartridges, the eluents will be evaporated slowly under nitrogen gas. The concentrated samples containing metolachlor will be analyzed by GC/MS while the fractions containing the polar metabolites will be analyzed by HPLC; both the GC and HPLC will be equipped with chiral columns. For GC/MS, a fused silica column coated with *tert*-butyldimethylsilyl-β-cyclodextrin will be used; this column has been shown to partially separate metolachlor isomers (*13*). Also, a nonbonded permethylated γ-cyclodextrin column will be tested to examine if enantiomeric separation of metolachlor could be improved. For the HPLC analysis of the polar metabolites, two columns will be examined. First, the silica-based chiral reversed-phase system with cellulose carbamate will be used. This column was shown to partially resolve the parent metolachlor (*13*), but its usefulness in separating the OXA and ESA compounds still needs to be tested. The other HPLC set-up that will be tested will use a conventional reversed-phase stationary phase and a chiral mobile-phase containing γ-cyclodextrin. Our preliminary work using capillary zone electrophoresis (CZE) proved γ-cyclodextrin to be an effective chiral selector for the separation of metolachlor oxanilic acids (see Figure 5) (*not included in excerpt*). Although only partial separation of the metolachlor OXA was achieved in CZE, we were able to show for the first time (in combination with NMR data) that metolachlor indeed exists as eight stable isomers (*14*). The additional interaction offered by the stationary phase in HPLC may result in increased separation efficiency.

Soil Degradation Study. A controlled degradation study will be conducted using soil samples taken from a field that has not been exposed to metolachlor and from a field that has been previously treated with metolachlor herbicide. The reason for taking samples from two different sources is to investigate the effect of different microbial populations on the stereoselectivity of metolachlor degradation. A sterilized soil sample will serve as a control. The soil samples will be divided into two sets: one will be treated with commercial formulation Dual® (racemic metolachlor herbicide formulation) and the other with Dual Magnum® (S-enriched herbicide formulation). Each set will be conducted in duplicate, and each sample will receive an application rate of **500 µg** of metolachlor per **200 g** soil. The fortified oil samples will be incubated in Petri dishes at **20–25 °C**, and the soil moisture will be controlled. Samples (about 10 g) will be taken periodically from the Petri dishes for analysis, the first one immediately after fortification and then after 3, 7, 15, 30, and **60 days**. The soil samples will be extracted and analyzed for the parent compound and the OXA and ESA metabolites. Enantiomeric ratios (ER) of the parent metolachlor will be calculated based on the peak areas of individual isomers and will be used to determine the stereoselectivity of metolachlor degradation. In addition, the ER of metabolites will be monitored.

Numbers and Units

Use correct abbreviations for units (e.g., µg, g, °C, days) and include a space between the number and its unit (e.g., 25 °C). (See appendix A and table 3.2.)

Exercise 13.2

Look back at excerpt 13A and complete the following tasks:

a. Aga addresses both prior accomplishments and preliminary results in this excerpt. Describe both and state how they help to establish Aga's research expertise.

b. Substantiate or refute the following statement: The author presents a logical progression of proposed experiments.

c. Find one example that illustrates how to do each of the following: (1) use an abbreviation for the first time, (2) report a number with units, and (3) format an in-line citation. Suggest rules for each convention.

d. Find one example that illustrates the appropriate use of the following: (1) commas, (2) hyphens, (3) semicolons, and (4) colons. What rules are followed in each case?

e. Comment on the verb tense(s) used to describe previous work and proposed work. Give an example of each.

Analyzing Organization

A move structure for the Experimental Approach section is shown in figure 13.1. The section is organized around three key moves: (1) Share Prior Accomplishments, (2) Share Preliminary Results, and (3) Describe Proposed Methodology. These moves parallel the information requested in many RFPs. For example, the ACS Analytical Chemistry Graduate Fellowship RFP (excerpt 11A) prompts applicants to summarize work already accomplished (i.e., prior accomplishments and preliminary results) and to summarize work planned for the term of the fellowship (i.e., proposed methodology). Similarly, the NSF CAREER award RFP (excerpt 11C) requires applicants to provide a summary of prior research accomplishments and an outline of the research plan, including the methods and procedures to be used. The Experimental Approach section is often the most technical section of the proposal.

Exercise 13.3

Consider the RFPs for two NSF program solicitations: the Faculty Early Career Development (CAREER) award (excerpt 11C) and the Collaborative Research in Chemistry (CRC) award (excerpt 11D). Explain how the moves in figure 13.1 are addressed in these RFPs.

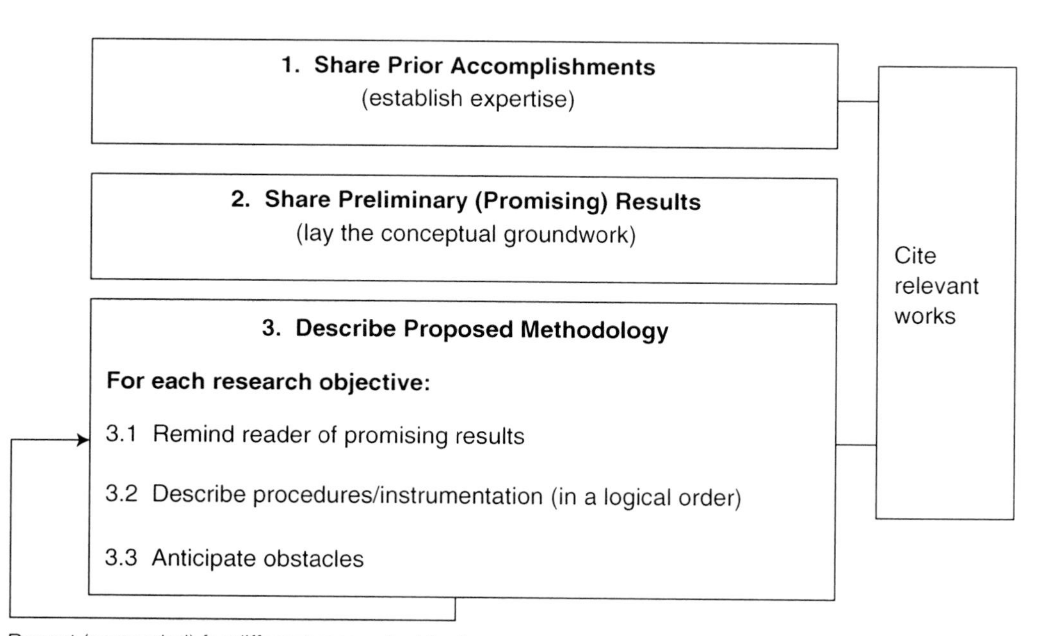

Figure 13.1 A visual representation of the suggested move structure for the Experimental Approach section of the Project Description.

Table 13.1 Common level 1 headings for the Experimental Approach section.

EXPERIMENTAL APPROACH	PROPOSED METHODS
EXPERIMENTAL SECTION	PROPOSED RESEARCH PLAN
EXPERIMENTAL DESIGN	PROPOSED RESEARCH ACTIVITIES

With the move structure for the Experimental Approach section in mind, we are ready to analyze the moves and submoves in more detail. The entire section often begins with a level 1 heading. Common level 1 headings are shown in table 13.1.

Move 1: Share Prior Accomplishments

The first move of the Experimental Approach section is to share prior accomplishments. The term "prior accomplishments" refers to completed works (e.g., published articles or otherwise disseminated results) and other accomplishments (e.g., awards, collaborations) that are related to the proposed work. (Unrelated accomplishments may be listed in a separate biographical statement but should not be mentioned in the proposal.) The purpose of move 1 is to establish expertise and convince reviewers that you have the necessary skills to complete the proposed work. Move 1 usually begins at or near the start of the Experimental Approach section. The length of move 1 varies with each proposal. Typically, the section is longer for experienced researchers (i.e., those with prior grant support) because they have more accomplishments to share. Move 1 is often demarked with a level 2 heading. Common level 2 headings for move 1 are shown in table 13.2.

Analyzing Excerpts

Let's examine move 1 in authentic proposals. Perhaps the most direct way authors share prior accomplishments is to cite their own published works. It is both appropriate and expected that authors call attention to their authored (or co-authored) works, provided that they are related to the proposed work. Excerpts 13B–13D illustrate this approach. Some authors, such as Lee and Lorigan (excerpts 13B

Table 13.2 Common level 2 headings for move 1 of the Experimental Approach section.

Prior Accomplishments	Summary of Prior Accomplishments
Previous Work	Previous Accomplishments

and 13C), mention explicitly that their papers were published in the *Journal of the American Chemical Society*. (Such journals as the *Journal of the American Chemical Society*, *Nature*, and *Science* are particularly prestigious; hence, authors will often mention them by name.) Other authors, such as Vyvyan (excerpt 13D), simply include a citation to their own work. Either way, we encourage you to cite your own works. The decision to include journal titles is up to you.

Excerpt 13B (Lee, 2001)

1. Nucleic Base Acidities. The N1 and N3 Acidities of Uracil. Published in *J. Am Chem. Soc.* **2000**, *122*, 6258–6262. Our progress to date involves calculations and experiments for the determination of nucleobase activities. We have learned that there is an enormous difference in the inherent stabilities of anions at the N1 (**7**) and N3 (**8**) positions in uracil (**5a**, Figure 1).

Excerpt 13C (Lorigan, 2002)

Logically, the first step for developing this method at lower magnetic fields (X-band for EPR studies) is to fully characterize the bicelle system without the added complications induced by protein-lipid interaction. Over the last two years, we have accomplished this goal and published several articles describing our results (*81*, *86–88*). The first paper was published in *J. Am. Chem. Soc.* and demonstrates the effects of magnetic phospholipids bilayer alignment by showing an EPR spectrum of a DMPC/DHPC/Yb^{3+} "bicelle" sample doped with a cholestane spin label (*81*). . . . We followed up our initial report with a second *J. Am. Chem. Soc.* paper that spectroscopically characterizes magnetically oriented phospholipid bilayers with EPR spectroscopy and discusses conditions for optimal bicelle alignment at X-band (*86*).

Excerpt 13D (Vyvyan, 2001)

Our general approach to the synthesis of the heliannuols is illustrated by our recently completed total synthesis of (±)-helliannuol D (**4**).[19]

A successful collaboration can also be mentioned as a prior accomplishment. Collaborations strengthen expertise (two heads are better than one) and add to the intellectual merit of a proposal by increasing the likelihood that the work will be completed on schedule. Excerpts 13E and 13F showcase two ways that authors describe collaborative relationships in their proposals.

Excerpt 13E (Aga, 2002)

Currently, the PI is collaborating with Prof. Bertold Hock's laboratory at the Technical University of Munich (TUM), Freising, Germany, to produce the antibodies against isoxaflutole and its metabolites. Two rabbits have been immunized with different immunogens that were prepared by the PI at the University of Nebraska. Antibody production is being carried out at the TUM because they have the expertise and facilities for polyclonal, monoclonal, and recombinant antibody production. Prof. Hock has agreed to produce recombinant antibodies for isoxaflutole in the future for **further** immunoassay development.

Farther vs. Further

Farther: more distant; at a greater distance

Further: additional; additionally

(See appendix A.)

Excerpt 13F (Kinsel, 1999)

In our initial studies, we have demonstrated, for the first time, that protein binding to surfaces has a significant impact on subsequent protein MALDI MS ion signals. This demonstration of principle was made possible through a number of important collaborations and student interactions between members of our research group and faculty in the Department of Chemistry and the Biomedical Engineering Program at the University of Texas at Arlington.

If you are at the start of your career, you may not yet have many prior accomplishments to share. In this case, you can establish expertise in move 2 (Share Preliminary Results) instead. Excerpt 13G illustrates this approach. Here, Lyon emphasizes what he has accomplished to date on the project, highlighting results and relevant skills that he and his group have mastered. Do not hesitate to point out, as Lyon does, that the results *demonstrate your ability* to perform tasks required for your proposed work.

Excerpt 13G (Lyon, 2000)

Preliminary Results

Colloid Synthesis. Our group has investigated the synthesis of hydrogel particles that can be derivatized covalently with proteins using standard aqueous cross-linking procedures. . . . The first synthetic efforts have been focused on lightly cross-linked (<1%) copolymers of N-isopropylacrylamide (NIPA) and acrylic (AA) or methacrylic acid (MAA).[51–53] These hydrogel spheres contain up to 95% water by volume, thereby allowing for diffusion of molecules into the particle with only moderate steric inhibition. The NIPA component of this material is thermoresponsive; poly-NIPA undergoes a first-order volume phase transition at ~32 °C.[27,30] This **phenomenon** is evident from the data presented in Figure 1. These are recently collected data from our group that demonstrate our ability to accurately measure polymer phase transitions via both turbidity and photon correlation spectroscopy (PCS). Furthermore, the PCS data also demonstrate our ability to measure the diameter of colloidal hydrogels in situ, without the use of invasive electron or probe microscopies.

Phenomenon/Phenomena

Phenomenon (singular)

Phenomena (plural)

(See appendix A for other plural and singular scientific words.)

Exercise 13.4

Refer back to the proposal in excerpt 11A written in response to the ACS Analytical Chemistry Graduate Fellowship. How does Amanda Haes, a graduate student at the time, establish her expertise?

Exercise 13.5

We include another excerpt from Lyon's proposal (excerpt 13H). In this excerpt, Lyon is responding directly to the CAREER prompt to "include a summary of prior research ... accomplishments." To respond to the prompt, he uses a chronological approach to describe his undergraduate, graduate, and postdoctoral experiences. What do you notice about his use of the third person (i.e., the PI) and the first person (i.e., *I*) when referring to himself?

Excerpt 13H (Lyon, 2000)

Summary of Prior Research Accomplishments

The PI has extensive and broad experience in inorganic, analytical, physical, and materials chemistry and has employed surface plasmon resonance spectroscopy (SPR), electrochemistry, transient absorbance spectroscopy, spectroelectrochemical quartz crystal microgravimetry, and photon correlation spectroscopy to the study of a wide range of nanostructured materials and colloidal systems. The PI has also investigated charge transfer in molecular and inorganic polymer systems using spectroscopic and electrochemical probes.[103–105] The PI's research experience began as an undergraduate in the lab of Professor Stephan S. Isied (1988–1992), where he synthesized small peptidic mimics of zinc-binding protein active sites. ...

Under the advisement of PhD mentor Professor Joseph T. Hupp, the PI successfully used spectroelectrochemical quartz crystal microgravimetry to elucidate the mechanism of charge transport ... for both aqueous and nonaqueous sytems.[106,107] This was the first demonstration of proton-coupled electron transfer at oxide semiconductor interfaces. These findings were then successfully applied to a new interpretation of photoinduced electron transfer at similar interfaces, which are of importance in the field of solar energy conversion.[108] ...

As a postdoctoral associate with Professor Michael J. Natan, the PI investigated plasmon coupling between noble metal films and metal colloidal particles using SPR in both scanning and imaging mode. It was discovered that SPR signals due to protein-protein interactions can be amplified ~100-fold through. . . .[1,2] This work eventually evolved into a particle-enhanced SPR immunoassay,[2] which is currently in commercial development. Also, as a part of this work, the PI designed and constructed an imaging SPR instrument for use in particle-enhanced high-throughput screening of combinatorial drug libraries.[109]

The PI has recently moved on to investigations of colloidal polymer systems. . . .

Analyzing Writing

Before we conclude our discussion of move 1, we call your attention to the pointers near selected excerpts above. These sidebars and excerpts together highlight correct usages of the following words, punctuation, or other writing features:

- numbers and units (excerpt 13A)
- farther vs. further (excerpt 13E)
- phenomenon/phenomena and other singular/plural words (excerpt 13G)

To test your knowledge of these features, complete exercise 13.6.

Exercise 13.6

Find and correct the mistakes in the following sentences, or indicate that the sentence is "correct as written".

a. To further develop the chemosensor design, we will focus on chelation-enhanced fluorescence.

b. We expect samples sizes to be about 100 g.

c. The expected phenomena was observed only in the single spectra collected at 300 K.

d. We will extend the synthesis farther than previous works.

We end this section by examining a few commonly used verb tense–voice combinations in move 1. As shown in table 13.3, prior accomplishments are typically written in active voice in either past tense or present perfect. Statements that establish expertise are typically written in present tense.

Table 13.3 Common functions of different verb tense–voice combinations in move 1 of the Experimental Approach section.

Function	Tense–Voice Combination	Example
To highlight prior accomplishments	Past–active	In previous work, we *demonstrated* that...
	Present perfect–active	We *have completed* a previous study that...
		Our previous efforts *have focused* on...
To establish expertise	Present–active	The data *demonstrate* our ability to measure...
	Present–passive	Our general approach *is illustrated* by our recently completed total synthesis of...

 **Exercise 13.7**

Identify the verb tense–voice combination in each sentence. What function does each sentence serve? Consult table 13.3.

a. In prior studies, we *demonstrated* that protein binding to surfaces has a significant impact on... ion signals. (Adapted from Kinsel, 1999)
b. In 2007, we *provided* the first example of a Lewis acid promoted C-C bond heterolysis of epoxides. (Adapted from Johnson, 2003)
c. Our previous works *have established* our ability to construct multilayered films.
d. Our ability to construct multilayed films *is evidenced* by our previous works.

13A Writing on Your Own: Share Prior Accomplishments

Make a list of research accomplishments that you have acquired to date that will help you with your proposed work (e.g., publications, conference presentations, collaborations, work-related experiences, instrumental expertise, technical training).

Based on this list, write a one-paragraph description of your skills and research accomplishments to establish expertise in your proposal.

Move 2: Share Preliminary Results

The second move in the Experimental Approach section is to share preliminary results. Preliminary results (unlike prior accomplishments) refer to results that are not yet published or disseminated. They suggest to reviewers that you have

Table 13.4 Common level 2 headings for move 2 of the Experimental Approach section.

Preliminary Results	**Current Work**
Initial Findings	**Work in Progress**

tested a few of the key ideas in your proposed work. Without at least some empirical evidence that your ideas will succeed, your proposal is unlikely to be funded. (Indeed, chemists often joke that proposals are only funded if 90% of the work is already done!) Like move 1, move 2 is often demarked with a level 2 heading. Some examples of level 2 headings are shown in table 13.4.

Analyzing Excerpts

Let's examine move 2 in authentic proposals. As authors share preliminary results, many authors also use move 2 to lay the conceptual groundwork for their proposed work. To this end, full color, computer-generated graphics are often included. Excerpts 13I and 13J illustrate this approach. In each case, the authors use full-color illustrations to share preliminary findings (the color is not reproduced here). We first saw the use of computer-generated graphics in chapter 12, to introduce fundamental concepts about the research area. They serve a similar purpose here. High-quality graphics effectively illustrate scientific concepts and, at the same time, attest to the author's skill and expertise as a writer. The familiar saying "a picture is worth a thousand words" is absolutely true in proposals.

Computer-Generated Graphics

Full-color, high-quality computer-generated graphics (drawings and illustrations) are widely used in proposals to illustrate scientific concepts.

In excerpt 13I, Lyon describes his group's progress in using hydrogels to fabricate thin films. As he shares his preliminary findings, he uses text and graphics to help readers better understand his plans to construct multilayered films. In Figure 2, he uses a drawing to illustrate how anionic hydrogel particles are attracted to an amine-coated surface. Three anionic hydrogels are taken up by four amines, suggesting that a "complete monolayer" does not form. In Figure 3, he presents empirical evidence from atomic force microscopy (*not included here*) to support this hypothesis. In Figure 4, he illustrates how a single surface can be extended to multiple layers, formed by alternating layers of hydrogel particles (negatively charged) and poly-allylamines (positively charged). In Figure 5 (*not included here*), he shares preliminary results that *show conclusively* that he and his group are able to construct multilayered nanoparticulate hydrogel films. After reading this section, his reviewers are better prepared to understand his work and are likely more convinced that it will succeed.

Exercise 13.8

Read excerpt 13I and answer these questions:

a. List two concepts that Lyon develops as he shares his preliminary results. Consider text and graphics.
b. What are the two ways in which Lyon refers to figures in the text?
c. How are Figures 2 and 4 labeled? Are the captions above or below the figures? Do Lyon's formatting conventions support those endorsed in chapter 16?
d. Would excerpt 13I be as effective without the figures? Explain.

Excerpt 13I (Lyon, 2000)

Preliminary Results

Thin Film Fabrication. The acrylic (AA) or methacrylic (MAA) acid portion of the copolymer also offers a group that will interact electrostatically with amines—a convenient method for immobilization of polymers to surfaces.[55–57] Figure 2 shows the general methods; modification of a solid surface with the appropriate alkylamine renders the surface amine-coated. Exposure of that surface to a suspension of carboxylated hydrogel spheres results in surface immobilization of the particles by electrostatic attachment. Figure 3 shows an atomic force microscopy (AFM) image of a representative hydrogel-modified surface recently prepared in our group. Due to interparticle repulsion, a complete monolayer is not formed under these conditions, resulting in a surface decorated with well-defined particles.

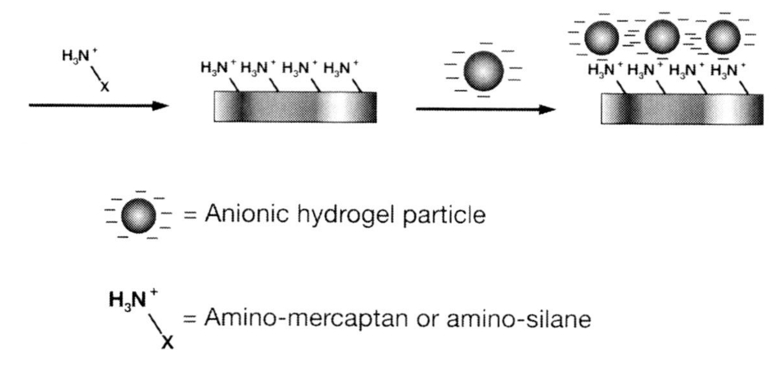

Figure 2. Electrostatic adsorption of colloidal hydrogels to functionalized surfaces.

This result suggests that polyelectrolyte multiplayer assembly methods can be applied to these materials. These methods (developed by Lvov and Decher)[58–60] allow for the construction of stable, reproducible, submicron-think, polymer films by alternating layer adsorption of polycations and polyanions from aqueous solutions. Accordingly, we have begun investigating the feasibility of this multiplayer film assembly method using hydrogel particle or linear polyelectrolytes (poly-styrenesulfonate) as the anion component and linear polyelectrolytes (poly-allylamine) as the cationic component in an alternate-layer deposition process (Figure 4).

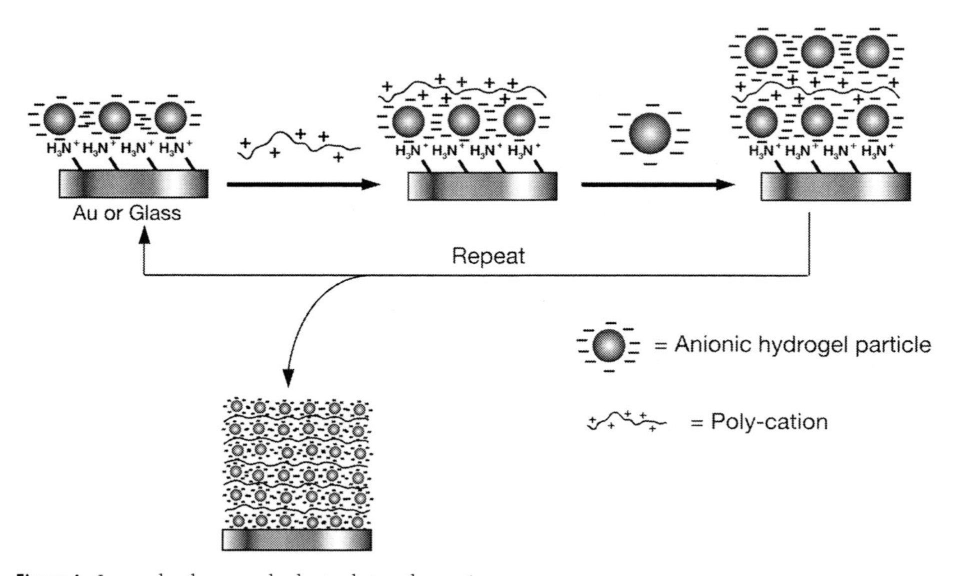

Figure 4. Layer-by-layer polyelectrolyte adsorption.

Figure 5 shows representative quartz crystal microgravimetry data recently collected in our group for the deposition of just such a multilayer film. . . . Despite the need for further investigation of the films, **these data show** conclusively that we are able to construct multilayered nanoparticulate hydrogel films via simple solution-based assembly methods. Films prepared in this fashion are currently being characterized with respect to their thickness and morphology (via profilometry and atomic force microscopy), porosity (voltammetry), viscoelasticity (crystal impedance), refractive index (surface plasmon resonance), and molecular structure (FTIR spectroscopy). These studies are elaborated upon in the research methodology.

Figures and Captions

See chapter 16.

Data/Datum

Data (plural)

Datum (singular, rarely used)

(See appendix A for other plural and singular scientific words.)

Excerpt 13J also involves the preparation of thin films at molecular-length scales but uses strategies of crystal engineering. Patrick proposes to investigate thin film growth on substrates submersed in a liquid crystal (LC) solvent/solute mixture under the influence of an external magnetic field. For example, in a preliminary experiment, they deposited a fatty acid (tetracosanoic acid) from an LC solvent onto a graphite substrate. In the presence of a magnetic field, the deposited film was uniformly oriented (crystalline); in the absence of the field, a randomly oriented (polycrystalline) film was observed. Patrick includes an illustration to depict these phenomena. By combining both text and illustration, Patrick presents his preliminary results in a way that establishes a clear conceptual framework for his proposed work.

Excerpt 13J (Patrick, 2000)

Summary of Preliminary Results

We have conducted preliminary investigations to assess the way LC solvents influence film formation in monolayers of small organic molecules on highly oriented pyrolytic graphite (HOPG) substrates. A number of different molecular solute/LC solvent combinations have been investigated over the last year. In most cases, samples were prepared by deeply immersing a graphite substrate into a reservoir of the LC/solute mixture at ~100 °C, then allowing the system to gradually cool to room temperature (Fig. 3).... The reservoir was located in a magnetic field, oriented with the field axis in the plane of the substrate surface. After preparation, samples were removed and analyzed with scanning tunneling microscopy (STM). The STM tip penetrated through the thick LC solution to image molecules in the monolayer at the graphite interface. Results from three representative systems are summarized in Table 1.

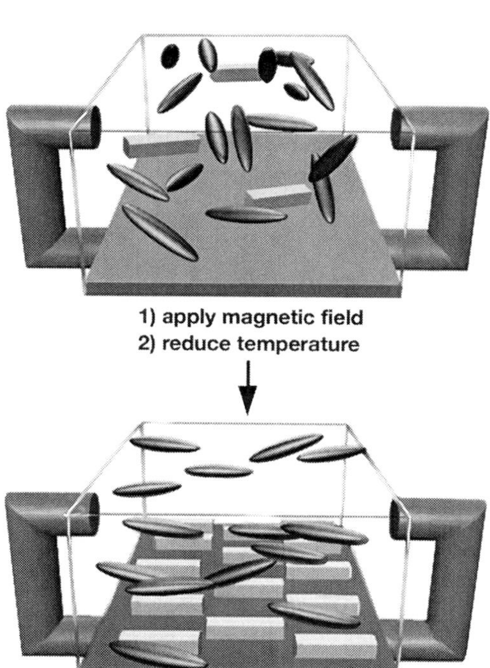

Figure 3. Preparation of oriented films.

The first entry in Table 1 represents the simplest implementation of the LCI method, in which the LC solvent and solute building block were identical. Because it is the simplest and best understood system studied so far, we will describe our results for 8CB in most detail. We wish to emphasize, however, that LCI is not restricted to 8CB; we have demonstrated it in other systems as well, including systems in which the building blocks differed from the LC solvent and were non-mesogens.

Exercise 13.9

Cover up the figure in excerpt 13J and complete the following tasks:

a. Read the excerpt (without looking at the figure).

b. Based solely on the text, try drawing a sketch on your own that illustrates the effect of the magnetic field.

c. Then look at the figure. How close were you?

d. In what ways did Patrick's illustration help you gain a better grasp of the proposed work?

e. Note how the text "wraps" around the figure in Excerpt 13J. In many instances, authors also "box" their figures. Explain why these practices are common in proposals but not in manuscripts submitted for publication.

Many authors use the word *promising* when sharing preliminary results, presumably because the word suggests that the proposed work will likely succeed. For example, 10 of the 22 CAREER proposals cited in this textbook included the word *promising* (and even *very promising*) in this move. Here are a few examples to give you an idea of how the word was incorporated:

The extension appears to be	**promising**...	(From Johnson, 2003)
The results...were very	**promising**...	(From Aga, 2002)
Employing a particularly	**promising** class of...ligands	(From Warren, 2002)
Our efforts...are very	**promising**.	(From Vyvyan, 2001)
Upon identification of	**promising** leads...	(From Johnson, 2003)
Early results are	**promising**...	(From Walker, 2001)
(make) imaging...ions very	**promising**.	(From Sanov, 2002)
...(are)	**promising** candidates...	(From Patrick, 2000)
A	**promising** new technique...	(From Lorigan, 2002)
A	**promising** lead structure...	(From Finney, 1999)

The word *recently* (or *recent*) is also common in move 2. *The ACS Style Guide* advises against using *recently* in journal articles (in part, because *recently* loses its meaning so shortly after a paper is published), but its use is less objectionable in research proposals, which are short-lived documents. Investigators like the word because it conjures up the image of a research group actively engaged in the proposed work. A few examples of how *recently* is used in move 2 are as follows:

Furthermore,	**recent** NMR data reveal...	(From Lee, 2001)
We have	**recently** prepared...	(From Warren, 2002)
Our	**recently** established methodology...	(From Lee, 2001)
Our	**recently** completed total synthesis...	(From Vyvyan, 2001)
...	**recently** collected data from our group.	(From Lyon, 2000)

Excerpt 13K illustrates how *promising* and *recently* are used in a complete paragraph. The author also uses the phrase "the *first* evidence of ____". The word *first* in this context is also common in move 2 because it underscores the originality of preliminary findings. Empirical evidence for the author's claims is provided in a figure. Note that the author uses a slightly reduced font size in the figure caption. This practice is sometimes allowed in proposals if space is tight.

Solvatochromism

Solvatochromism refers to changes in the electronic state of the solute (specifically, the solute's electronic state transition energy) caused by the solvent.

Excerpt 13K (Walker, 2001)

Early results are promising. We have recently completed a preliminary study that examined the solvatochromic behavior of 4-aminobenzophenone (4ABP) adsorbed to hydrophilic substrates from a variety of solvents. Hydrophilic substrates are polished quartz surfaces that demonstrate complete wetting when exposed to water.... Figure 5a shows the SHG spectrum of 4ABP adsorbed to a hydrophilic interface from cyclohexane.... To our knowledge, these data represent the first evidence of surface-induced **solvatochromism** at solid-liquid interfaces. Superimposed are UV–vis spectra of 4ABP in bulk cyclohexane and bulk diethyl-ether. As anticipated, data show that the hydrophilic, silanol terminated surface renders the interactial environment more polar than bulk cyclohexane.

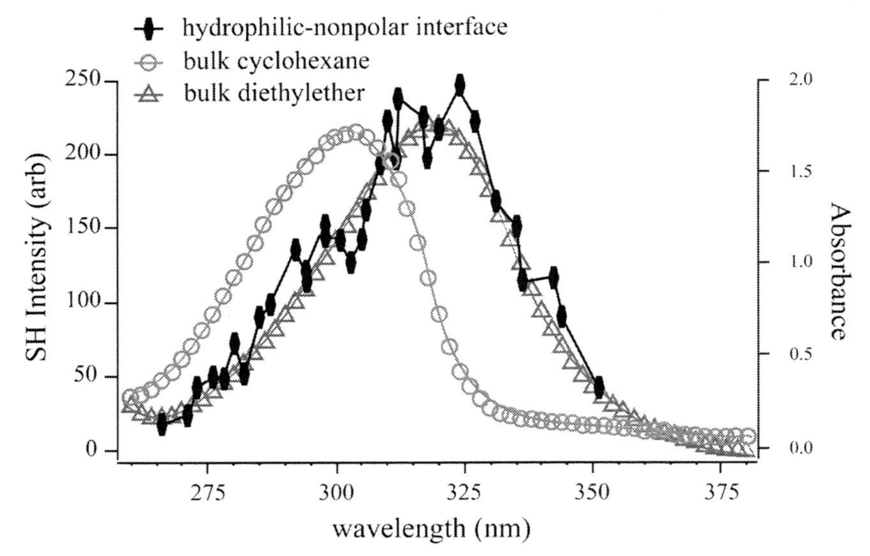

Figure 5a. UV–vis and SHG spectra of 4ABP. The two UV–vis spectra are for 4ABP in cyclohexane (open circles) and diethyl ether (open triangles). The SHG spectrum (filled diamonds) shows solvatochromic shift of 4ABP adsorbed to the interface between hydrophilic quartz and a cyclohexane solution.

Text Boxes, Wrapping, and Captions

Graphics in proposals are sometimes "boxed". In some cases, the text is wrapped around the graphic. To save space, slightly smaller font is sometimes allowed for figure captions.

Exercise 13.10

Read excerpt 13L and answer the following questions:

a. What words or phrases are used to emphasize that the preliminary results are both promising and original?
b. Note that Johnson describes the results of his synthesis in present tense (e.g., the thermal reaction *gives* a diastereomer ratio . . .) rather than in past tense (e.g., the thermal reaction *gave* a diastereomer ratio . . .). Suggest a reason for this choice.
c. Find examples of present perfect. For what purpose(s) does Johnson use present perfect? (See table 6.2 for more on present perfect construction.)

Excerpt 13L (Johnson, 2003)

Our preliminary experiments have provided the first example of Lewis acid promoted C-C bond heterolysis of epoxides and productive cycloaddition (eq 7). Under the influence of $TiCl_4\cdot(THF)_2$ (2 equiv), epoxide **26** reacts with methyl pyruvate to provide acetal **27** (52% isolated yield), along with C-O cleavage product **28** (23 °C, 3 h). The diastereoselectivity for formation of **27** is 2.3:1. We have performed the analogous reaction in the absence of a Lewis acid; the thermal reaction requires several days at 110 °C and gives a diastereomer ratio (dr) of ca. 1.3:1 ... **Although** not optimized from the standpoint of chemoselectivity, these results are promising because of the relatively low reaction temperature and potential for enhanced diastereocontrol.

While vs. Although

While (during the time that)

Although (even though)

(See appendix A.)

Exercise 13.11

Reexamine excerpts 13K and13L for ways in which authors share preliminary results in their proposals, and answer the following questions:

a. How much detail do the authors include in their preliminary results? What details are included?

b. Count the uses of *we* and *our* in these excerpts. Why do you think it is so common for authors to refer to themselves in move 2?

Analyzing Writing

Before we conclude our discussion of move 2, we call your attention to the pointers near selected excerpts above. These pointers and excerpts together highlight correct usages of the following words, punctuation, or other writing features:

- figures and captions (excerpt 13I)
- data/datum (excerpt 13I)
- while vs. although (excerpt 13L)

To test your knowledge of some of these features, complete exercise 13.12.

 **Exercise 13.12**

Find and correct the mistakes in the following sentences, or indicate that the sentence is "correct as written".

a. While it is clear that the first three reactions were solvent dependent, this may not be true for the next set of reactions.
b. The data clearly shows the expected relationship between time and temperature.
c. Although the word *recently* is discouraged in journal articles, it can be used in research proposals.

We end this section by examining a few commonly used verb tense–voice combinations in move 2 (table 13.5). Present perfect is commonly used to describe preliminary work (done in the past); present tense is commonly used to share preliminary findings (believed to be true over time). Active voice is also common. Note that because authors want to call attention to their own promising results in this move, personal pronouns such as *we* or *our* are often used.

Table 13.5 Common functions of different verb tense–voice combinations in move 2 of the Experimental Approach section.

Function	Tense–Voice Combination	Example
To describe preliminary work (done in the past)	Present perfect–active	We *have* recently *completed*... We *have conducted* preliminary investigations to assess...
	Past–passive	Films *were prepared* by... Samples *were analyzed* by...
To share preliminary findings (that will likely be true over time)	Present–active	Our preliminary results *demonstrate* that... To our knowledge, these data *represent* the first evidence of... Under the influence of $TiCl_4(THF)_2$ (2 equiv), epoxide **26** *reacts* with methyl pyruvate to...(From Johnson, 2003)
To refer to graphics (e.g., figures or illustrations)	Present–active	Figure 2 *shows* the general method.
	Present–passive	Results from three representative systems *are summarized* in Table 1.

Exercise 13.13

Consider the following sentences used to describe preliminary results. Which are written in active voice? Which are written in passive voice? Convert the sentences that are written in passive voice to active voice to place greater emphasis on the scientist/researcher.

a. A number of different molecular solute/LC solvent combinations have been investigated over the last year. (From Patrick, 2000)

b. In most cases, we prepared samples by deeply immersing a graphite substrate into a reservoir of the LC/solute mixture at ~100 °C. (From Patrick, 2000)

c. Our group has investigated the synthesis of hydrogel particles that can be derivatized covalently with proteins using standard aqueous crosslinking procedures. (From Lyon, 2000)

d. The first synthetic efforts have been focused on lightly crosslinked (<1%) copolymers of NIPA and acrylic (AA) or methacrylic acid (MAA).[51–53] (From Lyon, 2000)

Exercise 13.14

Consider the following sentences that describe preliminary results (adapted from Gudmundsdottir, 2001). For each sentence, choose the most appropriate verb form. Be prepared to explain your choices.

a. The IR spectra of **1a** before and after irradiation in an argon matrix *is/was/are/were* shown in Figure 8.

b. The transient spectrum of azide **1h** *is/was/are/were* taken immediately after the laser pulse over a 200 ns time window.

c. The difference spectrum in Figure 7 *has/had/have* a strong band between 270 and 320 nm.

d. As the temperature *is/was* lowered, more of compounds **4a**, **8a**, and **9a** *are/were* formed.

13B Writing on Your Own: Share Preliminary Results

Decide what preliminary (and promising) results you will highlight to convince your readers that your proposed work is feasible. For example, perhaps you have already collected your samples, conducted calibration experiments, or tested your instrument under background conditions. Decide if you need figures or tables to present the preliminary data or if they can be adequately reported in the text.

Write one to three paragraphs to share these results. If possible, share the results in such a way that builds a conceptual framework for your proposed work. If appropriate, include computer-generated illustrations to elucidate these concepts.

Move 3: Describe Proposed Methodology

In the third (and last) move of the Experimental Approach section, you describe how you will conduct your proposed work. A well-organized and logical progression of ideas is essential in this move. Most authors demark the start of this move with a level 2 heading, parallel to the level 2 headings used for moves 1 and 2. A few examples are shown in table 13.6.

As shown in figure 13.1, move 3 includes three submoves: remind reader of promising results, describe procedures/instrumentation, and anticipate obstacles. These submoves are typically accomplished in sequential order, and the sequence is reiterated for projects with more than one research objective. A level 3 heading is commonly used to demark each new objective; it may be generic (e.g., Objective 1, Objective 2) or descriptive (e.g., Quantification of Surface-Protein Binding Affinity). The language used in these level 3 headings should be parallel. Consider the parallelism in the following examples from Lyon (2000):

> **Bioconjugate Synthesis—Stage 1.** As described above, we have already demonstrated our ability to produce thermoresponsive copolymer hydrogel particles that possess groups for protein attachment.
>
> **Bioconjugate Synthesis—Stage 2.** The second stage of the bioconjugate project will involve more complex polymer structures and more challenging biological targets.

Parallelism

Headings should be parallel and begin with the same form of speech. (See appendix A.)

Table 13.6 Common level 2 headings for move 3 of the Experimental Approach section.

Proposed Methods	**Proposed Studies**
Proposed Research	**Proposed Work**
Research Plan	**Experimental Plan**

It is important that the proposed work for each research objective be presented in a logical order. Often, the section begins with lower risk experiments and moves progressively toward higher risk experiments. Ordinal language (e.g., first, second, last) is common and is used to reinforce a logical progression of ideas. (Note: The convention of using ordinal language in the research proposal is quite different from the journal article, where ordinal language is discouraged.)

Analyzing Excerpts

With that brief background in mind, let's examine how authors accomplish move 3 in authentic research proposals. Submove 3.1 (remind reader of promising results) serves as a transition between move 2 (Share Preliminary Results) and move 3 (Describe Proposed Methodology). Ideally, submove 3.1 is accomplished in a sentence or two at the start of each research objective. You want to remind the reader what you have already learned and then move on to the proposed work. Five examples are shown below (P1–P5). Each passage begins with a level 3 heading, to demark the start of a new objective. The text begins by referring to a preliminary finding and then shifts to the proposed work; the preliminary finding serves as the first step in a clear progression of planned experiments.

P1 **2A. IR studies of N^{15} labeled azide 1a.** We are concerned with providing further evidence that nitrene **2a** was formed upon the photolysis of azide **1a**. To this end, we plan to synthesize azide **1a** as an N^{15} isotope labeled in the N1 position (see Figure 15), obtain IR spectra before and after **irradiation** in an argon matrix, and compare the calculated shift for the C-N^{15} band in nitrene **2a** with the experimental value. Because isotope shifts in IR bands can be calculated very accurately, this will be an excellent proof of the formation of a nitrene intermediate. (From Gudmundsdottir, 2001)

P2 **1.1. Measurement of multiple acidic sites in thymine.** Because we have already embarked on the study of uracil (see Preliminary Results), we will begin with thymine (**4a**, Figure 1), which is the 5-methyl-derivative of uracil. (From Lee, 2001)

P3 **Bioconjugate Synthesis—Stage 1.** As described above, we have already demonstrated our ability to produce thermoresponsive copolymer hydrogel particles that possess groups for protein attachment. These syntheses will continue, with the goal of producing gels across a range of NIPA/AA copolymer ratios (protein attachment points). (From Lyon, 2000)

P4 **Heliannuols A, D, and K.** Our preliminary results show that the biomimetic phenol epoxide cyclization route is valid for the *exo* cyclization products such as heliannuol D (**4**), and we are hopeful that the planned acid-catalyzed cyclization of epoxides like **21** will produce *endo* products like heliannuol A (**1**). (From Vyvyan, 2001)

P5 **Effect of Sample Preparation Protocol.** Our preliminary studies strongly suggest that there is an inverse relationship between the MALDI ion signals observed for surface adsorbed proteins and the affinity of the surface for binding those proteins. Although this relationship appears to be well demonstrated above, numerous questions remain to be answered about this approach. For example, how does the sample preparation protocol impact the observed inverse relationship? In the first group of experiments, we will vary the order of sample and matrix deposition to determine if other sample preparation protocols lead to similar observations. (From Kinsel, 1999)

Irradiation vs. Radiation

Irradiation (n.) is the act of applying radiation (n.).

Exercise 13.15

Reread P1–P5 and complete the following tasks:

a. Identify the preliminary findings and the proposed work in each example.
b. How common is it for authors to incorporate a graphic (i.e., table or figure) into this submove?
c. How common is it for authors to cite relevant works in this submove?
d. How common is it for authors to use personal pronouns (*we*, *our*, *I*, and *my*) in this submove?

Exercise 13.16

Aga (2002) presents three research objectives in her proposal. One of these is shown in excerpt 13A. A second objective involves modeling the transport behavior of nitrates and antibiotics in the Central Platte watershed using the ArcView geographical information system (GIS) and the EPA software package BASINS. Given the following information (adapted from Aga's proposal), complete submove 3.1 for her second research objective. Include a level 3 heading, and write at least two sentences that transition from preliminary results to the proposed work.

- Assume as preliminary work, Aga has already built the project file using the BASINS software (i.e., she has input data related to watershed topography, point source discharges, flow rates, and property boundaries).

- She proposes to use these data to simulate the transport rates of nitrates and antibiotics from cattle and swine feedlot runoff into nearby waterways (primary, secondary, and tertiary streams).
- Samples will be collected at least once a month at the feedlots and in the nearby streams.
- The measured feedlot concentrations of nitrates and antibiotics will be imported into BASINS. Using these input values, BASINS will model the transport rates and final concentrations of nitrates and antibiotics in the nearby streams.
- Modeled results will be compared with experimental values.

After the transition from preliminary results to proposed work has been accomplished, you are ready to describe the procedures and instrumentation that will be used in the proposed work (submove 3.2). In essence, submove 3.2 offers a road map of the procedures to be followed. There are two conventional ways to organize this road map. The first approach presents the work chronologically, beginning with more straightforward and familiar experiments and then moving toward more challenging and unexplored methodologies. The second approach organizes the work according to underlying themes or desired outcomes. The first approach is preferred when experiments build on one another and need to be performed sequentially. The second approach can be used when the proposed methods are largely independent of each other. Of course, combinations of the two approaches are also possible.

Road Map for Proposed Work

Work plans can be organized (1) sequentially, (2) by underlying themes or desired outcomes, or (3) by combining the two approaches.

We start with examples of the sequential approach. With this approach, you begin with more routine experiments, ones that are reasonably likely to succeed (e.g., calibration or optimization procedures). The initial set of experiments can also serve as a test case and/or show that you can reproduce **literature values**. For example, Aga (P6) proposes first to explore conditions that will optimize immunoassay sensitivity, and Spain (P7) proposes to begin with a study of topography, using published methods and a self-assembled monolayer with a known structure.

Literature Values

Values reported in refereed publications.

P6 First, several parameters will be explored to determine the optimum conditions at which the immunoassay will have highest sensitivity for the target analytes. These parameters will include various preparations of hapten-enzyme conjugates, varying strengths of assay buffers, varying incubation times, and two signal-amplification techniques using avidin- or streptavidin-biotin complexes. (From Aga, 2002)

P7 We will begin with a study of the topography of an *n*-alkanethiol self-assembled monolayer (SAM) anchored to a Au(111) substrate. Alkanethiols will be attached to a Au(111) substate by published methods.[44] The SAM will be imaged by atomic force microscopy (AFM) to assure that the known structure is observed.[45] (From Spain, 1997)

Of course, a successful proposal must also forge ahead into less familiar territory. It is not enough to conduct the "easy" experiments; you must approach the "cutting edge" or forefront of your field. For this reason, Aga goes on to describe how the optimized immunoassay will eventually be used to test for analytes in more complex environmental samples, and Spain proposes a sequence of experiments that will culminate in the deposition of translationally "hot" metal atoms on a self-assembled monolayer system. The important point in these examples is how authors develop a clear and logical order for their proposed work.

A similar approach is used by Lyon (excerpt 13M). He divides his proposed methods into two stages; the second stage is described as more *complex* and *challenging* than the first. In his description, fewer specific details are included in stage 2 (e.g., stage 1 cites a specific synthetic procedure, whereas stage 2 refers to more general coupling chemistries). The lack of specificity in stage 2 is acceptable because of the difficulties associated with predicting specific approaches as you project your work farther into the future.

Excerpt 13M (Lyon, 2000)

Bioconjugate Synthesis—Stage 1. As described above, we have already demonstrated our ability to produce thermoresponsive copolymer hydrogel particles that possess groups for protein attachment. These syntheses will continue, with the goal of producing gels across a range of NIPA/AA copolymer ratios (protein attachment points)....

The bioresponsivity of the hydrogel conjugates will initially be investigated with simple protein-protein and protein-ligand pairs where one of the binding partners will be covalently attached to the hydrogel.... First, we will use carboiimide coupling methods[54] to covalently attach one of the binding partners to the carboxylate moieties in the hydrogel. Specifically, we will... [*goes on to describe more specific methods*].

Bioconjugate Synthesis—Stage 2. The second stage of the bioconjugate project will involve more complex polymer structures and more challenging biological targets. First, we will explore factors such as protein loading, particle size, polymer cross-linking density, and copolymer ratio/identity in a highly parallel fashion to evaluate the biosensitivity of a wide range of hydrogel bioconjugates. This will allow us to rapidly converge upon a highly optimized polymer-protein conjugate based on NIPA hydrogels. Second, we will expand our approach to other coupling chemistries (e.g., maleimide, biotin-streptavidin, photo-crosslinking). By broadening the range of chemistries available, these colloidal bioconjugates will have wider applicability to other biochemistries and film fabrication methods. We will also begin to explore other biological interactions such as ligand-receptor binding, DNA hybridization, enzyme-substrate interactions, or protein-DNA complexation to further evaluate the generality of our approach.

The sequential approach is also common in proposals written by synthetic chemists (a multistep synthesis is inherently step by step). Vyvyan (excerpt 13N), for example, proposes a strategy to synthesize a select group of heliannuols (alleopathic natural products isolated from the sunflower) in an optically pure form. One approach that he will explore involves enantioselective cross-coupling reactions between an alkyl zinc reagent and an aryl bromide. He begins with experiments that will utilize recently developed catalysts and produce products with known optical rotation data. Subsequent reactions are described that will lead potentially to the desired stereospecific heliannuols A and D.

Excerpt 13N (Vyvyan, 2001)

Asymmetric Cross-Coupling Reactions. An attractive method to attach the side chain and establish the benzylic stereocenter of our target compounds is an enantioselective cross-coupling of an alkyl zinc reagent and an aryl bromide.... Our approach will examine the reaction of secondary alkyl zinc species (e.g., **30**) with aryl halides (e.g., **31** and 4-bromotoluene) using chiral catalysts such as **45** and **46**. Simpler routes by which to prepare prospective chiral ferrocene-based catalysts have recently been developed, which bodes well for the future development of this process.[40] Chiral GC or HPLC and optical rotation data of the products will be compared to literature values for curcuhydroquinone **33** and curcumene **47**, **respectively**, to determine the sense and level of enantioselectivity (Scheme 6).[28,38a,41]

Once this process is explored with the model system to assess the level of enantioselectivity, we will then prepare alkyl zinc reagent **48** from **44** using standard methods[22,36,42] and cross couple **48** to aryl bromide **18** using the appropriate chiral catalysts (Scheme 7). Although the acetonide stereocenter in **48** is somewhat remote from the coupling site, the stereocenter may serve to enhance the stereoselectivity of the cross-coupling process because the two possible products are diastereomers, not simply enantiomers. This reaction will produce **49** from (*S*)-**48** and **50** from (*R*)-**48** that can then be converted to epoxides **51** and **52** using standard methods.[43] Epoxide **51** leads to heliannuol D **4** after base-promoted epoxide cyclization and deprotonation. Similarly, epoxide **52** leads to heliannuol A **1** after acid-promoted cyclization.

Respectively

See appendix A.

(*S*) and (*R*)

Use italic (not bolded) type for prefixes like (*S*) and (*R*) that denote stereochemistry. (See chapter 4.)

The last example of a sequential approach is from Sanov (excerpt 13O). A series of increasingly complex experiments is proposed to study the photochemistry of O_2^-, S_2^-, and OCS^-. Sanov begins with the "easier" diatomic anions (O_2^- and S_2^-), which will serve as prototypes for subsequent experiments. Next, he will study a larger, polyatomic anion (OCS^-) and its cluster ions, $OCS^-(H_2O)_k$. In the future, he will study even larger dimers and trimers $(OCS)_n^-$ ($n \geq 2$) and their hydrated counterparts.

Excerpt 13O (Sanov, 2002)

Electronic Structure via Photoelectron Imaging.... Initial experiments will focus on O_2^- and S_2^-.... Besides their importance, O_2^- and S_2^- will serve as prototype systems for developing the photoelectron imaging approach and...

The $(OCS)_2^-$ anion is the next logical system, taking us into the realm of polyatomic anions with rich photochemistry and more complex symmetry. $(OCS)_2^-$ is an example of how dramatically the properties of a compound (e.g., carbonyl sulfide) may change upon electron capture and/or solvation.[138] ...

Our next step will be to use "differential imaging" to study cluster chemistry and examine the effects of solvation (hydration) on electronic structure....

Building on the initial findings described in Section 4, we will acquire photoelectron images of $OCS^-(H_2O)_k$ cluster anions at different wavelengths in the visible and UV and investigate the dynamics of hydration and hydration-induced stabilization of OCS^-....

By examining the images of $OCS^-(H_2O)_k$ ions, we expect to attribute the changes with k to the **effects** of hydration.... In the future, we plan to extend these studies to $(OCS)_n^-$ ($n \geq 2$) cluster anions and their hydrated counterparts $[(OCS)_n(H_2O)_k]^-$.

Affect vs. Effect

Effect (n.): a consequence, result

Affect (v.): to influence, change

(See excerpt 13U for the proper use of *affect*. See also appendix A.)

Exercise 13.17

Dyer (2001), in his CAREER proposal, proposes six key experiments to evaluate polar order and polar stability of organic self-assembled monolayers. A gold-coated surface will be partitioned into two surfaces, allowing two different hydrogen-bonding moieties to be studied simultaneously. Dyer uses ordinal language (first, second, third, fourth, fifth, and sixth) to make the sequence of experiments clear. However, even without this language, the order is predictable because the experiments are arranged in order of increasing complexity.

We have listed the first three of the six experiments (a–c) below. Arrange them in order of complexity (and, hence, the order in which they will be performed), identifying the correct ordinal language (first, second, third) where indicated. (SHG = second harmonic generation; SAM = self-assembled monolayer.)

a. *First/Second/Third*, the SHG signal will be measured when both surfaces exhibit complementary H-bonding moieties. Ideally, we predict that the SHG signal should be twice that of the (*first, second, third*) experiment because both surfaces will contribute to the bulk polar orientation.

b. *First/Second/Third*, a control experiment will be performed... where the terminal functionality on both surfaces is an alkane.... In this instance, the SAM surface will not participate in H-bonding with the bulk liquid crystal (LC) and therefore the bulk material will not exhibit polar order; thus, the SHG signal should be very low.

c. *First/Second/Third*, the SHG signal will be measured when only one surface has a hydrogen bonding SAM and the other surface is terminated with an alkane. We expect the SHG signal to be significantly larger than for the *first/second/third* experiment where both surfaces were terminated with alkanes.

Exercise 13.18

Below are five excerpts (a–e) adapted from the "Proposed Studies" section of Kinsel's (1999) CAREER proposal. However, the excerpts are out of order, and we have omitted three words (*first*, *next*, and *collectively*) that were included in the original proposal. Using language clues and figure 13.1 as guides, arrange these excerpts in the correct order (1–5), and select the most appropriate missing word where indicated.

a. *First/Next/Collectively*, the above studies should allow us to determine the optimum sample preparation and MALDI conditions with which to characterize surface-protein interactions.

b. *First/Next/Collectively*, we will further extend our investigations to a broader range of biomolecules and to a wider array of pulsed RF plasma polymer modified surfaces. In these studies, the thrust of the research will be to further define the range of systems for which the MALDI methodology can be used to study surface-protein interactions. The selection of peptides and proteins will include larger proteins commonly used in MALDI studies (e.g., ubiquitin, chicken egg lysozyme, horse apomyoglobin, trypsin) as well as peptides and proteins that are readily radiolabeled (e.g., substance P, bovine serum albumin).

c. For example, our preliminary MALDI MS studies employed a somewhat unconventional sample preparation approach in which the protein deposition step preceded the matrix deposition step. (In the conventional approach, the protein is co-deposited with the MALDI matrix solution.)... Thus, in the *first/next/collectively* group of experiments, we will explore different sample preparation methods to determine if other protocols also lead to an inverse relationship between surface-protein binding affinity and the MALDI ion signal.

d. These studies will be followed with experiments that examine the influence of the matrix solvent on the inverse surface-protein binding affinity/MALDI ion signal relationship.

e. Our preliminary work strongly suggests that there is an intimate relationship between the MALDI ion signals for surface-adsorbed proteins and the affinity of the surface for binding those proteins. Although this relationship appears well demonstrated in the examples above, numerous questions remain to be answered.

We now consider two examples of proposed methods that are not organized sequentially. In excerpt 13P, Finney describes three near-term projects that will be conducted simultaneously. He offers an overview to describe the three different areas and then goes on to give more details about each area (not included). Ordinal language is still used, but in this case, it is used to convey the order in which the projects will be presented, rather than the order in which the experiments will be conducted.

Excerpt 13P (Finney, 1999)

Proposed Near-Term (0–24 month) Research

Overview. The proposed near-term research is divided into three areas, **which** will be pursued simultaneously. The first is the complete characterization of the current mercury-responsive fluorescent chemosensor system, including the measurement of fluorescence lifetimes, to discern the origin of the conformational control of fluorescence. The second is to develop two classes of substituted biaryl acetylenic fluorescent chemosensors, to move the observed fluorescence into the visible region and increase the magnitude of the fluorescence signal, **which** occurs upon conformational restriction.

The third is the construction of a 100-member combinatorial library of fluorescent chemosensors related to the established mercury binder, to identify a combination of recognition domains **that** will allow selective, high-affinity detection of mercury in the presence of other metal ions.

Which vs. That

Which signals extra information.

That signals essential information about the word or phrase that comes before it.

(See appendix A.)

A second example is shown in excerpt 13Q. Harpp's proposed work initially involves the collection of field data in two sampling campaigns. Following collection, the samples will be analyzed for trace and major element concentrations and isotopic ratios. In her case, the order in which she conducts these analyses is less important than *how* she will use the data to answer questions about plume-ridge interaction mechanisms. Thus, she organizes her proposed work (titled "Proposed Plan") not by the tests she will perform, but rather by the types of information the analyzed data will provide (i.e., information about formation mechanisms, melting dynamics, and spatial distributions).

Excerpt 13Q (Harpp, 1998)

Proposed Plan

To evaluate properly the plume-ridge interaction mechanisms described above, it is essential that we establish a thorough database from those poorly sampled regions of the northern Galápagos, Wolf, Darwin, and Genovesa Islands. We propose to carry out two field excursions, one to Wolf and Darwin Islands, and a second to Genovesa, to map and collect representative samples for geochemical analysis. All three islands remain essentially unmapped at this time. The geochemical data will then be examined in conjunction with the PLUM02 expedition data from Pinta Island (Cullen and McBirney, 1987; White et al., 1993) and Marchena Island (Vicenzi et al., 1990) for extensive modeling of mixing and melting parameters.

From each island, a suite of samples will be collected where the geologic relationships are well established. This will be done via careful flow-by-flow mapping, assessment of relative ages of flows, and paleomagnetic work to confirm correlation of field relationships. Emphasis will be placed on obtaining as extensive a stratigraphic section as possible. The samples will be examined for petrographic variations, trace element and major element concentrations, and isotopic ratios. With these data, we will evaluate the following:

(1) Formation mechanisms. How are the northern Galápagos Islands and seamounts related to the plume and the ridge? Is the WDL a channel to the ridge from the plume, a leaky fault, or related to near-ridge volcanism? How is Genovesa related to the ridge, and the adjacent islands of Pinta and Marchena? We will examine the geochemical evolution of Wolf, Darwin, and Genovesa lavas to address these questions. . . .

(2) Melting dynamics and source composition. Rare earth inversion using the path integral method of **White et al. (1992; McKenzie and O'Nions, 1991)** will be used to quantify the melting parameters and approximate source compositions (e.g., **Harpp, 1995; Mahoney et al., 1993**). . . .

Rare earth element data will also serve as the basis for a forward modeling study to better constrain melting systematics in the Galápagos. The melting model will invoke clinopyroxene-rare earth element partition coefficients, which vary with composition **(Gallahan and Nielsen, 1992)**, and a polybaric or column melting process. . . .

Isotopic ratios will be used to constrain source compositions and the contributions from the plume and ridge to the island lavas. Harpp (1995) showed that magma mixing occurs at the source level, prior to melting. The forward melting model will be applied to different proportions of plume and ridge compositions to reproduce trace element REE systematics in the erupted lavas. From this process, both the original source compositions as well as the range in melting conditions experienced by magma at each site will be constrained. This information will assist in developing a map of the thermal and compositional structure of the present-day Galápagos plume and its interaction with the GSC.

(3) Spatial distribution of geochemical variations. Existing data from Marchena and Pinta Islands, PLUM02 data from the WDL, and the new Wolf, Darwin, and Genovesa data will be compiled into a comprehensive geochemical database for further examination of the formation mechanisms described above. The spatial distribution of geochemical variations relative to the plume and the ridge will be examined in detail.

Author–Date Citations

The author–date format is one of three common ways to cite others' works. (See chapter 17.)

Exercise 13.19

Look back at excerpts 13P and 13Q and answer the following questions:

a. Which authors cite relevant works?

b. Based on these two excerpts, when do you think it is most appropriate to include citations in submove 3.2 (describe procedures/instrumentation)?

Before we leave submove 3.2, we want to address one more question: how much detail should be included when describing proposed methodology? In a journal article, the general rule of thumb is to include enough detail so that an expert in your field can repeat the experiment. Thus, a journal article Methods section would include details such as the names and locations of vendors, compound purity, gram weights for reagents, **operational parameters**, reaction conditions for a synthesis (time, temperature, solvent, etc.), and so on.

Operational Parameters

Unlike the methods section of a journal article, operational parameters (settings under which an instrument is operated) are not typically included in the methods section of a proposal. (See chapter 3.)

In a research proposal, however, such details are often omitted. This is true for several reasons. First, a journal article needs sufficient detail to allow others to repeat the work; a proposal needs only to convince your readers that you can do the work. Second, proposals typically describe multiple experiments that will culminate in numerous publications. You simply do not have enough space to describe each experiment in detail. Third, you most likely do not yet know the details for many of your proposed experiments, particularly those planned for the distant future; hence, it is customary to describe your work in more general terms. Finally, too many details can obscure the underlying purpose of your work. The emphasis should be on the general approach, not the minutiae; you do not want to bury your readers in details that may distract them from the importance of your project.

A journal article needs sufficient detail to allow others to replicate the work; a proposal, on the other hand, needs to convince your readers that you can do the work.

That said, let's consider a few examples. We begin with excerpt 13R, which includes a rather lengthy description of the proposed methods—but note that the general approach, not the experimental details, is emphasized. The author begins with a description of the general methods, focusing on the instrument that will be used, a Fourier transform ion cyclotron resonance mass spectrometer (FTMS). If this were a journal article, one would expect to find a formal description of the operational parameters, but instead Lee focuses on the novel way in which the instrument will be used (for acidity bracketing). Because this novel application is central to the success of the project, precious space is used to describe the technique. In a logical order, she describes two types of experiments; the first type (bimolecular reactivity) is simpler than the second (acidity bracketing). After explaining the general process of acidity bracketing, she goes on to describe how this method will be used to measure the multiple acidic sites in thymine. She mentions by name the reference acids that she will use, building on her experience with uracil and strengthening the reader's confidence that she knows what she is doing. She briefly points out that the acids are commercially available, but does not take valuable space to list potential vendors or compound purity, information that would be included in a journal article.

Excerpt 13R (Lee, 2001)

Research Design and Methods

General Methods. The instrument that will be used to execute the gas-phase experimental portion of the proposed research is a Finnigan 2001 **dual-cell** Fourier transform ion cyclotron resonance mass spectrometer (FTMS or FTICR), equipped with both electron impact (EI) and electrospray ionization (ESI). FTMS is a **high-resolution**, **high-sensitivity** technique that allows the entrapment and detection of **gas-phase** species.[17,18] Gas-phase ions are trapped in a magnetic field, much like a reactant sits in a flask in solution. The instrument is a mass spectrometer; therefore, we will often refer to the mass-to-charge (*m/z*) ratio of ions, which is the method we use to identify species. $(M-1)^-$ or $(M-H)^-$ refers to a molecule M that has been deprotonated; for example, H_2O has an $(M-1)^-$ ion of *m/z* 17 (HO^-).

Herein, we describe two main types of experimental techniques that we will utilize once the ion is trapped. First is simple bimolecular reactivity. An ion in the FTMS is "trapped" between plates of a "cell." One can introduce neutrals into that cell with which the ion can react in a bimolecular fashion.[17,18] The reactant and product ions can be detected with the mass spectrometer, allowing one to obtain qualitative information (i.e., what products are formed) as well as quantitative information (kinetics and product distributions). We have a "dual cell" setup, which **comprises** two interconnecting reaction regions. Ions can be transferred from one cell to another, but not neutrals. Therefore, if one produces an ion in one cell, and wishes to isolate that ion from any neutrals present, one can transfer that ion to the second cell.[17,18]

The second type of experiment is "acidity bracketing." This technique is used to measure the gas-phase acidity of a species of interest. In this technique, species of known acidities are allowed to react with the anion of unknown acidity. The ability of the anion of unknown acidity to deprotonate relatively stronger acids, and the inability of the anion to deprotonate weaker acids (stronger bases), allow one to bracket the acidity of the unknown acid.[19] We have also developed a new method for measuring multiple acidic sites.[3] To our knowledge, this is the first example of such measurement being carried out in an FTMS.[20] For example, uracil has two acidic sites, at the N1 and N3 positions. The N1 site is more acidic, so the $N1^-$ ion **7** is less reactive than the $N3^-$ ion **8**. Under normal bracketing conditions, deprotonation by hydroxide produces both the $N3^-$ and the $N1^-$ ions, but the $N3^-$ ion reacts with any neutral uracil present to deprotonate the N1-H and form the $N1^-$ ion. We call this neutral-catalyzed isomerization. Such isomerization has also been observed between the monoenolate of acetic acid and the acetate ion in a flowing afterglow apparatus.[20]

To bracket the N3 site, we must perform the deprotonation of uracil under conditions that will allow the $N3^-$ to be sustained; that is, we remove the $N1^-/N3^-$ mixture from the neutral uracil environment as quickly as possible. First, we allow hydroxide to deprotonate uracil, presumably at N1 *and* N3; then we immediately transfer the ions to our second cell, which is free of uracil neutral. We then allow the M-1 of uracil (m/z 111) to react with reference acids. In summary, by forming a mixture of ions deprotonated at the less and more acidic sites, and isolating the ions from the neutral precursor by transfer into the second cell, we avoid the neutral-catalyzed isomerization that results in loss of the ions deprotonated at the less acidic site. This technique is utilized extensively in this project....

Measurement of multiple acidic sites in thymine. Because we have already embarked on the study of uracil (see Preliminary Results), we will begin with thymine (**4a**, Figure 1), which is the 5-methyl-derivative of uracil. Our calculations at B3LYP/6–31+G* predict that the gas-phase acidities of thymine N1 and thymine N3 should be 331 and 343 kcal/mol, respectively. These values are in agreement with earlier calculations by Zeegers-Huyskens et al.[36]

To bracket the N1 site of thymine, we will use dichloroacetic acid (H_{acid} = 328.4 kcal/mol), difluoroacetic acid (331.0), hydrochloric acid (333.3), pyruvic acid (333.5), 2-bromo-butyric acid (336.8), 2-chloropropionic acid (337.0), and trifluoro-*m*-cresol (339.3). These acids have acidities in the range that should allow us to bracket the acidity of the N1H of thymine....

All these reference acids are commercially available liquids or gases; although some of the liquids are not very volatile (for example,...

Hyphenated Two-Word Modifiers

Two words that together describe a noun are usually hyphenated, unless the first word ends with –ly. (See appendix A.)

Comprise vs. Compose

See appendix A for more information on these commonly confused words.

Similarly, Aga (excerpt 13A) describes the types of columns that she will use to achieve enantiomeric separation (an essential feature of her proposed work), but she does not devote space to a description of GC/MS parameters (e.g., temperature program, carrier gas, flow rates). She describes the general approach that she will use to analyze soil samples in the soil degradation study but provides few details on how soil moisture will be controlled or how the soil samples will be extracted and analyzed (details we would expect to see in a journal article describing this work).

In excerpt 13S, Hergenrother, a combinatorial chemist, describes how he will synthesize a chemical library of 6,000 compounds. The synthesis is illustrated in a scheme (*not included here*) and described briefly in the text. To conserve space, he refers to reactions by name when possible (e.g., a Grignard alkylation and Horner–Emmons olefination) rather than describing the steps of the reaction sequence. He refers to general steps that will be taken (e.g., oxidation, deprotection, protection, and acylation) but does not include details such as stoichiometry, reaction times, and reaction temperatures. He concludes by suggesting that this method could be expanded 10-fold to create a library of 60,000 members.

Excerpt 13S (Hergenrother, 2002)

The library will be synthesized as follows (Scheme 5): To the silyltriflated 500 μm polystyrene beads will be attached 10 different heterocyclic alkenes to provide **1**; the synthesis of these alkenes will be described in the next section. After conversion of the olefin to the aldehyde, the electrophilic functionality will be appended. Thus, the beads will be split and the aldehyde will be subjected to three different types of reaction with 30 different building blocks: a Grignard alkylation followed by oxidation, a Horner–Emmons olefination, and a zinc triflate-mediated alkyne addition followed by oxidation.[22] This process will provide a collection of 300 compounds containing a ketone or a α,β-unsaturated electrophile (Scheme 5, **2**). The secondary alcohol will then be deprotected and subjected to 20 acylating agents, producing 6,000 compounds. Finally, deprotection of the carboxylic acid and simultaneous cleavage from the resin will provide compounds with the aspartic acid sidechain (Scheme 5, **3**). The carboxylic acid will be protected as the tris(2,6-diphenylbenzyl)silyl ester, a protecting group that is stable to Grignard reagents and other harsh conditions.[23] Of course, if this protecting group were to be a problem, the compounds would simply be protected as the alcohols and then oxidized at a later stage. . . . Once the synthesis of this 6,000 member library is successful, it could easily be expanded to 60,000 members.

Exercise 13.20

Consider the level of detail in excerpts 13T and 13U. What do the authors tell you about their general approach? Do they include enough detail for you to repeat their experiments? Propose an explanation for the amount of detail provided.

Excerpt 13T (Patrick, 2000)

General Methodology. Films will be prepared by deeply immersing substrates in a liquid crystal (LC) solution (several mm) to ensure that no memory of anchoring at the air-LC interface remains at the depth of the substrate.[36] In general, solutions will be prepared to give the smallest concentration possible such that the solute undergoes preferential adsorption (see below). In some experiments, films will be prepared in a magnetic field, using a variable strength (up to 1.3 T) electromagnet. Solution properties will be characterized in each case by separately measuring the bulk-phase anisotropic and mean magnetic susceptibility of the mixture. These measurements will be performed using the Faraday–Curie method.[37] **Because** solutes can alter the phase transition temperature of a LC, we will independently measure the phase transition temperature for each solution by optical analysis using a polarizing microscope equipped with a temperature-controlled sample stage.

Since vs. Because

Use *since* to connote the passage of time; otherwise, use *because*. (See appendix A.)

Excerpt 13U (Spain, 1997)

Film topography as a function of the metal atom kinetic energy will be measured. A beam of hyperthermal Au atoms will be directed onto the organic layer. If a polycrystalline film is obtained, as with the thermal Au atoms, we will use atomic force microscopy (AFM) to measure film morphology. With AFM, we will determine if the film is smooth or rough (with steps) and if it is amorphous or polycrystalline. If a polycrystalline film is produced with both low and high kinetic energy particles, we will measure the average size of the microcrystallites to see if larger grains are observed with the higher energy species.... In this way, our studies will explore whether or not increased Au atom energy can **affect** the grain size on amine-terminated surfaces.

Exercise 13.21

Reexamine excerpts 13R–13U and answer the following questions:

a. To what extent do the authors cite relevant works when describing their procedures and instrumentation?
b. For what purposes do the authors cite the primary literature (e.g., support claims)? Provide at least one example for each purpose.
c. Which citation format seems to be most common: superscript numbers, italic numbers in parentheses, or the author-date format?
d. Now glance back at excerpt 13Q. What citation format is used? Suggest a reason for the differences in citation formatting.

The third submove of move 3 (anticipate obstacles) stems from the fact that research proposals, to varying degrees, chart new territory. As with anything new, there will be obstacles that must be overcome. You have learned this lesson already if you have joined a research group. Unlike cookbook lab experiments, where most of the "bugs" have been worked out, research is full of bugs. Moreover, even well-established methods have limitations; indeed, one of the first lessons that you learn in a research group is the limitations of your technique. Such limitations can be straightforward, such as an instrument's detection limits, or more subtle, such as a side reaction that competes unknowingly with the desired reaction mechanism. It is best to address these limitations directly in your proposal. By doing so, it demonstrates that you have thought carefully about your project and that you have sufficient experience to recognize what might go wrong.

Let's examine how authors do this in their writing. Lyon (excerpt 13V) identifies *potential pitfalls* in his work (demarked by a level 3 heading). His anticipated obstacle is that the thermoresponsive polymers may lack sufficient bioresponsivity. He points out the problem, and also includes a **contingency plan**, or possible solution to the problem. In his case, he will draw on a large array of more complex responsive polymers, if the simpler ones do not work. Similarly, Lee (excerpt 13W) reassures her readers that she will address potential complications caused by "hot" ions in her instrument by using argon to cool the ions and will improve her ability to measure acidity of the less acidic site by using deuterated reference acids.

Contingency Plan

Plan to address unplanned events or obstacles. Effective research proposals anticipate obstacles and include contingency plans to solve them.

Excerpt 13V (Lyon, 2000)

Potential Pitfalls. The major pitfall in this work concerns the ability to induce bioresponsivity in thermoresponsive polymers. It is conceivable that while some thermodynamic perturbations may occur, they may not be large enough for the material to be of high utility or applicability. Fortunately, a tremendous array of more complex responsive polymers has been described, which implement backbones other than those found in the traditional NIPA gel (dextrans, carbohydrates, polyglycols, polystyrenes, polyamines, etc.)[24,29,32,45,100–102] These materials are often multiresponsive (temperature, pH, and/or ionic strength) and display much more complicated phase transition behavior. Although it would be unwise to use these materials in the early stages of the project due to their complexity, we will quickly move to explore these materials as colloidal bioconjugates if our initial choice of hydrogel proves to be unsuccessful. It is expected that we will learn enough of how proteins modulate gel solvation thermodynamics from the early work such that our choice of the second-generation hydrogel will be an educated, rational, and successful one.

Excerpt 13W (Lee, 2001)

A few caveats should be mentioned. First, one must worry that one's ions are not "hot"; that is, we are measuring thermochemical quantities and the ions should be at thermal temperature for an accurate determination of acidity. We use argon to cool our ions.

Second, a drawback with multiple-acidic site bracketing, as we learned with uracil, is that the more basic $N3^-$ ion can deprotonate a weak acid, which produces a strong base, which can then deprotonate the N1H of thymine. The end result is that one might never observe the $(M-1)^-$ of the weak acid indicating proton transfer between the $N3^-$ and the reference acid; instead, one sees the $N1^-$ of thymine, ***m/z*** 125. Therefore, while the most acidic site will have a very accurate measurement, the less acidic sites are, to be exact, a lower limit on acidity. Our proposed plan to circumvent this problem is to use deuterated acids. The $N3^-$ ion will deprotonate the deuterated acid; if the conjugate base deprotonates the N1H of thymine, then we will see the deuterated thymine anionic product as a mass-to-charge ratio one mass unit higher than deprotonated thymine! In this way, we hope to gain accurate values, even for the less acidic site.

m/z

This abbreviation for mass to charge ratio is italicized (not bolded).

It is also common to anticipate obstacles and propose contingency plans in synthetic work. In excerpt 13X, Johnson explains how he will deal with problems that might arise from a protic byproduct. Similarly, Vyvyan (excerpt 13Y) points out an alternative sequence if the initial reaction proves unfruitful. Contingency plans such as these suggest that the researcher is both farsighted and resourceful, qualities that make them a worthwhile investment for a funding agency.

Excerpt 13X (Johnson, 2003)

The effect of this protic by-product on the proposed reaction will need to be determined. If necessary, removal of the solvent and alcohol in vacuo and re-dissolution of the complexes (by means of a multichannel pipetter) should be feasible. Our glove box is equipped with a vacuum inlet, so this procedure could be executed under rigorously anhydrous conditions.

Excerpt 13Y (Vyvyan, 2001)

Should the proposed acid catalyzed 8-*endo* epoxide opening prove unfruitful, intermediate **49** could still be converted to heliannuol A through an alternative sequence (Scheme 7). Specifically, conversion of **49** to the corresponding aryl triflate, deprotection of the diol, and reprotection of the secondary hydroxyl set up formation of the aryl ether using the aryl etherification protocols developed by Buchwald[45] and Hartwig.[46] Heliannuol A **1** can be oxidized to produce heliannuol K **11** using any one of a number of possible reagents.

Exercise 13.22

A fluent writer avoids using the same word over and over again. Suggest at least three alternative choices for the italicized words in the following phrases. Look through excerpts 13V–13Y for some ideas.

a. If problems *arise*, we will...
b. A possible *problem* is...
c. If plan A proves *unsuccessful*, then we will...
d. *An alternative approach* is to...
e. To test the *feasibility*, we will...

Exercise 13.23

Ideally, submove 3.3 both identifies a potential obstacle and offers a contingency plan. Consider the following passages, all taken from submove 3.3 of Experimental Approach sections. In each passage, identify the problem and, if present, the possible solution.

a. If problems develop with the interpretation of the magnetically aligned nitroxide spin labeled-peptide bicelle data, we will incorporate the TOAC spin label onto different residues of the peptides and repeat the experiments. (From Lorigan, 2002)
b. Of course, if this protecting group were to be a problem, the compounds would simply be protected as the alcohols and then oxidized at a later stage. (From Hergenrother, 2002)
c. One drawback of immunoassays is the random occurrence of "false positives" (detection of the analyte when it is not really in the sample) and "false negatives" (analyte is not detected when it is actually present). False positive results are normally observed when there are interfering compounds in the matrix, which may be removed by a simple cleaning procedure. False negative results are observed when the concentration of the analyte in the sample is below the detection limit of the immunoassay. Sample preparation procedures such as SPE should prevent false negatives. False positive results may be eliminated by verifying dubious results with a more sophisticated instrumental method such as CZE/MS, which will be used only as a confirmatory test because it is more expensive and a more tedious method. (From Aga, 2002)
d. The self-assembly of the dicopper complex may require careful adjustment of conditions, such as solvent, order of addition, protonation state of the cysteine thiol, and/or oxidation state of the copper starting material(s). The formation of disulfide bonds and concomitant reduction of Cu(II) to Cu(I) through the well-known redox process may require retaining the protecting group on the cysteine thiol until after metal complex formation. (From Houser, 2001)

Analyzing Writing

Writing a research proposal is a good way for you to learn how chemists work—how they define meaningful problems, design experimental approaches to investigate these problems, and anticipate how their research will contribute reliable scientific knowledge to the community of chemists. Writing research proposals, in other words, asks you to do chemistry.

—Beal and Trimbur (2001)

We finish our discussion of move 3 by calling your attention to the pointers placed near selected excerpts above. These pointers and excerpts together highlight correct usages of the following words, punctuation, or other writing features:

- irradiation vs. radiation (passage P1)
- respectively (excerpt 13N)
- stereochemistry terms (*S*) and (*R*) (excerpt 13N)
- affect vs. effect (excerpt 13O)
- which vs. that (excerpt 13P)
- hyphenated two-word modifiers (excerpt 13R)
- comprise vs. compose (excerpt 13R)
- since vs. because (excerpt 13T)

To test your knowledge of these words and conventions, complete exercises 13.24–13.26.

Exercise 13.24

We have removed three hyphens from the following sentence. Where should the hyphens be placed and why?

> We will also begin to explore other biological interactions such as ligand receptor binding, DNA hybridization, enzyme substrate interactions, or protein DNA complexation to further evaluate the generality of our approach. (From Lyon, 2000).

Exercise 13.25

Make the following sentence more concise by using the word "respectively" correctly. Check units and punctuation as well.

> Our calculations predict that the gas-phase acidity of thymine N1 should be 331 kcal/mol and thymine N3 should be 343 kcal/mol. (Adapted from Lee, 2001)

Exercise 13.26

Consider the italicized words or letters in the sentences below. If incorrect, correct the error. If no changes are needed, indicate "correct as is".

a. *Because* SPE is also a pre-concentration technique, the method detection limit for the quantification of individual isomers could be improved by several orders of magnitude. (Adapted from Aga, 2002)

b. The proposed work is *comprised* of three separate studies.

c. We have already demonstrated our ability to produce thermoresponsive copolymer hydrogel particles *which* possess groups for protein attachment. (Adapted from Lyon, 2000)

d. We anticipate that only the (*S*)-amino acid will be formed.

e. Our studies will explore whether or not increased Au atom energy can *effect* the grain size on amine-terminated surfaces. (Adapted from Spain, 1997)

We conclude this section by analyzing common verb tense–voice combinations in move 3. Because move 3 describes work that will be done in the future, future tense is most common. Future tense is used with both active and passive voice:

Future–active	We will begin with a study of...
	Our approach will examine...
	We will measure the average size of...
Future–passive	Films will be prepared by...
	Several parameters will be explored.
	These syntheses will be continued until...

Exercise 13.27

Consider the following sentences, adapted from CAREER proposals. Rewrite each sentence in future tense and active voice.

a. "Differential imaging" is used as our next step to study cluster chemistry. (From Sanov, 2002)

b. XPS is used to monitor changes in surface composition. (From Fairbrother, 2000)

c. Our coupling strategies utilized standard peptide coupling procedures. (From Houser, 2001)

 Exercise 13.28

Consider the following passages, adapted from CAREER proposals. Each passage has one language or formatting error. Make corrections so that each passage follows standard writing conventions.

a. We followed up our initial report with a second *J. Am. Chem. Soc.* paper, which spectroscopically characterizes magnetically-oriented phospholipid bilayers with EPR spectroscopy and discusses conditions for optimal bicelle alignment at X-band (86). (Adapted from Lorigan, 2002)

b. A solid phase extraction (SPE) method that will isolate and fractionate metolachlor and its oxanilic acid (OXA) and ethanesulfonic acid (ESA) metabolites from soil and water will be developed. (Adapted from Aga, 2002)

c. Figure 4. **Layer-by-layer polyelectrolyte adsorption.** (Adapted from Lyon, 2000)

d. For example, hydrogen's three isotopes, protium (H), deuterium (D), and tritium (T), have substantial relative mass differences (1 amu, 2 amu, and 3 amu, respectively). (Adapted from Kohen, 2002)

e. For example, H_2O has an $(M-1)^-$ ion of m/z 17 (HO^-). (Adapted from Lee, 2001)

f. These findings were then successfully applied to a new interpretation of photoinduced electron transfer at similar interfaces, that are of importance in the field of solar energy conversion.[108] (Adapted from Lyon, 2000)

13C Writing on Your Own: Describe Your Proposed Methods

Write move 3 of your proposal. Be sure to (1) remind reader of preliminary results, (2) describe procedures and instrumentation in a logical manner (e.g., sequentially, by underlying themes, by desired outcomes, or some combination), and (3) anticipate obstacles and suggest alternatives with a contingency plan (that identifies both problems and possible solutions). Use appropriate headings and subheadings in your writing.

13D Writing on Your Own: Complete the Experimental Approach Section

You are ready to complete the first full draft of your Experimental Approach section (moves 1, 2, and 3). Add headings, subheadings, and ordinal language, as needed, to organize your ideas and to help your reader recognize the breadth and depth of your proposed work. You may need to make small changes when combining the three parts to be sure they fit seamlessly.

Review your written work (see chapter 18) before you submit it for peer review or to your instructor.

Chapter Review

To judge what you've learned from this chapter, define each of the following terms and explain its importance, in the context of this chapter, to a friend or chemistry colleague who is new to the field:

computer-generated graphics	literature values	prior accomplishments
conceptual framework	obstacles	promising results
contingency plan	preliminary results	text boxes

Also explain the following to a chemistry colleague who has not yet given much thought to writing the Experimental Approach section of a Project Description:

- Main purpose of the Experimental Approach section
- Differences between moves 1 and 2: share prior accomplishments and share preliminary results
- Connections between collaborative relationships and demonstration of expertise
- Ways to share your proposed methodology (move 3)
- Level of detail to include in your proposed methodology (move 3)
- Reasons for anticipating obstacles in your proposed research

Additional Exercises

Exercise 13.29

Reflect on what you have learned from this chapter. Select one of the reflection tasks below and write a thoughtful and thorough response:

a. Reflect on two underlying themes of the Experimental Approach section: to demonstrate (1) the expertise of the researcher and (2) the plausibility and promise of the proposed research. At this point in your professional development,
   - What is the best way to demonstrate your expertise?
   - How will you demonstrate the plausibility and promise of your proposed research?
   - What steps will you take to build reviewers' confidence in you and your research plan?

b. Research proposals require that writers (1) reveal potential obstacles in the design and methodology of the proposed work and (2) suggest contingency plans.

- How is it possible that a proposal can be strengthened by the inclusion of possible obstacles and/or weaknesses?
- How can drawbacks be introduced to establish a writer's expertise and abilities?
- How important do you think contingency plans are for reviewers of research proposals?

c. Reflect on the similarities and differences between the Methods section of a journal article and research proposal.

- What are the similarities?
- What are the differences?
- What will you do to make sure the Experimental Approach section of your proposal meets reviewers' expectations?

14 *Writing the Outcomes and Impacts Section*

A well-developed proposal shows that the investigator has grasped a problem well enough to justify second-party sponsorship.
—Paradis and Zimmerman (1997)

All good proposals must come to an end. In this chapter, we examine conventional ways in which authors summarize and conclude their Project Descriptions. We consider project timelines, lists of expected outcomes, and statements of broader impacts. By the end of this chapter, you should be able to

- Develop a project timeline
- Generate a list of expected outcomes
- Suggest broader impacts of your proposed work
- Reinforce the importance of your proposed work in concluding remarks

Staying on Track

This chapter covers what is normally presented in the last major section of the Project Description. Common level 1 headings in this section are

PROJECT TIMELINE

EXPECTED OUTCOMES

CONCLUSIONS

As you work through the chapter, you will write the closing section of your own Project Description. The Writing on Your Own tasks throughout the chapter guide you step by step as you do the following:

14A Create a project timeline

14B Create a list of expected outcomes

14C Conclude the proposed work

14D Complete the Outcomes and Impacts section

Like the previous sections of the Project Description (chapters 12 and 13), there is no one right way to end a proposal. However, proposal guidelines often instruct authors to include a projected timeline, a list of expected outcomes, a summary of objectives, and/or a statement of relevance or broader impacts in their concluding remarks. For example, the ACS Division of Analytical Chemistry Graduate Fellowship announcement (excerpt 11A) asks for a statement that links "the relevance of [the proposed] work to analytical chemistry." The NSF Grant Proposal Guide (see excerpt 15B) asks for "objectives for the period of the proposed work," their "expected significance," and their "relationship to longer-term goals of the PI's project." Moreover, the PI must describe "as an integral part of the narrative, the broader impacts of the proposed activities." Not surprisingly, each of the authors of our 22 CAREER proposals approached this task slightly differently. We examine several of their approaches in this chapter.

Reading and Analyzing Writing

We begin with an excerpt that you can read and analyze on your own (excerpt 14A), specifically, the conclusion to Harpp's proposal regarding plume-ridge interaction in the Galápagos. She includes a formal timeline (titled "Project Schedule") and conclusions for her work.

Exercise 14.1

Read excerpt 14A and answer the following questions:

a. How does the author preface the year-by-year timeline? What purpose does the preface serve?

b. How detailed is the timeline? What kinds of details are included and excluded?

c. Note that the timeline is written in a "to do" list fashion. Check to be sure that the author uses parallel language in her list. Find five phrases that are parallel to the opening phrase in the Year One description (i.e., Refinement of...).

d. Why do you think it is acceptable to write the timeline in a "to do" list fashion but the Conclusions section in complete sentences?

e. The author stresses importance again in her Conclusions section. Give two examples; what purpose is served by each example?

f. Suggest synonyms for the words "importance" and "important."

Excerpt 14A (Harpp, 1998)

PROJECT SCHEDULE

The schedule for this project is designed to permit participation by undergraduates such that students will have their own project, sufficient time to complete the analytical work, and the opportunity to present results at a national meeting. The fieldwork is divided into **two trips** to fit the academic calendar (travel during winter break), and to increase the number of opportunities for students to participate in a unique field experience.

Year One: Refinement of sample preparation and analytical techniques for trace element analysis in ocean island basalts using existing samples from the Galápagos. Compilation of existing geochemical data for Pinta and Marchena Islands. Training of students in analytical techniques, preparation for fieldwork.

Year Two: Field excursion to Wolf and Darwin Islands (**24 days** total; **6 days** on each island). Petrographic examination of samples at Lawrence University, initial analysis of trace elements by ICP-MS at Lawrence University (up to **100 samples**).

Year Three: Complete analysis of trace elements by ICP-MS at Lawrence University; analysis of major elements by XRF at Macalester College (up to 100 samples); determination of Sr, Nd, and Pb isotopic ratios of a selection of Wolf and Darwin samples by TIMS at Cornell (up to 30 samples). Interpretation of geochemical data, modeling of melting parameters. Presentation of results at Fall AGU meeting by undergraduate student(s). Preparation for fieldwork.

Year Four: Field excursion to Genovesa Island (24 days total; 12 on the island). Petrographic examination of samples at Lawrence University; initial analysis of trace elements by ICP-MS at Lawrence University (up to 100 samples).

Year Five: Complete analysis of trace elements by ICP-MS at Lawrence University; analysis of major elements by XRF at Macalester College (up to 100 samples); determination of Sr, Nd, and Pb isotopic ratios of a selection of Wolf and Darwin samples by TIMS at Cornell (up to 30 samples). Interpretation of geochemical data, modeling of melting parameters. Presentation of results at Fall AGU meeting by undergraduate student(s). Preparation of final plume-ridge interaction synthesis paper for publication with student authors.

CONCLUSIONS

Detailed geological and geochemical examination of Wolf, Darwin, and Genovesa Islands will provide important information regarding the volcanic evolution of these features, and their relationship to the main Galápagos archipelago via the Morgan pipeline hypothesis, the leaky fault model, or the mini-hotspot/near-ridge volcanism model. The collected data and mapping information will yield important information regarding shallow mantle flow, hotspot dynamics, and plume-ridge interactions. Finally, detailed information regarding Wolf, Darwin, and Genovesa Islands will serve as an important database for current biodiversity and evolutionary studies in the Galápagos.

Six or 6?

In most cases, use the number form rather than the word form for numbers with units of time and measure (e.g., 6 days, not six days). Use the word form with other units (e.g., two trips) unless the number is 10 or greater (e.g., 12 samples). (See Numbers and Units in appendix A.)

Analyzing Organization

The move structure for the Outcomes and Impacts section of the Project Description is shown in figure 14.1. The first two moves summarize the proposed work by highlighting expected achievements. Typically, no new information is provided in these moves; rather, their purpose is to summarize **deliverables** or tangible accomplishments that will result from the work. Examples include the delivery of a more efficient synthesis, an improved analytical procedure, or a novel application of an instrument. Deliverables also include the dissemination of findings through conference presentations, publications, and patents, allowing the larger scientific community to learn about your work.

Deliverables

The more tangible products of a research project, such as a novel synthesis.

Deliverables also include the dissemination of findings through publications, patents, and conference presentations.

1. Present Project Timeline

2. List Expected Outcomes

3. Conclude the Proposed Work

3.1 Reiterate goals and importance

3.2 Address broader impacts

Figure 14.1 A visual representation of the suggested move structure for an Outcomes and Impacts section of the Project Description.

Move 3 concludes the proposal and focuses on the project's broader impacts, allowing investigators one more chance to stress the importance of the work. The Outcomes and Impacts section, like the Conclusions section in a journal article, broadens its scope to address a scientific, rather than an expert, audience.

With the move structure for the Outcomes and Impacts section in mind, we are ready to analyze the three moves in more detail and consider headings for each move.

Move 1: Present a Project Timeline

A convenient way to summarize the more tangible accomplishments of your work is to present them in a timeline. Recall the specific-achievable-measurable (SAM) test for research objectives introduced in chapter 12; a timeline helps reviewers (and you) evaluate whether your project goals are achievable within the funding period. The start of this move is usually demarked with a level 1 heading. Common headings for move 1 are shown in table 14.1.

Analyzing Excerpts

A realistic time schedule becomes an effective argument. It suggests to readers that you understand how much time your project will take and that you are not promising miracles just to win approval.

—Adapted from Houp et al. (2006)

Let's consider a few examples to see how authors address timelines in their proposals. Harpp (excerpt 14A) presents a year-by-year synopsis of her project, listing tangible accomplishments in each time interval (e.g., compilation of existing data, analysis of up to 100 samples, presentation of results at a conference, and preparation of the final plume-ridge paper). Note the language in the timeline; it reads much like a "to do" list, using phrases rather than full sentences

Table 14.1 Common level 1 headings for move 1 of the Outcomes and Impacts section.

PROJECT TIMELINE	PLAN OF WORK
PROJECT SCHEDULE	PROPOSED WORK PLAN

(e.g., "Training of students in analytical techniques, preparation for fieldwork"). Many of the tasks begin with **nominalizations**:

refinement of ____	examination of ____	preparation of ____
analysis of ____	presentation of ____	interpretation of ____

Nominalizations

A noun formed from another part of speech. For example, the noun *interpretation* comes from the verb *interpret*. (See appendix A and table 2.2.)

Finney (excerpt 14B) and Sanov (excerpt 14C) also include timelines, but they integrate them into the running prose of their text. As you read their timelines, note their use of hedging language. (Indeed, it is difficult not to hedge when describing what you will be doing over the next 5 years!) Finney, for example, *anticipates* that the synthesis of the fluorophore will take 6 months (in light of his many other responsibilities), while the library synthesis and fluorescence characterization *are anticipated* to take 18 months. Sanov begins by acknowledging how *ambitious* his project is and points out that his progress will be *guided by* currently unknown intermediate results. Nevertheless, he does his best to provide *strategic landmarks*, and he describes his *hope* to *start tackling* bimolecular reactions in the second half of the award period.

Hedging

Hedging is used to soften a writer's stance. (See appendix A.)

Excerpt 14B (Finney, 1999)

ANTICIPATED TIMELINE FOR NEAR-TERM RESEARCH OBJECTIVES

The immediate goal of fully characterizing the parent Hg^{2+}-responsive fluorescent chemosensor, which will entail ^{1}H NMR and fluorescence titration experiments, isolation and characterization of the 1:1 Hg^{2+}:ligand complex, and fluorescence lifetime measurements of the simple diol/carbonate pair, is expected to be complete within 6 months. (This work will be the primary responsibility of the graduate student currently on the project, Sherri McFarland.)

The synthesis of biaryl fluorophores exhibiting visible fluorescence will be the responsibility of the PI (Nat Finney) **and** will begin within a month. The synthesis and complete characterization of the proposed fluorophores are anticipated to take 6 months, in light of teaching, advising, and grant-writing responsibilities.

The synthesis of the combinatorial library of fluorescent chemosensors will be assigned to an incoming graduate student. Initial work will focus on validation of the synthetic chemistry, which is anticipated to take 6 months. Upon completion of her studies on the parent system, Sherri McFarland will join the new graduate student on this aspect of the project. Library synthesis and fluorescence characterization are anticipated to be complete within 18 months.

"and" (with no comma)

No comma is needed before "and" when the sentence has a single subject and two verbs (e.g., I read and revise). (See Commas in appendix A.)

Excerpt 14C (Sanov, 2002)

TIMELINE

It is recognized that the range of the experiments described in this proposal is very ambitious. But this is a long-term program at the frontier of **science, and** I must be guided by intermediate results to decide which directions merit the greatest attention. Nonetheless, I envision the following strategic landmarks. Within several months, we will complete the diagnostic experiments on I^- and other atomic anions. By the end of 2001, we will have carried out the studies of O_2^- and S_2^- and should soon have the results on the OCS^- hydration. In 2002, we will begin the time-resolved experiments on $(OCS)_2^-$ and continue with reaction dynamics and alignment studies in other systems. Bimolecular reactions in clusters are our most ambitious goal, which I hope to start tackling in the second half of the award period.

"..., and" (with a comma)

A comma is needed before "and" when it joins two complete sentences (e.g., I read, and I revise). (See Commas in appendix A.)

Exercise 14.2

Reread excerpts 14B and 14C. Then answer the following questions:

a. How detailed are the two excerpts in specifying time periods and/or the passage of time? Give some examples.
b. What features distinguish these timelines from one another? In other words, what are the defining characteristics of each of these timelines?
c. Do the two researchers take advantage of the opportunity to stress the importance of their work one last time? If so, what language do they use to emphasize the value of their proposed work?

Timelines can also be portrayed in charts or figures, as illustrated in excerpt 14D. In fact, charts and figures represent excellent ways to illustrate how smaller, individual projects contribute to larger research goals and how smaller projects complement one another and overlap in time. (Note: In excerpt 14D, Kohen uses the following abbreviations in his chart, each defined previously in the proposal: hydrogen (H), tritium (T), deuterium (D), kinetic isotope effect (KIE), dihydrofolate reductase (a relatively small protein) (DHFR), and wild type (WT).)

Excerpt 14D (Kohen, 2002)

Timetable for work plan. The work plan spans **five years** and involves **eight** related **projects** (Figure 3). Enzyme preparation, syntheses of H/T- and D/T-labeled NADPH, and preparation of high resolution crystals have been started, or will be started, in the early stages of the project. Studies related to KIEs, the temperature dependence of KIEs, and DHFR dynamics will take place in years 1–5; modeling and interpretation of the results will take place in years 2–5.

Five or 5?

In a single sentence, refer to numbers (as either numerals or words) in the same way. In excerpt 14D, "five years" was used instead of "5 years" to be consistent with "eight projects." (See Numbers and Units in appendix A.)

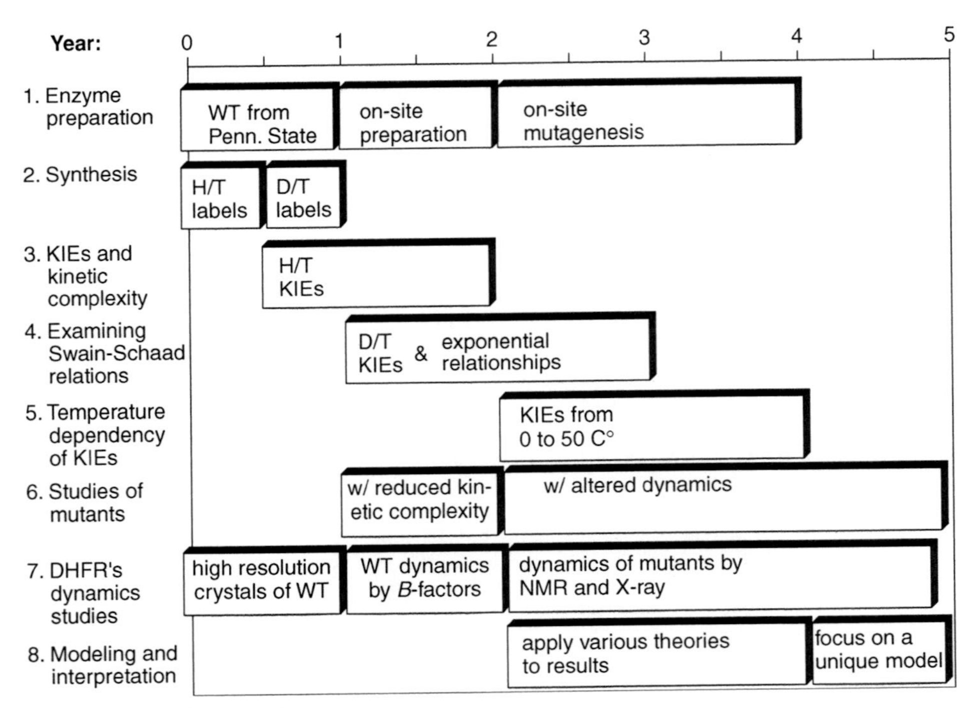

Figure 3. Timetable for work plan.

 Exercise 14.3

Excerpts 14A–14D illustrate three different ways to present a timeline: in a year-by-year list, formal prose, or a figure. All three approaches are appropriate. Which approach do you prefer and why?

Analyzing Writing

We used pointers near excerpts 14A–14D to call your attention to correct usages of the following writing features:

- numbers and units (excerpts 14A and 14D)
- nominalizations (excerpt 14A).
- hedging (excerpts 14B and 14C)
- commas (excerpts 14B and 14C)

To test your knowledge of some of these features, complete exercises 14.4 and 14.5. Refer back to the excerpts and pointers as needed.

 Exercise 14.4

The following sentences were adapted from project timelines. Correct the errors in each sentence.

a. In the first year, we will discover at least five new compounds.
b. The initial project is expected to take 2 years and will involve the synthesis of six new compounds.
c. In Year 2, we will collect fifty samples at the Grand Canyon site; in Year 3, we will collect fifty samples at the Yosemite site.
d. We will begin with time-resolved experiments, and continue with reaction dynamics and alignment studies.

 Exercise 14.5

Nominalize each of the following verbs; specify the noun that can be formed from the verb:

a. to prepare
b. to synthesize
c. to analyze
d. to present
e. to interpret

Table 14.2 Common functions of different verb tense–voice combinations in move 1 of the Outcomes and Impacts section of the Project Description.

Function	Tense–Voice Combination	Example
Present timeline (in narrative form)	Present–active	The work *spans* five years and involves eight related projects. (From Kohen, 2002)
	Present–passive	A comprehensive study of... ***has been presented***.
	Future–active	Studies related to... ***will be initiated*** in Year 3.
	Future perfect–active	By the end of 2009, we ***will have completed*** the studies on...

We conclude our discussion of move 1 by considering verb tense and voice combinations. Present and future tenses, with either active or passive voice, are commonly used in narrative timelines (e.g., excerpt 14B and 14C), as summarized in table 14.2. The timeline itself is often a list of fragments that need not contain verbs, as illustrated in the following timeline adapted from Harpp (1998):

Year One	Refinement of sample preparation and analytical techniques Training of students in analytical techniques
Year Two	Field excursion to Wolf and Darwin Islands

14A Writing on Your Own: Create a Project Timeline

Draft a timeline for your proposed research. Decide if you want to present your timeline in running prose, as a list, or as a graphic (chart or figure). Include a heading for your timeline.

Move 2: List Expected Outcomes

In addition to a timeline, a second way to summarize anticipated accomplishments is to list your expected outcomes. Expected outcomes are measurable achievements that link back to the measurable objectives listed at the start of the proposal. Recall that the Project Description often starts out with a list of specific, achievable, and measurable (SAM) objectives. Timelines attest to the achievability of the objectives; outcomes attest to their measurability.

Move 2 is often demarked with a level 1 heading (see table 14.3). In some headings, the term "outcome" is used explicitly (e.g., **EXPECTED OUTCOMES**).

Table 14.3 Common level 1 headings for move 2 of the Outcomes and Impacts section.

EXPECTED OUTCOMES	PROJECT OUTCOMES
EXPECTED FINDINGS	SUMMARY
PROJECT DELIVERABLES	SUMMARY OF RESEARCH PLAN

In other cases, the heading is more generic (e.g., **SUMMARY**). In the latter case, the outcomes are integrated into an overall summary of the proposed work.

Analyzing Excerpts

We include two excerpts for you to consider. Johnson (excerpt 14E) lists expected outcomes in his proposed Lewis acid–promoted [3+2] cycloaddition reactions. Projected outcomes include new *strategies* (for promoting cycloaddition reactions), a *diverse range* of hetereocycles (to be used as future chiral building blocks), and *new approaches* (for identifying additional catalysts). Patrick (excerpt 14F) lists expected outcomes for his liquid crystal imprinting investigations. His outcomes include *refinement* and *extensions* of current techniques, *improved fundamental understanding* (of liquid crystal imprinting mechanisms), and *preparations* of new types of films and film materials. Both lists are short and succinct. Moreover, both investigators preface their lists with a short paragraph that reiterates the overall goals of the proposed work.

Excerpt 14E (Johnson, 2003)

SUMMARY

Lewis acid-promoted [3+2] cycloadditions of aziridines and epoxides proceeding via carbon-carbon bond cleavage of three-membered ring heterocycles are demonstrated for the first time. This proposal details plans for extending these initial results into a general synthetic method for the enantioselective synthesis of structurally diverse pyrrolidine- and tetrahydrofuran-containing organic compounds. Expected outcomes of the proposed work will include

1. A variety of strategies for effecting dipolar cycloadditions from stable, neutral precursors, including carbon-carbon bond heterolysis induced by Lewis acidic metal complexes and chiral organic catalysts;
2. A diverse range of heterocycles and other chiral building blocks that can be generated by this cycloaddition strategy;
3. Focused approaches for rapidly evaluating metal complexes that are candidates to catalyze the cycloadditions.

Excerpt 14F (Patrick, 2000)

SUMMARY OF THE RESEARCH PLAN

The research plan is organized around a set of well-defined model systems and a limited number of detailed experiments. The experiments are designed to progress over a five-year period from fundamental studies on model systems to development of thin film materials with potential practical applications. Together they will form a coherent bond of film growth using LCI with a variety of molecular building and LC solvents. The expected outcomes of this work will be

1. Refinement and extension of the LCI technique to include additional building blocks and LC solvents.
2. Improved fundamental understanding of LC-surface interactions.
3. Preparation of a new type of nanostructured composite thin film in which one component possesses unidirectional order while the other component is randomly oriented.
4. Preparation of thin film materials for use as re-writable anchoring.

Parallelism in Enumerated Lists

Note the different, and correct, uses of parallel language, numbering, and punctuation in the lists of expected outcomes in excepts 14E and 14F. (See appendix A.)

Exercise 14.6

Examine excerpts 14E and 14F. What do you notice about the following features of the authors' writing? Generalize a rule for each feature that will assist you in your own writing.

a. Lack of punctuation after "will include" and "will be" just before lists begin
b. Punctuation at the end of listed items
c. Different forms of parallelism in lists

Exercise 14.7

In both excerpts 14E and 14F, the authors preface their lists with a short paragraph.

a. What kind of information is included in these paragraphs?
b. Do you notice more similarities or differences between these two paragraphs? Explain.
c. How does the information in these short paragraphs influence the choice of heading in each excerpt (e.g., "Summary" rather than "Expected Outcomes")?

Exercise 14.8

Dyer (excerpt 14G) uses a list of questions rather than a list of outcomes to accomplish move 2. Select four of his questions and rewrite them as expected outcomes. Be sure to use parallel language and formatting in your list.

Excerpt 14G (Dyer, 2001)

SUMMARY

The research activities presented here will represent a significant departure from previous approaches to polar order in organic thin films. This work will address several important questions: First, is it possible to induce thermodynamically stable polar order along the LC director axis? Second, how do surface interactions affect the global orientation of liquid crystals? Third, can hydrogen bonding force the growth and polar orientation of noncovalent polymer networks? Fourth, can we develop and utilize the synthetic methodology for growing densely packed polymer brushes? Fifth, will polymer brushes adopt conformations that yield a permanent polarization density? And sixth, to what extent may we control properties such as second order NLO, pyroelectricity, and piezoelectricity in organic thin films?

Analyzing Writing

We used a pointer to highlight parallelism in enumerated lists in excerpts 14E and 14F. Test your knowledge of parallelism by completing exercise 14.9.

Exercise 14.9

Consider the following first draft of expected outcomes (adapted from Hergenrother, 2002). Convert it to a final list of expected outcomes, as it might appear in a finished proposal. Be sure to use parallel language, punctuation, and numbering in the final list.

Expected outcomes for proposal:

- We will synthesize a 6,000-member focused library designed to inhibit the caspase family of cysteine proteases involved in apoptosis.
- If 6,000-member focused library is successful, extend it to a library of 60,000 members.
- synthesis of a 2,400-member dihydropyridone-based primary library containing four cyclic scaffolds, screened against apoptotic proteins
- We will extend the primary library to >30,000 compounds

We conclude our discussion of move 2 by considering verb tense and voice combinations used in this move. The sentence that introduces expected outcomes is usually written in future tense and active voice with "expected outcomes" as the subject of the sentence:

Future–active The expected outcomes of this work will be...
The expected outcomes of this work will include...

The expected outcomes themselves are often presented in an enumerated list. Like the timeline in move 1, the list often contains fragments with few or no verbs, as illustrated in the following list of expected outcomes adapted from Johnson (2003):

1. A variety of strategies for effecting dipolar cycloadditions
2. A diverse range of heterocycles and other chiral building blocks
3. Focused approaches for rapidly evaluating metal complexes

14B Writing on Your Own: Create a List of Expected Outcomes

Draft a list of expected outcomes for your proposed research. Strive to link your outcomes to the list of project goals and objectives that you created in the first section of the proposal (see Writing on Your Own task 12B). Make sure that you list outcomes in a parallel fashion. Include a heading and a brief introduction.

Move 3: Conclude the Proposed Work

The last move concludes the proposal. This is accomplished in two submoves. This first submove reiterates project goals and stresses (once again) the expected accomplishments and importance of the proposed work. In the second submove, authors take a step back to look at the broader impacts of the project. Who are the beneficiaries of this new knowledge, synthesis, methodology, or instrumentation? How does the work benefit the larger scientific community and general public? Most authors try to address questions such as these in their closing remarks. Move 3 is typically signaled with a level 1 heading. Common headings are shown in table 14.4.

Gender-Neutral Language

The ACS Style Guide urges writers to use gender-neutral language. When, for example, identifying broader impacts, refer to the general public, people, or humankind, rather than mankind.
See *The ACS Style Guide* for other suggestions.

Table 14.4 Common level 1 headings for move 3 of the Outcomes and Impacts section.

CONCLUSIONS	RESEARCH SUMMARY
CONCLUSIONS AND IMPACTS	SUMMARY
IMPACTS AND SIGNIFICANCE	

Analyzing Excerpts

To analyze how authors conclude their proposals, we consider three excerpts. In each case, the author begins with a reminder of project goals and objectives—reemphasizing importance—and concludes with a statement of broader impacts. For example, Harpp (excerpt 14A) begins her Conclusions section by reemphasizing the important information that will be gained by a detailed examination of islands in the Galápagos archipelago. She ends her Conclusions by emphasizing that her work will *serve as an important database* for others studying biodiversity and evolution in this area. Walker (excerpt 14H) uses his Research Summary section to remind us how his proposed work will elucidate dynamic solvation processes. He concludes by pointing out that his work can lead to *improved predictive models* with the potential to *significantly impact* the modeling of interfacial processes ranging from electron transfer reactions to protein folding. Rose-Petruck (excerpt 14I) begins by reiterating the immediate impacts of his research: a clearer understanding of ligand motions in the gas and liquid phases. He finishes by stating that his work may enable a *view* into the very center of chemical processes (ultrafast structural motion and dynamics), may *inspire others* to develop ultrafast X-ray systems, and will have *applications* in chemical and biomedical industries.

Excerpt 14H (Walker, 2001)

RESEARCH SUMMARY

Fruitful interplay between experiment and theory has led to an increasingly detailed understanding of equilibrium and dynamic solvation properties in bulk solution. However, applying these ideas to solvent-solute and surface-solute interactions at interfaces is not straightforward due to the inherent anisotropic, short-range forces found in these environments. Our research will examine how different solvents and substrates conspire to alter solution-phase surface chemistry from the bulk solution limit. In particular, we intend *to determine systematically and quantitatively* the origins of interfacial polarity at solid-liquid interfaces as well as identify how surface-induced polar ordering affects dynamic properties of interfacial environments. . . .

We hope that our results will aid in developing predictive models of interfacial permittivity. . . . One specific objective will be to search for evidence of dry layers at hydrophobic-aqueous boundaries. If such layers form (and if our experiments can successfully characterize their extent), our results will impact significantly the modeling of processes ranging from electron transfer to protein folding.

Split Infinitives

Avoid split infinitives unless needed for clarity.

Ambiguous: to develop completely new compounds

Unambiguous: to completely develop new compounds

(See appendix A.)

Excerpt 14I (Rose-Petruck, 2000)

IMPACT AND BROADER SIGNIFICANCE

The immediate impact of this research will be a clearer understanding of ligand motions during photoelimination reactions. In particular, comparative studies of molecular motions in the gas phase (using ultrafast electron diffraction) and in the liquid phase should become a source of very detailed understanding of the influence of solvation on chemical processes. Such combined studies in collaboration with Peter Weber, Dept. of Chemistry, Brown University are planned.

In general, UXAS opens a new spectroscopic window for the direct ultrafast time-resolved observation of molecular motions during chemical reactions. Its applications will range from physical to organometallic and bioinorganic chemistry. It will enable a "view" into the very center of chemical processes—ultrafast structural motion and dynamics. The demonstration that ultrafast **X-ray investigations** can be carried out with laboratory-based equipment will, hopefully, inspire others to build and apply such ultrafast systems themselves. This research is embedded in related investigations that are centered around the development of ultrashort and ultrabright laboratory based **X-ray sources** and their applications to chemical biomedical problems.

X-ray, UV–vis

Note that the "r" in X-ray is not capitalized. X-ray must be followed by a noun (e.g., X-ray sources).

UV–vis, too, must be followed by a noun (e.g., UV–vis spectroscopy).

Exercise 14.10

Reexamine excerpts 14H and 14I.

a. The author of excerpt 14H uses personal pronouns (*we*, *our*), while the author of excerpt 14I does not. Both writing styles are acceptable. Which style do you prefer and why?

b. What headings do the authors use to conclude their Project Descriptions?

Exercise 14.11

Another example of move 3 is shown in excerpt 14J. Read the excerpt and answer the following questions:

a. What are the primary goals of Tuckerman's research?
b. How does he emphasize the importance of his work?
c. What broader impacts does Tuckerman address? Who may potentially benefit from this project?
d. Explain the purpose of the author's last sentence.
e. Based on the heading used, how do you think Tuckerman formatted other headings in his proposal?

Excerpt 14J (Tuckerman, 1999)

7. SUMMARY

A proposal for the comprehensive study of chemical processes in a variety of important condensed-phase systems using modern theoretical methodology has been presented. The primary goals of the research are to provide microscopic information on the mechanisms and structural and dynamical properties of the chemical systems proposed for investigation, to test the applicability of modern **ab initio** molecular dynamics (MD) by comparison with experiment, and to develop and apply novel ab initio MD techniques in simulating complex chemical systems. The proposed research will contribute to the forefront of modern theoretical chemistry and address a number of important technological issues. The PI has carefully attempted to demonstrate his knowledge, ability, and resources to carry out the proposed research projects.

Ab initio

According to *The ACS Style Guide*, common Latin phrases such as "ab initio" should not be italicized or bolded (or placed inside quotation marks).

As mentioned in chapter 11, the term "broader impacts" also includes human impacts (i.e., how the proposed work will impact students and prepare them to be future scientists). For example, Aga (excerpt 14A) addresses students in her Project Schedule section ("Students will have their own project, sufficient time to complete the analytical work, and the opportunity to present results at a national meeting."). Similarly, Finney (excerpt 14B) mentions one graduate student by name and another "new" graduate student who will join the project. These are just two examples. In general, authors describe human impacts by (1) anticipating how many students will be involved in the project and for how long, (2) describing new lecture or laboratory curricula that will be developed, (3) listing specific skills that students will learn, and (4) identifying opportunities that students will have to share their results with the larger scientific community (e.g., through conference presentations and publications).

Analyzing Writing

We used pointers in excerpts 14H–14J to call your attention to the following writing features in this move (each described in more detail elsewhere in the textbook):

- use of split infinitives (excerpt 14H)
- use of UV–vis and X-ray (excerpt 14I)
- use of ab initio (excerpt 14J)

To test your knowledge of these features, complete exercises 14.12 and 14.13. Refer to the excerpts and pointers as needed.

Exercise 14.12

Rewrite the following sentences so they adhere to conventional writing practices in chemistry:

a. Our work is centered around the development of ultrashort and ultrabright X-Ray sources.

b. We will use *ab initio* molecular dynamics to simulate complex chemical systems.

c. FTIR and UV–VIS will be used to probe the mechanisms of the reaction.

Exercise 14.13

Identify the split infinitives in the following sentences. Rewrite each sentence without a split infinitive, unless the change will create unwanted ambiguity.

a. One purpose of move 3 is to clearly restate the importance of the proposed work.

b. Compounds will be chosen for their ability to efficiently catalyze the cycloaddition reactions.

c. We intend to systematically investigate the origins of interfacial polarity.

Table 14.5 Common functions of different verb tense–voice combinations in move 3 of the Outcomes and Impacts section of the Project Description.

Function	Tense–Voice Combination	Example
Reiterate goals and importance	Present–active	The primary goals of this research *are* to...
	Present perfect–passive	A comprehensive study of... ***has been presented***.
	Future–active	A detailed examination of... ***will*** possibly ***shed*** new light on...
		Our research ***will examine***...
Address broader impacts	Future–active	If successful, our results ***will contribute*** to...
		Our findings ***will*** potentially *impact*...

We conclude our discussion of move 3 by considering verb tense and voice combinations commonly used in this move (summarized in table 14.5). Present tense, present perfect, and future tense are used to reiterate goals. Future tense is used most often to state broader impacts. Active voice is common throughout the move, and personal pronouns (e.g., *we* and *our*) may be used at the authors' discretion.

Exercise 14.14

The following sentences were adapted from the concluding remarks of Project Descriptions. Select the most appropriate verb tense in each sentence. Explain your choice. (Note: There may be more than one correct answer.)

a. The proposed goals *represent/have represented/will represent* a significant departure from previous approaches to polar order in organic thin films. (Adapted from Dyer, 2001)

b. The proposed synthesis *benefits/has benefited/will benefit* biochemists in both academia and industry.

c. To contribute to students' research skills, the proposed work *involves/has involved/will involve* three undergraduate students in each year of the project.

14C Writing on Your Own: Conclude the Proposed Work

Write a conclusion for your Project Description that (1) reiterates project goals and importance and (2) addresses the broader impacts of the work. Take advantage of your last opportunity to stress the importance of your work.

14D Writing on Your Own: Complete the Outcomes and Impacts Section

Integrate your timeline, list of expected outcomes, and conclusions into a final draft of the Outcomes and Impacts section. Edit and proofread the section so that it is ready for peer review or final submission (see chapter 18).

Chapter Review

Check your understanding of what you've learned in this chapter by defining each of the following terms and explaining its importance, in the context of the chapter, to a friend or chemistry colleague who is new to the field:

broader impacts
deliverables
dissemination of results
expected outcomes
hedging language
nominalizations
parallelism
split infinitives
tangibles
timeline

Also explain the following to a chemistry colleague who has not yet given much thought to writing the Outcomes and Impacts section of a Project Description:

- Main purpose(s) of the Outcomes and Impacts section
- Purpose of the three moves in the section
- Role of hedging in the section
- Different ways to present a timeline
- Role of broader impacts in the section
- Ways to stress importance (again) in this section

Additional Exercises

Exercise 14.15

It is not often that the words *hope* and *hopefully* are seen in scientific writing, and many consider the use of *hopefully* grammatically incorrect. Yet, there are three instances in this chapter (see below). Based on these examples and what you know about the Outcomes and Impacts section of the Project Description, suggest circumstances in which words like *hope* and *hopefully* might be acceptable.

- Bimolecular reactions in clusters are our most ambitious goal, which I *hope* to start tackling in the second half of the award period. (From Sanov, 2002)
- We *hope* that our results will aid in developing predictive models of interfacial permittivity. (From Walker, 2001)
- The demonstration that ultrafast X-ray investigations can be carried out with laboratory-based equipment will, *hopefully*, inspire others to build and apply such ultrafast systems themselves. (From Rose-Petruck, 2000)

Exercise 14.16

Reflect on what you have learned from this chapter. Select one of the reflection tasks below and write a thoughtful and thorough response:

a. The Outcomes and Impacts section of the Project Description represents your final opportunity to emphasize the importance of your proposed work. Reflect on the different excerpts that you have read in this chapter.

- How do the authors emphasize the importance of their work? From your perspective, which approach is most effective?
- What tactics might you use in your own concluding remarks?
- Why is it permissible to reiterate importance, when chemists normally pride themselves in conciseness in their writing?

b. The Outcomes and Impacts section of the Project Description can greatly influence the views of reviewers, possibly making a difference in final decisions about funding and support.

- Why is the incorporation of broader impacts so important?
- In what ways might the broader impacts sway the opinion of reviewers?
- What are the most valuable lessons that you've learned from reading the excerpts in this chapter that will assist you in writing your Outcomes and Impacts section?

c. Although there is no single template for an effective proposal (because RFPs make different demands on writers), there is a lot to be gained from examining models of successful proposals.

- What have you gained from evaluating models of successful proposals?
- Considering the fundamental characteristics of the Outcomes and Impacts section, what strengths do you bring to the task?
- What weaknesses might you bring to the task? How will you overcome them?

15 *Writing the Project Summary and Title*

The Project Summary has a very long shelf life and is the only part of a research proposal that is available to the public. So write it carefully!

The Project Summary is a short description of the proposed work (one page or less), which, unlike the rest of the proposal, can be accessed by the public. It is usually the first page of the proposal (excluding the cover page, which provides institutional information) and, therefore, precedes the Project Description. In this chapter, we consider the Project Summary and highlight a few language features pertinent to the summary. We conclude the chapter (and module) with suggestions for writing a title for your proposal. By the end of this chapter, you will be able to do the following:

- Recognize the audience and purpose for the Project Summary
- Know what content should be included in (and excluded from) the Project Summary
- Follow the typical move structure of the Project Summary
- Write an appropriate title for your proposal

Staying on Track

The Project Summary is the first of three major divisions of the proposal: Project Summary, Project Description, References Cited.

The most common major division title is

PROJECT SUMMARY

As you work through the chapter, you will write your Project Summary and give your proposal a title. The Writing on Your Own tasks throughout the chapter will guide you step by step as you do the following:

15A Write the Project Summary

15B Write the proposal title

15C Complete the proposal

Most external funding agencies require a Project Summary. Funding agencies often share project summaries with their donors, boards of directors, prospective grantees, and others interested in the types of projects that they fund. The Project Summary is also used by **program officers** to determine the most appropriate review panel for the proposal. In complex agencies like the NSF and NIH, there are many related program initiatives, and it is ultimately the decision of the program officer to forward the proposal to the right division for consideration. As such, the Project Summary is typically written for a scientific audience, enabling a scientifically literate but nonexpert audience to understand the project.

Program Officer

Contact person assigned by funding agencies. Program officers coordinate the solicitation of proposals, answer questions about the funding initiative, and ultimately participate in award decisions.

Many RFPs recommend that investigators contact the program officer as proposals are being prepared.

The Project Summary is not the same as a journal article abstract (chapter 7) or a conference abstract (chapter 8), even though the Project Summary is sometimes called an abstract. The Project Summary summarizes work that has yet to be done and is written for a scientific audience. The journal article abstract summarizes work that has already been done and is written for an expert audience. The conference abstract describes work in progress and is written for a scientific audience. Because the Project Summary reiterates the major aspects of the proposed work, it is written last, after the Projection Description has been completed. For that reason, this chapter comes last in the research proposal module.

Abstracts

See chapters 7 and 8 for details on journal article abstracts and conference abstracts, respectively.

The Project Summary has a longer "shelf life" than the rest of the proposal. Thus, the summary lives on years after the funding has ended, serving as a record of funding for both the agency and recipient. As such, the Project Summary should be viewed as a stand-alone document. Some people will read only the Project Summary and no other section of the proposal; others will read the Project Description, but not the Project Summary. Thus, key concepts, major terms, definitions, and acronyms must be introduced and defined in both the Project Summary and the Project Description.

A Stand-Alone Document

The Project Summary lives on long after a project is completed. It should be written for a scientific audience and be able to stand on its own.

Reading and Analyzing Writing

Rather than begin this section with an excerpt from a Project Summary, we begin instead with excerpts from two guides for writing the Project Summary: the National Cancer Institute (NCI) Quick Guide for Grant Applications (excerpt 15A) and the NSF Grant Proposal Guide (excerpt 15B). Both guides highlight essential features of the Project Summary, including its purpose, intended audience, length, and content. Perhaps the most obvious difference between the two guides is what they call the summarizing document: the NCI calls it an Abstract, and the NSF calls it a Project Summary. Another difference is the emphasis on "broader" (i.e., human) impacts. The NSF requires the Project Summary to address how the project promotes the training and education of future scientists. The NCI does not include this requirement.

Exercise 15.1

Read excerpts 15A and 15B. While reading them, think about the difference(s) in section designations (Abstract and Project Summary) and funding agency expectations and answer the following questions:

a. Are the purposes of the NCI Abstract and NSF Project Summary the same or different? Why does the NSF make a distinction between an Abstract and Project Summary?
b. What similarities and differences do you notice in content and formatting expectations?

Excerpt 15A (from National Cancer Institute Grant Extramural Funding Opportunities, Quick Guide for Grant Applications. http://deainfo.nci.nih.gov/extra/extdocs/gntapp.htm (accessed January 9, 2008))

Abstract

1. **Purpose**: The purpose of the Abstract is to describe succinctly every major aspect of the proposed project except the budget. The Abstract is an important part of your application. It is used in the grant referral process, along with a few other parts of the application, to determine what study section is appropriate to review the application and to what institute at NIH it is most relevant. Members of the Study Section who are not primary reviewers may rely heavily on the Abstract to understand your proposal.
2. **Recommended Length**: The recommended length of the Abstract will vary among different funding agencies, but the NIH Abstract is a half-page, and confined to the designated space provided in the application.
3. **Content**

   The Abstract should include

   a. a brief background of the project
   b. specific aims or hypotheses
   c. the unique features of the project
   d. the methodology (action steps) to be used
   e. expected results
   f. evaluation methods
   g. description of how your results will affect other research areas
   h. the significance of the proposed research
4. **Suggestions**
   a. Be complete, but brief.
   b. Use all the space allotted.
   c. View the Abstract as your one-page (or half-page) advertisement.
   d. Write the Abstract last so that it reflects the entire proposal. Spend time reviewing it.
   e. Remember that the Abstract will have a longer shelf life than the rest of the proposal and may be used for purposes other than the review, such as to provide a brief description of the grant in annual reports, presentations, or in response to requests from top management at NIH.

Excerpt 15B (from NSF Grant Proposal Guide. http://www.nsf.gov/pubs/2002/nsf022/nsf0202_2.html (accessed January 9, 2008))

Project Summary

The proposal must contain a summary of the proposed activity suitable for publication, not more than one page in length. It should not be an abstract of the proposal, but rather a self-contained description of the activity that would result if the proposal were funded.

The summary should be written in the third person and include a statement of objectives and methods to be employed. It must clearly address in separate statements (within the one-page summary) (1) the intellectual merit of the proposed activity and (2) the broader impacts resulting from the proposed activity.... It should be informative to other persons working in the same or related fields and, insofar as possible, understandable to a scientifically or technically literate lay reader.

Proposals that are not consistent with these instructions may not be considered by NSF.... Both criteria must be addressed.

Exercise 15.2

Judge your understanding of excerpts 15A and 15B by answering the following questions:

a. Describe the intended audiences for the NCI Abstract and the NSF Project Summary. Are they the same or different?

b. Which opening phrase would you most likely find in a Project Summary? Explain your choice.

- The principal goal of my project will be to...
- Our proposed research has several related goals, including...
- The overarching goal of this research project is to...

c. Would the following be good advice to give to colleagues writing their first Project Summary? Why or why not?

> Do not exceed the page limit, and make your Summary even shorter than the specified page limit if you can.

With these guidelines in mind, consider an excerpt from Aga's CAREER proposal Project Summary (excerpt 15C). The excerpt summarizes the research component of the proposed work, thereby addressing the intellectual merit of the project. For completeness, the excerpt also includes Aga's broader impacts summary, highlighting the educational components of her work. (Educational impact summaries are not included elsewhere.)

Exercise 15.3

Although you have read other parts of Aga's Project Description (in chapters 11–14), read her Project Summary (excerpt 15C) with fresh eyes, as if it were the first time you learned of her work.

a. Do you understand what she is proposing? Explain.

b. Do you understand why her proposed work is important? Explain.

c. Can Aga's Project Summary be viewed as a half-page advertisement for the project? Why? Why not?

d. Based on what you have read, do you think that the proposal guide that Aga followed was more similar to the NCI guidelines (excerpt 15A) or the NSF guidelines (excerpt 15B)? Explain.

Excerpt 15C (Aga, 2002)

PROJECT SUMMARY

Intellectual Merit. The principal goal of this project is to study the occurrence, fate, and transport in the environment of agricultural chemicals **that** are widely used in crop production, such as herbicides and nitrates. In addition, the presence of antibiotics in the environment will be examined due to their widespread use in animal production, which may cause contamination of surface waters due to runoff from animal feedlots and confinement operations. This is an important issue because of the danger of promoting antimicrobial resistance among microorganisms that are exposed constantly to antibiotics. Assessment of herbicides, nitrates, and antibiotics in water resources is important for increasing the effectiveness of policies designed to protect water quality.

This project will involve development and applications of immunochemical techniques to measure residues of herbicides and antibiotics in soil and water. Immunochemical techniques, such as enzyme-linked immunosorbent assays (ELISAs), are sensitive, fast, and inexpensive, allowing long-term investigations to be feasible and economical. ELISAs will be developed using antibodies that will be produced in-house for new herbicides. In addition, commercially available ELISAs will be employed for the analysis of antibiotics and existing pesticides that are known to persist in soil and water. These ELISAs will be used in conjunction with instrumental methods such as capillary electrophoresis, liquid chromatography, and gas chromatography. Sample preparation procedures using solid-phase extraction will also be developed.

Broader Impacts. The educational component of this proposal will provide research experience for undergraduate students in Chemistry and enhance laboratory experiments in Analytical and Environmental Chemistry courses. . . . The investigations described in this proposal will allow undergraduate students to conduct research under close supervision by the principal investigator (PI). Students will learn to design experiments, collect and analyze environmental samples, interpret data, and present their results. Moreover, problems posed in this proposal will be incorporated into courses that the PI is assigned to teach during the academic year. For example, experiments that involve immunochemical detection and chiral separations of organic contaminants will be incorporated into Analytical Chemistry and Instrumental Analysis. In Environmental Chemistry, a course currently being developed by the PI, students can participate in open-ended investigations such as monitoring levels of pesticides and antibiotics in surface and ground waters. Students enrolled in these courses will experience unique opportunities to learn techniques used in the analysis of environmental samples. The data generated by the students in their laboratory experiments will become part of the database used to determine the seasonal pattern of agricultural chemical input into local rivers and reservoirs.

Which vs. That

Which signals extra information.

That signals essential information about the word or phrase that precedes it.

(See appendix A.)

Exercise 15.4

Review the list of content areas that should be included in an NCI Abstract (excerpt 15A). To what extent does Aga address these content areas? Explain.

Exercise 15.5

In your own words, what are the broader impacts of Aga's proposed research activities? Consider both the impacts of the work on society and on training future scientists.

Analyzing Organization

As illustrated in the move structure for a typical Project Summary (figure 15.1), there are four moves in a standard Project Summary. Because the Project Summary restates the major components of the Project Description, the moves will look

1. Introduce Topic and/or Project Goals

2. Emphasize Significance
(i.e., stress unique features and applications)

3. Summarize Methods
(i.e., highlight promising results)

4. Summarize Broader Impacts

Figure 15.1 A visual representation of the suggested move structure for a Project Summary of a research proposal.

familiar. In move 1, the author offers a brief introduction to the proposed work, identifying the research topic and project goals. Following the introduction, the significance of the project is emphasized (stressing unique features of the project and pointing out expected implications and applications of the work), and the research methodology is summarized (highlighting promising results). Moves 2 and 3 can be addressed in either order (significance/methods or methods/significance) or in an integrated manner. Move 4 addresses the broader impacts of the work (educational impacts are not included).

Analyzing Excerpts

Below we examine excerpts from several CAREER Project Summaries. Because the summaries are relatively short (one page or less), we consider the entire summary at once, rather than move by move. The Project Summary is written for a scientific audience or, as described in excerpt 15B, a "scientifically or technically literate lay reader." Consistent with this audience, there are no citations in the Project Summary.

In excerpt 15D, Walker begins with a statement of the topic (solvation at hydrophobic and hydrophilic solid-liquid interfaces) and then moves directly to the significance of the work. He emphasizes the need for information on interfacial phenomena and points out possible applications of his work for other areas of science (molecular recognitions, electron transfer, and macromolecular self-assembly). He goes on to describe his experimental methods, focusing on three aspects of his approach (in order of difficulty): equilibrium measurements, time-resolved studies, and distance-dependent measurements of solvation strength.

Excerpt 15D (Walker, 2001)

PROJECT SUMMARY

This career plan describes a research program that will experimentally investigate solvation at hydrophobic and hydrophilic solid-liquid interfaces and an educational initiative designed to introduce talented freshmen to independent research in physical and analytical chemistry.

Motivating the research is the need for systematic, quantitative information about how different surfaces and solvents affect the structure, orientation, and reactivity of adsorbed solutes. In particular, the question of how the anisotropy imposed by surfaces alters solvent-solute interactions from their bulk solution limit will be explored. Answers to this question promise to affect our understanding of broad classes of interfacial phenomena including electron transfer, molecular recognition, and macromolecular self assembly. By combining surface sensitive, nonlinear optical techniques with methods developed for bulk solution studies, experiments will examine how the interfacial environment experienced by a solute changes as a function of solvent properties and surface composition.

One aspect of the research will examine equilibrium aspects of solvation at hydrophobic and hydrophilic interfaces. In these experiments, solvent dependent shifts in chromophore absorption spectra will be used to infer interfacial polarity. Preliminary results from these studies are presented. The polarity of solid-liquid interfaces arises from a complicated balance of anisotropic, intermolecular forces. It is hoped that results from these studies can aid in developing a general, predictive understanding of dielectric properties in inhomogeneous environments.

Complementing the equilibrium measurements will be a series of time resolved studies. Dynamics experiments will measure solvent relaxation rates around chromophores adsorbed to different solid-liquid interfaces. Interfacial solvation dynamics will be compared to their bulk solution limits, and efforts to correlate the polar order found at liquid surfaces with interfacial mobility will be made. Experiments will test existing theories about surface solvation at hydrophobic and hydrophilic boundaries as well as recent models of dielectric friction at interfaces. Of particular interest is whether or not strong dipole-dipole forces at surfaces induce solid-like structure in an adjacent solvent. If so, **then** these interactions will have profound effects on interpretations of interfacial surface chemistry and relaxation.

A third goal of the research is to measure the distance over which surface effects extend into solution. These experiments face several technical challenges, but if experiments are properly designed, results would provide the very distance versus strength of interaction information necessary for incorporating molecular contributions into existing dielectric continuum theories of surface solvation.

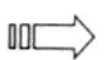

Then vs. Than

See appendix A for more information on these commonly confused words.

In excerpt 15E, Lorigan uses an approach that is similar to Walker's. He begins by describing the goal of his work and the topic of his research. Next, he stresses the significance of his work by pointing out gaps and challenges in the field. He concludes with a brief overview of his experimental methods, which will fill the gap. Abbreviations for EPR and NMR are defined in the text. The abbreviation for EPR is defined again in the Project Description, but NMR is not. According to *The ACS Style Guide*, common abbreviations such as NMR, IR, and DNA need never be defined. Lorigan opted to define NMR because he recognized that his Project Summary might be read by nonscientific audiences.

Excerpt 15E (Lorigan, 2002)

PROJECT SUMMARY

The goal of this proposal is to initiate an extensive research and teaching program in the emerging area of biophysical chemistry. The research focuses on the application of state-of-the-art magnetic resonance techniques specifically designed to investigate protein tems. The educational activities of this project are centered on the development of a modern biophysical chemistry curriculum for undergraduate and graduate students in both the classroom and laboratory.

Membrane proteins (which make up approximately one-third of the total number of known proteins) are responsible for many of the important properties and functions of biological systems. They transport ions and molecules across the **membrane**; they act as **receptors**; and they have roles in the **assembly**, **fusion**, and structure of cells and viruses. Presently, investigating membrane proteins is one of the most difficult challenges in the area of structural biology and biophysical chemistry. Our knowledge of membrane proteins is limited, primarily **because** it is very difficult to crystallize these protein systems due to the extreme hydrophobic interactions between the proteins and the membrane. New methods are needed and current techniques need to be extended to study the structural properties of membrane proteins.

This proposal concerns the use of magnetic resonance spectroscopy techniques to investigate novel oriented phospholipids membrane protein systems (bicelles). The general aim is to utilize a combination of electron paramagnetic resonance (EPR) spectroscopy and solid-state nuclear magnetic resonance (NMR) spectroscopy to investigate the structural and dynamic properties of the membrane proteins. These experiments will reveal new insights concerning the structural characteristics of the membrane-bound protein CREP-1. Specifically, (1) new methods will be developed for orienting lipid bicelle systems at lower magnetic fields for EPR studies at Q-band, (2) the orientation of the TM-A and TM-B peptides of CREP-1 will be determined with respect to the lipid bilayer, and (3) complementary dynamic information will be obtained on the same membrane protein in two different time scale regimes. This study has important applications to the fields of chemistry, biophysical chemistry, and structural biology.

Semicolons

Use semicolons rather than commas to separate items in a series that already contains commas. (See appendix A.)

Since vs. Because

Use *since* to connote the passage of time; use *because* to illustrate cause and effect relationships. (See appendix A.)

In excerpt 15F, Lyon uses a slightly different approach. He begins with the goal of the research but focuses next on his experimental approach, introduced as *the challenge*. He concludes with the significance of the work. Like Lorigan, acronyms are defined in the Project Summary (and again in the Project Description). Lyon uses bold font to call the reader's attention to the different moves in the Project Summary.

Excerpt 15F (Lyon, 2000)

PROJECT SUMMARY

The goal of this research is to develop a new class of bioresponsive materials that undergo rapid, large-magnitude, volume-phase transitions in response to specific biological stimuli. Rationally designed responsive hydrogel nanoparticle bioconjugates are the basis for these materials. The development of these unique biohybrids is based on two fundamental premises: (1) that the thermodynamics of hydrogel volume-phase transitions can be perturbed through localized polymer-protein interactions, and (2) that the kinetic limitations commonly associated with macroscopic hydrogel networks can be overcome by reduction to the sub-micron scale.

The challenge of this work is to develop responsive gel particles that can be covalently linked to proteins such that a polymer-confined protein-protein interaction event modulates the thermodynamics of solvation. Because the particulate hydrogel is a cooperative entity, local changes in polymer chemistry are expected to modulate hydration over a long-length scale, thereby resulting in a material that changes its volume in response to a biological interaction. Furthermore, reduction of the material to nanoscale dimensions is expected to allow for rapid equilibration of the particle with its surroundings, while simultaneously enhancing the ability of the biological analyte to diffuse into the material. Accordingly, this work will focus on the fundamental aspects of bioconjugate synthesis and hydrogel solvation thermodynamics and kinetics.

Bioconjugate hydrogel colloids (80–1000 nm dia) will be prepared by covalently linking the biological macromolecule of interest to the backbone of a particle derived from poly-N-isopropylacrylamide (p-NIPA). NIPA gels belong to a unique class of responsive materials with environmentally switchable volumes; these changes result from solvation or desolvation of the hydrogel network. Initial experiments will focus on carboiimide coupling of monoclonal antibodies (mAbs) to acid sites on the polymer backbone. . . . In addition to bioconjugate synthesis, parallel efforts will focus on methods for surface immobilization of hydrogel particles for thin film device (sensor) fabrications. . . . Finally, advanced-core shell hydrogel architectures will be designed and synthesized in order to create multiresponsive particles.

The significance of this proposal is multifold. First, this work will expand our fundamental knowledge concerning the chemistry and physics of colloidal hydrogels. A detailed understanding of the factors governing gel bioconjugate hydration kinetics and thermodynamics will be obtained. This will allow for the rational design of hydrogel matrices that are sensitive to biological stimuli. Also, synthetic routes for the biological modification of responsive gels, and methods for bioconjugate characterization, will be developed. Finally, control over the surface chemistry of these particles will lead to new methods for the construction of responsive hydrogel thin films. Colloidal responsive gels have never been investigated in this manner; hence, this work will significantly advance the fields of bioconjugates and hydrogel-based materials.

A second point of significance is that this work will lead to a new class of bioresponsive materials for a host of applications. Initial applications target biosensors and bioassays; colloidal materials provide tremendous versatility in biosensor design. For example, a film of responsive particles could be immobilized onto a transducer, which would then sense changes in film opacity, refractive index, mass, or elasticity as the particle volume changes.... Long term possibilities for these materials include drug delivery, artificial muscles, organ surrogates, skin grafts, affinity separations, and cell adhesion inhibitors.

Exercise 15.6

Read Kinsel's CAREER Project Summary (excerpt 15G).

a. Identify the parts of his Project Summary that match the moves illustrated in figure 15.1.
b. Do you notice any pattern in the author's verb tenses (present, past, future) for the different moves? If so, what pattern(s) do you see?

Excerpt 15G (adapted from Kinsel, 1999)

PROJECT SUMMARY

This proposal describes the development of a new, systematic approach for qualitatively and quantitatively studying surface-biomolecule interactions by matrix-assisted laser desorption ionization (MALDI) mass spectrometry (MS). This methodology is being developed because of the profound importance that surface-biomolecule interactions play in applications where biomaterials come into contact with complex biological fluids. It can readily be shown that undesired reactions occurring in response to surface-biomolecule contact (protein adsorption, biofouling, immune response activation, etc.) lead to enormous economic and human costs. Thus, the development of analytical methodologies that allow for efficient assessment of the properties of new biomaterials and/or the study of detailed fundamental processes initiated upon surface-biomolecule contact are of critical value....

Preliminary studies show that the effects of even small changes in surface chemistry on surface-peptide binding affinity can be examined by MALDI MS. This result suggests that the MALDI MS methodology can be used to gain greater insight into the relative influence of surface, protein, and solution properties on surface-biomolecule binding affinity. The preliminary studies will be expanded to encompass studies of the effect of changes in surface morphology, protein electrostatic charge, and protein solution environment on surface-protein interactions. These studies will be pursued through established collaborations that uniquely allow us to systematically alter surface properties (using pulsed radio frequency plasma polymerization), characterize the resultant surface (using atomic force microscopy), and quantitate the effect on surface-binding affinity (using MALDI MS and conventional protein-binding studies). The insights gained will lead to a deeper understanding of the mechanism(s) of surface-biomolecule interactions and will be of great value in the design of improved application-specific biomaterials.

Exercise 15.7

Look back at excerpts 15C–15G. Focus on the authors' use of language to accomplish the purposes of the Project Summary and answer the following questions:

a. Do any of these authors use personal pronouns? What rule might you hypothesize about personal pronoun use in the Project Summary?

b. What phrases do the authors use to refer to their proposed work (e.g., this project)?

c. How do the authors use the following words: goal, objective, aim?

d. Do any of the authors use ordinal language (first, second, third)? What purpose, if any, does the enumeration serve?

e. What words and phrases (besides *important*) are used to emphasize the importance of the research?

f. Do you see any gap and fill-the-gap statements? What moves are they associated with?

g. The author of excerpt 15F states that "this work will significantly advance the fields of bioconjugates and hydrogel-based materials." Is the word *significantly* used in a statistical sense or in another way?

Analyzing Writing

In our examination of the language in the Project Summary, we focus on easily confused words, punctuation, the use of personal pronouns, and common verb tense–voice combinations. To call your attention to confusing word pairs and punctuation, we placed pointers near numerous excerpts above:

- which vs. that (excerpt 15C)
- then vs. than (excerpt 15D)
- semicolons in a series with commas (excerpt 15E)
- since vs. because (excerpt 15E)

To test your knowledge of some of these features, complete exercises 15.8–15.10.

Exercise 15.8

Select the correct word in each sentence:

a. Side reactions will likely occur at temperatures higher *then/than* 75 °C.

b. Substances eluding from the first column will *then/than* be concentrated onto a second more polar column.

c. *Since/Because* our knowledge is limited, new approaches are needed.

d. *Since/Because* the early 1970s, agricultural chemicals have been used widely in crop production; their fate in the environment must be studied.

Exercise 15.9

Consider the following uses of *which* and *that*. Indicate if the sentence is correct as is, otherwise correct it. Modify punctuation, if necessary.

a. Antibiotics are used in animal products, *which* may lead to surface water contamination.

b. Membrane proteins (*that* make up approximately one-third of the total number of known proteins) are the focus of this study.

c. For example, a film of responsive particles could be immobilized onto a transducer, *which* would then sense changes in film opacity.

d. There is a danger of promoting antimicrobial resistance among microorganisms, *which* are exposed constantly to antibiotics.

e. ELISAs will be developed using antibodies *that* will be produced in-house for new herbicides.

f. This career plan describes a research program *which* will experimentally investigate solvation at hydrophobic and hydrophilic solid-liquid interfaces.

g. The goal of this research is to develop a new class of bioresponsive materials *which* undergo rapid, large-magnitude, volume-phase transitions.

h. The challenge is to develop gel particles *that* can be covalently linked to proteins.

Exercise 15.10

Rewrite this short passage (adapted from excerpt 15C) by correcting the punctuation. Everything else should remain the same.

Important for increasing the effectiveness of water-quality protection policies are the assessment of herbicides, nitrates, and antibiotics in water resources, the development of immunochemical techniques to measure herbicide residues, and the application of enzyme-linked immunosorbent assays.

We conclude this section by briefly considering personal pronouns and common verb tense and voice combinations used in the Project Summary. Personal pronouns (e.g., *I*, *our*, *we*), although common in other sections of the proposal, should not be used in the Project Summary. As noted in the NSF guidelines (excerpt 15B), the Project Summary should be written in the third person.

Not recommended *We hope* that results from these studies will aid in...
The principal goal of *our project* is to...
We are particularly interested in whether or not...

Recommended *It is hoped* that results from these studies will aid in...
The principal goal of *the project* is to...
Of particular interest is whether or not...

Note that *our* is acceptable in a Project Summary when it is used in a general sense, to connote scientists or society in general; it should not be used to refer to the PI's group.

Acceptable This work will expand *our* fundamental knowledge of...
Answers will affect *our* understanding of...
Our knowledge of membrane proteins is limited...

With respect to verb tense, most of the Project Summary is written in present and future tenses. Typical verb tense–voice combinations are listed in table 15.1.

Table 15.1 Common functions of different verb tense–voice combinations in the Project Summary.

Function	Tense–Voice Combination	Example
To introduce the research topic	Present–active	This proposal ***describes*** the development of a new, systematic approach...(From Kinsel, 1999) The research ***focuses*** on a state-of-the-art...
To present project goals	Present–active	The goal of this proposal **is** to initiate an extensive research program in the emerging area of biophysical chemistry. (From Lorigan, 2002) The general aim **is** to utilize a combination of...(From Lorigan, 2002)

continued

Table 15.1 (continued)

Function	Tense–Voice Combination	Example
To emphasize significance	Present–active	The significance of this proposal **is** multifold. (From Lyon, 2000)
		Assessment of herbicides... in water resources **is** important for increasing the effectiveness... (From Aga, 2002)
		Answers to this question ***promise*** to affect our understanding. (From Walker, 2001)
To summarize proposed methodology	Future–active	This project ***will involve*** development and applications of... (From Aga, 2002)
	Future–passive	These studies ***will be pursued*** through... (From Kinsel, 1999)
		Enzyme-linked immunosorbent assays ***will be employed*** for the analysis of antibiotics. (From Aga, 2002)
To highlight promising results	Future–active	These experiments ***will reveal*** new insights. (From Lorigan, 2002)
		The insights gained ***will lead*** to a deeper understanding. (From Kinsel, 1999)
To summarize broader impacts	Future–passive	The data... ***will be used*** to determine the seasonal pattern of agriculture chemical input into local rivers and reservoirs. (Adapted from Aga, 2002)

15A Writing on Your Own: Write the Project Summary

By now, you have written and revised the individual sections of your proposal (see Writing on Your Own tasks in chapters 11–14). It is time to write your Project Summary. Follow the guidelines in your RFP or use the NCI guidelines in excerpt 15A. Limit your Project Summary to one-half page. Be sure to introduce your topic, summarize your research methods, and emphasize the significance of your topic. Remember that your Project Summary should reflect every major aspect of your proposal.

When you proofread your Project Summary draft, make sure that you have written for a scientific audience, not an expert audience. Check to see that you have not used first-person pronouns and that you have used verb tenses appropriately. Finalize your Project Summary using suggested guidelines in chapter 18.

Writing a Title

The best time to determine the title is after the text is written; in this way, the title will reflect the contents and emphases of your written work accurately and clearly.

—Adapted from Coghill and Garson (2006)

The rules for writing a research proposal title are similar to those for the journal article and poster. Like journal article titles (see table 7.1), a proposal title should include keywords that will help the program officer decide where to send your proposal for review. The title should also include keywords that capture the need(s) targeted by the granting institution. Abbreviations and acronyms should be avoided. Although explicit restrictions on title length are rare, shorter titles are preferred, provided that they accurately depict the contents and emphases of the proposal.

Titles

See table 7.1.

Exercise 15.11

Read the list of CAREER titles below and answer the following questions:

a. What general patterns do you notice in the titles?
b. Are the titles complete sentences or fragments?
c. Are any titles made up of two parts, with a colon between the two parts?
d. What kinds of information come first? Second? Third?
e. Do any of the titles follow the X of Y by Z pattern found in journal article titles (see table 7.1)?

[*Note:* The original CAREER grant titles all began with "CAREER:" (as specified in an NSF directive), which has been omitted here.]

1. Immunochemical Techniques for Investigations of the Occurrence and Fate of Agrochemicals in the Environment (from Aga, 2002)

2. Career Plan for Research in Polar Organic Materials with an Integrated Emphasis on Industrial Problem Solving (from Dyer, 2001)
3. Exploring the Mechanisms of Organic Surface Modification Processes (from Fairbrother, 2000)
4. New Approaches to the Development of Fluorescent Chemosensors for Heavy Metal Ions (from Finney, 1999)
5. Photolysis of Alkylazides in Solution and in Crystals (from Gudmundsdottir, 2001)
6. Teaching through Research: Plume-Ridge Interactions in the Galápagos as Focus of an Integrated Interdisciplinary Chemistry Curricular Initiative (from Harpp, 1998)
7. Combinatorial Chemistry in the Classroom and Laboratory: Identification of Novel Small Molecule Ligands for Apoptotic Proteins (from Hergenrother, 2002)
8. Models for Metal Active Sites in Proteins (from Houser, 2001)
9. Discovery of New Enantioselective Dipolar Cycloadditions from Small Ring Heterocycles (from Johnson, 2003)
10. Analytical Applications of MALDI Mass Spectrometry to the Study of Surface-Protein Interactions (from Kinsel, 1999)
11. Protein Dynamics and Hydrogen Tunneling in Enzymatic Catalysis (from Kohen, 2002)
12. Mechanistic Studies of Nucleotide Reactivity (from Lee, 2001)
13. Investigating Membrane Proteins with Magnetic Resonance Spectroscopy (from Lorigan, 2002)
14. Bioresponsive Hydrogel Nanoparticles (from Lyon, 2000)
15. Liquid Crystal Imprinting (from Patrick, 2000)
16. Ultrafast X-ray Imaging of Molecular Dynamics in Solution: A Research Program that Enhances Students' Learning (from Rose-Petruck, 2000)
17. Structure and Dynamics of Negative Ions via Photoelectron Imaging Spectroscopy (from Sanov, 2002)
18. Atomic Force Microscopy Studies of Transition Metal Chalcogenide Deposition Using Translationally Hot Atoms (from Spain, 1997)
19. Theoretical Investigations of Chemical Processes in Bulk Crystals and on Surfaces (from Tuckerman, 1999)
20. Synthesis of Allelopathic Agents as Leads to New Agrochemicals (from Vyvyan, 2001)
21. Surface-mediated Solvation at Hydrophobic and Hydrophilic Interfaces (from Walker, 2001)
22. Metal-Ligand Multiple Bonding in Later, First Row Complexes (from Warren, 2002)

15B Writing on Your Own: Write the Proposal Title

Write your proposal title. Make sure that it (1) includes keywords that accurately depict the contents and emphases of the proposal and (2) communicates the needs targeted by the funding agency.

15C Writing on Your Own: Complete the Proposal

After you have written your title, it is time to put your proposal together so that it reads as a single, unified document:

- Combine the two major sections of the proposal: the Project Summary and the Project Description.
- Add your References Cited and finalize in-line citations (see chapter 17).
- Make sure headings and formatting are parallel throughout the document.
- Finalize your entire proposal (see chapter 18).

Congratulations! You have completed your research proposal.

Chapter Review

As a self-test of what you've learned in this chapter, define each of the following terms and explain its importance, in the context of this chapter, to a friend or chemistry colleague who is new to the field:

abstract	program officer	shelf life
broader impacts	Project Summary	stand-alone document

Also explain the following to a friend or colleague who has not yet given much thought to writing a Project Summary or title for a research proposal:

- Differences between a Project Summary, journal article abstract, and conference abstract
- Purpose, intended audience, and length of a Project Summary
- Content and organization (moves) of a Project Summary
- Use of personal pronouns in a Project Summary
- Relationship between a Project Summary and the rest of the research proposal

Additional Exercises

Exercise 15.12

Reflect on what you have learned from this chapter. Select one of the reflection tasks below and write a thoughtful and thorough response:

a. The Project Summary is said to have a longer shelf life than the rest of the research proposal.
   - In your own words, what exactly does that mean?
   - Why would the short Project Summary be so different from other parts of the proposal?
   - What role does the Project Summary play in the proposal review process?

b. Project Summaries, in general, span from one-half page to a full page.
   - What are the challenges associated with writing something so short?
   - How can a writer describe succinctly the major aspects of the project in such a short piece?
   - What principles can guide you in writing an effective Project Summary?

c. You have just completed your research proposal.
   - What have you learned about research proposals that you consider most important?
   - What advice would you give to someone writing a research proposal for the first time?
   - In your view, what are the keys to an effective proposal?

Section 2

Graphics, References, and Final Stages of Writing

16 *Formatting Figures, Tables, and Schemes*

Like a picture, a well-crafted graphic is worth a thousand words. (Of course, authors must avoid the temptation to include both the graphic and the thousand words.)
—Sharon R. Baker, University of Michigan

This chapter focuses on general formatting guidelines for three commonly used graphics in chemistry writing: figures, tables, and schemes. The major purposes and uses for each graphic are described, and common formatting expectations are shared. Before-correction and after-correction examples are used to identify common formatting errors and ways to correct them. Each section of the chapter ends with a table of useful guidelines. By the end of the chapter, you will be able to do the following:

- Know when it is appropriate to include a figure, table, or scheme
- Recognize common formatting mistakes in figures, tables, and schemes
- Format figures, tables, and schemes in appropriate and conventional ways

As you work through the chapter, you will format your own graphic, guided by the Formatting on Your Own task at the end of the chapter.

Formatting

A properly formatted graphic communicates scientific findings clearly and professionally.
Formatting involves selecting the appropriate font, text justification, and border style, among other details.

Graphics, in combination with the text, allow authors to communicate complex information efficiently. When done properly, text and graphics work together, reinforcing each other without duplicating information. Like the text, graphics must follow formatting conventions. In this chapter, we call your attention to some common formatting practices. Of course, we cannot address all of the formatting practices in chemistry, nor can we anticipate how these conventions will change over time. Thus, use this chapter for basic formatting information and for insights into the many details involved in a properly formatted graphic. As always, consult *The ACS Style Guide* and your targeted journal's Information for Authors for more detailed and current information.

Information for Authors

The technical formatting and submission requirements for graphics differ from journal to journal and from RFP to RFP. It is the author's responsibility to consult the relevant journal or RFP for specific requirements. See *The ACS Style Guide* for additional tips on achieving high-quality graphics.

Figures

Purpose and Use

Authors use **figures** (e.g., graphs, illustrations, photographs) to display scientific information. Examples of figures are included throughout the textbook, for instance, an ion source (excerpt 3S), a comet assay (excerpt 4E), a chromatogram (excerpt 9F), and an illustration of hydrogel adsorption (excerpt 13I). Figures are numbered consecutively throughout a paper (Figure 1, Figure 2, etc.) and mentioned by name and number in text preceding the figure. Although many figure types exist, by far the most common is the **graph**. Because of their frequency, we devote this section of the chapter solely to formatting graphs; however, the guidelines presented are applicable to many other figure types as well.

Figures

Figures display scientific information in, for example, graphs, illustrations, and drawings. By far, the most common figure is a graph; hence, we focus only on graphs in this chapter.

Graphs

Many types of graphs exist (e.g., bar graphs, contour plots, line graphs, pie charts, scatter plots).

- In journal articles, graphs are used to share information too complex for words.
- In posters, graphs are used to display even simple concepts visually.
- In proposals, illustrations and graphs are common.

The central purpose of a graph is to present, summarize, and/or highlight trends in data or sets of data. Graphs of various types (e.g., scatter plots, contour plots, two- and three-dimensional line graphs, and bar graphs) are used for different purposes; thus, authors must match their purpose with the appropriate type of graph.

Graphs give readers a chance to examine data independently, without being influenced by the authors' interpretations. To this end, effective graphs include navigational guideposts, such as clearly labeled axes and a descriptive caption. Graphs also help readers comprehend complex relationships between and among sets of data, relationships that would be difficult, if not impossible, to grasp with words alone. That said, care must be taken to use graphs only when necessary, and not when words alone would suffice. This is particularly true in journal articles, where conciseness is at a premium. Consider the following occasions when words alone, without graphs, are preferred:

- Use words to report a linear relationship between variables rather than graphing a single straight line, for example, "A linear relationship was observed between A and B ($r = 0.97$)."
- Use words to describe the results of calibration procedures rather than graphing calibration curves, for example, "All calibrations were accurate to ±0.05%."
- Use words to describe simple relative abundances rather than graphing such data in a pie or bar graph, for example, "Compounds **2**, **3**, **5**, and **7** were produced in 25, 30, 36, and 40% yield, respectively."
- Use words to report spectral data that verify compound identity, rather than include the spectrum (see the example of NMR data in excerpt 3K).
- Use words to describe related or similar results rather than plotting multiple sets of data that convey the same trend, for example, "Reaction 2 followed a decay curve like the one shown in Figure 2."

These rules are relaxed in posters. Poster guidelines often encourage authors to use graphs because of their visual appeal, even when words alone would suffice. In research proposals, graphs are used as they would be in journal articles; however, unlike journal articles, textbook-like drawings and illustrations are commonly used to illustrate new concepts and approaches described in the proposal.

A graph that contains too little information is a problem, but so is a graph that contains too much. Even in journal articles, graphs should not overwhelm the reader. Data points should be legible and easily distinguished (unless a trend rather than individual points is being stressed). If a graph is too crowded, consider graphing a representative sample of the data and describing similar or related trends in the text. Moreover, graphs should not be misleading. For example, the axes should not be scaled to make large differences look small or small differences look large. This problem is illustrated in exercise 16.1.

Exercise 16.1

The following fictitious dose–response curve for compound A is both misleading and unnecessary. To discover its problems, answer questions a and b. (Note: This dose–response curve is used to evaluate the toxicity of a chemical. Cells were treated with increasing doses of the chemical. The dose is plotted on the x axis, and percent cell survival is plotted on the y axis.)

a. Why is this graph misleading? (Hint: First look at the data points alone; then examine the data points and axis values together with the error terms in the figure caption. Compare your two observations.)

b. Why is the graph unnecessary? Suggest a way to state the same information in the text.

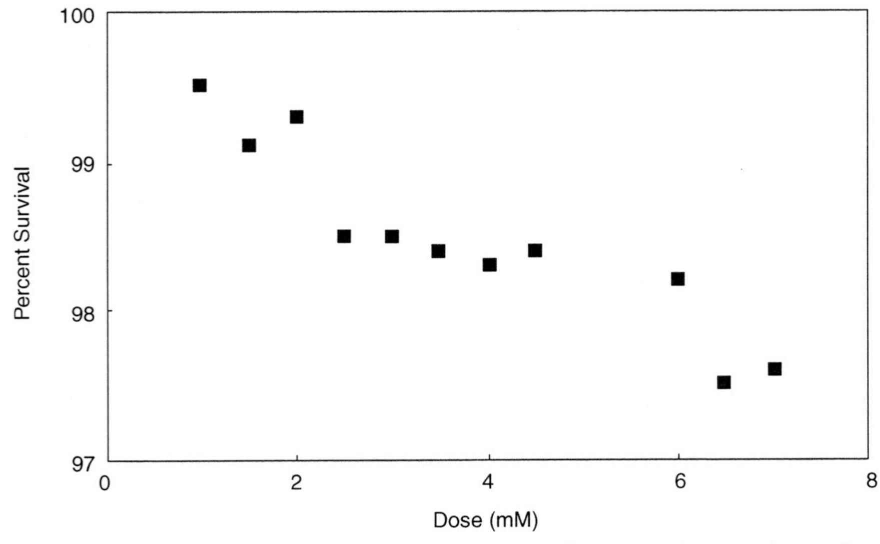

Figure 1. Dose–response curve showing the cytotoxicity of compound A to rat liver cells. Error bars (not shown) are ±10%.

Graph Formatting

With these considerations in mind, we examine how to format a graph properly. We begin by examining two graphs that contain numerous formatting errors: the before-correction graphs in figure 16.1. Exercise 16.2 helps you identify these errors and then directs you to figure 16.2 to view the after-correction graphs. Important graph-formatting tips are summarized in table 16.1.

Table 16.1 Some generally accepted guidelines for formatting graphs. (Because formatting conventions vary, consult the appropriate Information for Authors for specific guidelines.)

Attribute	Description
Axis labels	Label all axes. A common format is to use unbolded font and to capitalize only the first letter of words (e.g., Reaction Time, not REACTION TIME). Place units in parentheses, using standard abbreviations (e.g., min, s) and special characters (e.g., α, μ, m^3, Cl_2). Place labels outside and parallel to the axes.
Axis scaling	Choose axis scales that accurately represent the data; do not mislead readers by using scaling that makes small changes look large or large changes look small. When possible, minimize empty space.
Background color and lines	Use a white background, not a gray or colored background. (Colored backgrounds may be used in posters to add visual appeal.) Remove horizontal and vertical gridlines unless they are needed to convey scientific information.
Boxes	In general, do not "box" graphs. Boxes are sometimes used in research proposals, to help the graph stand out, and in posters, to add visual appeal.
Captions	Place a caption below the graph, aligned with the left-hand margin. Do not include a title above the graph (except in posters, which sometimes have a caption and a title). A common format is to begin the caption with an identifier in bolded font (e.g., **Figure 2.**). Continue with a short informative descriptor (often a fragment) that can be understood apart from the text. The descriptor is usually written in sentence case and often ends with a period (e.g., **Figure 2.** Effect of heating on reaction rate.). The descriptor can be followed by additional information (e.g., see Legends below) in one or more sentences or fragments. In general, treat captions as text, and use the same font and font size as in the text. (A slightly smaller font size is sometimes used in proposals.)

continued

Table 16.1 (continued)

Attribute	Description
Color	In journal articles, use color only if needed to convey scientific information and allowed by the journal. In posters and proposals, color may also be used to add visual appeal. (Proposal graphs should also be legible in gray scale, in case reviewers print out a hard copy to read.)
Curves	Avoid including more than five curves per graph. Label all curves clearly. Leave sufficient space between curves so that they are easily distinguishable.
Font sizes	Choose font sizes that are roughly the same size as the text. Axis labels may be slightly larger than labels for tick marks on the axes. Avoid using more than two font sizes.
Legends	Use legends with two or more data sets. If possible, make the legend part of the caption (e.g., **Figure 2.** Yields with (□) and without (■) heating). Otherwise, place the legend inside the graph (without a box), or, if it overlaps data, place it to the right of the graph. Do not rescale the graph to make room for the legend.
Symbols and bars	Choose colors and patterns for symbols and bars that maximize contrast (e.g., hatch marks or white- and black-filled shapes). When appropriate, follow a theme (e.g., filled shapes for experimental data, unfilled shapes for theoretical data). Avoid grays that are too close in color. Judge contrast from a printed version of the graph, not from the computer screen.
Variables on x and y axes	Most graphs typically plot two variables: the independent variable (or "cause") and the dependent variable (or "effect"). Plot the independent variable on the x axis and the dependent variable on the y axis.

Exercise 16.2

Examine the two before-correction graphs in figure 16.1 and answer the following questions.

a. What formatting errors do you observe in the two graphs? Consider errors in backgrounds, use of color, axis labels, titles, and legends.
b. Are the graphs misleading because of the formatting errors? Explain.
c. Identify at least three features of the before-correction graphs that are correct.
d. Do you think the graphs are necessary? Explain.

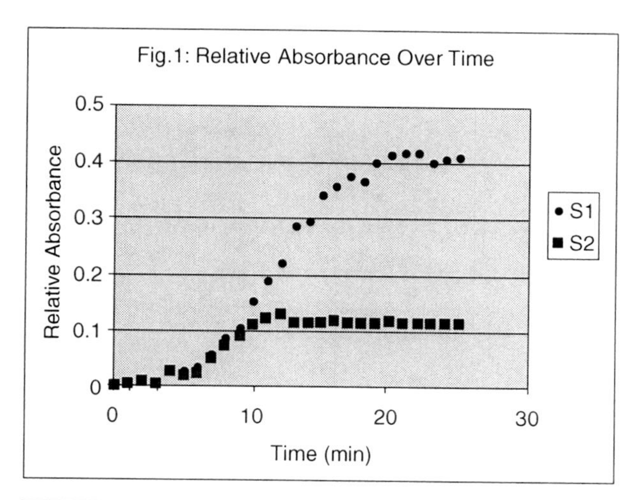

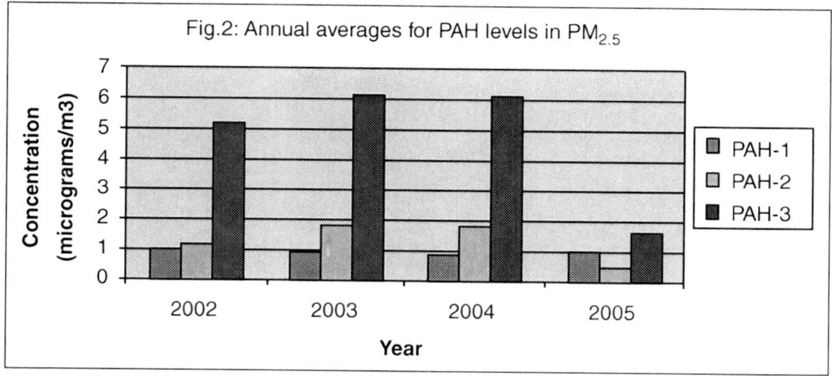

Figure 16.1 (Before correction) Examples of common formatting errors in line graphs (Figure 1) and bar graphs (Figure 2). (Note: The authors used red and blue symbols in Figure 1; blue, red, and purple bars in Figure 2; and gray backgrounds.)

e. Now look at figure 16.2. What formatting errors in the before-correction graphs (figure 16.1) have been corrected in the after-correction graphs (figure 16.2)?

Exercise 16.3

Examine Figure 3 in excerpt 16A. How well does the graph adhere to the guidelines in table 16.1? Specifically, consider the following:

a. Is the legend appropriately formatted and placed? Explain.

b. Is the caption appropriately formatted and placed? Explain.

c. Are the symbols chosen for the six curves appropriate? Explain.

d. Have the authors placed the independent variable (the "cause") on the x axis and the dependent variable (the "effect") on the y axis?

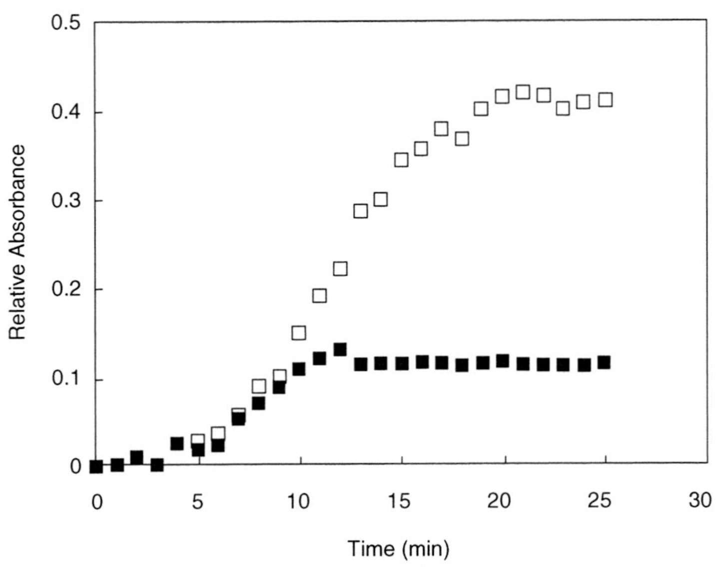

Figure 1. Absorbance over time for uncatalyzed (■) and catalyzed (□) products.

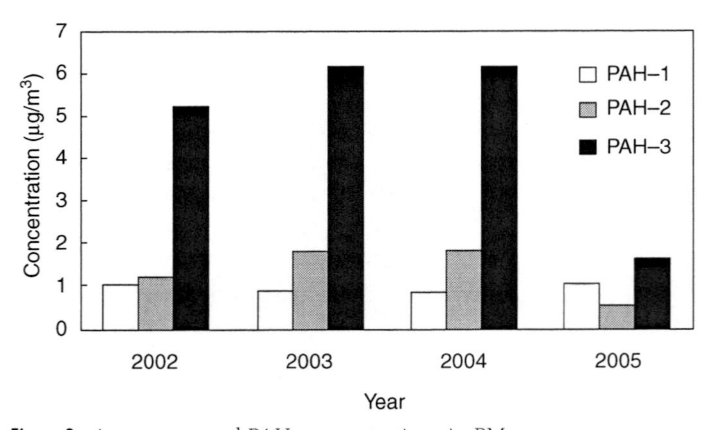

Figure 2. Average annual PAH concentrations in $PM_{2.5}$.

Figure 16.2 (After correction) The line graph (Figure 1) and bar graph (Figure 2) from Figure 16.1 after formatting errors have been corrected.

e. Comment on the presentation of the data. Are the data points easily distinguished? Are the axes scaled appropriately, or are they misleading?

f. Would you suggest any changes to the figure? If so, what would you suggest? Why?

Excerpt 16A (adapted from Satoh et al., 2003)

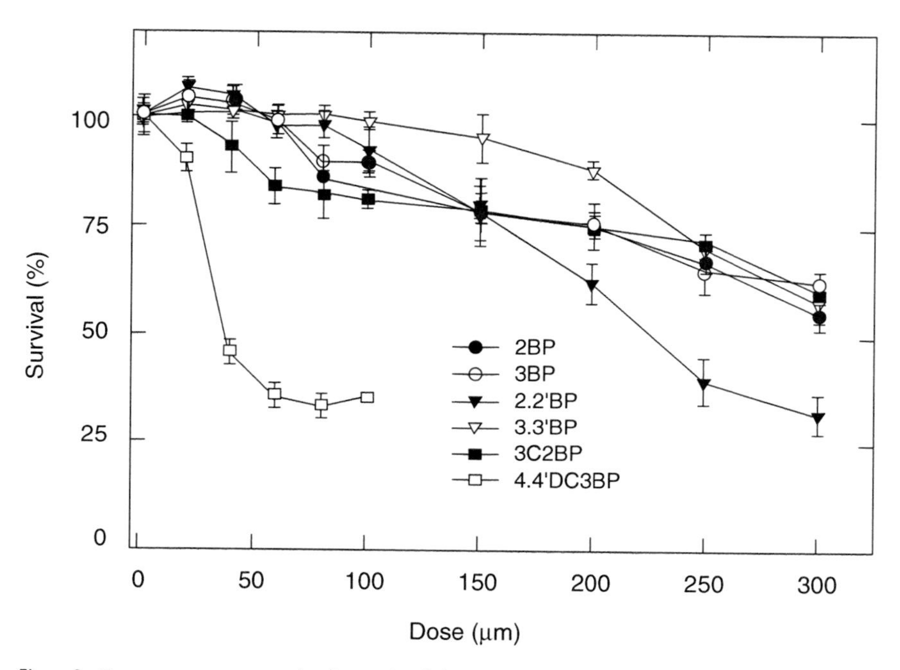

Figure 3. Dose–response results for each of the six chemicals. Cell cultures were exposed to the chemicals for 30 min. Each chemical was tested for a dose range of 0–300 μM. Each data point is representative of the results for a set of chemically treated triplicates reported as an average % survival ± standard deviation determined at the 95% confidence interval.

Exercise 16.4

The following four graphs have formatting problems.

a. Find at least four problems in each graph.

b. How would you correct these formatting problems? Propose solutions that are in line with standard formatting practices.

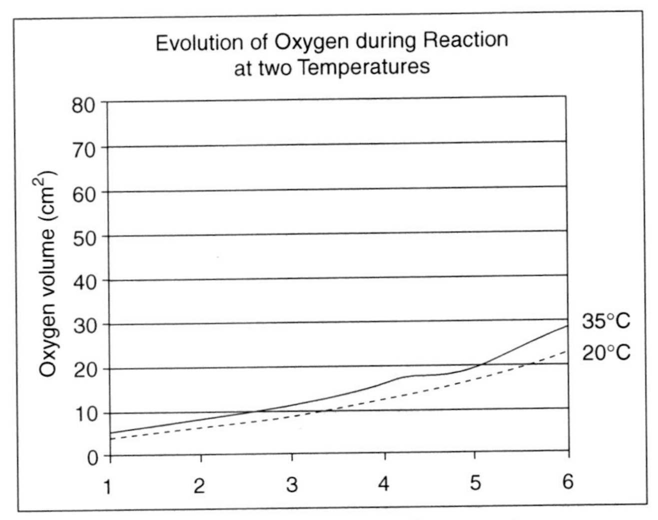

Figure 1. The Evolution of Oxygen (in cm^3) during the Reaction at Two Temperatures.

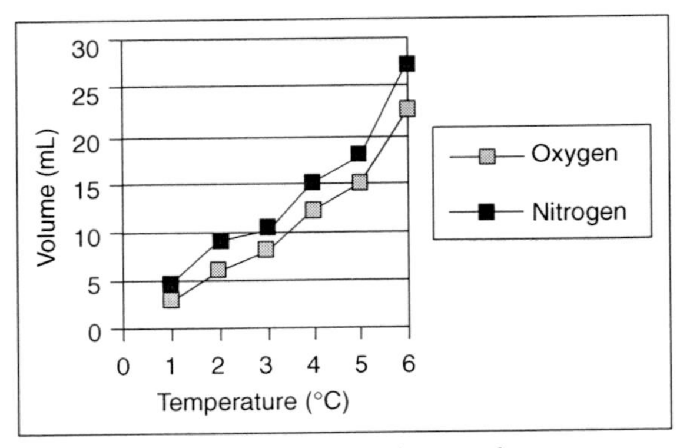

Figure 2. Relationship between volume and temperature for two gases.

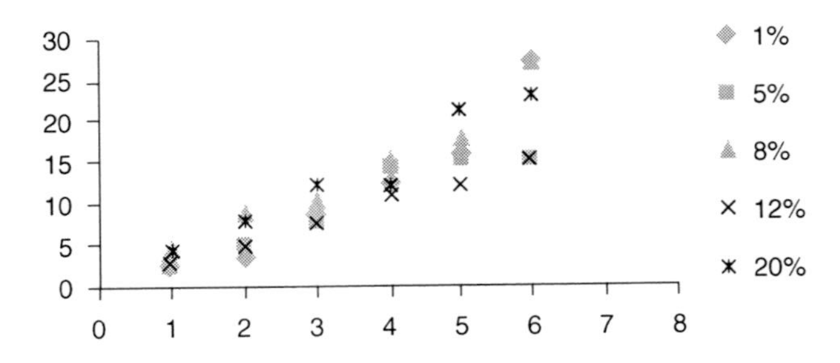

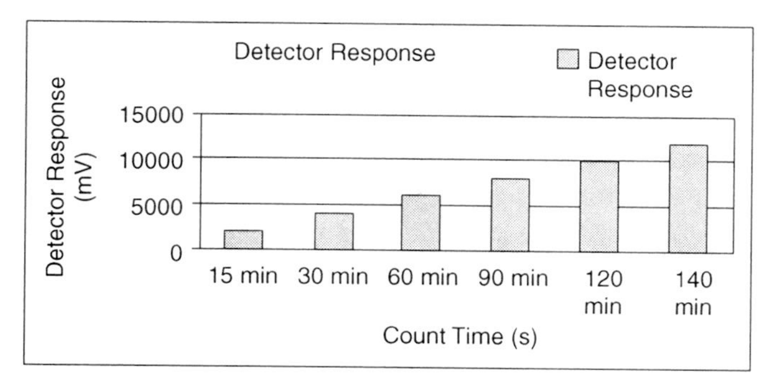

Figure 4. Detector response (mV) for various count times (s).

 **Exercise 16.5**

Reexamine the drawing of instrumentation, the chromatograms of GC data, and the photographs of comet assays in excerpts 3S, 4C, and 4E, respectively. Which guidelines, if any, in Table 16.1 offer insights on the formatting of such figures?

Tables

Purpose and Use

Tables are used to summarize, group, and highlight numerical data so that they can be viewed collectively. Tables are numbered consecutively throughout the paper (Table 1, Table 2, etc.) and mentioned by name and number in text preceding the table. Like figures, tables are designed to conserve space, not waste it. A general rule of thumb is to use a table only if you have enough data to fill at least three rows and three columns; otherwise, describe the data in the text. Also, be careful not to report the same data in both a figure and a table; typically, data should be reported in either a table or a figure, not both. Exceptions do occur; for example, a table could be used to present a comprehensive set of data and then one or more graphs could be used to highlight important features or trends in these data. But do not simply repeat the data in two formats.

Tables

In general, tables summarize, group, and highlight numerical data so that the data can be viewed collectively.

A bare minimum for a table is three rows and three columns.

Be sure to call out (i.e., mention) the table in text preceding the actual table.

Table Formatting

We now consider how to format a table. You may be surprised at what a properly formatted table looks like. Novice authors often include unnecessary frills in their tables, such as bolding, gridlines, and italics. Hence, a properly formatted table may be simpler than you expect. Once again, we use before-correction and after-correction tables (figures 16.3 and 16.4, respectively) to walk you through the formatting process. Exercise 16.6 guides you in this activity. Essential guidelines for formatting a table are summarized in table 16.2.

Table 16.2 Some generally accepted guidelines for formatting tables. (Because formatting conventions vary, consult the appropriate Information for Authors for specific guidelines.)

Attribute	Description
Alignment, columns	Use one type of alignment (left-justified or centered) per column. Within a single table, columns may be aligned differently.
Alignment, numbers	Align numbers on the decimal. Include a leading zero for numbers <1 (e.g., 0.5). Include zeroes to the right of the decimal point only if they are significant figures (e.g., 10.0 when the tenths place is known).
Alignment, text	Left-justify word entries (e.g., ethanol or EtOH).
Color	Use black and white for tables, unless the table is in a poster.
Column entries	Use lowercase for words in columns (e.g., ethanol not Ethanol) unless the word is normally capitalized (e.g., EtOH). Use unbolded font unless the entry is normally bolded (e.g., **2**, when **2** is a compound label).
Column headings and units	Give every column a brief heading. In general, headings are unbolded, except for normally bolded items (e.g., **2** as a compound label). Include units in the headings, in parentheses, not with the values in the columns. (The unit alone as a column heading is not acceptable.) Format units and symbols properly (e.g., °C, μg, m^3). A second line may be used for the unit. In this case, drop single-line headings to the second line.

continued

Table 16.2 (continued)

Attribute	Description
Column headings, capitalization	Consult guidelines because practices vary. Some journal articles use lowercase for headings (e.g., reaction time); others use title case (e.g., Reaction Time) or sentence case (e.g., Reaction time). Title case is often used in posters. Capitalized abbreviations (e.g., T for temperature) are capitalized even in lowercase headings. Capitalize headings consistently throughout your work.
Column order	Place independent variables (i.e., what you controlled) in the leftmost columns, in a logical order. Place dependent variables (i.e., what you measured) in the rightmost columns. Hence, tables are read from left to right, from controlled variables to measured values.
Entry column	Include an entry column (leftmost column) to list sample numbers or trials. Include this column only if you refer to specific entries in the text. Order entries according to their presentation in the text, and number them sequentially (1, 2, 3, etc.). Do not use labels that were used in the work (e.g., S0101208) because they are meaningless to readers. Center entry numbers under the heading "entry".
Font and font size	Strive to use the same font in the table as in the text. If necessary, reduce the table font by one size. Use the same font and font size in the entire table (i.e., for titles, headings, entries, and footnotes).
Footnotes	Use footnotes, if necessary, to share information about the whole table, explain column headings, define abbreviations, or provide additional information. Footnotes must be referenced in the table (using, e.g., a superscript a, b, c); footnotes are placed beneath the table (either above or below the bottom horizontal line). Use the same font size in the footnote as in the rest of the table.
Gridlines and boxes	Avoid gridlines, boxes, and vertical lines. (Such lines may be visible on your computer but should not be visible in the printed document.) Use only horizontal lines; two or three are most common. In the two-line format, horizontal lines are placed above the table (and table title) and below the table; in the three-line format, horizontal lines are placed below the table title, below the column headings, and below the table.

continued

Table 16.2 (continued)

Attribute	Description
Repeated values	Do not include a column that repeats the same value for each row. Instead, state the value in the table title or in a footnote (e.g., The temperature was 25 °C).
Size and orientation	Try to fit your table to the width of one page. If necessary, exchange rows and columns, or break a long table into smaller ones. Avoid horizontal tables (landscape orientation).
Spacing and empty cells	Single-space table titles and table entries to minimize unused space. Similarly, plan your table so that there are no empty cells (or very few).
Titles	Place a title above the table. Begin with an identifier (e.g., **Table 2.**). Continue with a brief, informative descriptor that can be understood apart from the text (e.g., Reaction conditions for synthesis of **3**). The descriptor is usually not a complete sentence. Consult guidelines for formatting titles because practices vary. The identifier is usually bolded (e.g., **Table 2.**), but the descriptor is sometimes bolded and sometimes not. Descriptors can be in sentence case (ending with a period) or in title case (omitting the period). Some titles are left-justified; others are centered. Three examples are included below: **Table 1.** Reaction conditions for acid synthesis. **Table 1.** Reaction Conditions for Acid Synthesis **Table 1. Reaction Conditions for Acid Synthesis**

Exercise 16.6

Examine the before-correction table in figure 16.3.

a. What formatting features in the table in figure 16.3 do you think are correct? For example, should there be a column for temperature (75 °C), or is it correct to state the temperature only once in a footnote? Should the title be positioned above the table (as shown) or be placed below the table? The dependent variable (what is measured) should be in the last column. Is it?

b. What formatting features in the table in figure 16.3 do you think are incorrect?

c. Now compare the tables in figures 16.3 and 16.4. Identify the ways in which inappropriate formatting in the table in figure 16.3 has been corrected in the table in figure 16.4.

Table 1. Yield of **3** in reaction of **1a** and **2a**.

NO.	SOLVENT	TIME	2a (equiv)	YIELD OF 3
S2T1A4	Ethanol	40 min	.5	10%
S2T1A5	Ethanol	60 min	1.1	16%
S2T1A6	Ethanol	90 min	2.2	20%
S3T1A1	CH_3CN	40 min	.5	35%
S3T1A2	CH_3CN	60 min	1.5	52%
S3T1A3	CH_3CN	90 min	2.0	65%

All reactions were conducted at 75 °C.

Figure 16.3 (Before correction) An example of a table with common formatting errors.

Table 1. Yield of **3** in the reaction of **1a** and **2a.**[a]

entry	solvent	time (min)	**2a** (equiv)	yield of **3** (%)
1	ethanol	40	0.5	10
2	ethanol	60	1.1	16
3	ethanol	90	2.2	20
4	CH_3CN	40	0.5	35
5	CH_3CN	60	1.5	52
6	CH_3CN	90	2.0	65

[a]All reactions were conducted at 75 °C.

Figure 16.4 (After correction) The same table as in figure 16.3 but with the formatting errors corrected.

Exercise 16.7

The following three tables have formatting problems.

a. Find at least three problems in each table.

b. How would you correct these formatting problems? Propose solutions that are in line with standard formatting expectations.

Table 1. Reaction conditions for the formation of product 3.

COMPD.	ATM. PRESSURE	TEMP. (deg C)	Yield after 24 hours
BENZENE	1	50	10%
TOLUENE	1	50	50%
XYLENE	1	50	20%
TOLUENE	.5	50	80%
TOLUENE	1.5	50	40%
TOLUENE	.5	100	95%

Table 2. Reaction Conditions for Formation of **1a**

Sample ID #	SOLVENT	RXN TIME	YIELD 1a (%)	PRESS. (atm)
B1_20_1	benzene	20 min	10	1
T1_20_1	toluene	20 min	20	1
X1_20_1	xylene	20 min	30	1
X2_40_1	xylene	40 min	70	1
X3_60_1	xylene	60 min	60	1
X4_40_0.5	xylene	40 min	95	0.5
X5_40_1.5	xylene	40 min	75	1.5

Table 3. Reaction conditions for the formation of **2a**.

Entry	solvent	T (°C)	P (atm)	Yield (%)
1	benzene	50	1	60
2	toluene	50	1	30
3	xylene	50	1	10
4	benzene	75	1	50
5	benzene	100	1	75
6	benzene	100	.5	60

Exercise 16.8

Convert the following information into a properly formatted table, with an appropriate table title. Assume that you are preparing the table for an expert audience.

- Lead was extracted from nine replicates of NIST standard reference material 2709 (a contaminated soil) using various methods. The efficiencies of the various extraction methods were compared.
- Six samples (1–6) were digested using microwave-assisted acid procedures. Each sample was run in triplicate.
    - Samples 1–3 used a 3:1 nitric acid (HNO_3)/hydrogen peroxide (H_2O_2) mixture and were extracted for 120, 60, and 30 min, respectively.
    - Samples 4–6 used a 2:1 HNO_3/hydrochloric acid (HCl) mixture and were extracted for 120, 60, and 30 min, respectively.
    - The extraction efficiencies (averaged over three runs) for samples 1–6 were 66, 56, 59, 80, 75, and 73%, respectively.

- Three samples (7–9) were extracted using sonication in a 10:1 methanol (MeOH)/water (H_2O) mixture for 120, 60, and 30 min, respectively. Each sample was run in triplicate. Extraction efficiencies (averaged over three runs) were 5, 8, and 13%, respectively.

Exercise 16.9

Glance through the tables in excerpts 4A, 4D, 4E, 4F, and 4G in chapter 4. How well do the tables adhere to the formatting guidelines presented in this chapter? What obvious deviations do you see?

Schemes

Purpose and Use

Schemes are used to depict a series of steps that progress in time. (Note that schemes differ from charts, which list groups of compounds or structures that do not change in time.) Most commonly, schemes are used to illustrate chemical reactions. In such cases, schemes often include arrows (e.g., to denote a forward reaction, resonance, equilibrium, and/or electron movement), intermediates, transition states, reactants, and products. Schemes are numbered in order of appearance in the text (Scheme 1, Scheme 2, etc.). As with tables and figures, the scheme is mentioned in the text before the scheme is encountered. Schemes are perhaps most common in Discussion sections of journal articles (e.g., to illustrate proposed mechanisms) but can appear most anywhere in journal articles, posters, and proposals.

Schemes

In general, schemes depict a series of steps that progress in time.

Scheme Formatting

The formatting of schemes is quite straightforward; hence, we include only one example of a properly formatted scheme. (Often the most challenging part of a scheme is to correctly draw the reaction using chemistry-drawing tools, skills that are beyond the scope of this textbook.) An example of a correctly formatted scheme is shown in scheme 16.1, which depicts the addition of Br_2 to butane to give (2*R*, 3*R*)-2,3-dibromobutane **1**. The reaction proceeds through a bromonium carbocation intermediate. What types of formatting do you notice in this scheme? Exercise 16.10 helps you answer this question. Essential guidelines for formatting a scheme are summarized in table 16.3.

Table 16.3 Some generally accepted guidelines for formatting schemes. (Because formatting conventions vary, consult the appropriate Information for Authors for specific guidelines.)

Attribute	Description
Compound labels	Label only those compounds (and other structures) in a scheme that are referred to in the text. Do not label all compounds. Use bolded numbers and/or letters (e.g., **2**, **3a**). Center the labels just below the compounds.
Titles	Place a title above the scheme. Minimally, the title includes an identifier in bolded font with no period (e.g., **Scheme 2**). Sometimes a short descriptor follows the identifier, but this is not required. If included, the descriptor is in title case (with no period) and may be bolded or not. The title is usually centered but may be left-justified. Three examples follow: **Scheme 1** **Scheme 1. Formation of Carbocation Intermediate** **Scheme 1.** Formation of Carbocation Intermediate

Scheme 16.1

Exercise 16.10

Consider five schemes in this textbook: Scheme 16.1 (above), Schemes 1 and 2 in Boesten et al. (2001) at the end of chapter 2, scheme 4.1 in chapter 4, and Scheme 1 in excerpt 5D (Demko and Sharpless, 2001). Use these schemes to answer the following true or false questions:

a. Every compound in a scheme should have an alphanumeric label.
b. The title goes above the scheme.
c. Schemes should depict at least two different reactions.
d. It is customary to include reactants above the arrows in schemes.
e. Most scheme titles include a brief description.
f. Scheme titles use bolded font.
g. Periods are included after the scheme title (e.g., **Scheme 1**.).

Exercise 16.11

Reexamine Schemes 1 and 2 in Boesten et al. (2001) at the end of chapter 2. At the same time, consider Figure 2 in the same article. Could Figure 2 have been a scheme? If so, does this suggest that authors have a choice to use a figure or a scheme?

16 Formatting on Your Own: Format a Figure, Table, or Scheme

Format your graphics (figures, tables, and/or schemes) according to the general guidelines presented in this chapter. It is advisable to consult other sources (e.g., Information for Authors for specific journals, funding agencies, or conferences; *The ACS Style Guide*) to find out if (and how) their expectations differ from the general guidelines presented here.

Pay attention to the formatting expectations specified in this chapter, as small or as large as they may be. In all cases, check your punctuation (periods, commas, semicolons, parentheses, etc.), capitalization, fonts, bolding, sequential numbering, and, of course, clarity of presentation. Make sure that you mention the graphic(s) by type and number (e.g., Figure 2, Table 1, Scheme 3) in the text of your journal article, research proposal, or poster (unless you do not have room in your poster) before the reader encounters the actual graphic.

For figures, pay special attention to axes, boxes, captions, colors, legends, lines, and symbols.

For tables, pay careful attention to footnotes, gridlines, units, titles, and column headings, entries, and sequencing.

For schemes, pay close attention to scheme titles and structure labels.

After you have formatted your graphic(s), set aside time to proofread your work. Consult chapter 18 for guidance.

17 *Formatting Citations and References*

An author should cite those publications that have been influential in determining the nature of the reported work and that will guide the reader quickly to earlier works that are essential for understanding the present investigation.

—Adapted from the American Chemical Society, *Ethical Guidelines to Publication in Chemical Research* (https://paragon.acs.org/)

In this chapter, we introduce you to formatting conventions for **citations** and **references**. Writers are obligated to cite others' works that have significantly influenced or are relevant to their own work. It goes without saying that to include another's original ideas without proper acknowledgment is plagiarism. Consequently, this stage of the writing process is critical both professionally and ethically. Although citations and references are inextricably linked, for clarity we deal with them separately. By the end of the chapter, you will be able to do the following:

- Know what information should be cited
- Know what information need not be cited
- Use in-text citations appropriately
- Recognize different reference formats (numerical and alphabetical)
- Determine the appropriate reference format for your work

Citations and References

Citation: An authoritative source referred to in the text to indicate work that has influenced the current work or that provides essential background information.

Reference: Bibliographical information for in-text citations.

Two Citing on Your Own tasks will guide you in preparing proper citations and references as you do the following:

17A Finalize citations

17B Compile and format references

The seriousness of properly compiling and formatting citations and references is predictably emphasized in Information for Authors documents. The excerpts below serve as vital reminders of the importance of checking relevant guidelines before finalizing citations and references in your written work:

> Authors should be judicious in citing the literature; unnecessarily long lists of references should be avoided. (Author Information, *J. Org. Chem.* **2007**, *72*, 16A)
>
> The accuracy and completeness of the references are the authors' responsibility. (Authors' Guide, *Anal. Chem.* **2007**, *79*, 390)
>
> Avoid references to works that have not been peer-reviewed. (Instructions to Authors, *Environ. Sci. Technol.* **2007**, 30)
>
> The citation of references in text . . . varies widely from journal to journal and publisher to publisher. . . . Authors are encouraged to check the author guidelines for a specific publication to find information on citing references. (*The ACS Style Guide*: Coghill and Garson, 2006)

Most authors attend to citations and references when their work is nearly complete; useful reminders may be inserted into the text as they write—such as "cite Kopinski here" or "add ref"—but properly formatted citations and references are often added in the last stages of writing. Such timing is prudent because the text typically undergoes changes until it is considered done, which often require additions and deletions of citations. When you are ready to finalize your citations and references, do so with great care. Be certain that each citation is correct and perfectly matched to its corresponding reference. Pay careful attention to punctuation (e.g., periods, commas, semicolons, parentheses), fonts (e.g., italic, bold), and information sequencing. Although a simple comma (or perhaps a bold comma) and an italic number in parentheses might seem trivial, their appropriate use signals a concern for detail that is expected of good writers (and good chemists) and by the individuals who will read and review the work.

Citations

Many chemistry students, particularly second language students, are intimidated by the polished writing styles of some renowned chemists. Rather than paraphrasing what the experts have said, many students are tempted to use quotations to retain the excellent writing style. However, because my goal is to teach our students to become better writers, I essentially forbid the use of quotations in their papers. I tell them that "one quote is two too many."

—Donald Paulson, California State University–Los Angeles

Citations, inserted into the text of your written work, identify publications that have influenced your thinking and that provide pertinent background information for your readers. Although citations are formatted in different ways, they serve the same purposes, such as acknowledging others' works, enhancing the writer's credibility, placing the current work into a broader context, and making it easy for readers to locate cited materials.

You are likely already familiar with citations and have used them countless times in other written works. Nonetheless, we must point out one important difference between citations in chemistry and those in many other fields. The difference pertains to **direct quotations**, in which you quote verbatim the words of others. Although commonplace in many genres, direct quotations are exceedingly rare in chemistry writing. They are virtually nonexistent in journal articles and uncommon in posters and research proposals. Thus, no examples of direct quotations are included in textbook excerpts. Instead, we include examples of how chemistry writers paraphrase the pertinent ideas of others, citing sources appropriately, without quoting them directly. We encourage you to adopt the practice of **paraphrasing** in your own chemistry-specific writing.

Direct Quotes

Although common in other genres, direct quotes should be avoided in chemistry journal articles and used sparingly in scientific posters and research proposals.

direct quotes

Paraphrasing

See Writing on Your Own task 6A.

The largest number of citations is found in the Introduction sections of journal articles, posters, and research proposals, consistent with the purposes of the section. Introductions of both journal articles and research proposals often include 15 or more citations in opening paragraphs (often with multiple citations in a single sentence). Far fewer citations (sometimes even none) are included in poster Introductions because of space limitations and the poster's role in emphasizing new results.

New citations, as well as repeated citations to works first mentioned in the Introduction, appear in other sections of the journal article, poster, and proposal, but with far less frequency. For journal articles, the second largest number of citations (after the Introduction) typically occurs in the Discussion section; here, citations are needed as authors interpret their data in light of the prevailing literature. In posters and proposals, citations are often second most abundant in the Methods section, reflecting the author's need to cite precedent methodology and instrumentation. Not surprisingly, the smallest number of citations in all three genres is typically found in the Results (or Preliminary Results) section, where authors focus largely on their own work.

How many works total should you cite in your written work? The answer to this question varies for each genre. Typically, posters have the fewest citations, often less than five; hence, only the most essential works are cited. The number increases for journal articles and proposals and is influenced by what is judged as appropriate to cite in each genre.

How Many Works Should You Cite?

Authors normally cite fewer works in journal articles than in research proposals. *Environmental Science & Technology* recommends fewer than 20 references, while *Analytical Chemistry* specifies no more than 50.

Authors often cite 50–100 works in research proposals.

In journal articles, only essential works (i.e., those that have critically influenced your research) are cited; marginally relevant or tangential works should not be cited. The goal is to read widely, but cite only the most pertinent, precedent works. In general, only information (e.g., methodologies, instrumentation, results) from the primary literature needs to be cited. The primary literature includes journal articles, technical reports, and other scholarly works but normally does not include sources of general knowledge, such as textbooks. This is not to say that you should not read textbooks or allow textbook-like materials to influence your writing. By all means, read widely in your field, both at introductory and more advanced levels. However, if you decide to include general knowledge in your manuscript, rephrase the information in your own words and omit the citation.

Citing the Primary Literature

The primary literature includes peer-reviewed journals, technical reports, patents, edited and scholarly books, and other publications that present original ideas and research.

In contrast, general information (gleaned, e.g., from textbooks) is not part of the primary literature.

To illustrate content that is too general to cite, we include excerpts from four journal articles (excerpts 17A–17D). For contrast, we also include examples of cited content from these same articles.

Excerpt 17A (adapted from Beck et al., 2004)

Not cited

Heterocycles are among the most important structural classes of chemical substances and are particularly well represented among natural products and pharmaceuticals. It is estimated that far more than 50% of the published chemical literature concerns heterocyclic structures. One striking structural feature inherent to heterocycles, which continues to be exploited to great advantage by the drug industry, lies in their ability to manifest substituents around a core scaffold in a defined three-dimensional representation, thereby allowing for far fewer degrees of conformational freedom than the corresponding conceivable acyclic structures.

Cited

An especially effective and fruitful way to synthesize heterocycles is by isocyanide-based multicomponent reaction (MCR).[1]

Excerpt 17B (adapted from Celis et al., 2000)

Not cited

The existing methods for the removal of heavy metals from the environment can be grouped as biotic and abiotic. Biotic methods are based on the accumulation of heavy metals by plants or microorganisms; abiotic methods include physicochemical processes such as precipitation, coprecipitation, and adsorption of heavy metals by a suitable adsorbent.

Cited

Among the different adsorptive materials that have conventionally been used to capture metal ions from solution are activated charcoal (*2*), zeolites (*3*, *4*), and clays (*5*).

Excerpt 17C (adapted from Huang and Chen, 2002)

Not cited

The Mukaiyama aldol reaction is one of the most important means for C–C bond formation. Silyl enol ethers react with aldehydes in the presence of Lewis acids to give β-hydroxy carboxylates.

Cited

Using chiral catalysts, not only various enantioselective Mukaiyama[1] and vinylogous Mukaiyama[2] aldol reactions have been developed but also asymmetric reactions of α,α-difluoro silyl enol ethers (**1**) with carbonyl compounds have been reported.[3]

Excerpt 17D (adapted from Daglia et al., 2002)

Not cited

However, coffee beverages are complex mixtures of several hundred chemicals that either occur naturally or are later induced in coffee by the roasting process in the form of nicotinic acid or melanoidins.

Cited

Coffee, in particular roasted coffee, has been found to act as a potent antioxidant and to inhibit lipid peroxidation both in chemical (*1*) and in biological systems in rat liver microsomal fractions (*2*).

Exercise 17.1

Consider the following passage, adapted from a journal article Introduction section. Which sentences should include citations? Explain how you arrived at your decisions.

Seaweed was once widely used as a fertilizer in coastal regions of the Atlantic seaboard of Europe and elsewhere access to this resource was available and arable farming was conducted. Seaweed is rich in a wide range of nutrients, improves soil texture and quality, and is a renewable and sustainable resource as it is deposited routinely in large quantities on many beaches. For coastal farms, this resource is on their doorstep and requires only limited transportation.

With the advent of modern fertilizers, seaweed fell out of use because of the ease of application of formulated pesticides and because the contamination of seaweed washed up on beaches with plastic waste (fishing lines and ropes, bottles, etc.). With the onset of green agriculture and the increasing demand for organically farmed produce, seaweed use as a fertilizer has come back into fashion. . . .

One factor has been overlooked in this reversion to seaweed use. Seaweed naturally contains high levels of arsenic, typically between 20 and 100 mg kg^{-1} dry weight (dw). Thus, sustained use of seaweed may lead to the buildup of arsenic in soils. The dominant species of arsenic in these seaweeds are in the form of arsenoribofuranosides (arsenosugars). These are assumed to be relatively nontoxic to humans and animals as compared to inorganic species. The arsenosugars are metabolized to different organoarsenic species but mainly to DMA(V) (dimethylarsinic acid) when consumed as a food source. (Adapted from Castlehouse et al., 2003)

Research proposals typically have the largest number of citations; guidelines regarding what is appropriate to cite are less stringent for proposals than for journal articles, for several reasons:

1. Reviewers want to see that you are familiar with the literature in your field; hence, you need to cite more than only the most pertinent works.
2. The research proposal is not a published document; hence, you may cite references that are not part of the primary literature.
3. In a proposal, references are often included in a separate section (without a page limit); hence, space is not an issue.
4. Although it is unwise to pad your references with marginally relevant or overly general works, because a proposal is often read by nonexperts, relevant sources of general information may be cited.

For these reasons, a 15-page proposal will often contain 50–100 cited works.

Writers must not only know *what* to cite, but also *how* and *where* to cite. Most chemistry genres employ one of three citation formats:

- superscript numbers
- italic numbers in parentheses
- author name and year of publication, known as author–date format

All three formats are approved by the ACS, but the vast majority of ACS publications employ one of the two numerical formats (the first two options listed above).

In-Text Citations

Most chemistry-related genres use one of three citation formats:

Superscript numbers

Italic numbers in parentheses

Author–date format

We examine all three formats in this chapter, but before we do, we note an important writing convention in chemistry: in general, authors are not referred to by name in the actual sentences that describe their work. Consider the following examples, illustrated with the superscript number and author–date formats:

Not preferred	Jung et al.[1] have shown that ozonolysis can be used to effectively control the taste and odor of drinking water.
Preferred	Ozonolysis has been used to effectively control the taste and odor of drinking waters.[1]

Not preferred	Jung et al. (2004) have shown that ozonolysis can be used to effectively control the taste and odor of drinking water.
Preferred	Ozonolysis has been used to effectively control the taste and odor of drinking waters (Jung et al., 2004).

In the preferred sentences, the emphasis is placed on the science (ozonolysis), rather than the scientists (Jung et al.), and on conciseness. On occasion, names are used in formal science writing (last names only), but this practice should be kept to a minimum.

Note the use of "et al." in the preceding examples. The abbreviation "et al." (from the Latin *et alia*) means "and others" and is used to refer to works that have three or more authors. In the author–date system, the abbreviation "et al." follows the last name of the first author every time the work is cited. In most ACS journals, "et al.", if it is used, is used without capital letters, quotation marks, or italics. In some non-ACS journals, "et al." is italicized ("*et al.*"), but the rules regarding capital letters and quotation marks are the same. Because "et al." means "and others," the verb following "et al." must agree with the plural subject (e.g., "Snow et al.[2] show…"). Consider the following incorrect and correct examples (based on ACS guidelines):

Incorrect
(a) Snow, Creus, and Kretzer[2] show…
(b) Snow, et al. (2006) show…
(c) Snow et al.[2] shows…
(d) Snow *et al.*[2] show…
(e) Snow et. al.[2] show…
(f) Both metals show similar results (Snow, et al., 2006).

Correct
(g) Snow et al.[2] show…
(h) Snow et al. (2006) show…
(i) Both metals show similar results (Snow et al., 2006).

et al.: Do's and Don'ts

Do use
- a period after "al."
- a space between "et" and "al."
- the plural form of the verb after "et al."

Don't use
- capital letters
- a comma between the name and "et"
- quotation marks
- italics (in most ACS journals)
- a period after "et"

Every time a work is cited, the complete and correct citation must be used. Thus, "(Snow et al., 2000)" or "[2]" or "(2)" must be used every time the work by Snow and others is cited. If you are using the name in the text, the phrase "and co-workers" should be used instead of "et al." when you cite two or more references with the same principal author but different coauthors. For example, "Snow and co-workers[2,3]" is used when citation 2 refers to an article by Snow, Fernandez, and Yazzie and citation 3 refers to an article by Snow, Garcia, and Myint.

Exercise 17.2

Identify what is wrong with the use of "et al." in the incorrect examples a–f above.

Superscript Numbers

The most common citation format utilizes superscript numbers. The first citation is assigned the number 1; subsequent citations continue in consecutive order throughout the text ([1], [2], [3], etc.). The same number is used each time the identical work is cited. Thus, every time the work that is assigned the number 1 is cited, the number 1 is used, even when it is cited after works assigned higher numbers. Some useful guidelines include the following:

Superscript Numbers

Superscript numbers should be placed as close as possible to cited information. Thus, they may appear anywhere in the sentence.

Superscript numbers are placed immediately after the punctuation in the text.

- The superscript number is placed as close as possible to the cited information; consequently, superscript numbers can occur anywhere in the sentence. Thus, a superscript citation should not be placed at the end of a sentence if it refers to information mentioned earlier in the sentence.
- When the cited authors are named in the sentence, the superscript number is placed immediately after the name of the last cited author (with no space between the name and number).

  Al-Dahbi[3] Kim and Jeon[7] Soussi et al.[8]

- Superscript numbers are placed immediately after any punctuation (e.g., commas, periods, semicolons), with no space between the number and punctuation:

  . . . demonstrated.[6] This suggests . . .
  . . . demonstrated,[6] suggesting that . . .

- Within the superscript, commas (without spaces) are used to separate consecutive or nonconsecutive superscript numbers; en dashes (–), rather than hyphens (-), are used to indicate a range of three or more consecutive numbers. For example,

  . . .[8,9]

  . . . ,[1,3,5]

  . . . ;[10–13,15,21]

  . . . .[8,10–12,14]

Some authentic examples from the literature are provided in excerpts 17E and 17F.

Excerpt 17E (Guo et al., 2002)

In 1988, Tanaka et al.[8] first reported the use of ultrafine metal powder in protein analysis. Since then, many inorganic materials, including graphite particles,[9–12] fine metal or metal oxide powder,[13] silver thin-film substrates or particles,[14] and silica gel,[15] have been used in the MALDI-TOF analysis of low-mass molecules.

Excerpt 17F (adapted from Monteiro and Hervé du Penhoat, 2001)

To date, D_t coefficients of carbohydrates established with the PFGSE approach[13,17,22–26] have been undertaken to (1) validate the theoretical self-diffusion coefficients calculated from MD trajectories,[13,25] (2) demonstrate the complexation of lanthanide cations by sugars,[17] (3) probe the geometry of a molecular capsule formed by electrostatic interactions between oppositely charged β-cyclodextrins,[23] (4) study the influence of concentration and temperature dependence on the hydrodynamic properties of disaccharides,[24] and (5) discriminate between extended and folded conformations of nucleotide-sugars.[26]

Exercise 17.3

Examine the superscript citation formatting in this passage and correct formatting errors.

Carbon nanotubes (CNTs) have potential applications in fields such as molecular electronics[8], conductive polymers[8,9,10], and energy storage.[2–3,5, 7]

Italic Numbers in Parentheses

Another common citation method is to use italic in-line (not superscript) numbers placed inside parentheses. Like superscript citations, these begin with (*1*) and continue in consecutive order. Numbers are repeated when the same reference is being cited. Other guidelines include the following:

Italic Numbers in Parentheses

Like superscript numbers, italic numbers in parentheses should be placed as close as possible to cited information. Thus, they may appear anywhere in the sentence.
Italic numbers are placed inside parentheses and before punctuation.

- These numbers are always italicized so that they can be distinguished from other numbers in the text.
- A space is included between words in the sentence and the left and right parentheses. No space is inserted after the right parenthesis when punctuation follows it:

  Results for nanoparticles (*4*) have been reported.
  Results have been reported for nanoparticles (*4*).

- As with superscript numbers, the citation is placed as close as possible to the cited information or the last cited author's name.
- Unlike superscript numbers, the italic in-line citation comes before, rather than after, punctuation.
- Multiple citations are separated by commas (with spaces after the commas); a range of three or more citations in sequence are separated by en dashes. For example,

  . . . (*8, 9*) . . .
  . . . (*1, 3, 5*), . . .
  . . . (*10–13*); . . .
  . . . (*8, 10–12, 14*).

Some authentic examples from the literature are provided in excerpts 17G–17I.

Excerpt 17G (adapted from Bogialli et al., 2004)

We have proposed rapid methods for determining 12 sulfonamide antibacterial in bovine and fish tissues (*14, 15*), milk, and eggs (*16*). These methods are based on analyte extraction from the matrix dispersed on sand by hot water followed by injection, directly (*14*) or after little manipulation (*15, 16*), of a large aliquot of the extract on an LC column.

Excerpt 17H (adapted from Filippova and Duerksen-Hughes, 2003)

There are reports indicating that arsenic compounds can damage cellular DNA, thereby acting as genotoxins, as assessed by the comet or single-cell gel assay. Some reports indicate that arsenite possesses this activity (*5–7*), whereas another indicates that although the methylated trivalent species are genotoxic, arsenite itself is not (*8*).

Excerpt 17I (adapted from Chouchane and Snow, 2001)

The arsenic–GSH complex [$As^{III}(GS)_3$] was synthesized using previously described methods (*25, 26*).

Exercise 17.4

Correct the italic in-text numbers in parentheses in the following passage.

Carbon nanotubes (CNTs) have potential applications in fields such as molecular electronics, (8) conductive polymers,(8,9,10) and energy storage (2-3,5, 7).

Author–Date

The third ACS-approved approach to citing the literature is to use authors' last names and year of publication (author name, year). This format is relatively uncommon in ACS journals but is frequently used in journals of other scientific disciplines, such as *Ecology* and *Reviews of Modern Physics*. With the author–date citation technique, the authors' last names, followed by a comma, a space, and the year of publication are placed in parentheses (or brackets in some journals). Publications with three or more authors are referred to by the last name of the first author, followed by et al., a comma, and the year. Multiple citations are separated by semicolons and listed in alphabetical order (by first author).

One author	(Phyu, 2008)
Two authors	(Uchiyama and Onel, 2008)
Three or more authors	(Abdulaziz et al., 2007)
Multiple citations	(Huang and Eble, 1999; Ureta et al., 2008)

If the authors are named in the sentence, only the year is placed in parentheses:

One author	Mencuccini (2008)...
Two authors	Horn and Patanasorn (2008)...
Three or more authors	Morgensen et al. (2007)...

Two or more publications by the same author(s) in the same year are distinguished by lowercase letters (a, b, c, etc.) after the year:

(Sukhinina et al., 2008a, 2008b, 2008c)

Sukhinina et al. (2008a, 2008b, 2008c)...

To illustrate more fully, excerpts 17G–17I have been rewritten using the author–date format.

Adaptation of Excerpt 17G (adapted from Bogialli et al., 2004)

We have proposed rapid methods for determining 12 sulfonamide antibacterials in bovine and fish tissues (Bogialli et al., 2003a, 2003b), milk, and eggs (Bogialli et al., 2003c).

Adaptation of Excerpt 17H (adapted from Filippova and Duerksen-Hughes, 2003)

There are reports indicating that arsenic compounds can damage cellular DNA, thereby acting as genotoxins, as assessed by the comet or single-cell gel assay. Some reports indicate that arsenite possesses this activity (Hartmann and Speit, 1994; Schaumloffel and Gebel, 1998; Wang et al., 2001), whereas another indicates that although the methylated trivalent species are genotoxic, arsenite itself is not (Mass et al., 2001).

Adaptation of Excerpt 17I (adapted from Chouchane and Snow, 2001)

The arsenic–GSH complex $[As^{III}(GS)_3]$ was synthesized using previously described methods (Scott et al., 1993; Serves et al., 1995).

Exercise 17.5

Examine the author–date citations in the following passage. Correct inappropriate formatting.

Saratz (et al., 2005) reviewed carbon nanotubes (CNTs) and described potential applications in fields such as molecular electronics (Saratz et al.; 2005), conductive polymers (Saratz et al.; 2005; Soussi et al., 2001; Dalǵlu, Yilmaz, and Alptikin 2004), and energy storage (Soussi et al. 2001; Dalǵlu et al., 2004, Mach and Tardy, 2001; Mach and Tardy, 2001).

Exercise 17.6

Novice writers often ask if they can cite a publication more than once in the same paragraph. Review excerpts 17F and 17G. What is the answer? Find at least two examples in each excerpt that illustrate the correct answer to this question.

Exercise 17.7

The author–date format was formerly used in the *Journal of Agricultural and Food Chemistry*. By 2004, the journal was using italic in-line numbers in parentheses, as shown in excerpt 17G. Rewrite the first two sentences in excerpt 17J (published in 2000) so that they follow the italic in-line numbers in parentheses format:

Excerpt 17J (adapted from Gurrieri et al., 2000)

Introduction

The prickly pear cactus (*Opuntia ficus indica*) is a plant highly distributed in the Mediterranean area, Central and South America, and South Africa (Barbera and Inglese, 1993; Munoz de Chavez et al., 1995). Owing to its crassulacean acid metabolism, this plant is characterized by a high potential of biomass production with low water consumption (Barbera and Inglese, 1993; De Cortazar and Nobel, 1992; Dominguez-Lopez, 1995). It is therefore extremely drought tolerant and grows abundantly under semiarid conditions. The content of proteins, carbohydrates, minerals, and vitamins (mostly vitamins A and C) in the fleshy stems (cladodes) is nutritionally significant, and in Mexico people eat them cooked as vegetables with meat or beans.

Exercise 17.8

Passages a–e illustrate three different citation formats; however, there is at least one mistake in each passage. Without converting to another citation format, revise the passages to correct the citations and related language.

a. Surface-enhanced neat desorption (SEND) is another matrix-free method for small molecule analysis[5]. (Adapted from Guo et al., 2002)

b. Organosilicon reagents have played an increasingly important role in Pd(0)-catalyzed cross-coupling with organohalides.[1,2,3,4,7] (Adapted from Riggleman et al., 2003)

c. Buriak et al. (2007) suggests that laser desorption/ionization can be used successfully on porous silicon. (Adapted from Guo et al., 2002)

d. Arsenic groundwater levels in some areas of Bangladesh groundwater arsenic concentrations approach 2 mg L^{-1}. (Tondel, Rahman, Magnuson, and Chowdhury, 1998) (Adapted from Meharg et al., 2003)

e. [*From the beginning of the Introduction section*] *Cereus peruvianus* (L.) Miller (apple cactus, known also as koubo) is a large thorny columnar cactus, native to the subtropical southeastern coast of South America (3). *C. peruvianus* is also common as an ornamental plant (*1, 2*). It has been recently introduced to cultivation in Israel in the framework of our program for developing new crops suitable for the Negev Desert (*3, 4*). (Adapted from Ninio et al., 2003)

17A Citing on Your Own: Finalize Citations

Finalize who and what you are going to cite. Consider your audience. For expert and most scientific audiences, general information (found in textbooks) is not cited.

Determine the citation format that you will use throughout your written work (superscript numbers, italic numbers in parentheses, author–date, or some variation). Find out if your targeted journal, funding agency, or conference has a preference.

Format your citations so that they follow all expected conventions. Don't guess if you are unsure; consult the appropriate style guide.

Proofread your in-text citations carefully, paying special attention to periods, commas, semicolons, en dashes, italics, parentheses, and the use of et al. Double-check your work.

References

The last required section of most written work is the References. This section goes by various names. In journal article and posters, it is often called References or Literature Cited. In proposals, it is often called References Cited. Whatever the name, its purpose is to offer more detailed information about the publications cited within the text, making it easy for readers to retrieve each cited source, if so desired. As the name implies, only the literature cited in the text should be included in the references. The format that you use for your references should parallel the format that you use with in-text citations:

- References for numerical citations (superscript numbers or italic numbers in parentheses) are listed in numerical order (1, 2, 3, etc.).
- References for author–date citations are listed in alphabetical order.

Formatting References

References should parallel the formatting of in-text citations.

With in-text numerical citations, list references in numerical order.

With in-text author–date citations, list references in alphabetical order by last name of the first author.

Like in-text citations, the formatting of references requires great care. The appropriate use of punctuation, fonts, parentheses, in addition to the inclusion, exclusion, and sequencing of information (e.g., authors, title of article, title of journal, year of publication, page numbers), reveals your attention to detail, expected of good writers and by expert and scientific readers.

For brevity, we focus only on references in journal articles. References in scientific posters are normally truncated versions of journal article references (to conserve space), whereas references in research proposals typically mirror the formats used by journals in the author's field. Moreover, we focus on references to only two types of publications: periodicals (i.e., journal articles) and books. (We refer you to *The ACS Style Guide* for information on other types of publications such as patents, theses, technical reports, and online documents.)

When preparing references, certain miminal information is required. References to journal articles must include the authors' surnames and initials, year of publication, journal name, journal volume number, and page numbers (inclusive page numbers are preferred). References to books must include the authors' surnames and initials, editors' surnames and initials (when applicable), book title, publisher, city of publication, and year of publication. It is the author's responsibility to be sure this information is complete and accurate.

In most chemistry journals, journal names are abbreviated according to the *Chemical Abstracts Service Source Index* (*CASSI*). A few commonly used abbreviations are listed in table 17.1. Note that only journal titles with at least two words are abbreviated. One-word journal names (e.g., *Science*) are not abbreviated.

CASSI

Acronym for the *Chemical Abstracts Service Source Index.*

CASSI abbreviations are used to refer to journal names in reference lists. (See *The ACS Style Guide* for *CASSI* abbreviations.)

In the rest of this chapter, we consider examples of references to journal articles and books. Both numerical and alphabetical references are addressed. These examples are included to increase your awareness of the level of detail that is needed to format references correctly and to showcase the many differences in formatting that exist across journals. If you are preparing a manuscript for submission, do not rely on these examples, but consult a recent issue of the target journal and its Information for Authors before you prepare your references. It is ultimately your responsibility to check that your references adhere to the journal's current practices.

Table 17.1 Sample journal-title abbreviations according to *CASSI*.

Acc. Chem. Res.	*Chem. Res. Toxicol.*	*J. Phys. Chem. A*
AIChE J.	*Environ. Sci. Technol.*	*Langmuir*
Anal. Chem.	*Inorg. Chem.*	*Nano Lett.*
Biochemistry	*J. Agric. Food Chem.*	*Nature*
Biochem. J.	*J. Am. Chem. Soc.*	*Org. Lett.*
Can. J. Chem.	*J. Org. Chem.*	*Science*

Numerical References

When authors use numbered in-text citations (i.e., superscript numbers or italic numbers in parentheses), references are numbered and listed in citation order at the end of the text (1, 2, 3, ...). Publications that are cited more than once in the written work are listed only once in the references. In all cases, the numbers in the reference list must match the numbers cited in the text exactly.

Examples of numerical references to journal articles are shown in table 17.2. Some ACS journals require authors to include the title of the article in the

Table 17.2 Examples of numerical references for journal articles.

Source	Example
Without article title	
The ACS Style Guide	(8) Klinger, J. *Chem. Mater.* **2005**, *17*, 2755–2768.
Anal. Chem.	(8) Kamholz, A. E.; Weigl, B. H.; Finlayson, B. A.; Yager, P. *Anal. Chem.* **1999**, *71*, 5340–5347.
J. Org. Chem.	(8) Keipert, J. S. J.; Knobler, C. B.; Cram, D. J. *Tetrahedron* **1987**, *43*, 4861–4874.
J. Phys. Chem. C	(8) Valiullin, R.; Furo, I. *J. Chem. Phys.* **2002**, *116*, 1072–1076.
With article title	
The ACS Style Guide	(8) Klinger, J. Influence of Pretreatment on Sodium Powder. *Chem. Mater.* **2005**, *17*, 2755–2768.
Biochemistry	8. Gilles-Gonzalez, M. A., and Gonzalez, G. (2005) Heme-based sensors: defining characteristics, recent developments, and regulatory hypotheses, *J. Inorg. Biochem. 99*, 1–22.
Chem. Res. Toxicol.	(8) Evans, D. C., Watt, A. P., Nicoll-Griffith, D. A., and Baillie, T. A. (2004) Drug-protein adducts: an industry perspective on minimizing the potential for drug bioactivation in drug discovery and development. *Chem. Res. Toxicol. 17*, 3–16.
Environ. Sci. Technol.	(8) Shoeib, M.; Harner, T.; Vlahos, P. Perfluorinated chemicals in the Arctic atmosphere. *Environ. Sci. Technol.* **2006**, *40*, 7577–7583.
J. Agric. Food Chem.	(8) Sata, N. U.; Fusetani, N. Amaminols A and B, new bicyclic amino alcohols from an unidentified tunicate of the family Polyclinidae. *Tetrahedron Lett.* **2000**, *41*, 489–492.

reference; others prefer that authors omit the titles. Examples of both practices are shown. In each case, we include the generic format recommended in *The ACS Style Guide* and several examples from actual ACS journals. To emphasize that these are numerical references, each reference is given the number 8.

As shown in table 17.2, ACS journals do not necessarily follow the format suggested in *The ACS Style Guide*. This is because *The ACS Style Guide* is just that, a guide; each journal is allowed to establish its own formatting practice. Note the following similarities and differences among the references in table 17.2:

- Indentation styles differ across journals.
- Authors' names are listed last name first, followed by initials only (no first names). Initials are followed by a period and separated by a single space (e.g., Adams, C. B.). Authors' names are usually separated by semicolons. In some instances, the word "and" precedes the name of the last author.
- *The ACS Style Guide* recommends title case for journal article titles (i.e., capitalizing the first letters of all major words), but the ACS journals examined use sentence case (i.e., capitalizing only the first letter of the first word and proper nouns).
- Journal names are italicized and abbreviated according to *CASSI*. No punctuation is included after the journal name, unless the name ends with an abbreviation (e.g., *Biochemistry* but *Anal. Chem.*).
- The journal volume number is italicized in all examples.
- The placement and formatting of the publication year differ across journals.
- Page numbers are inclusive, that is, they include the first and last page number of the article (e.g., 117–128 or 117–28). (Most ACS journals require inclusive page numbers; a few permit first page only.) No commas are used for page numbers ≥1000, and an en dash (–) is used between the first and last page.
- All references end with a period.

A Bold or Italic Comma?

The ACS Style Guide recommends that punctuation in references take on the formatting (bolding or italics) of the preceding number, for example,

bold year and comma: **2008,**

italic volume number and comma: *72,*

This practice is observed in some but not all ACS journals. For consistency, we follow this practice in table 17.2.

Exercise 17.9

Format the following information as reference 8 for three different journals: *Environmental Science & Technology*, *Chemical Research in Toxicology*, and *Analytical Chemistry*. Follow the formatting examples in table 17.2.

Authors	Gary R. List and Roberta G. Reeves
Title	Overview of infrared methods for alkenes
Journal	*The Journal of Organic Chemistry*
Volume/Issue/Year	Volume 72, Issue 12, 2007
Pages	3980 to 3985

Exercise 17.10

Non-ACS chemistry journals require different reference formatting. Consider the following citation, adapted from a reference in *Annals of Occupational Hygiene*. What differences do you notice when compared to the recommended format in *The ACS Style Guide* (table 17.2)?

(8) Benoit FM, Davidson WR, Lovett AM, Ngo A. Breath analysis by API/MS—human exposure to volatile organic solvents. Int. Arch. Occup. Envir. Health 1985;55:113–20.

We next consider reference formats for books (table 17.3). Two types of books are addressed: books with and without editors. We again show the generic formatting that is recommended in *The ACS Style Guide* and examples of book citations from actual ACS journals. As you read the table, note the following:

- Author lists for both journal articles and books are formatted in the same way in a single journal, though they differ across journals (e.g., compare the *Biochemistry* and *Anal. Chem.* entries in tables 17.2 and 17.3).

Table 17.3 Examples of numerical references for books.

Source	Example
Without editors	
The ACS Style Guide	(8) Le Couteur, P.; Burreson, J. *Napoleon's Buttons: How 17 Molecules Changed History*; Jeremy P. Tarcher/Putnam: New York, 2003; pp 32–47.
Anal. Chem.	(8) Bard, A. J.; Faulkner L. R. *Electrochemical Methods*, 2nd ed.; Wiley: New York; 2001.

continued

Table 17.3 (continued)

Source	Example
Chem. Res. Toxicol.	(8) Aldridge, W. N., and Reiner, R. E. (1972) *Enzyme Inhibitors as Substrates: Interactions of Esterases with Esters of Organophosphorus and Carbamic Acids*. North-Holland Publishing, Amsterdam.
Environ. Sci. Technol.	(8) Mackay, D. *Multimedia Environmental Models: The Fugacity Approach,* 2nd ed.; Lewis Publishers: Boca Raton, FL, 2001.
With editors	
The ACS Style Guide	(8) Almlof, J.; Gropen, O. Relativistic Effects in Chemistry. In *Reviews in Computational Chemistry*; Lipkowitz, K. B., Boyd, D. B. Eds.; VCH: New York, 1996; Vol. 8, pp 206–210.
Anal. Chem.	(8) Francesconi, K. A.; Kuehnelt, D. In *Environmental Chemistry of Arsenic*; Frankenberger, W. T., Jr., Ed.; Marcel Dekker: New York, 2002; pp 51–94.
Biochemistry	8. Fierke, C. A., and Hammes, G. G. (1996) Transient Kinetic Approaches to Enzyme Mechanisms, in *Contemporary Enzyme Kinetics and Mechanisms* (Purich, D., Ed.) 2nd ed., pp. 1–35, Academic Press, New York.
J. Phys. Chem. C	(8) Boschloo, G.; Edvinsson, T.; Hagfeldt, A. Dye-sensitized nanostructured ZnO electrodes for solar cell applications. In *Nanostructured Materials for Solar Energy Conversion*; Soga, T., Ed.; Elsevier: Amsterdam, 2007; pp 227–254.

- The book title is italicized and in title case. The chapter title (when included) is not italic. In some journals, the chapter title is in title case; in others, it is in sentence case.
- Editors' names are written last name first, with initials only. Names are followed by an abbreviation for editor(s): Ed. (for books with one editor) or Eds. (for books with multiple editors).
- The punctuation and formatting of the publisher, place of publication, and year vary across journals.
- Page numbers (when included) are inclusive. Most often, "pp" is used without a period. (*Biochemistry* is an exception.)
- All references end with a period.

Exercise 17.11

Format the following book information for a reference in *Environmental Science & Technology* and *Chemical Research in Toxicology*. Follow the formatting examples in table 17.3.

Author	R. B. Darlington
Book title	Regression and Linear Models
Publisher	McGraw-Hill
Year	1990
Place	New York, NY

Exercise 17.12

The following book reference appeared in *Analytical Chemistry*. Reformat the reference for *Chemical Research in Toxicology*. Follow the formatting examples in table 17.3.

(8) Bard, A. J.; Faulkner L. R. *Electrochemical Methods*; Wiley: New York; 1980.

With ever-increasing frequency, authors also need to know how to format references to electronic sources (e.g., general Web sites, online periodicals, and online books). When this textbook was printed, such references were relatively uncommon. However, we briefly address references to general Web sites in exercise 17.13. For examples of other types of electronic sources and formatting expectations, consult a recent issue of your targeted journal or the most recent edition of *The ACS Style Guide*.

Exercise 17.13

The ACS Style Guide recommends the following format for general Web sites:

ACS Publications Division Home Page. http://pubs.acs.org/ (accessed Jan 9, 2008).

Compare this recommended format to the following examples found in three different ACS journals. What differences and similarities do you notice?

http://www.ams.usda.gov/science/pdp/SOPs.htm. (accessed Feb 8, 2007).

Facts about Cambodia; worldfacts.us; http://worldfacts.us/Cambodia.htm.

http://www.ams.usda.gov/science/pdp/SOPs.htm. (accessed 02/08/07).

 Exercise 17.14

Consider these hastily written references. Reformat them so that they are suitable for the indicated journal. Follow the examples in tables 17.2 and 17.3.

a. Reformat for *The Journal of Organic Chemistry*:

> (8) Duarte, I. F.; Barros, A.; Belton, P. S.; Righelato, R.; Spraul, M.; Humpfer, E.; Gil, A. M. High-resolution NMR spectroscopy and multivariate analysis for the Characterization of Beer. *J. Agric. Food Chem.*, **2002**, *50*, 2475–2481.

b. Reformat for *Chemical Research in Toxicology*:

> (8) Alison S. Bateman, Simon D. Kelly, and Mark Woolfe. Nitrogen Isotope Composition of Organically and Conventionally Grown Crops *Journal of Agric. Food Chemistry*, 55 (7), 2664–2670, **2007.**

c. Reformat for *Environmental Science & Technology*:

> (8) George Cacalano, John Lee, Karen Kikly: *Neutrophil and B Cell Expansion in Mice That Lack the Murine IL-8 Receptor Homolog;* **Science, 1994**, *265(200)*, pp. 657–682

d. Reformat for the *Journal of Physical Chemistry C*:

> (8) Perlmutter, P. (1992). Conjugate Addition Reactions in Organic Synthesis; *New York: Pergamon Press.*

Alphabetical References

When the author–date citation format is used in the body of the work, references are listed in alphabetical order (based on the last name of first author) at the end of the work. Although rare in ACS journals, which use largely numerical citations and references, many journals prefer this approach. Just like numbered references, the correct format for alphabetical references varies among journals. We illustrate these differences for journal articles (table 17.4) and books (table 17.5). All examples were taken from non-ACS journals. A few similarities and differences are highlighted below:

- All use a hanging indent, with the first line left-justified and subsequent lines indented one tab space.
- The author lists vary in punctuation, spacing, sequencing of authors' first and last names, and the use of "and" before the last author.

Table 17.4 Examples of alphabetical references for journal articles in three non-ACS journals.

Source	Example
Ann. Occup. Hyg.	Carey WP, Beebe KR, Kowalski BR. Multicomponent analysis using an array of piezoelectric crystal sensors. Anal. Chem. 1987;59:1529–34.
Ecology	Davis, M. A., P. Grime, and K. Thompson. 2000. Fluctuating resources in plant communities: a general theory of invasibility. Journal of Ecology 88: 528–534.
J. Geophys. Res.	Gardner, M. W., and S. R. Dorling (2000), Meteorologically adjusted trends in UK daily maximum surface ozone concentrations, *Atmos. Environ., 34,* 171–176.

Table 17.5 Examples of alphabetical references for books in three non-ACS journals.

Source	Example
Ann. Occup. Hyg.	Ballantine DS. Acoustic wave sensors theory, design, and physico-chemical applications. San Diego, CA: Academic Press, 1997.
Ecology	Grime, J. P. 2001. Plant strategies, vegetation processes, and ecosystem properties. John Wiley and Sons, Chichester, UK.
J. Geophys. Res.	Darlington, R. B. (1990), *Regression and Linear Models,* McGraw-Hill, New York.

- Sentence case is used for journal article titles. Sentence case without italics, and title case with italics are used for book titles.
- The formatting of journal names varies in terms of font (e.g., italics) and abbreviated journal names.
- Years of publication are formatted and placed differently.
- Page numbers are inclusive (with en dashes), but in one example, a truncated form is used (1529–34 rather than 1529–1534).
- All references end with a period.

Exercise 17.15

Convert the following hastily prepared numerical reference list to a reference list appropriate for the *Journal of Geophysical Research*. Follow the examples in tables 17.4 and 17.5.

1. Mackay, D. *Multimedia Environmental Models: The Fugacity Approach*, 2nd ed.; Lewis Publishers: Boca Raton, FL, 2001.
2. Shoeib, M.; Harner, T.; Vlahos, P. Perfluorinated chemicals in the Arctic atmosphere. *Environ. Sci. Technol.* **2006**, *40*, 7577–7583.
3. Gilles-Gonzalez, M. A., and Gonzalez, G. (2005) Heme-based sensors: defining characteristics, recent developments, and regulatory hypotheses, *J. Inorg. Biochem.* *99*, 1–22.
4. Evans, D. C., Watt, A. P., Nicoll-Griffith, D. A., and Baillie, T. A. (2004) Drug-protein adducts: an industry perspective on minimizing the potential for drug bioactivation in drug discovery and development. *Chem. Res. Toxicol.* *17*, 3–16.

Exercise 17.16

Beginning writers sometimes incorrectly combine numerical and alphabetical reference formats. Consider the example below. The in-text citations are not numbered sequentially (i.e., 1, 2, 3, . . .), and the literature cited is in alphabetical order rather than in citation order.

a. Revise both the in-text citations and references to demonstrate the correct use of the italic number in parentheses format. Assume that these are the first three citations in the written work.
b. Revise both the in-text citations and references to demonstrate the correct use of the author–date format. Assume that these are the first three citations in the written work.

The stabilization of the complex was first accomplished by Munroy (*3*); reaction rates were studied by Baker (2), and the mechanism was elucidated by Adams (*1*).

Literature Cited

1. Adams, J. K. *J. Org. Chem.* **2007**, *8*, 18–22.
2. Baker, B. L. *J. Org. Chem.* **2006**, *12*, 1188–1195.
3. Munroy, H. H. *J. Org. Chem.* **2005**, *5*, 110–115.

We conclude this chapter by summarizing a few tips that will assist you in preparing your references, whether you use a numerical or alphabetical format.

For all reference sections

- The heading **References** or **Literature Cited**, if required, should match the other headings in your work in terms of font, case, and placement.
- There must be an exact match between in-text citations and references. All cited works (but only cited works) must be included.
- Publications that are cited more than once in the written text should be listed only once in the References.
- References must include all authors listed on a paper, in the exact order in which they appear on the original publication. Hence, et al. should not be used in references. Note that a few journals allow et al. to be used after listing the first 10 authors of an article that is written by 11 or more authors.
- Authors are almost always listed in inverted form: last name, first initial, middle initial, and qualifiers (e.g., Jr., II).

For journal article reference sections

- References in submitted manuscripts are double-spaced (like the rest of the text), even though the references will be single-spaced in the final publication.
- References should be placed at the end of the manuscript, after the text or Acknowledgments section (if acknowledgments are included). Some (but not all) journals ask that you start the references on a new page.
- A recent issue of your targeted journal and its Information for Authors should be consulted before you begin your references. Be sure to examine the printable (**PDF**) forms of recent articles, not the **HTML** forms. (The formatting of HTML references is often unconventional, as a result of hot links to cited articles.)

For scientific poster reference sections

- A highly abbreviated format may be used in posters if space is limited (e.g., Gordon et al. *JOC,* **2007,** *12,* 13.); formatting should be consistent for all entries.

For research proposal reference sections

- References are often single-spaced (like the rest of the proposal) but with an additional space between references.
- In most proposals, references are placed in a separate section and do not count against the page limit for the Project Description section.
- Full-reference formatting (with titles of articles and chapters included for each reference) is customary (and often required) in proposals.
- If proposal guidelines do not specify a citation/reference format, use a format common in your field. Numerical formats are preferred because they save space.

PDF or HTML?

When viewing a journal electronically, view only PDF files. HTML files do not display the appropriate formatting.

There is no single way to format citations and references. Authors are responsible for determining and then following the appropriate formatting conventions for their work.

17B Citing on Your Own: Compile and Format References

Use a reference format that complements the in-text citation format that you've used.

- For numerical citations, list your references numerically.
- For author–date citations, list your references alphabetically.

If you haven't yet checked the citation/reference expectations of your intended audience (journal, conference, or funding agency), attend to this vital task now, before you begin formatting your references.

Fill in all mandatory information so that you have a complete reference list. Format each reference, paying careful attention to information sequencing, abbreviations, punctuation, fonts, and parentheses. Don't guess if you are unsure; consult the appropriate style guide or Information for Authors.

Proofread your references before claiming your written work complete. Because it is easy to overlook the bolding of a comma, an italic number, a missing pair of parentheses, a period, a semicolon, and so on, it is always a good idea to read through your references one last time!

18 *Finalizing Your Written Work*

Writing is a lot like composing a piece of music. Random notes do not make a symphony, nor do random assemblies of words tell a cogent story. It is critical that writers learn how to put ideas together, through the very final stages of composing, using tried-and-true methods that are common to the field of chemistry. Only in this way can their ideas unfold clearly for the reader.
—Timm Knoerzer, Nazareth College

Writing is a process, not a one-time event.
—Ellen R. Fisher, Colorado State University

In this chapter, we guide you through the final stages of completing your written work. Too often novice writers skip this vital stage of writing, thinking that their earlier drafts and revisions are sufficient. But experienced writers know that one last careful review is needed to assure the highest quality of work possible. By the end of this chapter, you will be able to do the following:

- Identify the different features of effective writing
- Examine your written work from various perspectives
- Take systematic steps to improve the quality of your writing
- Finalize your written work so that it meets the expectations of your target audience

As you work through the chapter, the Revising and Editing on Your Own tasks will guide you as you do the following:

18A Prepare to finalize your work

18B Check organization

18C Verify follow-through

18D Eliminate redundancy

18E Maximize conciseness

18F Check tense and voice

18G Correct common errors

18H Review science content

18I Reconfirm adherence to external guidelines

18J Check formatting and overall appearance

By the time you consult this chapter, you have most likely revised and edited your work countless times. You have written separate sections, modified them, and then merged them, altering your writing throughout the process in an attempt to fit all the pieces together seamlessly. You have probably lost count of all the changes, large and small, that you have made (or been asked to make) in an effort to perfect your writing. Indeed, for most writers, more time is spent rewriting than writing.

In this chapter, we ask you to look again at your written work, but this time as an entire document. Rather than write and rewrite section by section, we ask you to take a step back and reflect on your work as a whole. The goal is to achieve a coherent and fluent piece of writing that reads well from start to finish. To facilitate this process, we have created a checklist of nine items (listed below). We present each item and ask you to review your entire work from the perspective of this item, making changes and improvements as needed. As you become familiar with these items, you will become a better and more efficient editor of your own written work, a necessary prerequisite for becoming an effective writer.

1. organization	4. conciseness	7. science content
2. follow-through	5. tense & voice	8. adherence to external guidelines
3. redundancy	6. common errors	9. formatting and overall appearance

Because the goal of this chapter is to help you finish your work, we forgo the usual read-analyze-write tasks, exercises, and chapter review. Likewise, we include only a few excerpts, and we have replaced the Writing on Your Own tasks with Revising and Editing on Your Own tasks. The latter are designed to guide you in the final revision process.

18A Revising and Editing on Your Own: Prepare to Finalize Your Work

Finalizing your written work requires that you reflect on your work as a whole and read through it many times, each time focusing on a different area.

Glance at the nine items above. Which areas are likely to be strengths in your writing? Which are likely to be weaknesses and require special attention?

Decide on a strategy that you will use to review your work holistically. The goal is to read your work slowly and deliberately. Many authors read their work aloud, so as not to skip any parts of their written work. If you decide to read your work silently, avoid skimming or you will miss areas that require attention.

Check Organization

Writers of all abilities struggle with organization. The move structures presented throughout the textbook, and reproduced in appendix B, were developed to guide writers in organizing individual sections of their work. Effective writing, however, requires that individual sections fit together to create a cohesive whole. The goal is for each section of your work to flow and transition seamlessly into the next, leading your readers effortlessly through a logical progression of ideas.

One way to evaluate the overall organization of your written work is to examine the headings, subheadings, and opening sentences in key paragraphs, rather than reading the work in its entirety. If you can follow the presentation of ideas in these few phrases and sentences, chances are good that the broad organizational structure of your written work is sound.

To see how this approach works, let's reexamine the aldehydes-in-beer journal article. Instead of reading the full article, this time we look only at the broad framework of the paper (excerpt 18A), limiting ourselves to key opening sentences and subheadings, to evaluate the paper's overall organization. A quick glance at these important parts of the article reveals a well-organized paper with a logical progression of ideas. Only one essential ingredient is missing, the conclusions, because these were embedded (appropriately) in the last few sentences of the paper.

Excerpt 18A (adapted from Vesely et al., 2003)

[First sentences in the four paragraphs comprising the Introduction]

- Carbonyl compounds, particular aldehydes, are considered to play an important role in the deterioration of beer flavor and aroma during storage.
- Several analytical methods for the determination of aldehydes in beer have been developed, and good results have been obtained using liquid-liquid extraction (*2*), distillation (*3*), or sorbent extraction (*4*).
- A simple way to increase the selectivity of extraction techniques is to derivatize the carbonyl compounds.
- In this work, we adapted a method for the analysis of beer aldehydes using solid-phase microextraction (SPME) with on-fiber derivatization.

[*Subheadings from the Materials and Methods section*]

- **Chemicals.**
- **Beer Samples.**
- **SPME Fiber.**
- **Derivatization Procedure.**
- **GC Conditions.**

[*Subheadings from the Results segment of the combined R&D section*]

- **Identification.**
- **Optimization of Derivatization Procedure.**
- **Calibration.**
- **Method Validation.**

[*First sentences from the first three paragraphs in the Discussion segment of the combined R&D section*]

Beer Analysis

- Nine aldehydes were detected in analyzed beer (Figure 5).
- The aldehydes 2-methylpropanal, 2-methylbutanal, 3-methylbutanal, methional, and phenylacetaldehyde are so-called Strecker aldehydes, formed as a result of a reaction between dicarbonyl products of the Amadori pathway and amino acids, having one less carbon atom than the amino acid (*1*).
- During long-term storage at elevated temperature, American-style beers develop a stale flavor (*10*).

18B Revising and Editing on Your Own: Check Organization

Check the organization of your written work by making an outline from its headings, subheadings, and first sentences of key paragraphs. Is the organizational structure clear? Logical?

Based on the outline, how well does the organizational structure of your written work adhere to the move structures introduced in the textbook (see appendix B)? Does each section contain the appropriate information? Is the information sequenced as your audience would expect it to be? Are any required moves missing? Revise your work to improve its organization.

Read over your work in its entirety. Does it flow? Are transitions in place that will assist your readers in moving from one section to the next? Make changes, large or small, to improve your organization so that it is in line with your audience's expectations and the move structures suggested in this textbook.

Verify Follow-Through

The opening section of most written work (typically the Introduction) sets the stage for the rest of the work; hence, more than any other section, the early part of one's writing foreshadows what is to come in subsequent sections. Foreshadowing can be viewed as making a promise to your readers. In a journal article, as an example, the promise typically appears in the fill-the-gap statement ("In this work, we demonstrate . . ."). As you finish your written work, make sure that you have fulfilled any promises that you have made. If you have either overstated or understated your case, be sure to bring these statements in line with what your work ultimately achieves.

18C Revising and Editing on Your Own: Verify Follow-Through

Reread the opening section of your written work while considering questions such as these:

1. What promises have you made to your readers? What have you told your readers that you will accomplish? Report? Demonstrate?
2. Have you overstated or understated your intentions?
3. How well have you foreshadowed the information presented later in the written work (e.g., results, future plans) with language, gap statements, and fill-the-gap statements?

Revise your opening section, if necessary, to strengthen foreshadowing and follow-through.

Eliminate Redundancy

When we write, the intentional repetition of some content is natural. This is particularly true when writing for general audiences, who often benefit from having an idea expressed in several different ways. Sometimes, however, identical phrases or sentences are repeated unnecessarily in a work, often unintentionally. We call this **redundancy**. During the final stages of writing, you must look for redundancies in your work and either alter the redundant passages or delete them entirely.

Redundancy

The unnecessary repetition of words or phrases.

Check for redundancy as part of the final writing process.

One common place to find redundancy is after headings or subheadings. In this case, words in the heading are unnecessarily repeated in the sentence after the heading. The redundant words should be deleted. Consider the following examples:

> **Sampling Procedure.** ~~The sampling procedure was as follows:~~ A sample (10 mL) was placed in a septum-sealed glass vial (30 mL).
>
> **Site of Investigation.** ~~The site of investigation was the Coconino National Forest.~~ The Coconino National Forest is the largest contiguous ponderosa pine forest in North America.
>
> **Conclusions.** ~~In conclusion, t~~[T]he concentrations of PAHs in...

Note that the phrase "In conclusion," is redundant only if it follows the heading "Conclusions." Without the heading, the phrase is appropriate and serves as a useful signal to the reader.

Another common place for redundancy is in transitional sentences. In journal articles, there are two transitional moves: at the start of the Results section (to remind readers about methods) and at the start of the Discussion section (to remind readers about results). Caution is needed not to repeat the same sentence in these moves. For example, if you wrote in the Results section "As shown in Figure 1, the rate increased with the addition of the catalyst," you could not use this sentence again at the start of the Discussion section. To avoid repetition, you must either state the repeated information differently, or combine it with new information. An example of the latter approach is shown in excerpt 18B, where the authors combine repeated information from the Methods section (that rats were fed 3 μg Se/g of diet) with new information from the Results section (that this diet significantly reduced the incidence of mammary tumors).

Excerpt 18B (adapted from Finley et al., 2001)

[From the Results]

Experiment 1. When fed the 3 μg Se/g of diet, high-Se broccoli significantly reduced the incidence and total number of mammary tumors... (Table 1).

Redundancy is also common in sections of a work that have overlapping purposes. Examples include the abstract of a journal article or the Project Summary of a research proposal; each has the purpose of summarizing key points from the accompanying document. Although information can be repeated in the abstract or Project Summary, authors are required to state the information differently, resisting the temptation to merely copy sentences from other sections of their work. Another vulnerable place for redundancy is in the Discussion section of a journal article, which shares several overlapping purposes with the Introduction section. For example, beginning writers sometimes repeat the fill-the-gap statement of the Introduction (e.g., "In this work, we measured ___") in the summarizing remarks of the Discussion section. Similarly, authors sometimes repeat words used to establish the importance of their work in both the Introduction and Discussion sections. Now is the time to look for such redundancies in your writing and correct them as needed.

18D Revising and Editing on Your Own: Eliminate Redundancy

Reread your work with an eye toward locating (and then adjusting) unnecessary repetitions. Keep your audience in mind when making decisions about what to keep and what to eliminate.

1. Check the sentences immediately following your headings and subheadings. Eliminate repeated words as necessary.
2. Examine sentences that transition the reader from one section to the next. Minimize redundancy, when appropriate.
3. Compare the different sections of your written work. Revise sections to eliminate unnecessary duplication of information.

Maximize Conciseness

Conciseness, a cornerstone of writing in chemistry, is expected by expert and scientific audiences. Revising and editing at this final stage offers additional opportunities to make your writing more concise. Consider the changes made in this example to achieve conciseness:

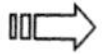

Conciseness

See appendix A and table 2.1.

Wordy version (adapted from Carroll et al., 2001)

There are multiple pathways that are currently known to inactivate lipoxygenases, ranging from competitive to allosteric to reductive inhibition. The competitive pathways have been studied by Zherebtsov, Popova, and Zyablova (2000). The allosteric process has been studied by Mogul Johansen, and Holman (2000). Reductive inhibition has been studied by Kemal Louis-Flamberg, Krupinski-Olsen, and Shorter (1987). (57 words)

More concise version

Multiple pathways are currently known to inactivate lipoxygenases, ranging from competitive[1] to allosteric[2] to reductive inhibition.[3] (16 words)

18E Revising and Editing on Your Own: Maximize Conciseness

Reread your written work. Look for (and then delete) wordiness. Take steps, such as the following, to maximize conciseness:

1. Eliminate unnecessary words and phrases.
2. Replace multiple-word phrases with more concise single-word alternatives that have the same meaning (e.g., "to" instead of "in order to," "because" instead of "due to the fact that").
3. Delete information that can be assumed (e.g., "In our judgment").
4. Use nominalizations (e.g., "after distillation" instead of "After we distilled the product").
5. Focus on the science by omitting authors' names.

Check Tense and Voice

Throughout the textbook, we have described how verb tense and voice are used to convey meaning. By now, you understand that different tense–voice combinations have particular functions. In general, rules for tense and voice hold true across all the genres addressed in this textbook. For example, past–passive is used to describe specific steps taken in your work (e.g., "initial HSSPME experiments were performed"), and present–active is used to state scientific "truths" and knowledge (e.g., "Cr^{6+} induces strand breaks"). One exception is that future tense is used in research proposals, including both future–active (e.g., "We will address several objectives.") and future–passive (e.g., "This goal will be pursued in two parts."). Checking tense and voice in your writing is an important fine-tuning step. Consult tense–voice tables throughout the textbook (see tables 4.1, 5.1, 6.5, 12.3, 12.5, 12.6, 13.3, 13.5, 14.2, 14.5, 15.1) to be certain that your intended message matches reader expectations.

18F Revising and Editing on Your Own: Check Tense and Voice

Read through your written work and look for unconventional uses of tense and voice.

1. Consider the function of individual sentences to determine if tense and voice are used appropriately. Consult tables 4.1, 5.1, 6.5, 12.3, 12.5, 12.6, 13.3, 13.5, 14.2, 14.5, and 15.1.
2. Check that past tense is used to describe work completed in the past and that present tense is used to report scientific "truths."
3. Reread your Methods section and verify that it is written largely in passive voice.
4. When using active voice, check that you used personal pronouns (e.g., *we*) sparingly and in conventional ways (e.g., to signal a decision or the start of the fill-the-gap statement).

Correct Common Errors

Possessing a mastery of the mechanics of writing (e.g., grammar, punctuation) is not trivial, but rather it indicates an attention to detail that reflects well on the seriousness of the writer.

—Joseph H. Aldstadt, University of Wisconsin–Milwaukee

A careful proofreading of your written work is necessary to locate (and then correct) the common errors that we all make in our writing. Look for problems in spelling, subject–verb agreement, punctuation, grammatical accuracy, and parallelism (in headings, subheadings, lists, and series). Revising and editing are a bit easier these days because computers can find common typos (e.g., chemsitry) and grammatical errors (e.g., the reactions was . . .). But one should never rely entirely on a computer. The computer cannot, for example, recognize common writing conventions in chemistry (e.g., mL), nor will it consistently find words that are spelled correctly but used improperly (e.g., affect/effect, principal/principle, their/there). Thus, it is important that authors read over their own written work with great care.

Common Errors

See appendix A for proofreading tips on the following:

Grammar (e.g., subject–verb agreement, parallelism)

Punctuation (e.g., capitalization, commas, semicolons)

Easily confused words (e.g., affect/effect, comprise/compose)

18G Revising and Editing on Your Own: Correct Common Errors

Reread your written work to locate common errors. Pay particular attention to the following:

1. acronyms and abbreviations
2. capitalization
3. numbers and units
4. parallelism in series and lists
5. placement of figure captions and table titles
6. plural words (e.g., data, spectra, syntheses)
7. punctuation (especially colons, semicolons, and commas)
8. spelling and typos

9. subject–verb agreement
10. troublesome words (e.g., affect/effect, comprise/compose, fewer/less, since/because, then/than, which/that, while/although)
11. use of the word respectively
12. words to avoid or use cautiously (e.g., look, prove, recently, see, this, very, we)

Review Science Content

It goes without saying that your work will be judged first and foremost on science content. In general, high-quality writing and high-quality science go hand in hand. Your mastery of the science content is revealed directly through your writing. Pay particular attention to the following areas where your writing skills and knowledge of science content directly overlap:

1. Your description of importance and implications: Have you explained how your work fills a gap, furthers scientific knowledge, and benefits society?
2. Your references to others' works: Have you acknowledged all pivotal works and paraphrased and cited others' works correctly and accurately?
3. Your presentation of methods: Have you described your methods accurately, in enough detail that an expert could repeat them? Have you paid appropriate attention to detection limits and quality assurance/quality control?
4. Your presentation of results: Have you double-checked the accuracy of all data, units, and error terms? Have you told your story of scientific discovery in a correct, logical, coherent way?
5. Your interpretation of findings: Have you hedged your claims and supported your interpretations with scientific evidence, avoiding hand-waving arguments?

18H Revising and Editing on Your Own: Review Science Content

Reread your written work, paying attention to the science content. Think about the following questions while reading your written work and correct any inaccuracies:

1. Have you conveyed the importance of your work?
2. If asked, could you explain the words and concepts in your work?
3. Are your methods described in sufficient detail for an expert? Could you state their limitations?
4. Is your presentation of science complete? Logical?

5. Have you double-checked all numbers, units, error bars, citations, and references for accuracy?
6. Have you hedged your claims and supported them with evidence?
7. Have you paraphrased relevant works and cited them correctly?

Reconfirm Adherence to External Guidelines

An extremely important step is to check the specific requirements of the publication and to follow them. Journals often specify a format, the number of pages, what software package or file formats are acceptable, how to cite references, and many other aspects of manuscript preparation. Requirements can vary from journal to journal even if the same publisher publishes them. . . . Understanding the requirements for the manuscript cannot be overemphasized.

—*The ACS Style Guide* (Coghill and Garson, 2006)

This textbook emphasizes writing conventions that are common in four writing genres of chemistry, but each and every work must conform to external conventions, as well. When proofreading, keep in mind that your work must meet external evaluator expectations:

- For journal articles, expectations are specified in the journal's Information for Authors.
- For conference abstracts and posters, expectations are specified on the conference Web site.
- For research proposals, expectations are specified in Requests for Proposals (RFPs) and supplementary grant proposal guidelines.

These expectations are often presented in a convenient checklist that can be consulted as you finish your work. Ideally, you checked for specifications early in the writing process. If not, consult the guidelines now to make sure that you have followed them precisely. Now is the time (and your last chance) to make changes so that your written work conforms to the expectations of these external sources.

18I Revising and Editing on Your Own: Reconfirm Adherence to External Guidelines

Consult relevant external sources (e.g., Information for Authors for journal articles, conference Web sites for poster guidelines, or RFPs for research proposals) to confirm that your written work conforms to specified expectations. Make changes that are necessary.

Examine Formatting and Overall Appearance

Few of us are 100% consistent as we write. For example, some section headings will be bolded, others will be italicized; we will use both Fig. and Figure, forget page numbers, use nonuniform indentations and spacing, and have two Table 3s. And who can blame us? We were concentrating so hard on the content of our written work that we rightfully ignored such pesky details.

Now it is time to make things right. Tables and figures should be numbered consecutively, in the order in which they are referenced in the text. Tables, figures, and schemes also need to be formatted in conventional ways (see chapter 16). Headings and subheadings require consistent formatting (capitalization, punctuation, bolding, font size, etc.); and margins, indentations, and spacing should be consistent. Headings should have at least two lines of text following them on the same page; headings that don't ("**dangling headings**") should be moved to the next page. Citations and references need to be formatted in a parallel fashion, following standard conventions (see chapter 17).

Margins, Indentations, and Spacing

Common page setup conventions are as follows:

Margins: At least one inch on all sides

Indentations: Standard tab at about one-half inch

Spacing: Double spacing required throughout most of the manuscript

Confirm these expectations in the appropriate journal or RFP.

Dangling Headings

Headings with fewer than two lines of associated text on the same page; such headings should be moved to the following page.

18J Revising and Editing on Your Own: Check Formatting and Overall Appearance

Look over your written work, this one last time. Consider questions such as these, and make whatever changes are necessary to improve the presentation of your written work:

1. Are tables and figures numbered consecutively and in the proper order? Are they formatted in a conventional manner?
2. Are subheadings parallel and consistently formatted? Do you have any dangling headings?

3. Are margins, indentations, fonts, and spacing consistent throughout your work?
4. Are citations and references formatted in a parallel fashion? Have you used only one citation format throughout (superscript numbers, italic numbers in parentheses, or author–date format)?
5. Do the visual elements (e.g., of your poster) follow standard practices?
6. How professional does your written work look?

Appendix A *Language Tips*

Appendix Contents

Audience and Purpose

- Concise Writing 584
- Fluid Writing 586
- Formal Vocabulary 588
- Hedging 590
- Nominalizations 594
- Respectively 597
- Unambiguous Writing 599

Writing Conventions

- Abbreviations, Acronyms, and Compound Labels 601
- Active and Passive Voice 604
- Numbers and Units 608

Grammar and Mechanics

- Grammar
  - Parallelism 612
  - Plural and Singular Scientific Words 616
  - Split Infinitives 619
  - Subject–Verb Agreement 620
- Punctuation
  - Capitalization 624
  - Colons and Semicolons 626
  - Commas 629
  - Hyphenated Two-Word Modifiers 632

Word Usage

Affect/Effect 634

Between/Among 636

Comprise/Compose 638

Farther/Further 641

Fewer/Less 643

Its/It's 644

Precede/Proceed 645

Principle/Principal 648

Since/Because and While/Although 650

Then/Than 653

Which/That 654

Audience and Purpose

Concise Writing

The importance of concise writing is stressed throughout this textbook. Unnecessary words clutter your writing and make the content more difficult to understand. Here are three rules that will help make your writing more concise. (See also the language tips for *respectively* and *nominalizations*.)

Rule 1: Eliminate unnecessary words or phrases, including references to scientists.

1. ~~As stated earlier in this paper,~~ TCE demonstrated specific antiproliferative activity against the more progressive and metastatic SW620 cells. (Adapted from Seeram et al., 2004)
2. ~~The cancer cell lines~~ SW480 and SW620 are two colon cancer cell lines ~~that were obtained~~ from the same patient; ~~the~~ SW480 ~~cell line~~ was ~~obtained~~ from the primary tumor, and ~~the~~ SW620 ~~cell line~~ was ~~obtained~~ from a metastatic site. (Adapted from Seeram et al., 2004)
3. ~~As shown in a paper by Conderoy et al. (2008), there is an~~ increasing ~~amount of~~ evidence ~~that~~ suggests an association between cancer and the cyclooxygenase (COX) enzyme (5); hence, the COX-expressing cell line HT-29 was studied ~~in this present work~~.

Rule 2: Replace wordy phrases with more concise words or phrases that have the same meaning.

1. ~~Similar to this~~ [Similarly], fractions enriched in organic acids also inhibited cell growth.
2. ~~As an example of what we mean~~ [For example], the detection limit increased 3-fold.
3. The rate increased ~~due to the fact that~~ [because] the temperature increased.

Rule 3: Use parentheses to report information such as vendors, reaction conditions, amounts, or product yield.

1. All reagents were research grade ~~and purchased from Aldrich~~ (Aldrich).
2. The highest yield ~~of 92%~~ was observed in THF ~~after 1 h of reaction time~~ (1 h, 92%).

Proofreading Tip: One of the hallmarks of writing in chemistry is conciseness. As you proofread your work, look for places where you can make your writing more concise. With practice, concise writing will become easier.

Exercise

Rewrite these passages to make them more concise. Delete unnecessary words, edit wordy phrases, and use parentheses when appropriate.

1. In order to determine the rate of the reaction, the temperature was increased.
2. Despite the fact that the temperature was increased, the rate of the reaction remained unchanged.
3. After increasing the solvent to 25 mL, the reaction proceeded quickly at room temperature with a reaction time of 30 min and a percent yield of 77%.
4. The polyphenol fraction, which was the most active of all the cranberry fractions, showed the highest antiproliferative activity against HCT116 of 92.1%, against SW620 of 63%, against HT-29 of 61.1%, and against SW480 of 60.1%. (Adapted from Liberty et al., 2007)

Concise Writing Answers (other answers possible)

1. ~~In order~~ to determine the rate of the reaction, the temperature was increased.
2. ~~Despite the fact that~~ [Although] the temperature was increased, the rate of the reaction remained unchanged.
3. After increasing the solvent (25 mL), the reaction proceeded quickly at room temperature (30 min, 77%).

4. The polyphenol fraction, ~~which was~~ the most active ~~of all the~~ cranberry fractions, showed the highest antiproliferative activity against HCT116 ~~of~~ (92.1%), ~~against~~ SW620 ~~of~~ (63%), ~~against~~ HT-29 ~~of~~ (61.1%), and ~~against~~ SW480 ~~of~~ (60.1%).

Fluid Writing

Sometimes you may find sentences in your writing that seem awkward or choppy, even though they are grammatically correct. The following rules will help make your writing flow more smoothly.

Rule 1: The part of the sentence that comes before the verb should be shorter than the part of the sentence that comes after the verb.

Awkward One of the most common and costly diseases in poultry *is* coccidiosis.

Better Coccidiosis *is* one of the most common and costly diseases in poultry. (Adapted from Peippo et al., 2004)

Awkward A rapid and simple method for the screening of residues of these two coccidiostatic compounds in poultry and eggs *is presented* in this paper.

Better This paper *presents* a rapid and simple method for the screening of residues of these two coccidiostatic compounds in poultry and eggs. (Adapted from Peippo et al., 2004)

Rule 2: In a sentence that contains both new information and information that has been mentioned previously (given information), state the given information first. In this way, the reader encounters familiar ideas first, making the sentence more fluid and easier to read.

Consider the following examples. In the awkward example, the given information (italicized) comes after the new information, making the second sentence harder to understand.

Awkward Anticoccidial drugs are extensively used in the poultry industry to control infection. The polyether ionophores, such as narasin and salinomycin, are *the most commonly used coccidiostats in poultry.*

Better Anticoccidial drugs are extensively used in the poultry industry to control infection. *The most commonly used coccidiostats in poultry* are the polyether ionophores such as narasin and salinomycin. (Adapted from Peippo et al., 2004)

Note, too, that different words are used to refer to the given information in each example: "coccidiostats" is another way of saying "anticoccidial drugs". By using different words with the same meaning, you can create linkages between sentences without becoming overly repetitive.

Proofreading Tip: Writing that flows is easier to read. To increase fluidity, start sentences with the shorter or more familiar concepts.

Exercise A

Compare the following sets of passages. Which passage is more fluid in each set? Identify the changes that have been made in the more fluid version.

Set A (adapted from Dóka et al., 2004)

A1. Varying concentrations of phenolics that may be determinants of nutritional quality in foods for both humans and animals are contained in sorghum grain (*3*). Digestibility and palatability may be hindered by phenolics that interact with proteins (*4*). Macronutrients, such as proteins and carbohydrates, can form indigestible complexes when linked with tannins, a special group of high molecular weight phenolic compounds.

A2. Sorghum grain contains varying concentrations of phenolics that may be determinants of nutritional quality in foods for both humans and animals (*3*). Phenolics that interact with proteins may hinder digestibility and palatability (*4*). Tannins, a special group of high molecular weight phenolic compounds, can form indigestible complexes when linked with macronutrients, such as proteins and carbohydrates.

Set B (adapted from Kozukue et al., 2004)

B1. Tomatoes, a major food source for humans, accumulate a variety of secondary metabolites including glycoalkaloids (*1*). These metabolites protect against the adverse effects of pathogens and predators, including fungi, bacteria, viruses, and insects.

B2. Tomatoes, a major food source for humans, accumulate a variety of secondary metabolites including glycoalkaloids (*1*). Pathogens and predators, including fungi, bacteria, viruses, and insects, have adverse effects that these metabolites protect against.

Revise the second sentence in the following pairs of sentences to make it more fluid (adapted from Ranatunge et al., 2004).

1. Nonsteroidal anti-inflammatory drugs (NSAIDs) are widely used to treat the signs and symptoms of inflammation, particularly arthritic pain.[1,2] It is mainly through the inhibition of cyclooxygenases (COXs), key enzymes in prostaglandin (PG) biosynthesis from arachidonic acid, that NSAIDs exert their anti-inflammatory effect.[3–5]
2. COX-1 and COX-2 are two of the mammalian COX isoforms.[6,7] The chronic, low-level production of cytoprotective PGs in the gastrointestinal (GI) tract is caused by constitutive COX-1, whereas the generation of PGs in inflammatory cells is the main function of inducible, short-lived COX-2.

Fluid Writing Answers

Exercise A

Set A: A2 is more fluid. In the first sentence, the part before the verb is shorter than the part after the verb. The second and third sentences have been reordered to front given, rather than new, information.

Set B: B1 is more fluid. The given and new information are sequenced so that the given information comes first. The word "metabolites" is added to the third sentence to remind readers what glycoalkaloids are.

Exercise B

1. The anti-inflammatory effect of NSAIDs is achieved mainly through the inhibition of cyclooxygenases (COXs), key enzymes in prostaglandin (PG) biosynthesis from arachidonic acid.[3–5]
2. Constitutive COX-1 causes the chronic, low-level production of cytoprotective PGs in the gastrointestinal (GI) tract, whereas the main function of the inducible, short-lived COX-2 is the generation of PGs in inflammatory cells.

Formal Vocabulary

Writing in chemistry requires authors to use more formal vocabulary than they would use in speaking. For example, you could say "we cut back on the solvent," but you would write "the solvent was decreased (or reduced)."

Consider the following examples:

Less Formal	Mutant forms were studied to *find out* what changes, if any, occurred in carbohydrate-binding specificity.
More Formal	Mutant forms were studied to *determine* what changes, if any, occurred in carbohydrate-binding specificity.

Proofreading Tip: Watch for informal "spoken" vocabulary in your writing. Substitute informal "spoken" vocabulary with more formal "written" vocabulary.

Exercise

Replace the informal verbs italicized in sentences 1–17 with more formal verbs (from the list below) to increase the formality of each sentence. You may need to change the form of verb so that the sentence remains grammatically correct. There may be more than one possible answer.

accumulate	discover	extract	reduce
analyze	eliminate	imply	remove
conduct	enable	include	report
conserve	ensure	investigate	suggest
consider	enter	limit	utilize
decrease	establish	obtain	withdraw
determine	examine	originate	

1. Because of its anticipated easy removal via catalytic hydrogenolysis, we decided to *see* the application of (*R*)-phenylglycine amide as a chiral auxiliary in asymmetric synthesis. (Adapted from Boesten et al., 2001)
2. Cr^{3+} also increased DNA polymerase processivity and *cut down* its fidelity during DNA replication in vitro (*15–17*) (Adapted from Plaper et al., 2002)
3. Chromate may *go into* cells through the general anion channel, leading to rapid intracellular accumulation (*10*). (Adapted from Plaper et al., 2002)
4. Several countries have *set up* standards for PCBs in dietary products such as fish, meats, and eggs. (Adapted from Llompart et al., 2001)
5. During the study, it was *picked up* that Cr^{3+} influences DNA topology. (Adapted from Plaper et al., 2002)
6. The general analytical procedure for the determination of PCBs in full-fat milk *is made up of* four main steps: (Adapted from Llompart et al., 2001)

7. SPME *makes use of* a small segment of fused-silica fiber coated with a polymeric phase to *take out* the analytes from the sample and to introduce them into a chromatographic system. Initially, SPME was used to *look at* pollutants in water[20,21] via direct extraction. (Adapted from Llompart et al., 2001)
8. We *found out* that the time to reach equilibrium between stationary phase and sample headspace was 90 min (Figure 3). (Adapted from Vesely et al., 2003)
9. Once inhaled deep into the lungs, $PM_{2.5}$ are difficult to *get rid of* and may pose significant health risks. (Adapted from Dellinger et al., 2001)
10. The Strecker reaction is historically one of the most versatile methods for *getting* α-amino acids in a cost-effective manner (Adapted from Boesten et al., 2001)
11. β-CD was also *said* to improve the biodegradation of a single hydrocarbon (dodecane) (*15*). (Adapted from Jozefaciuk et al., 2003)
12. In this manuscript, we *tell* the results of studies that indicate that $PM_{2.5}$ does indeed contain semiquinone-type radicals. (Adapted from Dellinger et al., 2001)
13. However, Se intake is not the only factor to *look at* for the reduction of carcinogenesis. (Adapted from Finley et al., 2001)
14. Our mechanistic studies to date *mean* that the role of zinc is not simply that of a Lewis acid. (Adapted from Demko and Sharpless, 2001)
15. A substantial fraction of the fine particles in the atmosphere *comes* from combustion sources (*9*). (Adapted from Dellinger et al., 2001)
16. Beer was also analyzed by gas chromatography/mass spectrometry (GC/MS) without being derivatized by PFBOA to *make sure* that there were no other sources of *m/z* 181 besides the derivatization agent. (Adapted from Vesely et al., 2003)
17. Hydrogen ions are added to the system or are *taken away* to hold the pH constant. (Adapted from Alberty, 2005)

Formal Vocabulary Answers

(1) investigate/examine/consider; (2) decreased/reduced; (3) enter; (4) established; (5) discovered/determined; (6) includes; (7) utilizes, extract, analyze/examine/investigate; (8) determined/discovered; (9) remove/eliminate; (10) obtaining; (11) reported; (12) report; (13) consider; (14) imply/suggest; (15) originates; (16) ensure; (17) withdrawn/removed

Hedging

Hedging is a mechanism used by writers to indicate some degree of caution, tentativeness, or restraint. Hedging prevents writers from overstating a claim or over interpreting their results. Without hedging, your writing may sound exaggerated or nonscientific, resulting in a loss of credibility. Imagine a weather forecaster uttering these statements:

Without hedging	Sunday will be windy with 60 mile per hour winds between 2:00–4:00 p.m.
With hedging	Sunday is likely to be windy; gusts up to 60 miles per hour might be observed in the afternoon.
Without hedging	A winter storm is headed our way and will dump 6 inches of snow between noon and 2:00 p.m.
With hedging	A winter storm appears to be headed our way; it may dump up to 6 inches of snow over the course of the day.

Forecasters who fail to hedge may lose listeners' trust because they will be wrong over and over again. More credible forecasters predict the weather with caution and restraint, using hedging words such as *may*, *might*, *likely*, and *appears*.

Table A1 lists hedging words commonly used by chemists to soften their interpretations and temper their claims.

Proofreading Tip: Avoid making bold assertions in your writing. Communicating some degree of caution and restraint (through the use of hedging words) will strengthen your position rather than weaken it. While proofreading, look for opportunities to use hedging words to soften your interpretations, claims, or expected outcomes and impacts.

Table A1 Common hedges in chemistry writing.

Verbs	Helping Verbs	Adjectives	Adverbs
appear	can	possible	apparently
indicate	could	probable	generally
seem	may		largely
suggest	might		mainly
support	should		possibly
	would		potentially
			presumably
			probably
			typically

Exercise A

Examine these pairs of sentences for hedging. In each pair, identify the sentence with hedging and without hedging. What words and phrases led you to your decisions?

1. The lack of *lacZ* induction in the case of chromium oxalate is possibly due to the inability of that compound to enter the bacterial cells.

   The lack of *lacZ* induction in the case of chromium oxalate is due to the inability of that compound to enter the bacterial cells. (Adapted from Plaper et al., 2002)

2. The thermal stability of the metal-containing DNA oligonucleotides proves that the incorporation of **9** into DNA oligonucleotides causes no destabilization of the duplex.

   The thermal stability of the metal-containing DNA oligonucleotides suggests that the incorporation of **9** into DNA oligonucleotides causes little or no destabilization of the duplex. (Adapted from Yu et al., 2001)

Exercise B

Read these passages and identify the authors' attempts to hedge. Look for words and phrases that reveal the authors' restraint.

1. Firm conclusions cannot be drawn because the concentrations of these compounds were not determined in the broccoli used in this experiment and because a direct comparison of high-Se garlic and high-Se broccoli was not made. (Adapted from Finley et al., 2001)
2. The immediate goal of fully characterizing the parent Hg^{2+}-responsive fluorescent chemosensor... is expected to be complete within 6 months.... The synthesis and complete characterization of the proposed fluorophores are anticipated to take 6 months. (Adapted from Finney, 1999)
3. In particular, we intend to systematically and quantitatively determine the origins of interfacial polarity at solid-liquid interfaces as well as identify how surface induced polar ordering affects dynamic properties of interfacial environments. (From Walker, 2001)
4. In general, UXAS opens a new spectroscopic window for the direct ultrafast time-resolved observation of molecular motions during chemical reactions. (From Rose-Petruck, 2000)

Exercise C

Rewrite these passages with hedging words to communicate more caution and tentativeness. Consult table A1 as needed.

1. The fluorescence lifetime measurement of the diol/carbonate pair will be complete in 2 months. (Adapted from Finney, 1999)
2. Se-enriched broccoli was not more effective (Table 1) than enriched garlic (3) in reducing the number of tumors; this finding proves that the combination of sulforaphane, indole carbinol, and chlorophyll with Se did not provide additional protection against mammary tumors. (Adapted from Finley et al., 2001)
3. Because it is known that humans exposed to different Cr^{3+} species accumulate high levels of Cr^{3+} intracellularly (*17*), results will affect human intake of Cr^{3+} as a nutrition additive. (Adapted from Plaper et al., 2002)

Hedging Answers

Exercise A

(1) With hedging	The lack of *lacZ* induction in the case of chromium oxalate is *possibly* due to the inability of that compound to enter the bacterial cells.
Without hedging	The lack of *lacZ* induction in the case of chromium oxalate *is due* to the inability of that compound to enter the bacterial cells. (Adapted from Plaper et al., 2002)
(2) Without hedging	The thermal stability of the metal-containing DNA oligonucleotides *proves* that the incorporation of **9** into DNA oligonucleotides causes *no* destabilization of the duplex.
With hedging	The thermal stability of the metal-containing DNA oligonucleotides *suggests* that the incorporation of **9** into DNA oligonucleotides causes *little or no* destabilization of the duplex. (Adapted from Yu et al., 2001)

Exercise B

1. *Firm conclusions cannot be drawn* because the concentrations of these compounds were not determined in the broccoli used in this experiment and because a direct comparison of high-Se garlic and high-Se broccoli was not made. (Adapted from Finley et al., 2001)

2. The immediate goal of fully characterizing the parent Hg^{2+}-responsive fluorescent chemosensor . . . *is expected to be complete* within 6 months. . . . The synthesis and complete characterization of the proposed fluorophores *are anticipated to take* 6 months. (Adapted from Finney, 1999)
3. In particular, we *intend to* systematically and quantitatively determine the origins of interfacial polarity at solid-liquid interfaces as well as identify how surface induced polar ordering affects dynamic properties of interfacial environments. (From Walker, 2001)
4. *In general*, UXAS opens a new spectroscopic window for the direct ultrafast time-resolved observation of molecular motions during chemical reactions. (From Rose-Petruck, 2000)

Exercise C (other answers are possible)

1. The fluorescence lifetime measurement of the diol/carbonate pair *is expected to be complete* in 2 months. (Adapted from Finney, 1999).
2. Se-enriched broccoli was not more effective (Table 1) than enriched garlic (*3*) in reducing the number of tumors; this finding *suggests/indicates* that the combination of sulforaphane, indole carbinol, and chlorophyll with Se did not provide additional protection against mammary tumors. (Adapted from Finley et al., 2001)
3. Because it is known that humans exposed to different Cr^{3+} species accumulate high levels of Cr^{3+} intracellularly (*17*), results *may/might* affect human intake of Cr^{3+} as a nutrition additive. (Adapted from Plaper et al., 2002)

Nominalizations

Nominalizations are nouns that are formed from other parts of speech, usually by adding an ending such as -ment, -tion, or -ity to a verb or an adjective.

Other Parts of Speech	→ *Nominalizations*
React (verb)	→ Reaction (noun)
Concentrate (verb)	→ Concentration (noun)
Measure (verb)	→ Measurement (noun)
Intense (adj.)	→ Intensity (noun)

Nominalizations are common in scientific writing because, in most instances, they are more concise.

Without nominalization	The heterogeneous property of ricin is generally not considered in studies of how toxic it is. (16 words)
With nominalization	The *heterogeneity* of ricin is generally not considered in *toxicity* studies. (11 words) (Adapted from Fredriksson et al., 2005)

Nominalizations can also minimize awkwardness, adding to the fluidity and clarity of your writing.

Awkward	The ability to deprotect the carboxylic acid and to simultaneously cleave from the resin will provide compounds with the desired side chain.
Less awkward	The *deprotection* of the carboxylic acid and the simultaneous *cleavage* from the resin will provide compounds with the desired side chain. (Adapted from Hergenrother, 2002)

Note: English has many pairs of verbs and nouns that look identical, for example, conduct (v.)/conduct (n.), decrease (v.)/decrease (n.), structure (v.)/structure (n.), and transfer (v.)/transfer (n.). The nouns in these pairs are not examples of nominalizations. Nominalizations require a change in word form, for instance, convert (v.) → conversion (n.).

Proofreading Tip: When proofreading your work, look for opportunities to make your writing more concise by using nominalizations. Refer to table 2.2 for a list of nominalizations common in chemistry writing.

Exercise A

Consider the italicized nouns in the following passages. Which are examples of nominalizations? What word do they stem from?

1. The binding *interactions* of the CD-capped nanoparticles with a *series* of five alkyldimethyl(ferrocenylmethyl)ammonium *ions* can be utilized for the phase *transfer* of the hydrophilic, CD-capped nanoparticles into a nonpolar chloroform *phase*. (Adapted from Liu et al., 2001)
2. The *structure* of the aggregates formed upon transfer of the CD-capped nanoparticles to the chloroform solution is postulated to resemble that of reverse micelles, as the nanoparticles direct the peripheral *arrangement* of the

cationic ferrocene amphiphiles, counterions, and water *molecules* around their *surfaces*. (Adapted from Liu et al., 2001)

3. A heat *treatment* results in the *reaction* of the *surface* of Si particles to form SiO_2. The *decrease* of particle *size* is monitored by a blue *shift* of the absorption maximum as a function of heating time. (From Bol and Meijerink, 2001)
4. Clearly, a *correlation* between Hg···C distance versus the ^{13}C chemical *shift* exists. Three *observations* may be made from the *data* in Figure 8. First, compounds **1–5** are indeed best considered to have η^1 *coordination* of the arene in the solid *state*, whereas coordination in compound **7** is best described as η^2. (From Borovik et al., 2001)

Exercise B

Rewrite these sentences so that they are more concise and in line with chemistry writing conventions. Use nominalizations whenever possible.

1. When we identify trimethylsiloxy-1,2-dioxetane and assign trimethylsiloxymethyl formate as a reaction product, we demonstrate how feasible the trimethylsilyl group migration is. (Adapted from Fajgar et al., 2001)
2. We describe how we prepared and characterized gold nanoparticles (~3 nm in diameter) capped with thiolated cyclodextrins. The CD-capped nanoparticles are hydrophilic, and they bind compounds derivitized from ferrocene as evidenced by values measured by ^{1}H NMR spectroscopy. (Adapted from Liu et al., 2001)

Nominalizations Answers

Exercise A

1. *interactions* from *interact* (v.); (2) *arrangement* from arrange (v.); (3) *treatment* from *treat* (v.), *reaction* from *react* (v.); (4) *correlation* from *correlate* (v.), *observations* from *observe* (v.), *coordination* from *coordinate* (v.)

Exercise B (other answers possible)

1. The identification of trimethylsiloxy-1,2-dioxetane and the assignment of trimethylsiloxymethyl formate as a reaction product demonstrate the feasibility of a trimethylsilyl group migration.
2. The preparation and characterization of gold nanoparticles (~3 nm in diameter) capped with thiolated cyclodextrins are described. The CD-capped

nanoparticles are hydrophilic, and they bind ferrocene derivatives as evidenced by ^{1}H NMR spectroscopic measurements.

Respectively

The word *respectively* is an adverb that means *in the order specified, each separately in the order mentioned*. When used correctly, the word can make your writing clearer and more concise.

Set A

Less concise	The injector temperature was 260 °C. The ECD temperature was 280 °C.
More concise	Injector and ECD temperatures were 260 and 280 °C, respectively. (Adapted from Llompart et al., 2001)

Set B

Less concise	Arsenite was dissolved in water; arsenobetaine was dissolved in methanol; arsenocholine was dissolved in chloroform.
More concise	Arsenite, arsenobetaine, and arsenocholine were dissolved in water, methanol, and chloroform, respectively.

Notice how *respectively* is used in the examples above and below:

- *Respectively* is preceded by a comma in all examples.
- The ordering of the sequenced items linked by *respectively* match. The first item in the first series is linked to the first item in the second series, the second item in the first series is linked to the second item in the second series, and so forth.
- *Respectively* most often appears at the end of a sentence, though it can appear within a sentence (e.g., *X* and *Y* are coordinates parallel and perpendicular to the surface, *respectively*, and *t* represents time).
- When units for items in a series are the same, they are repeated only once at the end of the series (e.g., The aldol reactions required 30 and 60 min, respectively.)
- *Respectively* is only used to relate two or more series (or sequences) in the same sentence. *Respectively* is not used in a simple list, as in this example:

  Flavonoids are of particular interest because of their potential anti-inflammatory, antiallergic, and antiviral activities.

Because there is only one series (anti-inflammatory, antiallergic, and antiviral activities), the use of *respectively* would be inappropriate.

Proofreading Tip: Look for opportunities to use *respectively* in your writing. Its proper use also requires attention to commas and units.

Exercise

Use *respectively* to improve the conciseness and clarity of the following sentences. If *respectively* cannot be used appropriately, indicate "No change needed."

1. CHO cell strain 32-3 was grown on α-MEM. Strain 31-8 was grown on MSB.
2. A 100 μg/L Ge stock solution was prepared in deionized water. A 1000 μg/L As stock solution was prepared in dilute nitric acid (5%).
3. The solid circles show the cross sections for the parent cluster ions having a temperature of 300 K. The open circles show the cross sections for the parent cluster ions having a temperature of 77 K. (Adapted from Hanmura et al., 2002)
4. It triggers conformational changes in retinoid proteins, such as rhodopsin, relevant to vision, and bacteriorhodopsin, relevant to ATP synthesis. (Adapted from Saito and Kobayashi, 2002)
5. Common extraction methods include liquid-liquid extraction and Soxhlet extraction.
6. Note the low water solubility of 1.75 mg/L for HHCB and the low water solubility of 1.25 mg/L for HTN (2). (Adapted from Buerge et al., 2003)
7. All bond lengths were optimized to better than 0.001 Å, while bond angles were optimized to better than 0.1°. (Adapted from Tsipis and Karipidis, 2003)
8. The experiments were conducted at 50, 70, 90, and 100 °C.

Respectively Answers

1. CHO cell strains 32-3 and 31-8 were grown on α-MEM and MSB, respectively.
2. Stock solutions of Ge (100 μg/L) and As (1000 μg/L) were prepared in deionized water and dilute nitric acid (5%), respectively.
3. The solid and open circles show the cross sections for the parent cluster ions having temperatures of 300 and 77 K, respectively.

4. It triggers conformational changes in retinoid proteins, such as rhodopsin and bacteriorhodopsin, relevant to vision and ATP synthesis, respectively.
5. No change needed.
6. Note the low water solubilities of 1.75 and 1.25 mg/L for HHCB and HTN, respectively (*2*).
7. All bond lengths and bond angles were optimized to better than 0.001 Å and 0.1°, respectively.
8. No change needed.

Unambiguous Writing

Novice writers often use the pronouns *this*, *that*, and *it* in their writing, but their use often leads to ambiguity because readers cannot be certain what *this*, *that*, or *it* refers to. To minimize ambiguity in your writing, avoid the use of such pronouns by themselves.

Set A

Ambiguous	As a mosquito repellant, eucamalol has been reported to outperform *N,N*-diethyl-*m*-toluamide in terms of potency.[12,13] The presence of eucamalol in *S. insularis* explains the widespread use of *this* to fend off mosquitoes[4] and further exemplifies the potential of plants to provide biodegradable alternatives to synthetic insecticides.
Unambiguous	As a mosquito repellant, eucamalol has been reported to outperform *N,N*-diethyl-*m*-toluamide in terms of potency.[12,13] The presence of eucamalol in *S. insularis* explains the widespread use of *this plant* to fend off mosquitoes[4] and further exemplifies the potential of plants to provide biodegradable alternatives to synthetic insecticides. (Adapted from Fattorusso et al., 2004)

Set B

Ambiguous	A 65 µm poly(dimethylsiloxane)/divinyl benzene (PDMS/DVB) fiber coating (Supelco, Bellefonte, PA) was used in this method. *This* was selected for its ability to retain the derivatizing agent and for its affinity for the PFBOA–aldehyde oxime.
Unambiguous	A 65 µm poly(dimethylsiloxane)/divinyl benzene (PDMS/DVB) fiber coating (Supelco, Bellefonte, PA) was used in this method. *This fiber coating* was selected for its ability to retain the derivatizing agent and for its affinity for the PFBOA–aldehyde oxime. (Adapted from Vesely et al., 2003)

Set C

Ambiguous	A qualitative model aimed at explaining the observed field-induced adsorption of negatively charged gold particles on a similarly charged silica surface is proposed. *It* is based on the idea that the adsorbed ion distribution at the particle surface is influenced by the applied electric field.
Unambiguous	A qualitative model aimed at explaining the observed field-induced adsorption of negatively charged gold particles on a similarly charged silica surface is proposed. *The model* is based on the idea that the adsorbed ion distribution at the particle surface is influenced by the applied electric field. (Adapted from Kloepper et al., 2004)

Proofreading Tip: No one wants to produce ambiguous writing, but it occasionally happens. Very often, the culprit is one of three simple words: *this*, *that*, or *it*. When proofreading your work, make sure that your reader can easily determine what *this*, *that*, and *it* refer to.

Exercise

Read the following sentences. All have the potential for ambiguity because of the ways in which *this*, *that*, or *it* is used. Suggest a way to minimize ambiguity in each sentence.

1. (*R*)- and (*S*)-warfarin are in their bound forms. *This* takes place within the protein human serum albumin. (Adapted from Clarke et al., 2001)
2. The iron solution in ethanol develops an intense black color that fades to yellow-brown within minutes. From *this*, the cage complex can be isolated in 80% yield. (Adapted from Venkateswara Rao et al., 2004)
3. The α,β-unsaturated aldehyde moiety in these molecules was responsible for their fungicidal action (5). *This* was described to result from the structural disruption of the cell membrane (6). (Adapted from Kubo et al., 2001)
4. One of the characteristics of the LSPR biosensor is its generality. *This* is a powerful attribute for fast, high throughput screening of adsorbates. (Adapted from Haes, 2003)
5. The extraction was performed with a liquid-liquid extractor; 50 mL of dichloromethane (Merck) and 50 mL of sample were used. *It* was maintained for 24 h. (Adapted from Fraile et al., 2000)

6. Once this process is explored with the model system to assess the level of enantioselectivity, we will then prepare alkyl zinc reagent **48** using standard methods[22,36,42] and cross couple *it* to aryl bromide **18** using the appropriate chiral catalysts (Scheme 7). (Adapted from Vyvyan, 2001)

Unambiguous Writing Answers

(1) This binding; (2) This solution; (3) The fungicidal action; (4) This generality OR... its generality, which is a powerful...; (5) The extraction; (6) **48**

Writing Conventions

Abbreviations, Acronyms, and Compound Labels

Abbreviations and acronyms are short forms of single words (e.g., M for molar) or multiple words (e.g., NMR for nuclear magnetic resonance). In abbreviations, the individual letters are usually pronounced (e.g., A-C-S for American Chemical Society); in acronyms, the letters form a new word (e.g., *CASSI* for *Chemical Abstracts Service Source Index*). Compound labels are used to represent chemical compounds. (See also chapter 3 and *The ACS Style Guide* for more information on abbreviations and acronyms.)

Some rules for using abbreviated forms are as follows:

Rule 1: To define an abbreviation or acronym, use the full term first, followed by the abbreviated form in parentheses:

> Detection and characterization of a kinetic product of *deoxyadenosine (dA)* alkylation helps to reconcile the apparent contradiction between the strength of nucleophiles in DNA and their propensity for addition to a model quinone methide. (Adapted from Veldhuyzen et al., 2001)

Rule 2: Do not define abbreviations for elements (e.g., Fe), empirical formulas (e.g., CH_3CH_3), units (e.g., mL, g, µm, °C), or a few other commonly used chemical abbreviations (see table A2). (If you are writing for a nonexpert audience, you may want to define even these commonly used terms.)

Rule 3: Do not use an abbreviated term before it has been defined (unless it is a term that does not require definition).

Table A2 A partial list of abbreviations that do not need to be defined for a chemistry audience.

at. wt	atomic weight	NMR	nuclear magnetic resonance
bp	boiling point	RNA	ribonucleic acid
DNA	deoxyribonucleic acid	U.S.	United States (as adjective)
equiv	equivalent(s)	UV	ultraviolet
fp	freezing point	v/v	volume per volume
IR	infrared	w/v	weight per volume
mp	melting point	w/w	weight per weight
		wt	weight

Rule 4: Define an abbreviated term only once in the body of the work. After it has been defined, you may use it in the remainder of the work:

> The experimental desorption isotherms for *randomly methylated* β-*cyclodextrin (RAMEB)* and *RAMEB*-enriched minerals are presented in Figure 1. The isotherms were measured for the *RAMEB* in the forms of powder and crystals; however, these were the same within the range of experimental error. (From Jozefaciuk et al., 2001)

Rule 5: After an abbreviated term has been defined, you may still use the unabbreviated form if it seems more appropriate.

Rule 6: Avoid abbreviations and acronyms in titles.

Rule 7: If you define an abbreviation or acronym in an abstract or Project Summary, define it again in the body of the work.

Rule 8: Preface abbreviations and acronyms, when appropriate, with articles "a" and "an" according to the pronunciation (not the letter) of the first sound of the abbreviation or acronym:

> *a* nuclear magnetic resonance spectrum
>
> *an* NMR spectrum

Rule 9: Abbreviate units of measure when they follow a number. Without a number, spell them out:

> 9 V/s
>
> measured in volts per second

Rule 10: Form the plural of multiple-letter, all-capital abbreviations and abbreviations ending in a capital letter by adding a lowercase "s" only. Do not put an apostrophe before the "s"; do not add an "s" to units of measure:

PCBs

pHs

PAHs

CFCs

10 mL

Rule 11: Use bolded numbers (e.g., **1**, **2**, **3**,...) and/or numbers and letters (e.g., **5a**, **5b**, **5c**,...) to represent chemical compounds.

Like other abbreviations, compound labels must be defined before they are used alone. In the text, the label is defined by placing it immediately after the fully named compound, usually without parentheses. (Some authors use parentheses, especially for compound labels in subheadings.) Compound labels may also be defined in graphical illustrations of the compound (e.g., equations, tables, figures, schemes). Once defined, the label may be used alone, without parentheses. Consider these examples:

> We report a safer and exceptionally efficient process for transforming nitriles **1** into tetrazoles **2** in water (Adapted from Demko and Sharpless, 2001).

> The asymmetric Strecker reaction of (*R*)-phenylglycine amide **1**, pyvaldehyde **2**, and HCN generated in situ from NaCN and AcOH was studied. (From Boesten et al., 2001)

> Compounds **5–8** are shown in Figure 2. Our first project will involve the synthesis of **5**.

> **2-(*p*-Toluenesulfonyl)-4'-methoxyacetophenone (2a).** A mixture of 2-bromo-4'-methoxyacetophenone (45.8 g, 200 mmol) and *p*-toluenesulfinic acid sodium hydrate (35.6 g, 200 mmol) in ethanol (1 L) was heated at reflux for 1.5 h. The mixture was stirred, cooled to room temperature, and dried to give 54.6 g (90%) of pure **2a**. (Adapted from Swenson et al., 2002)

Proofreading Tip: Abbreviations and acronyms are common in chemistry writing. Don't guess if you are uncertain about them. Consult *The ACS Style Guide* for a comprehensive list of abbreviations and acronyms.

Exercise

Consider the abbreviations, acronyms, and compound labels used in the following excerpts. Correct those that do not follow standard writing conventions. (Consult chapter 3 and *The ACS Style Guide* for additional hints.) For those excerpts that are correct, indicate "correct as is".

1. Particles were less than 10 microns (µm) in diameter.
2. Twelve polycyclic aromatic hydrocarbons (PAH's) were detected.
3. The most stable isomer agrees with experimental nuclear magnetic resonance and X-ray data. (Adapted from Laurencin et al., 2006)
4. **Analysis of Deoxyribonucleic Acid Damage Using the Comet Assay**. Two methods were employed. The first determined the comet moment . . . (Adapted from Dellinger et al., 2001)
5. An aqueous solution of the derivatization agent *O*-(2,3,4,5,6-pentafluorobenzyl)hydroxylamine was prepared at a concentration of 6 g/L. The PFBOA solution was prepared every 3 months. (Adapted from Vesely et al., 2003)
6. The foam-negative effects of lipids can also be counteracted with the addition of a lipid-binding protein, wheat puroindoline (PIN), to beer. The PIN may bind the residual free lipids in such a way that they can no longer destabilize the foam. (Adapted from Cooper et al., 2002)
7. Alkylation of sodium tolylsulfinate with bromomethyl or chloromethyl ketones (1a–c) under reflux for 2 h in ethanol gave β-keto sulfones (2a–c) in 62–90% yields after crystallization. (Adapted from Swenson et al., 2002)

Abbreviations, Acronyms, and Compound Labels Answers

(1) less than 10 µm (no need to define µm); (2) PAHs (not PAH's); (3) NMR (instead of nuclear magnetic resonance); (4) DNA (instead of Deoxyribonucleic Acid); (5) . . . the derivatization agent *O*-(2,3,4,5,6-pentafluorobenzyl)hydroxylamine (PFBOA) was prepared at a concentration . . . ; (6) correct as is; (7) chloromethyl ketones **1a–c** under reflux for 2 h in ethanol gave β-keto sulfones **2a–c** in 62–90% yields . . . (bolded numbers without parentheses)

Active and Passive Voice

In scientific writing, passive voice is often used to describe experimental procedures (see chapter 3). It is also used to shift the emphasis from the researcher to the research (i.e., from the scientist to the science). The hints provided here will help you avoid some common mistakes made by novice writers when using passive voice.

Note that in an active sentence, the subject is the person or thing that performs the action of the verb and the object is the person or thing that receives the action of the verb. In a passive sentence, the subject receives the action of the verb and there is no object.

	Subject	*Verb*	*Object*
Active	We	stirred	the solution.
Passive	The solution	was stirred.	

Passive voice is formed using the following formula:

(object of active sentence) + (is/was/were/will be/has been/have been) + (past participle)

In this formula, *is/was/were/will be/has been/have been* are forms of "to be" that (1) match the tense of the active verb and (2) agree with the singular or plural form of the object of the active sentence. The past participle is usually the same as the past tense verb form.

Consider the examples in table A3.

Table A3 Examples of active and passive voice in different verb tenses.

Tense	Active Voice			Passive Voice		
	Subject	Verb	Object	Object of Active Sentence	Form of "to be"	Past Participle Form of Verb[a]
Present	We	stir	the mixture.	The mixture	is	stirred.
Past	We	stirred	the mixture.	The mixture	was	stirred.
Past	We	stirred	the mixtures.	The mixtures	were	stirred.
Future	We	will stir	the mixtures.	The mixtures	will be	stirred.
Present Perfect	We	have stirred	the mixture.	The mixture	has been	stirred.
Present Perfect	We	have stirred	the mixtures.	The mixtures	have been	stirred.

[a]Passive voice constructions require the use of the past participle (e.g., shown, become), which is sometimes different from the past tense verb (e.g., showed, became). See the footnote in table 6.2 for more details.

Mistake 1: Novice writers, accustomed to lab manual instructions, sometimes describe procedures using "command language" rather than passive voice.

Incorrect	Distill DMF before use.
Correct	DMF *was distilled* before use.

Mistake 2: Novice writers often forget to use a form of "to be" (*is*, *was*, *were*, *will be*, *has been*, *have been*) before the verb.

Incorrect	Raman shifts calibrated with indene.
Correct	Raman shifts *were calibrated* with indene.

Mistake 3: Novice writers often repeat the subject and/or form of "to be" (*is*, *was*, *were*, *will be*, *has been*, *have been*) when the sentence has two passive verbs with the same subject.

Incorrect	The product was decanted, and it was purified by recrystallization.
Incorrect	The product was decanted and was purified by recrystallization.
Correct	The product *was decanted* and *purified* by recrystallization.

Mistake 4: Novice writers often forget to include the form of "to be" (*is*, *was*, *were*, *will be*, *has been*, *have been*) when two sentences, joined with "and," have different subjects and two passive verbs.

To correct this mistake, the second verb must be preceded by the appropriate form of "to be."

Incorrect	After 12 h at 83 °C, the solvent was evaporated and the residue dissolved in MTBE.
Correct	After 12 h at 83 °C, the solvent *was evaporated* and the residue *was dissolved* in MTBE.

Mistake 5: Novice writers often phrase passive sentences inappropriately when using the verb "added."

Sentences with "added" often begin with the word "To," with the added item(s) coming after the verb (e.g., "To a flask was added 10 mL ethanol"). This sentence construction is preferred when the added items include numerical amounts (e.g., 10 mL) because it prevents the awkwardness of starting the sentence with a number. When numbers are not involved, the more standard passive sentence pattern can be used.

Incorrect	17.86 mmol of *N*-benzylhydroxylamine and 16.93 mmol of $MgSO_4$ *were added* to this mixture.
Correct	To this mixture *were added* 17.86 mmol of *N*-benzylhydroxylamine and 16.93 mmol of $MgSO_4$.
Correct	*cis*-2-Bromo-3-methylcyclohexane *was added* to (*R*)-lactic acid.

Proofreading Tip: To meet the expectations of a chemistry audience, do your best to use active and passive voice following conventions in the field. Refer to tables 4.1, 5.1, 6.5, 12.3, 12.5, 12.6, 13.3, 14.2, 14.5, 15.1 for summaries of common tense-voice combinations and their functions.

Exercise A

Rewrite these sentences so that they correctly use passive voice to describe experimental procedures.

1. We heated the mixture to 65 °C for 15 min.
2. We added 0.31 mmol of Me_4ZnLi_2 to a solution of **1a** (30 mg, 0.21 mmol) in THF (4 mL).
3. We attached the in situ NIR probe to a steel holder on the fermentation tank, and we immersed the probe into milk broth.
4. We added several drops of water to quench the reaction, and we stirred the mixture vigorously for 10 min to allow the precipitation of zinc salts.

Exercise B

The Experimental section below (adapted from Pal et al., 2004) contains several errors in the use of passive voice. Identify and correct the errors.

Experimental Section

Materials. Boron trichloride (Aldrich), sodium tetraphenylborate (Aldrich), and cobaltocene (Strem) used as received. 9-Hydroxy-1-oxophenalene synthesized according to literature procedures.[31] Toluene was distilled from sodium benzophenone ketyl immediately before use. Acetonitrile was distilled from P_2O_5 and redistilled from CaH_2 immediately before use.

9-N-Benzyl-1-oxophenalene. A mixture of 9-hydroxy-1-oxophenalene (0.98 g, 0.005 mol) and benzylamine (10 mL) was refluxed for 10 h in argon. After cooling, yellow crystals were formed and were separated by filtration. The crude product was purified by column chromatography on Al_2O_3 with $CHCl_3$ to give a yellow solid (1.3 g, 92%) and further purification done by crystallization from hexane.

Active and Passive Voice Answers

Exercise A

1. The mixture was heated to 65 °C for 15 min.
2. To a solution of **1a** (30 mg, 0.21 mmol) in THF (4 mL) was added 0.31 mmol of Me_4ZnLi_2.

3. The in situ NIR probe was attached to a steel holder on the fermentation tank and immersed into milk broth. [Note that it is permissible to include "was" with "immersed," but it is less concise that way.]
4. Several drops of water were added to quench the reaction, and the mixture was stirred vigorously for 10 min to allow the precipitation of zinc salts.

Exercise B

Experimental Section

Materials. Boron trichloride (Aldrich), sodium tetraphenylborate (Aldrich), and cobaltocene (Strem) *were* used as received. 9-Hydroxy-1-oxophenalene *was* synthesized according to literature procedures.[31] Toluene was distilled from sodium benzophenone ketyl immediately before use. Acetonitrile was distilled from P_2O_5 and redistilled from CaH_2 immediately before use.

9-N-Benzyl-1-oxophenalene. A mixture of 9-hydroxy-1-oxophenalene (0.98 g, 0.005 mol) and benzylamine (10 mL) was refluxed for 10 h in argon. After cooling, yellow crystals were formed and ~~were~~ separated by filtration. The crude product was purified by column chromatography on Al_2O_3 with $CHCl_3$ to give a yellow solid (1.3 g, 92%) and further purification *was* done by crystallization from hexane.

Numbers and Units

This section highlights a few rules regarding the conventional formatting of numbers and units in chemistry writing, for example, 15%, 15 mL, three samples (not 15 %, 15mL, 3 samples). (See chapter 3 for more examples; refer to *The ACS Style Guide* for a comprehensive examination of numbers and units.)

Rule 1: Abbreviate units of measure (e.g., mL, cm, g, K), including units of time (e.g., s, h, min), when they follow a numeral. Do not use a period with an abbreviated unit (unless you are using the abbreviation for "inch": in.).

Inappropriate	5 seconds	16 milliliters	13.5 in
Appropriate	5 s	16 mL	13.5 in.

Rule 2: Spell out units of measure if they do not follow a numeral.

Inappropriate	several mg	a few min
Appropriate	several milligrams	a few minutes

Rule 3: Use the numerical form of numbers (e.g., 7) rather than the word form (e.g., seven) with units of measure and time, unless the number is at the beginning of a sentence. Make sure you leave a space between the number and unit, unless you are using the % sign.

Inappropriate	seven mL	five cm	thirty percent	11 hours
Appropriate	7 mL	5 cm	30%	11 h

Rule 4: Use °C with a space after the number, but no space between the degree symbol and the capital C. For K (kelvin), include a space between the number and the K; do not use the degree symbol with kelvin.

Inappropriate	10°C	10 ° C	10K	10 °K
Appropriate	10 °C	10 K		

Rule 5: Do not add a plural "s" to make an abbreviated unit plural (unless you are using the plural form of spelled-out units, e.g., milligrams).

Inappropriate	4 mgs	3 mols
Appropriate	4 mg	4 mol

Rule 6: Use the word form of numbers less than 10, except when referring to units of measure.

Inappropriate	4 sites	five mg	thirteen sites	8 samples
Appropriate	four sites	5 mg	13 sites	eight samples

Rule 7: Do not define abbreviated units of measure (unless you are writing for a nonexpert audience).

Inappropriate	3 min (minutes)	300 mL (milliliters)
Appropriate	3 min	300 mL

Rule 8: Use numerals (not the word form) in a series or range containing numbers 10 or greater to maintain parallelism.

Inappropriate	one, five, 10, and 15 mg, respectively
Appropriate	1, 5, 10, and 15 mg, respectively

Rule 9: In a series or range with a span of three or more numerals, include the unit of measure only once. Use the en dash to mean "to" or "through." (See *The ACS Style Guide* for additional information on when and where to use en dashes.)

Inappropriate	3 mL–10 mL	1 kg–4 kg
Appropriate	3–10 mL	1–4 kg

Exception 1 to Rule 9: When one or both of the numbers are negative or include a symbol that modifies the number, use the words "to" or "through" rather than the en dash.

Inappropriate	–3–10 °C	<2–4 kg
Appropriate	–3 to 10 °C	<2 to 4 kg

Exception 2 to Rule 9: Do not use the en dash when the words "from" or "between" are used.

Inappropriate	from 200–400 mL	between 1–3 h
Appropriate	from 200 to 400 mL	between 1 and 3 h

Rule 10: Use a slash (/), not the word "per," for units of concentration when both units are abbreviated.

Inappropriate	20 μg per mL
Appropriate	20 μg/mL

Proofreading Tip: When you finalize your work, examine all units of measure included in your writing. Make sure that your abbreviations and spacing (between numbers and units) are correct. Check that you are using numerals and words appropriately. If you have any doubts, refer to the rules above and *The ACS Style Guide*. Do not guess!

Exercise

Complete these sentences with the proper units of measure and time.

1. The fat content in full-fat milk is ___.
   a. 3.61 percent
   b. 3.61 %
   c. 3.61%
2. The flask was heated for ___ and then cooled to room temperature.
   a. 6 min.
   b. 6 minutes
   c. 6 min
   d. 6 mins.
   e. 6min.
3. The temperature was held at ___ for 2 min.
   a. 90° C
   b. 90 °C
   c. 90 deg. C
   d. 90 °C.
4. The carrier flow rate was ___ in helium.
   a. 1.2 mL/min
   b. 1.2 mL/minute
   c. 1.2 mL/min.
   d. 1.2 ML/min
5. Water isotherms were measured at 294 ± ___.
   a. 0.1K
   b. 0.1 Kelvin

c. 0.1 °K

d. 0.1 K

6. The sample of tree trunk contained ___ well-defined rings.

   a. seventeen

   b. 17

7. The wood pulp was soaked for ___ in dimethylchlorine.

   a. twelve hours

   b. 12 hrs.

   c. 12 h.

   d. 12 h

8. The dry mass ranged from ___.

   a. 1.5–5.4mg

   b. 1.5 mg–5.4 mg

   c. 1.5–5.4 mg

9. The 9-mm test tube contained ___ of zinc chloride.

   a. 200mg

   b. 200 mg

   c. 200 m.g.

10. ___ were obtained....

    a. Four different starches

    b. 4 different starches

11. They were immersed in an acidic solution for ___, respectively.

    a. 4, 6, and 12 h

    b. four, six, and 12 h

    c. 4 h, 6 h, and 12 h

12. The crude product was recrystallized from ethanol at ___.

    a. −10–5 °C

    b. −10 °C to 5 °C

    c. −10 to 5 °C

Numbers and Units Answers

1. c; 2. c; 3. b; 4. a; 5. d; 6. b; 7. d; 8. c; 9. b; 10. a; 11. a; 12. c

Grammar

Parallelism

The need for parallelism in writing is ever-present. We see instances of parallelism in words and phrases, lists, headings, and subheadings. Furthermore, words and phrases joined by *and* or *or* should be parallel; they should have the same grammatical form (grammatical parallelism), and the concepts should be of similar scale and importance (conceptual parallelism).

Examples of Grammatical Parallelism

Set A

Not Grammatically Parallel	In the reaction of 4-fluorophenylboronic acid, the hydrolysis was suppressed to some extent by *a reduction in* the amount of water and *lowering* the reaction temperature.
Grammatically Parallel	In the reaction of 4-fluorophenylboronic acid, the hydrolysis was suppressed to some extent by *reducing* the amount of water and *lowering* the reaction temperature. (Adapted from Senda et al., 2001)

Set B

Not Grammatically Parallel	The next step was *to collect* the fraction eluting in the corresponding time interval and *subjecting* it to prolonged aerial oxidation.
Grammatically Parallel	The next step was *collecting* the fraction eluting in the corresponding time interval and *subjecting* it to prolonged aerial oxidation.
Grammatically Parallel	The next step was *to collect* the fraction eluting in the corresponding time interval and *(to) subject* it to prolonged aerial oxidation. (Adapted from Napolitano et al., 2001)

Set C

Not Grammatically Parallel	The 15-HLO screen is both *a robust tool* and *reliable for reporting* lipoxygenase inhibition.
Grammatically Parallel	The 15-HLO screen is both *a robust tool* and *a reliable reporter* of lipoxygenase inhibition. (Adapted from Carroll et al., 2001)

Examples of Conceptual Parallelism

Just as words and phrases joined by *and* or *or* should be grammatically parallel, they should also be *conceptually parallel*; that is, the words and phrases on either side of *and* or *or* should have the same level of importance. Revising sentences with segments that are not conceptually parallel may require more work on your part; instead of changing one or two words, as you would to make a sentence grammatically parallel, you may need to make more extensive changes to ensure conceptual parallelism.

Set D

Not Conceptually Parallel This reaction has applications in *industry*, *synthesis*, and *the creation of amino acid derivatives*.

Explanation: The three items in this series are not conceptually parallel. Both "industry" and "synthesis" are broad terms, whereas the "creation of amino acid derivatives" is a much narrower, specific task.

Conceptually Parallel This reaction, which can be used to create amino acid derivatives, has other applications in *industry* and *synthesis*.

Explanation: The entire sentence had to be revised to retain the same content but express it in a parallel manner.

Lists

Lists are commonly used in research proposals (not journal articles) to (1) identify project objectives and goals and (2) describe proposed methods. Lists are often used in posters to describe methods and procedures, summarize results, discuss findings, and offer concluding remarks. When used, lists need to be formatted in a parallel fashion. The following components of each list must be parallel:

Numbering scheme

In continuous text, numbers are enclosed in parentheses: (1) (2) (3). In a displayed list, numbers can be enclosed in parentheses: (1) (2) (3) or followed by a period:

1.
2.
3.

Capitalization

Listed items start with either lowercase letters or uppercase letters, not both.

Punctuation

Listed items end with the same form of punctuation (e.g., comma, period, or no punctuation).

Language

Listed items start with the same form of speech (e.g., nouns, verbs). Headings and subheadings also are parallel to one another in terms of grammar, punctuation, and capitalization.

Not Parallel

Summarizing Project.

Description of Project:
1. Giving Background and Stating Objectives
2. Design of Research and Methods
3. scheduling of project
4. Conclusions

List of References

Parallel

Project Summary

Project Description

Background and Objectives

Research Design and Methods

Project Schedule

Conclusions

References Cited

Proofreading Tip: When proofreading your writing, pause at the words *and* and *or*. Check the words on either side of *and* and *or* to ascertain whether the items are grammatically and conceptually parallel.

If you have lists in your writing, make sure that the numbering scheme, punctuation, capitalization, and language are parallel. It is the writer's responsibility to check parallelism. When it is lacking, it reflects poorly on the writer.

Exercise A

Rewrite these sentences so that they are parallel grammatically (and conceptually). If the sentence does not need to be rewritten, write "correct as is".

1. The low yields in Friedel-Crafts reactions are due to the polyalkylation of aromatic compounds, the production of racemic mixtures, and forming other undesirable products.
2. Stable free radicals such as nitroxides have been used in correlation with EPR to study the effects of changes in pH and the determination of rate constants for spin exchange.
3. Pedal motion, a key process of photoreaction in crystals, is considered to occur only in crystals that have an orientational disorder or a large void around the molecules.
4. The present reaction allows various types of silylated carbocycles to be prepared safely, inexpensively, and in an efficient manner.

Exercise B

Consider the following list from a research proposal. Make necessary corrections so that the numbering scheme, capitalization, punctuation, and language are parallel in all instances.

The project will involve the following discrete but interrelated facets:

1) We will synthesize a chemical library of small molecules with maximal structural diversity and a format compatible with protein binding assays.
2) screening of the library against defined protein targets via small molecule microarrays
3) assess the biological use of the identified compounds in vivo to help elucidate various aspects of the apoptosis pathway.

Exercise C

Consider the following subheadings from a draft of the Materials and Methods section of a journal article. Make changes, where appropriate, so that the headings are parallel.

Collection Of The Samples.

STANDARDS:

Sample Preparation.

Analysis using GC/MS:

Parallelism Answers

Exercise A

(1) the formation of [instead of *forming*]; (2) to determine [instead of *the determination of*]; (3) correct as is; (4) efficiently [instead of *in an efficient manner*]

Exercise B (other answers possible)

The project will involve the following discrete but interrelated facets:

1. synthesizing a chemical library of small molecules with maximal structural diversity and a format compatible with protein binding assays
2. screening the library against defined protein targets via small molecule microarrays
3. assessing the biological use of the identified compounds in vivo to help elucidate various aspects of the apoptosis pathway

Exercise C

Sample Collection.

Standards.

Sample Preparation.

GC/MS Analysis.

Plural and Singular Scientific Words

Several chemistry words have tricky plural forms, and the plural form must be accompanied by a plural verb. A few particularly troublesome singular/plural word combinations are described below.

Data/Datum

The word *datum* (singular) is rarely used. The word *data* is plural; it must be accompanied by verb forms that are used with plural nouns. Note that if the term *data set* is used, the singular verb form is used. (For examples of *data* as a singular noun, consult table 4.2).

> Although the IR *data* of **B** *are* limited, the absorption at 2023 cm^{-1} suggests that it is **5F**. (Adapted from Nicolaides et al., 2001)

> The *data suggest* that electron-withdrawing groups increase the cyclization barrier.

Spectra/Spectrum

The word *spectra* is the plural form of *spectrum*. Use *spectra* with plural verb forms and *spectrum* with singular verb forms:

The photoelectron *spectrum is* shown on the *y* axis.

The NMR *spectrum* in the ring proton region *confirms* this assignment.

The electronic absorption *spectra* of **4** and **5** *were* found to shift slightly upon addition of a metal ion.

Similar absorption *spectra have* been reported for Co(II) derivatives of natural blue copper proteins.

Analyses/Analysis

The word *analyses* is the plural form of *analysis*. Use *analyses* with plural verb forms and *analysis* with singular verb forms.

Our *analyses confirm* the formation of an ice Ih layer.

Further *analysis suggests* that the organic layer contains the dissolved ions.

(Note: When we use the present tense in sentences such as these, we must be careful that we are offering interpretations that we believe will be true over time.)

Syntheses/Synthesis

The word *syntheses* is the plural form of *synthesis*. Use *syntheses* with plural verb forms and *synthesis* with singular verb forms.

The *syntheses were* most successful when a nonpolar protic solvent was used.

In this work, a *synthesis* for the formation of alkyl nitriles with a hydroxy group at the α-position *is* presented.

Scientific Disciplines That End in "s"

The names of numerous scientific disciplines end in "s", even though they are singular words (e.g., ballistics, chemometrics, dynamics, genetics, genomics, kinetics, mathematics, physics, quantum mechanics, thermodynamics). These words should be followed by singular verb forms.

Nonequilibrium *thermodynamics*—with applications to physical, chemical, and biological systems—*has received* much attention in recent years.

The *kinetics* of reactions in the atmosphere *depends* on the concentration of the hydroxyl radical.

et al. (for Multiple Authors)

The abbreviation *et al.* (from the Latin *et alia*) means *and others*. It is used to refer to a work that has multiple authors. Instead of referring to a work by writing out all of the authors' names, writers use *et al.* after the surname of the first author. The "surname + et al." combination (without a comma or italics) should be followed by a plural verb form.

Felmatti et al.[4] *have shown* that a compound in smoke promotes seed germination.

Logsdon et al.[16] *have used* quantum mechanics to explain how the Stark effect influences the orientation of polar molecules in the gas phase.

Proofreading Tip: The singular/plural words introduced in this language tip are used frequently in chemistry writing. If you are unfamiliar with the two forms, memorize them. Remember that it is rare to use the singular word *datum*. Because you will most likely use the plural form, *data*, make sure you use it with a plural verb form.

Exercise

The sentences below are taken from student papers. For each one, decide whether the correct verb form is used; make corrections if they are needed. If the sentence is correct, indicate "correct as is".

1. The data does not seem to suggest a practical application of the equation.
2. All data has been organized chronologically.
3. The data included in these tables are only those necessary for the calculation.
4. Hatakiyama et al.[3] note that forensic toxicological analyses has increased in the past decade.
5. Elemental analyses were performed by Detroit Laboratories.
6. Our analysis of the reaction rate coefficients suggest that the product is formed through an S_N2 mechanism.
7. Spectra were collected in triplicate.
8. FT-Raman spectra, collected in triplicate, was obtained using a Nicolet 870 spectrometer.
9. The synthesis of tetrazole proceed through a one-step concerted mechanism.
10. Chemometrics play an important role in analytical chemistry and are involved in both monitoring and control applications.
11. Whisnant et al.[1] has developed an online project in computation chemistry that can be used to investigate the role of Cl_2O_4 in stratospheric ozone depletion.

Plural and Singular Scientific Words Answers

(1) do [instead of *does*]; (2) have [instead of *has*]; (3) correct as is; (4) *note* correct as is; have [instead of *has*]; (5) correct as is; (6) suggests [instead of *suggest*]; (7) correct as is; (8) were [instead of *was*]; (9) proceeds [instead of *proceed*]; (10) plays... is [instead of *play... are*]; (11) have [instead of *has*]

Split Infinitives

Infinitives comprise the word "to" plus the base form of a verb (e.g., to contribute, to indicate, to transfer). A split infinitive occurs when a word (or words) is placed between the two parts of the infinitive (e.g., to fully investigate, to further complicate).

Although split infinitives are common in spoken language, they are less frequent in formal written language. A common belief is that infinitives should not be split. Yet, according to *The ACS Style Guide*, it is acceptable to split infinitives if doing so will prevent awkwardness or ambiguity.

Ambiguous The goal of this research is *to develop completely* new classes of bioresponsive materials.

Explanation: What is the source of ambiguity? Is the author's goal completely new classes of bioresponsive materials *or* development that is thorough and complete?

Unambiguous The goal of this research is *to completely develop* new classes of bioresponsive materials (Adapted from Lyon, 2000)

Proofreading Tip: To avoid infringing upon a traditional rule that is engrained in so many readers' minds, it is better not to split infinitives unless doing so minimizes awkwardness and ambiguity.

Exercise

Consider the sentences below. Does the split infinitive prevent awkwardness and/or ambiguity? If so, indicate "leave as is." If not, indicate "rewrite without a split infinitive."

1. The calculations allowed us *to fully investigate* dissociated states.
2. Co adsorption to Si(100) has also been shown *to entirely depend* on Co translational energy. (Adapted from Spain, 1997)
3. The immediate goal is *to fully characterize* the parent Hg^{2+}-responsive fluorescent chemosensor. (Adapted from Finney, 1999)

4. First, we will use carboiimide coupling methods[54] *to covalently attach* one of the binding partners to the carboxylate moieties in the hydrogel. (Adapted from Lyon, 2000)
5. This will allow us *to rapidly converge* upon a highly optimized polymer–protein conjugate based on NIPA hydrogels. (Adapted from Lyon, 2000)
6. In particular, we intend *to systematically and quantitatively determine* the origins of interfacial polarity at solid–liquid interfaces. (Adapted from Walker, 2001)
7. We will explore other biological interactions *to further evaluate* the generality of our approach. (Adapted from Lyon, 2000)
8. There are three main obstacles to any attempt *to experimentally evaluate* the effect of enzyme dynamics on covalent bond activation. (Adapted from Kohen, 2002)
9. The pharmaceutical industry quickly adopted this burgeoning technology as a method *to rapidly derivatize* and evaluate promising new compounds. (Adapted from Hergenrother, 2002)

Split Infinitives Answers

1. Leave as is
2. Rewrite without a split infinitive: to depend entirely
3. Rewrite without a split infinitive: to characterize fully
4. Rewrite without a split infinitive: to attach covalently
5. Rewrite without a split infinitive: to converge rapidly
6. Rewrite without a split infinitive: to determine systematically and quantitatively
7. Rewrite without a split infinitive: to evaluate further
8. Rewrite without a split infinitive: to evaluate experimentally
9. Leave as is

Subject–Verb Agreement

Subject–verb agreement occurs when the subject and verb of a sentence agree in number; that is, a singular subject must be accompanied by a singular verb form and a plural subject must be accompanied by a plural verb form. With simple sentences, we usually don't make subject–verb agreement mistakes. However, with more complex sentences, such as those in scientific writing, errors are more common. The tips below will help you to avoid the most common types of subject–verb agreement errors.

Tip 1: The most common subject–verb agreement mistake is made when the subject of a sentence is separated from its verb by a phrase or clause.

The ratio of the intensities of these two doublets *is* ~5:1. (From Cantrill et al., 2001)

As more words come between the subject and the verb, it becomes even more difficult to check for agreement:

The importance of nonadditive contributions for the accurate description of the intermolecular interactions *is* well documented. (Adapted from Tongraar et al., 2001)

GRACE implements a cutoff distance d_c such that *any lattice mismatch* between a pair of overlayer and substrate lattice points exceeding d_c *is ignored*. (From Mitchell et al., 2001)

Tip 2: When two or more subjects are joined with the word *and*, use a plural verb form. When two or more subjects are joined by *or*, the verb should take the number (i.e., singular or plural) of the closest subject.

Foam *and* flavor stability *are* important considerations for a brewer. (Adapted from Cooper et al., 2002)

Cesium iodide *or polyethylene glycol was* employed as a reference compound. (From Cantrill et al., 2001)

The appropriate metal ion concentration *or* the *rate constants were* used. (From Dodd, 1997)

Water *or oxygen is* required for these photochemical reactions. (From Bol and Meijerink, 2001)

We suggest that V-O-V-O-V domains *or clusters are* interrupted by incorporation of Sb to form V-O-Sb-O-V species. (From Spengler et al., 2001)

Tip 3: Units of measurement take singular verb forms.

Approximately *2 mg* of the sample *was* placed in aluminum sample cups and crimped with a cover. (From Arulsamy and Bohle, 2001)

To this mixture *was* added *25 mL* of water . . .

Tip 4: *Each* and *every* are followed by singular verb forms.

Each chemical *was* tested for a dose range of 0–300 μM. (From Satoh et al., 2003)

Each point in the new series *corresponds* to a position, a momentum, and a time variable, whereas in Mayer's theory, *each* point *corresponds* to a position. (From Andersen, 2003)

In Figure 6, *every* molecular species *is* represented by a single letter for simplicity. (Adapted from Yasuda et al., 2002)

Tip 5: *All*, *none*, *some*, *most*, and *any* can be followed by singular or plural verb forms, depending on the object of the preposition (or noun) following it.

All aldehyde analyses were run in the single-ion monitoring (SIM) mode. (From Vesely et al., 2003)

All of the solutions also *contain* 0.05 M phosphate buffer (pH 7.5). (Adapted from Cameron and Fielding, 2001)

All of the precipitate was transferred to the flask.

None of the analyzed aldehydes exceed their flavor threshold in beer. (Adapted from Vesely et al., 2003)

Some esters play a central role in flavor changes during beer aging. (Adapted from Vesely et al., 2003)

Most aldehydes, except formaldehyde, *form* two geometrical isomers. (Adapted from Vesely et al., 2003)

Tip 6: It is common to make subject–verb agreement errors with certain nouns, such as *data* and *spectra*. Refer to "Plural and Singular Scientific Words," above, for help with errors of this type.

Proofreading Tip: Make sure that subjects and verbs agree in your final written work: singular subjects with singular verb forms and plural subjects with plural verb forms. Because it is easy to make an agreement error while writing early drafts, it is in the final proofreading stage that you should check subject–verb agreement.

Exercise A

Check these sentences for subject–verb agreement. Identify the subject and main verb. If agreement is faulty, correct it. Otherwise, indicate "correct as is".

1. American lager beer samples used for the aldehyde analysis was stored at 30 °C for 4, 8, or 12 weeks. (Adapted from Vesely et al., 2003)
2. A stock solution containing a mixture of the standard compounds in ethanol was prepared in the concentration 100 ppb each. (Adapted from Vesely et al., 2003)

3. ^{1}H NMR and ^{13}C NMR spectra was recorded in ppm on a 300 MHz instrument using TMS as internal standard. (Adapted from Swenson et al., 2002)
4. Isooctane, acetone, and sodium hydroxide were obtained from Merck.
5. Blends of olive oil and hazelnut oil was prepared by mixing these oils. (Adapted from Ozen and Mauer, 2002)
6. In the new series, each diagram contains points, and each point has associated with it a phase-point variable and a time. (Adapted from Andersen, 2003)
7. Every rate coefficient for the forward reaction is expressed by a complex rate coefficient. (Adapted from Yasuda et al., 2002)
8. Some electron-rich aromatic nitriles requires higher temperatures. (Adapted from Demko and Sharpless, 2001)
9. All specimens were exposed to rainfall for an additional 5 days. (Adapted from Lebow et al., 2003)
10. All of the mass were collected on one filter.
11. Unpaved road dust or aerosols contributes to haze in class I airsheds.
12. Each particle, from dust, soot, or soil, react with light in a unique way.
13. Minutes or even hours are required to complete the separation process.
14. $PM_{2.5}$ or PM_{10} were monitored at each site.
15. A single core sample or multiple surface samples were collected at each site.

Exercise B

Consider the following passages adapted from CAREER proposals. Identify the subject. Correct the verb, if it does not agree with the subject, or indicate "correct as is".

1. Most of the theoretical work regarding the utility of time-resolved photoelectron angular distributions to probe electronic, vibrational, and rotational dynamics *have* also *concerned* neutral photoionization.[6–13,17–23] (Adapted from Sanov, 2002)
2. In contrast, metabolism of pesticides by microorganisms in water and soil and by enzymes in higher organisms often *proceed* stereoselectively. (Adapted from Aga, 2002)

Exercise C

Consider the following questions adapted from CAREER proposals, formed by a verb and auxiliary verb (do/does). Identify the subject. Correct the verb, the auxiliary, or both if they do not agree with the subject, or indicate "correct as is".

1. *Does* the ultrafast dissociation process of $Fe(CO)_5$ in solution, and that of related compounds, *occur* in a concerted or a rapid sequential way? (Adapted from Rose-Petruck, 2000)
2. If the dissociation process is concerted, which ligand motions *do* the caging solvation shell *inhibit*? (Adapted from Rose-Petruck, 2000)

Subject–Verb Agreement Answers

Exercise A

(1) *samples*... ~~was~~ were; (2) *solution*... *was*; correct as is; (3) *spectra* ~~was~~ were; (4) *Isooctane, acetone, and sodium hydroxide were*; correct as is; (5) *Blends*... ~~was~~ were; (6) *each*... *contains*... *each*... *has*; correct as is; (7) *Every rate coefficient*... is; correct as is; (8) *Some electron-rich aromatic nitriles* ~~requires~~ require; (9) *specimens were*; correct as is; (10) *mass* ~~were~~ was; (11) *dust or aerosols* ~~contributes~~ contribute; (12) *Each particle*... ~~react~~ reacts; (13) *Minutes or even hours are*; correct as is; (14) $PM_{2.5}$ or PM_{10} ~~were~~ was; (15) *A single core sample or multiple surface samples were*; correct as is.

Exercise B

(1) Most of the theoretical work... ~~have~~ *has* also concerned...
(2) ...metabolism of pesticides... ~~proceed~~ *proceeds*...

Exercise C

(1) Subject: the ultrafast dissociation process. Verb: correct as is.
(2) Subject: the caging solvation shell. Verb: ~~do~~ *does* the caging solvation shell *inhibit*?

Punctuation

Capitalization

Chemists are particular about what they capitalize and what they do not capitalize. Written work directed to an expert audience should follow standard capitalization rules.

One common mistake is to capitalize the names of molecules or atoms. Except at the beginning of a sentence, molecules and atoms should be written in lowercase:

- Methane was prepared by reacting hydrogen and carbon.
- A synthesis was proposed for 3-methylbutane.

- The most common gases in the atmosphere are nitrogen and oxygen.
- Helium was used as the carrier gas.
- 2,3-Chlorobenzene was detected in 94% of the wells tested.

There are numerous other capitalization rules to remember. The following letters are capitalized:

- the unit K (for kelvin), following a number (e.g., 100 K)
- the unit C (for Celsius), following a number (e.g., 100 °C)
- only the H in pH
- only the first letter in two-letter symbols for elements (Na, Fe, Cl, $FeCl_3$)
- FTIR (*but* Fourier transform infrared spectroscopy)
- NMR (*but* nuclear magnetic resonance spectroscopy)
- the first letter of a molecule at the beginning of a sentence (e.g., "3-Butanol was reacted with ethane.")
- all letters, except the plural "s," in most abbreviations (e.g., PCBs, CFCs, FTIR, NMR)
- the "T" and "F" in Table and Figure when referring to specific graphics (e.g., As shown in Figure 2 and Table 1 . . .)

Proofreading Tip: To communicate with your target audience, follow the capitalization conventions of the field. For more information on capitalization, refer to tables 3.1 and 3.2 and *The ACS Style Guide*.

Exercise A

Which letters should be capitalized in the following sentences?

1. High-purity helium was added to the gas mixture, lowering the temperature to 400 k.
2. The reaction of 4-butyl-(3-methylethyl)nonane and HO^- was monitored using fourier transform infrared spectroscopy (ftir) with nitrogen as the background gas.
3. The injector temperature was held constant at 270 °c.
4. 3-methylpropane produces several free radical halogenation products when reacted with cl_2 and light. The branching ratios are shown in figure 1.

Exercise B

Proofread the following passages. Look for capitalization errors. Find letters that need to be capitalized and capitalize them. Find letters that do not need to be capitalized, and change them to lowercase.

1. The kinetic behavior and microstructure of citrus pectin gels in 60% Sucrose were investigated by transmission electron microscopy (TeM). Ca^{2+} addition at Ph 3.0 resulted in faster gel formation. (Adapted from Löfgren et al., 2005)
2. To work with a homogeneous series of chitosans with the same distribution of degrees of polymerization but different DAs, a starting sample of chitosan of DA 5.2% was reacetylated. Acetylation was performed with Acetic Anhydride in a hydroalcoholic medium. Characteristics of the various prepared chitosan samples are given in table 1. (Adapted from Montembault et al., 2005)

Capitalization Answers

Exercise A

1. K
2. Fourier, FTIR
3. C
4. Methylpropane, Cl_2, Figure

Exercise B

1. The kinetic behavior and microstructure of citrus pectin gels in 60% *sucrose* were investigated by transmission electron microscopy (*TEM*). Ca^{2+} addition at *pH* 3.0 resulted in faster gel formation. (Adapted from Löfgren et al., 2005)
2. To work with a homogeneous series of chitosans with the same distribution of degrees of polymerization but different DAs, a starting sample of chitosan of DA 5.2% was reacetylated. Acetylation was performed with *acetic anhydride* in a hydroalcoholic medium. Characteristics of the various prepared chitosan samples are given in *Table* 1. (Adapted from Montembault et al., 2005)

Colons and Semicolons

Punctuation marks (e.g., colons and semicolons) are small in size, but they communicate a lot. It is important to use them correctly. Many writers confuse colons (:) and semicolons (;), though they serve different purposes in writing.

Colons are used at the end of a complete statement to introduce examples that directly relate to the statement. A key to using colons correctly is as follows: whatever comes before the colon must be able to stand alone as a complete sentence.

Incorrect Among 1–3, the relative Ag^+ selectivities to Hg_2^+ were: 34, 36, and 41%, respectively.

Correct At least three mechanisms should be taken into consideration: carbanion mechanism, hydride mechanism, and addition-elimination mechanism.

Correct Figure 1 shows the electronic spectra that result during laser photolysis of three different precursors: diphenyl carbonate, deuterated nitrosobenzene, and 1-^{13}C-phenol. (From Spanget-Larsen et al., 2001)

Explanation: Why is the first statement incorrect? What comes before the colon is not a complete sentence. The incorrect statement could be modified by simply changing it to read as follows: "Among 1–3, the relative Ag^+ selectivities to Hg_2^+ were as follows: 34, 36, and 41%, respectively." But this correction leads to a wordy solution. The original sentence is fine without the colon: "Among 1–3, the relative Ag^+ selectivities to Hg_2^+ were 34, 36, and 41%, respectively."

Semicolons have two primary uses:

(1) Semicolons are used to connect two related independent clauses (i.e., sentences that can stand on their own).

What comes before and after the semicolon must be able to stand alone as complete sentences:

Both first and fourth harmonics gave identical results; however, the ACV spectra have different appearances. (From Yu et al., 2001)

Mechanistic studies imply that zinc is not simply acting as a Lewis acid; a number of other Lewis acids were tested and caused little to no acceleration of the reaction.[23] (Adapted from Demko and Sharpless, 2001)

The semicolon rule does not apply to sentences joined with *and*, *but*, or *or*:

Incorrect Each diagram contains points; and each point has associated with it a single phase point variable and time.

Correct Each diagram contains points, and each point has associated with it a single phase point variable and time.

Correct Each diagram contains points; each point has associated with it a single phase point variable and time. (Adapted from Andersen, 2003)

(See "Commas," Rule 2, below for more information on sentences that are joined by words like *and*, *but*, *or*.)

(2) Semicolons, not commas, should be used to separate items in a series when one or more of the series contain commas within them.

The last item in the series is usually preceded by the word *and*, except in a series that reports physical properties and spectral data (see last example below).

Correct In each structure, 01 is coordinated by two M1, one M2, and one Si; 02 by one M1, one M2, and one Si; and 03 by one M2 and two Si. (Adapted from Ashbrook et al., 2002)

Correct **Figure 1**. EPR spectra at room temperature for $PM_{2.5}$ collected at five cities in the U.S. (trace A, Rubidoux, CA; trace B, Phoenix, AZ; trace C, Philadelphia, PA; trace D, Durham, NC; and trace E, Baton Rouge, LA). (From Dellinger et al., 2001)

Correct The mixture was cooled to room temperature; the resulting solid was collected, washed with ethanol (2 × 50 mL), and dried to give 54.6 g (90%) of pure **2a**: mp 126.0–127.0 °C; IR 2951, 2906, 1676, 1599, 1572 cm^{-1}. (Adapted from Swenson et al., 2002)

Proofreading Tip: It is amazing that punctuation marks as small as colons and semicolons can change the meaning of a sentence. Make sure that you use them correctly. Remember the following: (1) Colons should only be used after complete sentences, that is, sentences that can stand on their own. (2) When you join two *related* sentences with a semicolon, check that you've placed the semicolon where the period could be and that the first letter of the second sentence is lowercase.

Exercise

Consider the following sentences. Decide if the colons and semicolons are used correctly. If they are not used correctly, correct the punctuation. If they are correct, write "correct as is".

1. Although there are more than 4000 different components in cigarettes; nicotine has been identified as a key contributor to the long-term health problems associated with smoking.
2. Two aromatic systems were investigated for vesicle insertion: 5,5'-dithiobis(2-nitrobenzoic acid) and *p*-nitrophenylacetate.
3. The four samples from each location were: (1) nonfiltered, nonstabilized water for alkalinity testing, (2) field-filtered, nonstabilized water for nitrate and chloride testing, (3) field-filtered water stabilized with nitric acid for calcium, magnesium, and sodium testing, and (4) field-filtered water stabilized with sulfuric acid for ammonia and phosphate testing.
4. The two different soil sources utilized were; soil "A," a commercial topsoil, and soil "B," a top horizon soil from Flagstaff, AZ.

5. Boc was used as a protecting group, but it was not soluble in water; therefore, it was dissolved in dioxane prior to use.
6. Cantilever sensors have been shown to detect changes in surface tension; ligand–receptor binding; and molecular conformation.
7. The solutions were prepared with different ionic strengths; in different concentrations (10, 20, and 30 mM); in Tris-HCl and HEPES buffers; and at pH 7, 8, and 9.

Colons and Semicolons Answers

(1) replace semicolon with comma; (2) correct as is; (3) remove colon, replace commas with semicolons after the three instances of "testing", and keep the last period in the series; (4) remove semicolon; (5) correct as is; (6) replace semicolons with commas; (7) correct as is

Commas

Where to place commas can be confusing because of the different rules that specify comma placement. Use the rules specified below to help you. At first, you may need to consult these rules often, but as you write more, the rules will become second nature.

Rule 1: Put a comma after introductory words, phrases, or clauses that are followed by a complete sentence.

> *Nevertheless*, a trend in the enhancement of the quantum efficiency is observed for samples coated with different polymers. (From Bol and Meijerink, 2001)
>
> *On the other hand*, the diffraction pattern in Figure 12b exhibits distinct spots on diffuse lines. (From Spengler et al., 2001)
>
> *As previously described*, the bridging iodine atom is located at the midpoint between two diplatinum units. (Adapted from Mitsumi et al., 2001)

Rule 2: If two clauses that could be separate sentences are joined by a linking word, such as *and*, *but*, *or*, *so*, or *yet*, put a comma before the linking word.

> Only a minor fraction of the N1 alkylated derivative hydrolyzed under acidic conditions, *and* all loss generated an equivalent quantity of dA. (Adapted from Veldhuyzen et al., 2001)
>
> A very weak band is also observed at 114 cm^{-1} for **3**, *but* the energy of this band shows little temperature dependence. (From Mitsumi et al., 2001)

[Note the use of the word *very* in the example above. Although the word *very* should be used sparingly in writing for expert and scientific audiences, the use of *very* here—as in *a very weak* or *a very strong* band—is appropriate. In instances such as this, *very strong* and *very weak* are semiquantitative terms used to define absorbance.]

Rule 3: When a sentence contains a list or series of three or more items, put a comma before the *and* or *or* that comes before the last item.

> Samples passivated with PVB, MA, *and* PVA exhibit a substantial increase in quantum efficiency upon UV irradiation. (From Bol and Meijerink, 2001)

Rule 4: Use a comma between two or more adjectives that come before a noun only if you can reverse the order of the adjectives.

With commas	A *novel, inexpensive biosensor* was used to measure calcium.
Without commas	There are no reports on the *optical storage luminescence* exhibited by this *important industrial phosphor*. (From Dhanaraj et al., 2001)

Rule 5: Separate words such as *for example*, *however*, or *therefore* from the rest of the sentence with commas. (See table 6.6 for other words like these.)

> The band at 17 000 cm^{-1}, *however*, disappears in solution. (From Mitsumi et al., 2001)
>
> This may occur, *for example*, when the overlayer unit cells are significantly larger than the substrate unit cells. (From Mitchell et al., 2001)

Rule 6: If a clause or phrase adds extra information about the preceding word or phrase, information that could essentially be deleted without changing the sentence's main message, set the clause or phrase off from the rest of the sentence with commas.

> Concurrently, the α-chloroamine **9a**, *in a solution of THF*, was added to the NaH/Celite column. (From Hafez et al., 2001)

Rule 7: If the word *which* is used to introduce a nonrestrictive clause (i.e., extra information that does not limit the meaning of the sentence's main message), place a comma before the *which* and after the additional information. (See "Which/That", under "Word Usage" below, for more information.)

> This transformation, *which was verified by Raman spectroscopy*, was apparently induced by the mechanical pressure exerted by the AFM tip in contact mode. (From Mitchell et al., 2001)

Proofreading Tip: When commas are used properly, your writing will be more easily understood. Take commas seriously! With attention and care over time, you will begin to understand the rules that govern their use.

Exercise

All commas have been removed from the sentences below. Insert commas in the correct places.

1. In contrast the N1 derivative was much less stable under mild alkaline conditions. (Adapted from Veldhuyzen et al., 2001)
2. Like silylenes and carbenes germylenes are isolobal with phosphines and may function as ligands to transition metal complexes. (Adapted from Bazinet et al., 2001)
3. Polymorph selectivity can be achieved through two-dimensional epitaxy which allows efficient screening of substrates for polymorph control through geometric lattice modeling prior to performing experiments with actual libraries. (Adapted from Mitchell et al., 2001)
4. The assembly consists of three fritted jacketed columns (each 2 cm wide) including two top columns (type A) for reagent synthesis a catalytic column (type B) into which the reagent columns feed and a scavenger resin column (type C) below the catalytic column. (Adapted from Hafez et al., 2001)
5. The catalyst-loaded resin column was washed with methanol dichloromethane and diethyl ether and dried under high vacuum. (Adapted from Hafez et al., 2001)

Commas Answers

1. In contrast, the N1 derivative was much less stable under mild alkaline conditions. (From Veldhuyzen et al., 2001)
2. Like silylenes and carbenes, germylenes are isolobal with phosphines and may function as ligands to transition metal complexes. (From Bazinet et al., 2001)
3. Polymorph selectivity can be achieved through two-dimensional epitaxy, which allows efficient screening of substrates for polymorph control through geometric lattice modeling, prior to performing experiments with actual libraries. (From Mitchell et al., 2001)
4. The assembly consists of three fritted jacketed columns (each 2 cm wide), including two top columns (type A) for reagent synthesis, a catalytic column

(type B) into which the reagent columns feed, and a scavenger resin column (type C) below the catalytic column. (Adapted from Hafez et al., 2001)

5. The catalyst-loaded resin column was washed with methanol, dichloromethane, and diethyl ether and dried under high vacuum. (From Hafez et al., 2001)

Hyphenated Two-Word Modifiers

Chemistry writing includes many two-word modifiers, two words that together describe a noun:

Two-Word Modifier	*Noun*
air-dried	flask
solid-state	material
radiation-sensitive	compound
above-average	results
high-frequency	sound

There are many types of two-word modifiers (noun + adjective, adjective + noun, noun + noun), but almost all are hyphenated when used to modify a noun (see table A4).

Two- and three-word modifiers can be used with compound nouns (i.e., two or more terms used to express a single idea). Some compound nouns are hyphenated (e.g., half-life), and some are not (e.g., redox couple). The compound noun should not be connected to the hyphenated modifier with an additional hyphen.

Hyphenated Modifier	*Compound Noun*
temperature-dependent	absorption cross section
moisture-sensitive	iodine monochloride
three-dimensional	NMR spectra
light-induced	charge redistribution
light-catalyzed	miniaturized solar batteries
high-resolution	crystal structures
aqueous-phase	half-life
solvent-dependent	back-reaction

Some exceptions to the hyphenation rule include the following:

- Do not hyphenate two-word units if the first word is an adverb ending in "ly":

Incorrect	generally-accepted belief commonly-accepted practice
Correct	generally accepted belief commonly accepted practice

Table A4 Different types of two-word modifiers.

Type	Examples
Noun + adjective	time-dependent changes, mass-selective detector
Adjective + noun	long-term storage, high-density lipoprotein
Adjective + participle	fast-growing bacteria, far-reaching implications
Noun + participle	time-consuming calculation, phospholipid-coated capillaries
Adverb + adjective	below-average results
Noun + noun	solvent-solute interaction, room-temperature solution

- Do not hyphenate two-word modifiers that are chemical names (e.g., amino acid level, barium sulfate precipitate).
- Do not hyphenate two-word modifiers if one of the words is a proper name (e.g., Lewis acid catalysis).
- When two or more two-word modifiers have the same base (the second word in the modifier), and modify the same noun, use a hyphen after each modifier, without repeating the base:

Base	*Two-Word Modifiers with Same Base, Plus Noun*
range	medium- and long-range forces
order	first- and second-order kinetics

- Hyphenate three (or more) word modifiers unless other hyphenation rules are broken by doing so.

Correct	general-acid-catalyzed reaction
Correct	Lewis acid catalyzed reaction

Proofreading Tip: Consult *The ACS Style Guide* for more detailed information on hyphenated modifiers (see Unit Modifiers).

Exercise

Read the following sentences, all taken from the chemical literature but with most hyphens removed. Locate two- or three-word modifiers. Decide if they require hyphens. Add hyphens where necessary. If the sentence has no two- or three-word modifiers, write "correct as is".

1. A five point calibration curve for eight carbonyl compounds was measured.
2. Therefore, the pH of standard mixtures was adjusted to 4.5 using 0.1% phosphoric acid. (Adapted from Vesely et al., 2003)
3. Some electron rich aromatic nitriles require higher temperatures, which are achieved using a sealed glass pressure reactor. (Adapted from Demko and Sharpless, 2001)
4. A detailed cross sectional view of the source is shown in Figure 1. (Adapted from Takats et al., 2003a)
5. An output power of 0.77 W was used, which was low enough to prevent possible laser induced damage. (Adapted from Kizil et al., 2002)
6. Therefore, a normalizing and variance stabilizing logarithmic transformation was applied. (Adapted from Dellinger et al., 2001)
7. We conclude that these nanoparticle centered assemblies are conceptually similar to gold filled reverse micelles. (Adapted from Liu et al., 2001)
8. The experiment was adapted from a commonly accepted practice.
9. We have reported a gas phase and solution phase study comparing mono- and dinuclear Cr^{III} complexes as catalysts for the ring opening polymerization of propylene oxide. (Adapted from Schön et al., 2004)
10. Two three-dimensional models were prepared, as shown in Figure 2.

Hyphenated Two-Word Modifiers Answers

(1) five-point calibration; (2) correct as is; (3) electron-rich, sealed-glass; (4) cross-sectional; (5) laser-induced; (6) variance-stabilizing; (7) nanoparticle-centered, gold-filled; (8) correct as is; (9) gas-phase, solution-phase, ring-opening; (10) correct as is.

Word Usage

Affect/Effect

Numerous word pairs are commonly confused by writers. The confusion often stems from similar pronunciations or spellings. For example, the words *affect* and *effect* sound similar, so they are often used incorrectly, confusing the reader. Using the correct word and spelling promote unambiguous communication. In chemistry writing, *affect* and *effect* are typically used as follows:

Affect (verb): to influence, to modify, to change
Effect (noun): a result, a consequence, an outcome
Effect (verb): to bring about; to cause to happen

Consider the following examples:

Possible *Effects* on DNA Topology (a journal title from Plaper et al., 2002)

Before processing the spectral data, area normalization and baseline correction of the spectra were done to eliminate certain unwanted instrumental and cosmetic *effects* on the data set. (From Kizil et al., 2002)

Saponification of fats to their corresponding glycerols and carboxylates facilitates the release of PCBs from fatty matrixes and also can selectively degrade many other interfering substances without *affecting* the PCBs. (From Llompart et al., 2001)

The new procedure *effected* a 50% increase in yield. (From Dodd, 1997)

Proofreading Tip: Even the most skilled writers sometimes confuse *affect* and *effect*. When you are proofreading, pause at each instance of *affect* and *effect* to make sure that you are using it correctly.

Exercise A

Fill in the blanks with the correct form of either *affect* or *effect*.

1. Moreover, the ___ should increase for rotationally hot H_3^+ target ions.
2. We examined how a thymine analog ___ the ability of the DNA polymerase to make DNA.
3. The ab initio calculations show that the torsion does not significantly ___ the electronic properties of the polymer. (Aadpted from Weinberger et al., 2001)
4. The application of an inside negative membrane potential had a profound ___ on this comparably poor rod organization in unpolarized EYPC-SUVs. (Adapted from Sakai et al., 2001)

Exercise B

For each sentence, decide whether the word *affect* or *effect* is used correctly. Make corrections when necessary. Write "correct as is" for sentences that require no changes.

1. Because EH calculations show that these factors strongly *effect* the HOMO-LUMO gaps, different absorption and emission energies are expected to exist. (Adapted from Rawashdeh-Omary et al., 2001)
2. In these systems, the resonances of protons closer to the metal surface are more seriously broadened, whereas the *effect* is less pronounced for the protons that are farther away. (Adapted from Liu et al., 2001)

3. Our hypothesis was that the linkers would *affect* the reaction chemistry in terms of rate and selectivity. (Adapted from Hafez et al., 2001)
4. Such a quantum size *affect* is observable even in an insulator system with a band gap of 6 eV. (Adapted from Dhanaraj et al., 2001)

Affect/Effect Answers

Exercise A

(1) effect; (2) affects/affected; (3) affect; (4) effect

Exercise B

(1) affect; (2) correct as is; (3) correct as is; (4) effect

Between/Among

The words *between* and *among* are often used incorrectly. Fortunately, the rules that distinguish one term from the other are easy (and important) to remember.

Rule 1: *Between* is used with two, and only two, named objects.

> Spectra were obtained in the Raman shift range *between 400* and *4000* cm^{-1}. (From Kizil et al., 2002)
>
> Previous research has established a strong association *between the dietary form of Se* and *the cancer-preventive properties* of this element (5, 28). (From Finley et al., 2001)

Rule 2: *Among* is used with three or more named objects.

> In this study, we present evidence for the first time of a three-way interaction *among starch, protein, and lipid* that alters starch paste viscosity profiles. (Adapted from Zhang and Hamaker, 2003)
>
> On the other hand, significant differences ($p < 0.05$) were observed *among the three organic cultivations*, with the highest value observed in the plums grown on soil covered with trifolium. (From Lombardi-Boccia et al., 2004)

Rule 3: Rules 1 and 2 apply even if the objects are not explicitly stated.

> The idea that out-of-plane undulations could lead to repulsive interactions *between lipid bilayers* was first proposed by Helfrich in the late 1970s. (From Walz and Ruckenstein, 1999)

Rosell et al.[3] reported the highest lauric acid concentrations *among coconut oil samples* from Sri Lanka and the Philippines. (Adapted from Laureles et al., 2002)

Proofreading Tip: When you are proofreading, pause at each instance of *between* and *among* to make sure that you are using the terms correctly.

Exercise A

Consider the following titles of ACS journal articles. Fill in the blanks with either *between* or *among*.

1. Characterization of the Binding Interface ___ the E-Domain of *Staphylococcal* Protein A and an Antibody Fv-Fragment (From Meininger et al., 2000)
2. Phenolic Acids and Derivatives: Studies on the Relationship ___ Structure, Radical Scavenging Activity, and Physicochemical Parameters (From Silva et al., 2000)
3. What Accounts for the Remarkable Difference ___ Silabenzene and Phosphabenzene in Stability toward Dimerization? (From Brown and Borden, 2000)
4. Variation in Diosgenin Levels ___ 10 Accessions of Fenugreek Seeds Produced in Western Canada (From Taylor et al., 2002)
5. The Interface ___ a Protein Crystal and an Aqueous Solution and Its Effects on Nucleation and Crystal Growth (From Haas and Drent, 2000)

Exercise B

Read the following excerpts from the chemical literature and fill in the blanks with either *between* or *among*.

1. The surface force ___ a silica sphere and a copper electrode was measured in concentrated solutions of $MgSO_4$ with an atomic force microscope as a function of the electrode potential. The interaction ___ the two surfaces was compared with the DLVO theory. (From Dedeloudis et al., 2000)
2. The origins of these differences ___ the ring expansion reactions of **1b** and **1c** have been elucidated through the calculation of the energies of relevant isodesmic reactions. (From Galbraith et al., 2002)
3. These concentrations were established by interlaboratory calibration exercises ___ several laboratories. (Adapted from Huggett et al., 2003)

4. Starch, proteins, and lipids are the three major food components in cereal-based food products, and interactions ___ them in a food system are of importance to functionality and quality. The impetus for our research on a three-way interaction ___ these components arose from an observation of an unusual amylogram profile found in aged sorghum flour pastes. (Adapted from Zhang and Hamaker, 2003)
5. The comparisons were performed either ___ conventionally and organically grown plums (both grown on tilled soil) or ___ the three types of organic cultivations, each characterized by a different type of soil management (tilled soil, trifolium, meadow). (From Lombardi-Boccia et al., 2004)

Between/Among Answers

Exercise A

(1) between; (2) among; (3) between; (4) among; (5) between

Exercise B

(1) between, between; (2) between; (3) among; (4) among, among; (5) between, among

Comprise/Compose

The words *comprise* and *compose* are often misused in the chemical literature. Although the two words look and sound somewhat similar, they are used differently, with different word combinations. Most writers make mistakes when using *comprise*, confusing it with the ways in which *compose* is used.

Tip 1: The only time *comprise* and *compose* have the same meanings—"to contain" or "to consist of"—is when they are used with these word combinations: (1) comprise (by itself); (2) (be) composed of (with the verb "be" and the preposition "of"):

> The former *is composed of* an anionic MMX chain necessitating countercations, whereas the latter *comprises* an electrically neutral chain. (Adapted from Mitsumi et al., 2001)

In other words, *comprise* and *(be) composed of* can be used synonymously. Consider the following sentences, which have the same meaning:

> The human body *is composed of* 206 bones.
>
> The human body *comprises* 206 bones.

Tip 2: *Comprise* is a free-standing verb. It is used by itself, never with the word "of" following it or some form of the verb "be" preceding it.

Incorrect	This results in decreased conjugation of the oligothienyl units that *are comprised of* the backbones of these ligands.
Correct	This results in decreased conjugation of the oligothienyl units that *comprise* the backbones of these ligands. (Adapted from Weinberger et al., 2001)
Incorrect	The images show that the Au films *are comprised of* large (approximately 1000 Å wide) crystalline (111)-terminated grains.
Correct	The images show that the Au films *comprise* large (approximately 1000 Å wide) crystalline (111)-terminated grains. (Adapted from Han et al., 2001)

Tip 3: In contrast to *comprise*, *compose* is used with other words, not as a stand-alone word.

Keep in mind the following word combination so that you use *compose* correctly: *(be) composed of*. Notice that *composed* (ending in "ed") is preceded by some form of the verb "be" and followed by "of."

Incorrect	The mixtures *compose* molecules that are macromolecular (>500 Da), hydrophilic, and highly polar.
Correct	The mixtures *are composed of* molecules that are macromolecular (>500 Da), hydrophilic, and highly polar. (Adapted from Kujawinski et al., 2002)
Incorrect	Each compound *composes* an amino sugar group and a nitrogen-free sugar group bound directly to a 14-membered-ring aglycon moiety.
Correct	Each compound *is composed of* an amino sugar group and a nitrogen-free sugar group bound directly to a 14-membered-ring aglycon moiety. (Adapted from McClellan et al., 2002)

Tip 4: The *(be) composed of* word combination can be altered slightly when the "be" form is not the main verb of the sentence. In these cases, the "be" form can be dropped, but it is always understood.

Be is implied	In this paper, we report on a method based on hexachloroplatinic acid impregnation on carbon followed by an electrochemical reduction step *composed of* several current pulses.[7,8]
Be is included	In this paper, we report on a method based on hexachloroplatinic acid impregnation on carbon followed by

	an electrochemical reduction step *that is composed of* several current pulses.[7,8] (Adapted from Adora et al., 2001)
Be is implied	The proven validity of the differences method for this kind of system will enable more complex systems to be studied, like those *composed of* particular metal-IDA complexes with some resin and oxyanions such as selenite and tellurite.
Be is included	The proven validity of the differences method for this kind of system will enable more complex systems to be studied, like those *that are composed of* particular metal-IDA complexes with some resin and oxyanions such as selenite and tellurite. (Adapted from Atzei et al., 2001)

Proofreading Tip: The improper use of *comprise* and *compose* is a pet peeve of many editors and skilled writers. Make sure that you use the terms correctly. Improper use signals a novice writer who has not taken the time to proofread carefully!

Exercise A

Identify the incorrect uses of *comprise* and make corrections as needed.

1. The Na^+ or K^+ cation interaction has been experimentally probed by using synthetic receptors that *comprise* diaza-18-crown-6 lariat ethers having ethylene side arms attached to aromatic donors. (Adapted from Meadows et al., 2001)
2. The photoelectrons with energies from 0 to 0.20 eV are *comprised* of the third region, corresponding to dissociation. (Adapted from Alconcel et al., 2001)
3. These serine proteases process the capsid's scaffolding, *comprised* of roughly 103 assembly protein (AP) molecules. (Adapted from Pray et al., 2002)
4. Estimates of the number of water shells surrounding the DNA and the membranes suggest that one to four layers of water molecules *comprise* the interfacial water signal detected. (Adapted from Ruffle et al., 2002)

Exercise B

Complete the following sentences with the proper form of *comprise* or *compose*, leaving the rest of the sentence unchanged.

1. For TTF, the calculations of the vertical excitation energies ___ valence singlet and triplet states as well as the lowest members of the Rydberg series converging to the first ionization limit. (Adapted from Pou-Amerigo et al., 2002)

2. Not surprisingly, the structures of these different annexin gene products are nearly superimposable, ___ of individual repeats that themselves are nearly identical in structure. (From Isas et al., 2002)
3. They occur in nearly all igneous rocks, some metamorphic settings, and ___ up to 25% of the earth's upper mantle. (Adapted from Ashbrook et al., 2002)
4. Note that both terms, A and B, depend on the solvent, because they ___ the relative permittivity of the solvent and its viscosity. (From Muzikar et al., 2002)
5. Catalysts are often ___ of a single metal with nickel and cobalt.
6. The initial structure ___ four C-terminal amino acids.
7. The initial stage of the reaction is ___ of a two-step rearrangement mechanism (Scheme 1).

Comprise/Compose Answers

Exercise A

(1) correct as is; (2) comprise, are composed of; (3) composed of; (4) correct as is.

Exercise B

(1) comprise; (2) composed; (3) comprise; (4) comprise; (5) composed; (6) comprised, comprises; (7) composed

Farther/Further

Farther and *further* are commonly confused words. Using the ACS Journals Search, we found that *further* was used roughly 60 times more often than *farther*; nevertheless, it is important to learn to distinguish these two words and use them correctly. Most often, *further* signals quantity or degree and *farther* signals literal or figurative distance.

Tip 1: *Further* (adjective) indicates *additional* or *more*

> *Further* proof is given by the observation of two titrating groups of equal contributions in the heme redox titration curve. (From Mileni et al., 2005)
>
> Thus, it is difficult to assign contributions of any protonated heme propionates without *further* distinction such as ^{13}C-labeling. (From Mileni et al., 2005)

Tip 2: *Further* (adverb) indicates *additionally* or *to a greater degree*

The precursor and mature BACE1 proteins were *further* characterized for their ability to interact with a substrate-based transition state BACE1 peptide inhibitor. (From Wang et al., 2005)

The pH was immediately adjusted to 8.7, and the refolding mixture was *further* incubated with slow stirring at room temperature for 4 h. (From Wang et al., 2005)

Tip 3: *Further* (verb) is also used to mean *to help* or *to advance* (something)

To further our fundamental understanding of the underlying science,...

Tip 4: *Furthermore* is often used as a transition devise to indicate "in addition" (see table 6.6).

Tip 5: *Farther* (adjective) means *more distant*

In the third case, the upper surface was moved a *farther* 0.74 Å to the right. (From Neitola and Pakkanen, 2001)

Tip 6: *Farther* (adverb) means *at* or *to a greater distance* or *more distant point*. It is often accompanied by words such as these: apart, away, from, and toward.

When the concentration is higher, the metal ions move *farther* toward the maximum field. (From Fujiwara et al., 2004)

The magnitude becomes smaller because the chromophore is *farther* apart from the rim of the cavity of the chiral macrocyclic host.[6,7] (Adapted from Park et al., 2002)

Proofreading Tip: Although not foolproof, most instances of *further* and *farther* can be distinguished as follows: *further* signifies quantity or degree; *farther* designates distance.

Exercise

Fill in the blanks with the correct word, either *further* or *farther*.

1. When the surfaces were moved ___ apart, the energies slowly approached zero. (Adapted from Neitola and Pakkanen, 2001)
2. ___ mutagenesis work is required to clarify the precise origin of this distinct spectral feature. (From Mileni et al., 2005)

3. Our study provides ___ insight into the molecular mechanisms for the conversion of precursor BACE1 to active proteolytic enzyme. (From Wang et al., 2005)
4. The spectra from the NDO-naphthalene complex also revealed a second binding conformation (denoted as B), in which the substrate is located ~0.5 Å ___ from the Fe atom. (From Yang et al., 2003)
5. Importantly, ___ probing with anti-C-terminal antiserum revealed that the carboxyl terminus remains shielded. (Adapted from Lecchi et al., 2005)

Farther/Further Answers

(1) farther; (2) Further; (3) further; (4) farther; (5) further

Fewer/Less

The words *fewer* and *less* are commonly confused. A few simple rules will help you use these words correctly.

Rule 1: Use *fewer* with items that can be counted.

> These crystals have *fewer defects* than those grown from solution.
>
> *Fewer* than *four chlorine atoms* surround each platinum atom.

Rule 2: Use *less* with quantities that cannot be counted.

> The preferred addition requires *less activation* than the HCl addition to ethylene.
>
> The other compounds possessed significantly *less potency*.

Rule 3: Use *less* with units of measure.

> Data collection time was *less than 3 h* per crystal.
>
> Barriers of *less than 7 kcal/mol* are predicted.
>
> The oxide was isolated in high yield but with *less than 5% ee*.

Proofreading Tip: If you feel uncertain about the use of *fewer* and *less*, consult this language tip when proofreading your work to be sure that you are using the terms appropriately.

Exercise

Some of the sentences below use *less* and *fewer* correctly, but others do not. Identify the incorrect uses of *less* and *fewer*, and make corrections. Indicate "correct as is" for sentences that require no changes.

1. The extra quenching sites have less influence on samples with low quantum efficiency.
2. In the IR spectrum of **4**, there are less than three strong bands between 1279 and 1221 cm^{-1}.
3. The average spectrum recording lasted fewer than 2 s.
4. Fewer samples were used in the second series.
5. The modified catalyst yielded significantly less product.
6. Only one peak was apparent with concentrations of fewer than 2 mM.

Fewer/Less than Answers

(1) correct as is; (2) fewer; (3) less; (4) correct as is; (5) correct as is; (6) less

Its/It's

Its and *It's* are commonly used words, with different meanings but similar spellings, that are often confused by writers. The difference between *its* and *it's* is one that you simply need to memorize.

Tip 1: *Its* is the possessive form of it.

> Conversion of the CT-glyceryl ester to *its* corresponding aldehyde can be achieved at acidic pH. (Adapted from Tam et al., 2001)
>
> Changes in the film's composition after heat treatment were indicated by changes in *its* infrared absorption pattern. (Adapted from Xu et al., 2001)

Tip 2: *It's* is the contracted form of *it is*. It is inappropriate to use the contracted form in writing for an expert audience. *It's* is sometimes acceptable in writing for a general audience, as in these examples:

> "I think *it's* beautiful," says Stephen Weber, a chemist at the University of Pittsburgh. (From Boyd, 1998)
>
> *It's* still too early to know if these or other approaches to high-speed natural products discovery will be adopted throughout the industry. (From Service, 1999)

Proofreading Tip: When proofreading, look for contractions (e.g., it's, doesn't, don't, they'll); they have no place in formal written genres of chemistry (e.g., journal articles, scientific posters, research proposals). Check to make sure that you are using *its* and *it is* correctly. Although the difference between the two looks small, the difference in meaning is large.

Exercise A

Consider the use of *its* and *it's* in the sentences below. Decide whether the sentences follow the writing conventions expected by an expert audience. If the sentence is not appropriate, correct it. Indicate "correct as is" for sentences that require no changes.

1. The protonated nature of the Nt-cysteinyl thiol inhibits thioester ligation but permits imine capture of the Ct-glycoaldehyde ester and *it's* tautomerization to a stable thiazolidine ester **4**. (Adapted from Tam et al., 2001)
2. The power of the method is *its* ability to predict which bonds are most subject to being severed and which bonds are likely to be formed. (Adapted from Zimmerman and Alabugin, 2001)
3. Although the IR data of **B** are limited, the absorption at 2023 cm^{-1} suggests that *it's* **5-F**, the azidodiazo isomer of **4b**. (Adapted from Nicolaides et al., 2001)
4. Water is an important molecule on earth. *It's* shape is bent even when *it's* ionized.
5. MPA, a small thiol, is a useful additive because *it's* a water-soluble reducing agent and does not have a strong stench. (Adapted from Tam et al., 2001)

Its/It's Answers

(1) its; (2) correct as is; (3) it is (no contraction); (4) Its, it is (no contraction); (5) it is (no contraction)

Precede/Proceed

Precede and *proceed* are sometimes confused by writers, even though the words are not interchangeable. Using the ACS Journals Search, we found that *proceed* appears about 1.5 times more often than *precede*, but both words are used regularly; hence, it is important to use them correctly.

Tip 1: *Precede* (verb): To come, exist, or occur before in time, order, rank, position, or place:

A mechanism in which reductive activation of dioxygen *precedes* the C–H bond cleavage cannot explain such a tight coupling of oxygen and substrate consumption. (From Crespo et al., 2006)

The decrease in the concentration of IMP *preceded* a simultaneous increase in the concentrations of both inosine and hypoxanthine in meat samples. (From Tikk et al., 2006)

The oxidative addition is *preceded* by decoordination of the anion.[11] (From Ahlquist et al., 2006)

Tip 2: *Preceding* (adjective): Previous; existing or coming before something in time, order, rank, position, or place:

As mentioned in the *preceding* section, the reaction of...

In agreement with the *preceding* experiment, the average concentration...

Note that the adjective *preceding* commonly precedes words such as those in table A5.

Table A5 Words and phrases that commonly follow the adjective *preceding*.

conformational alteration	paper	section
conformational change	paragraph	stage
cycle	position	step(s)
discussion	residue	study
example	results	treatment
experiment	scheme	

Tip 3: *Proceed* (verb): To go forward or onward; to continue; to begin an action or a process:

The thermal decomposition of 2,1-benzisoxazole will *proceed* at the lower temperature. (Adapted from Lifshitz et al., 2006)

Both are positive, which indicates that these reactions *proceed* via loose transition states, especially the second channel (Si–Si bond breaking step). (From Dávalos and Baer, 2006)

The major fragmentation processes correspond to several competing radical-driven pathways that must necessarily *proceed* from different tautomeric forms of the peptide radical cation. (From Barlow et al., 2006)

Proofreading Tip: Stop and think for a moment when you come across *precede* or *proceed* in your writing. Does the word oblige you to think "back" (precede) or "forward" (proceed)? To ensure that your ideas are interpreted as you want them to be, make sure you use these words correctly.

Exercise

Fill in the blanks with the correct form of either *precede* or *proceed*.

1. Rain conditions were characterized by a wet winter ___ each growth period, indicating that soils were close to field capacity at budbreak. (From Koundouras et al., 2006).
2. Because simple ligand elimination from a metal center is expected to ___ via a barrierless or "loose" transition state,[3] one can reasonably expect the single ligand losses to be the channels having the lowest energy barriers to dissociation. (Adapted from Dunbar et al., 2006)
3. These dissociations ___ in an entirely different way depending on whether X is a hydrogen or a halogen atom. (From Gridelet et al., 2006)
4. In the case of rare earth metals, the reaction did not ___ at all. (From Saito and Kobayashi, 2006)
5. The kinetic values are consistent with Fdx-to-CYP51 electron transfer... (see the ___ section). (From McLean et al., 2006)
6. To achieve a polymer with 100 repeat units, the reaction must ___ to 99% completion. (From Wright et al., 2006)
7. It is sometimes difficult to implement this method. This is especially the case for coordination polymers that often collapse or decompose during the ___ desorption treatment. (From Daiguebonne et al., 2006)
8. This raises the question about the order of events in the active site of EF-Tu: either Pi release ___ and limits the rate of the conformational change of EF-Tu, or the conformational change takes place first and limits the rate of subsequent Pi release. (From Kothe and Rodnina, 2006)

Precede/Proceed Answers

(1) preceding; (2) proceed; (3) proceed; (4) proceed; (5) preceding; (6) proceed; (7) preceding; (8) precedes

Principle/Principal

Principle and *principal* are commonly misused. Using the ACS Journals Search, we found that *principle* is used more than twice as often as *principal*. However, *principal* occurred in more than 20,000 documents; hence, both words are common, and it is important to learn to distinguish between the two.

Tip 1: *Principle* (a noun): a basic truth or assumption; a standard; a guiding rule, code of conduct, method of operation; a rule or law concerning the functioning of natural phenomena or mechanical processes:

> The remarkable selectivity of ions by macrocycles illustrates the *principle* of molecular recognition. (From Parra et al., 2006)
>
> The basic *principle* of semiconductor photocatalysis involves photogenerated electrons and holes migrating to the surface. (From Wang et al., 2006)

Tip 2: *In principle*: with regard to the basics; according to what is supposed to be true, though not proven; in theory; in general, but not necessarily in all details:

> More calculations are, *in principle*, needed to corroborate this assumption.
>
> *In principle*, $CH_3COCH_2COCH_3$ may exist as an asymmetric (C_s) or symmetric (C_{2v}) structure in the ground state. (From Chen et al., 2006)

Tip 3: *First principles*: a calculation is said to be from first principles (or ab initio) if it relies on basic and established laws of nature without additional assumptions or special models.

The term is often used in titles and as a two-word modifier:

first-principles calculations	first-principles molecular dynamics study
first-principles kinetics	first-principles simulations
first-principles methods	first-principles study

Tip 4: *Principal* (an adjective): most important; main; chief; first in rank, worth, or degree:

The elimination of HO_2 from the alkylperoxy radicals is the *principal* reaction pathway for the R + O_2 reactions at temperatures above the transition region. (From Taatjes, 2006)

Principal is often used with words such as these:

principal advantages
principal characteristics
principal driving force
principal electronic parameters
principal emission sources
principal factors
principal goal
principal indices
principal ions
principal product
principal reaction
principal role
principal setup
principal step

Tip 5: *Principal components analysis (PCA)*: the name of a statistical procedure: Partitioning was based on *principal* components analysis.[11]

Proofreading Tip: Don't count on your computer spellchecker to catch commonly confused words. When you read over your final draft and run across *principle* and/or *principal*, stop for a moment. Make sure you've used the words correctly. Ultimately, you are responsible for getting them right!

Exercise

Fill in the blanks with the correct word(s): principle, in principle, principles, or principal.

1. In the most favorable case, this can lead to a balanced treatment of these states, but more accurate calculations are, ___, needed to corroborate this assumption. (Adapted from Dreuw, 2006)
2. ___, it is possible to determine absolute mole fractions of every single isomer in each flame. (From Hansen et al., 2006)
3. **Table 1.** ___ characteristics of three apple cultivars harvested at three commercial maturity stages. (Adapted from Mehinagic et al., 2006)
4. From the viewpoint of the microscopic reversibility ___, elimination of a Pd–Pd bonded moiety from the dipalladium(II) complexes is of great interest in relation to the dinuclear addition reactions discussed above. (Adapted from Murahashi et al., 2006)
5. The ___ components analysis (PCA) was realized on the averaged data (only for apples harvested in 2003) according to the cultivars. (Adapted from Mehinagic et al., 2006)

6. The structural stability and bonding mechanisms of $LiAlH_4$ have been thoroughly investigated with first-___ electronic structure and total-energy calculations. (From Song, Y., et al., 2006)
7. The three ___ pharmacological factors affecting the cytotoxicity of platinum drugs are cellular uptake and efflux, the frequency and structure of target (DNA) adducts, and the extent of metabolizing reactions of sulfur-containing proteins and peptides. (From Liu et al., 2006)
8. Simulating Temperature Programmed Desorption of Water on Hydrated γ-Alumina from First-___ Calculations (From Joubert et al., 2006)

Principle/Principal Answers

(1) in principle; (2) In principle; (3) Principal; (4) principle; (5) principal; (6) principles; (7) principal; (8) Principles

Since/Because and While/Although

Two of the most frequently misused words encountered in formal written language are *since* and *while*. Although these words are commonly used to mean *because* and *although*, respectively, *The ACS Style Guide* recommends that they be used only to connote time:

Incorrect That position is important *since*, as shown previously, crystallization kinetics are maximized at temperatures slightly exceeding the fluid–fluid critical temperature.

Correct That position is important *because*, as shown previously, crystallization kinetics are maximized at temperatures slightly exceeding the fluid–fluid critical temperature. (Adapted from Curtis et al., 2001)

Incorrect *While* there exist empirical rules for determining conditions favorable for protein crystallization, protein crystallization remains predominantly a trial-and-error process.

Correct *Although* there exist empirical rules for determining conditions favorable for protein crystallization, protein crystallization remains predominantly a trial-and-error process. (Adapted from Curtis et al., 2001)

Tip 1: *Because* is generally used to indicate a cause-and-effect relationship.

Knowledge of the SO_2 level in beer is important *because* SO_2 complexes with aldehydes and only "free" aldehydes are measured by the described method. (From Vesely et al., 2003)

Tip 2: *Since* is generally used to indicate "during the period after" and "from a point in the past until the present."

> There is a need to verify whether ozone-depleting compound emissions have truly decreased *since* the passage of the Montreal Protocol.

Tip 3: *While* is generally used to indicate "during the time that" and "at the same time that."

> The NCs should be dispersed in the lower density nonpolar hexane phase, *while* the HF, butanol, and water remain in the acetonitrile phase. (Adapted from Langof et al., 2002)

Tip 4: *Although* is generally used to indicate a contrast of some sort.

> At the beginning of this study, concentrations of benzene and trichloroethene (TCE), *although* low, were well above drinking water standards, and implementation of some treatment strategy was necessary. (From Richmond et al., 2001)

Proofreading Tip: Do not be surprised if you see the words *since, because, while,* and *although* improperly used in professional writing. It occasionally happens. Do your best to use them properly.

Exercise A

Some of the sentences below use *while* and *since* correctly (to indicate time), but others do not. Identify the incorrect uses of *while* and *since*, and make corrections. Indicate "correct as is" for sentences that do not require a change.

1. *Since* these calculations do not take fluctuations into account, the obtained values are only approximations.
2. *Since* those early studies, X-ray crystallography has confirmed this structure.
3. *Since* the rapidly changing intensities of the individual cells are difficult to control, a large variance within each experiment is an unavoidable consequence of using this method. (Adapted from Dellinger et al., 2001)
4. Despite the fact that these molecules have been synthesized *since* 1901, their preparation is still difficult.
5. The mixture was stirred *while* it cooled.

6. *While* the IR data of **B** are limited, the absorption at 2023 cm^{-1} suggests that it is **5-F**, the azidodiazo isomer of **4b**. (Adapted from Nicolaides et al., 2001)
7. This assumes equal concentrations of CO and CO_2, *while* from our experiments, it is clear that CO is in greater abundance.

Exercise B

Consider the following pairs of sentences. How do the sentences differ in meaning? What is the intended meaning of each sentence?

1. (a) The second reaction produced a very sharp peak at 2200 cm^{-1}, *while* the second peak disappeared.

   (b) The second reaction produced a very sharp peak at 2200 cm^{-1}, *although* the second peak disappeared.
2. (a) Gifted with novel chemical features and extraordinary biological activity, sceptrin has remained a prominent unanswered synthetic challenge *since* its characterization in 1981 by Faulkner and Clardy.

   (b) Gifted with novel chemical features and extraordinary biological activity, sceptrin has remained a prominent unanswered synthetic challenge *because* of its characterization in 1981 by Faulkner and Clardy.

Since/Because and While/Although Answers

Exercise A

(1) Because; (2) correct as is (note that some may argue "because of" could work—to indicate a cause–effect relationship; nonetheless, the word "early" suggests that a temporal meaning is intended); (3) Because; (4) correct as is; (5) correct as is; (6) Although; (7) although

Exercise B

1. (a) With *while*, the simulation produced a sharp peak at the same time a second peak disappeared.

   (b) With *although*, one can infer that the disappearance of the second peak was unexpected during the simulation.
2. (a) With *since*, there is the straightforward indication that sceptrin has remained a challenge in the time that has lapsed between 1981 and the present time.

   (b) With *because*, the sentence makes no sense: The sentence is not describing a cause–effect relationship.

Then/Than

Then and *than* have very different meanings; nonetheless, novice writers often mistakenly use them interchangeably. To communicate without ambiguity, it is important that these two words be used properly. Essentially, you need to memorize which one is which and then use them correctly in your written work. Note that *then* is used only sparingly in expert scientific writing because the order of events (e.g., first, second, then) is generally implied and does not need to be stated explicitly.

Tip 1: *Then* (adverb): next or after that:

> The solvent was *then* refluxed, whereupon benzophenone and metallic sodium in an atmosphere of nitrogen were added. (From Danil de Namor et al., 2002)

[Remember that *then*, as used above, is used infrequently in expert writing.]

Tip 2: *Than* (preposition, conjunction): used to join two parts of a comparison:

> This behavior can be justified given that the size of the former is greater *than* that of the uncomplexed cation. (From Danil de Namor et al., 2002)

Proofreading Tip: Make sure that you use *then* and *than* correctly. The meanings of the two terms are distinct even though the spellings are similar. When writers misuse *then* and *than*, it conveys a sloppiness and disregard for accuracy and may cause the reader to dismiss the work.

Exercise

Fill in the blanks with either *then* or *than*.

1. The PDMS/DVB SPME fiber was ___ placed in the headspace of the PFBOA solution for 10 min at 50 °C. (From Vesely et al., 2003)
2. The N–Na–N angle is 122.2° rather ___ 180° as expected for a nearly planar macroring conformation.
3. The new precipitate was ___ filtered, washed with 2 × 200 mL of 1 N HCl, and dried in a drying oven at 90 °C overnight. (Adapted from Demko and Sharpless, 2001)
4. Cells were plated into 60 mm diameter dishes at several different densities (250–500 cells per dish, three dishes per density) to obtain more ___ one set

in which the number of surviving colonies ranged from 100 to 200. (Adapted from Ou et al., 2000)

5. An earlier study from this group[10] found that such purified water contains less ___ ca. 1 ppb of typical UV-absorbing impurities. (Adapted from Quickenden et al., 1996)

Then/Than Answers

(1) then (remember that "then" is generally not stated explicitly in expert scientific writing.); (2) than; (3) then (again, "then" is often deleted from expert scientific writing.); (4) than; (5) than

Which/That

The difference between *that* and *which* may appear to be subtle, but the improper use of these two vital words can alter the interpretation of a sentence. You will often see and hear *that* and *which* used interchangeably. Nonetheless, you should strive to use the words as endorsed by *The ACS Style Guide* to ensure a correct interpretation of your written work. Remember the following rules:

Rule 1: A *that* clause (referred to as a "restrictive clause" in *The ACS Style Guide*) introduces information (about the word or phrase just before it) that is considered essential to the meaning of the sentence. If the *that* clause were deleted, the sentence would lose its intended meaning:

Meaningful	This has led to the expectation *that* smaller molecules may be capable of inducing their activation. (Adapted from Goldberg et al., 2002)
Not Meaningful	This has led to the expectation.

Explanation: In the meaningful sentence, the *that* clause answers the question "Which *expectation*?" or "Which specific *expectation*?" Without the *that* clause, the meaning of the sentence is lost.

Rule 2: A *which* clause (referred to as a "nonrestrictive clause" in *The ACS Style Guide*) signals *extra, though not necessary*, information about the word or phrase before it:

With *which* clause	Thus, the dimeric compounds, *which* contain two sets of binding groups, were found to be sufficient for functional activity.

Without *which* clause	Thus, the dimeric compounds were found to be sufficient for functional activity. (Adapted from Goldberg et al., 2002)

Explanation: In the sentence with the *which* clause, *which* introduces extra information about *the dimeric compounds* (the phrase before it). That information is not essential to the meaning of the sentence. Thus, the *which* clause could be deleted without sacrificing the authors' intended message. Notice that the extra information, introduced by *which*, is set off by commas.

There are a couple tests that you can use to check whether you are using *that* and *which* correctly.

Test 1

Check to verify that the word you have chosen (*that* or *which*) works with a comma immediately before it (and possibly at the end of the clause). *Which* usually comes after a comma.

Incorrect	Pennslvania, that is my home state, is known as the Keystone State.
Correct	Pennsylvania, which is my home state, is known as the Keystone State.

Test 2

Try to imagine the words *by the way* after every instance of *which*. If it sounds right, then you are probably using *which* correctly to signal extra information; if it sounds wrong, you probably should use *that*.

Incorrect	Linoleic acid was selected as the fatty acid, which *(by the way)* would be employed to simulate lipid-damaged beer.
Correct	Linoleic acid was selected as the fatty acid that would be employed to simulate lipid-damaged beer. (Adapted from Cooper et al., 2002)

Test 3

Ask yourself if the information following *that* or *which* is extra or essential. If the information could be deleted without changing your principal message, use *which*. If the information is essential for communicating your intended message, use *that*.

Extra Information	Some electron-rich aromatic nitriles require higher temperatures, *which* are achieved using a sealed-glass pressure reactor. (Adapted from Demko and Sharpless, 2001)

Explanation: The reader can understand "Some electron-rich aromatic nitriles require higher temperatures" without the information following *which*.

Essential Information Most aldehydes, except formaldehyde, form two geometrical isomers of the derivatives *that* are represented by two peaks in the chromatogram. (From Vesely et al., 2003)

Explanation: The *that* clause is needed for the authors to make their point.

Proofreading Tip: Check your use of *which* and *that* when proofreading. Their proper use will result in a more precise interpretation of your written work.

Exercise A

Fill in the blanks with *that* or *which*.

1. This methodology can be extended to other receptor systems, providing a general method for identifying small molecules ___ modulate protein–protein interactions. (Adapted from Goldberg et al., 2002)
2. Effective, yet technically nondemanding conditions, ___ are practical for a large library synthesis, were developed to free base the resulting formic acid salts.
3. Many substitutions on the aromatic ring are permitted, although none of the new carbamates led to activity ___ exceeded that of the original benzyl carbamate. (From Goldberg et al., 2002)
4. Mixture sublibraries were deconvoluted by synthesizing the individual components of the mixtures ___ displayed the highest competitive binding activity. (From Goldberg et al., 2002)
5. Markert and Weckert (*33–35*) have given detailed descriptions of the moss *P. formosum* in this area, ___ means that a large amount of background information was available. (From Kunert et al., 1999)
6. A laser output power of 0.77 W was used, ___ was low enough to prevent possible laser-induced sample damage yet provided a high signal-to-noise ratio. (From Kizil et al., 2002)
7. The samples of $PM_{2.5}$ ___ were used in the mechanistic studies were collected at the Louisiana Department of Environmental Quality station. (Adapted from Dellinger et al., 2001)
8. The cells were treated with increasing doses (20–160 μM) of butachlor, ___ was previously dissolved in ethanol, and the treated cells were left overnight. (From Ou et al., 2000)

 Exercise B

What is the difference in meaning between these two sentences (adapted from Van Berkel et al., 2002)?

1. The metal sampling/sprayer tube that extends the length of the device is a 10 cm long, 31 gauge stainless steel tube.
2. The metal sampling/sprayer tube, which extends the length of the device, is a 10 cm long, 31 gauge stainless steel tube.

Which/That Answers

Exercise A

(1) that; (2) which; (3) that; (4) that; (5) which; (6) which; (7) that; (8) which

Exercise B

1. The writers are providing information about the tube that they consider essential. The authors are only referring to the tube that extends the length of the device; they are not referring to other tubes.
2. The writers are providing what they consider to be additional, but not obligatory, information about the tube. That the tube "extends the length of the device" is not considered essential in the eyes of the authors. If the *which* clause (and commas) were deleted, readers would still have access to the information that matters to the authors: the metal sampling/sprayer tube is a 10 cm long, 31 gauge stainless steel tube.

Appendix B *Move Structures*

The move structures presented throughout the book are repeated here to serve as convenient references for readers. These move structures depict common organizational frameworks for the various sections of four chemistry genres: the journal article, the conference abstract, the scientific poster, and the research proposal.

Journal Article Move Structures

Abstract

1. State What Was Done

1.1 Identify the research area and its importance (optional)

1.2 Mention a gap addressed by the work (optional)

1.3 State purpose and/or accomplishment(s) of work

2. Identify Methods Used
(i.e., procedures and/or instumentation)

3. Report Principal Findings

3.1 Highlight major results (quantitatively or qualitatively)

3.2 Offer a concluding remark (optional)

Introduction

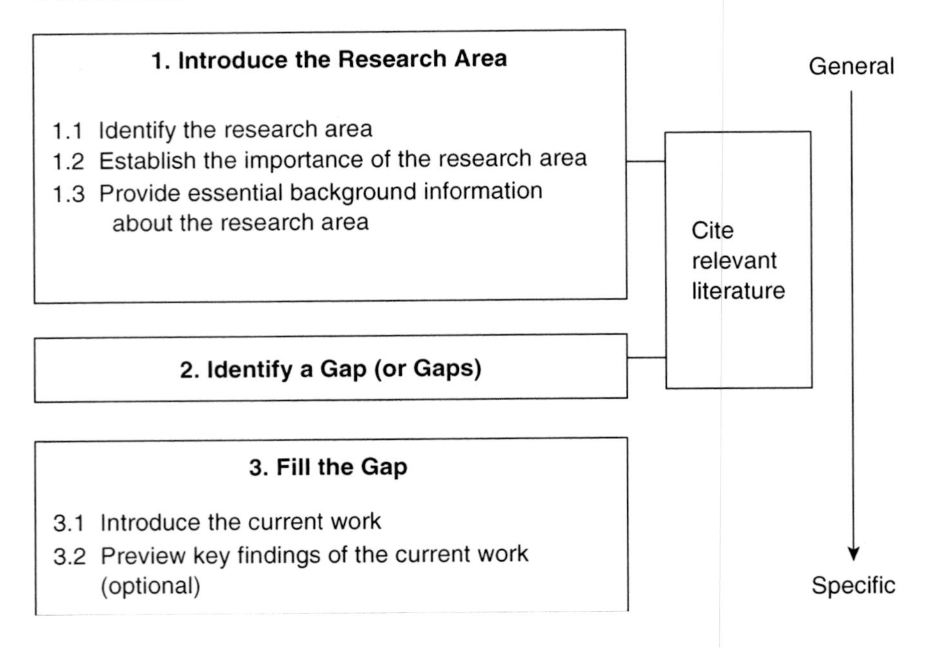

Methods

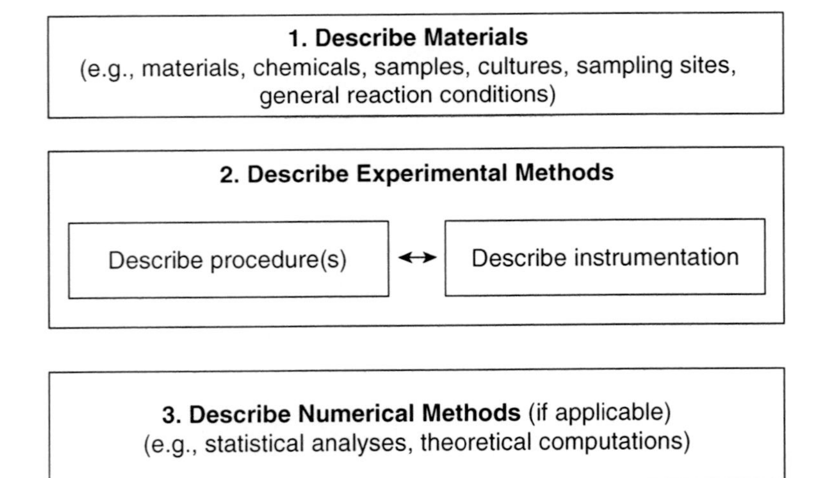

Results

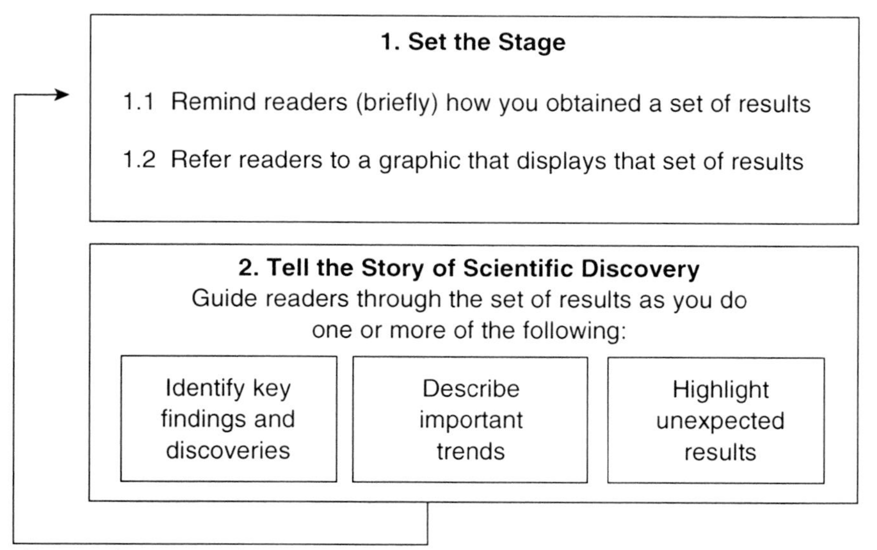

Discussion

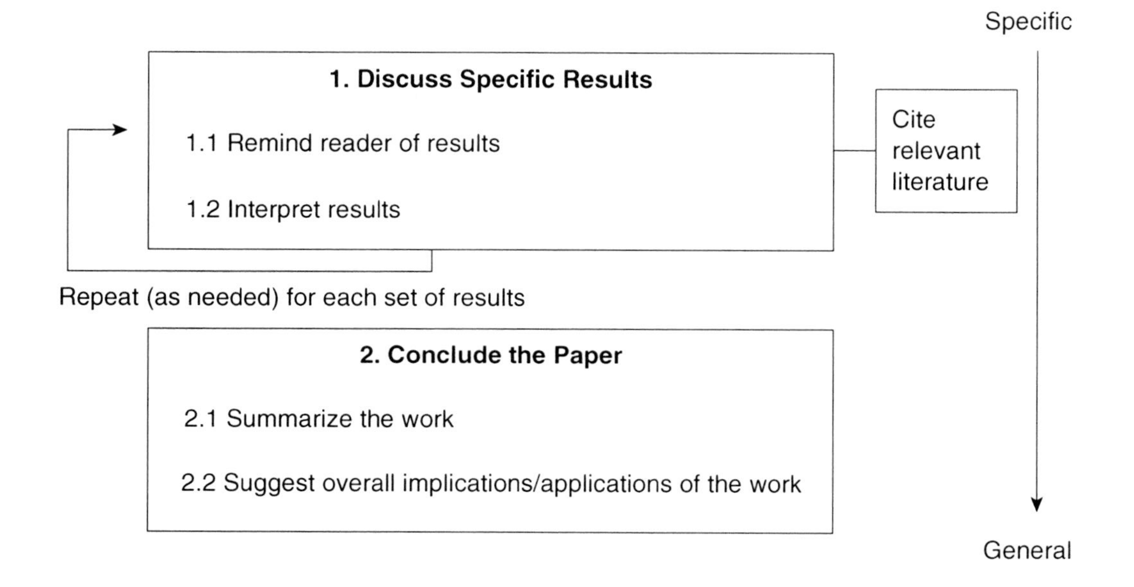

Conference Abstract Move Structure

1. Introduce the Research Area

1.1 Identify the topic

1.2 Highlight the importance of the research

2. Suggest a Gap in the Field

3. Describe the Work to be Presented

(i.e., the methods used and the results obtained or to be obtained)

Scientific Poster Move Structures

Introduction

1. Establish Importance of Research Area

1.1 Introduce the research area

1.2 Emphasize importance (through background infromation, gap statements)

Cite essential works (optional)

2. Preview Accomplishments

(Identify major goals that have been achieved)

Methods

1. Highlight Essential or Novel Materials
(e.g., chemicals, reagents, samples)

2. Summarize Essential or Novel Methods

- Identify major instrumentation
- Identify experimental and/or numerical procedures

Results

1. Share Preliminary Results
Prepare viewers for principal findings by doing one or more of the following:

- Share results that build confidence in your approach
- Share results that motivated your study
- Share results that lay groundwork for your principal findings

2. Share Principal Results
(i.e., share key findings; identify and summarize key trends)

3. Share Related Results (optional)
(i.e., share results that support, extend, or strengthen your principal findings)

Discussion

1. Interpret or Explain Results

2. Conclude with a Take-Home Message

Research Proposal Move Structures

Project Summary

1. Introduce Topic and/or Project Goals

2. Emphasize Significance
(i.e., stress unique features and applications)

↔

3. Summarize Methods
(i.e., highlight promising results)

4. Summarize Broader Impacts

Goals and Importance

1. State Goals and Objectives

2. Establish Importance

2.1 Identify the research area

2.2 Develop the research story

- Explain fundamental concepts
- Provide essential background information

3. Introduce the Proposed Work

3.1 Identify gap(s) in the field

3.2 Introduce your project to fill the gap(s)

Cite relevant literature

Experimental Approach

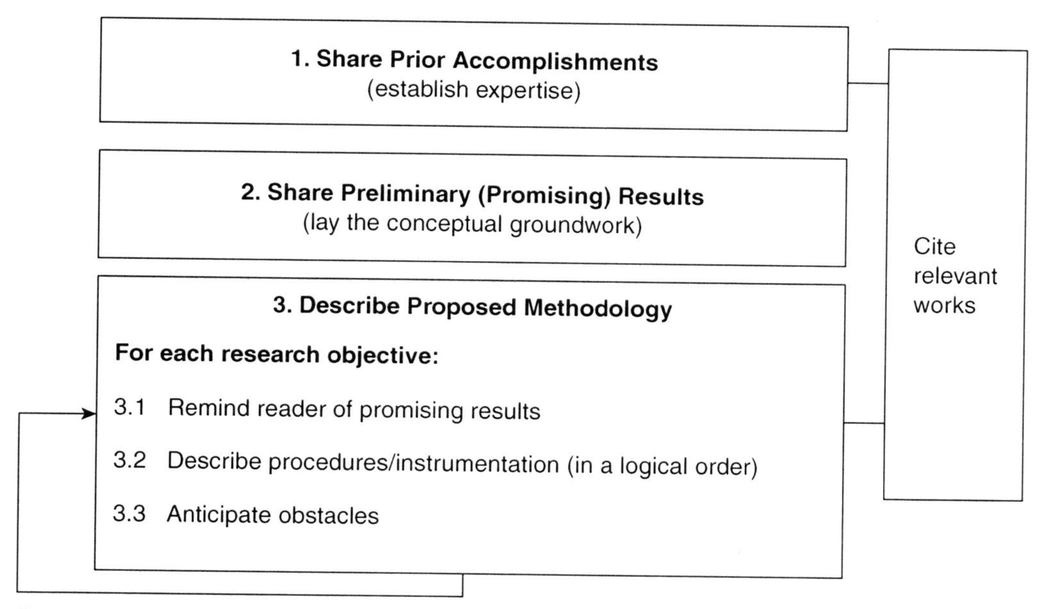

Outcomes and Impacts

1. Present a Project Timeline

2. List Expected Outcomes

3. Conclude the Proposed Work

3.1 Reiterate goals and importance

3.2 Address broader impacts

Sources of Excerpts

Adora, S.; Soldo-Olivier, Y.; Faure, R.; Durand, R.; Dartyge, E.; Baudelet, F. Electrochemical Preparation of Platinum Nanocrystallites on Activated Carbon Studied by X-ray Absorption Spectroscopy. *J. Phys. Chem. B* **2001**, *105*, 10489–10495.

Aga, D. CAREER: Immunochemical Techniques for Investigations on the Occurrence and Fate of Agrochemicals in the Environment, 2002 (NSF CHE 0233700).

Agnihotri, S.; Mota, J. P. B.; Vadlaman, L. N. Gas Adsorption To Estimate the Fraction of Open-Ended Nanotubes in Samples. *Proceedings of the 233rd American Chemical Society National Meeting,* Chicago, IL, March 25–29, 2007.

Aguilera, A.; Brotons, M.; Rodriguez, M.; Valverde, A. Supercritical Fluid Extraction of Pesticides from a Table-Ready Food Composite of Plant Origin (Gazpacho). *J. Agric. Food Chem.* **2003,** *51,* 5616–5621.

Ahlquist, M.; Fabrizi, G.; Cacchi, S.; Norrby, P.-O. The Mechanism of the Phosphine-Free Palladium-Catalyzed Hydroarylation of Alkynes. *J. Am. Chem. Soc.* **2006,** *128,* 12785–12793.

Ahmed, S.; Jabeen, N.; Rehman, E. Determination of Lithium Isotopic Composition by Thermal Ionization Mass Spectrometry. *Anal. Chem.* **2002,** *74,* 4133–4135.

Alberty, R. A. Components and Coupling in Enzyme-Catalyzed Reactions. *J. Phys. Chem. B.* **2005,** *109,* 2021–2026.

Alconcel, L. S.; Deyerl, H.-J.; DeClue, M.; Continetti, R. E. Dissociation Dynamics and Stability of Cyclic Alkoxy Radicals and Alkoxide Anions. *J. Am. Chem. Soc.* **2001,** *123,* 3125–3132.

Almela, C.; Laparra, J. M.; Velez, D.; Barbera, R.; Farre, R.; Montoro, R. Arsenosugars in Raw and Cooked Edible Seaweed: Characterization and Bioaccessibility. *J. Agric. Food Chem.* **2005,** *53,* 7344–7351.

Andersen, H. C. Diagrammatic Formulation of the Kinetic Theory of Fluctuations in Equilibrium Classical Fluids. III. Cluster Analysis of the Renormalized Interactions and a Second Diagrammatic Representation of the Correlation Functions. *J. Phys. Chem. B* **2003,** *107,* 10234–10242.

Aranda, A.; Jimenez-Marti, E.; Orozco, H.; Matallana, E.; del Olmo, M. Sulfur and Adenine Metabolisms Are Linked, and Both Modulate Sulfite Resistance in Wine Yeast. *J. Agric. Food Chem.* **2006,** *54,* 5839–5846.

Arimitsu, S.; Jacobsen, J. M.; Hammond, G. B. Synthesis of Multi Substituted 3-Fluoro Furans and 3,3-Difluoro Hydrofurans. *Proceedings of the 233rd American Chemical Society National Meeting,* Chicago, IL, March 25–29, 2007.

Arulsamy, N.; Bohle, D. S. Multiplicity Control in the Polygeminal Diazeniumdiolation of Active Hydrogen Bearing Carbons: Chemistry of a New Type of Trianionic Molecular Propeller. *J. Am. Chem. Soc.* **2001,** *123,* 10860–10869.

Ashbrook, S. E.; Berry, A. J.; Wimperis, S. Multiple-Quantum MAS NMR Study of Pyroxenes. *J. Phys. Chem. B* **2002**, *106*, 773–778.

Atzei, D.; Ferri, T.; Sadun, C.; Sangiorgio, P.; Caminiti, R. Structural Characterization of Complexes between Iminodiacetate Blocked on Styrene-Divinylbenzene Matrix (Chelex 100 Resin) and Fe(III), Cr(III), and Zn(II) in Solid Phase by Energy-Dispersive X-ray Diffraction. *J. Am. Chem. Soc.* **2001**, *123*, 2552–2558.

Banwell, M. G.; McRae, K. J. A Chemoenzymatic Total Synthesis of ent-Bengamide E. *J. Org. Chem.* **2001**, *66*, 6768–6774.

Barlow, C. K.; Moran, D.; Radom, L.; McFadyen, W. D.; O'Hair, R. A. J. Metal-Mediated Formation of Gas-Phase Amino Acid Radical Cations. *J. Phys. Chem. A* **2006**, *110*, 8304–8315.

Bazinet, P.; Yap, G. P.; Richeson, D. S. Synthesis and Properties of a Germanium(II) Metalloheterocycle Derived from 1,8-Di(isopropylamino)naphthalene. A Novel Ligand Leading to Formation of $Ni\{Ge[(iPrN)_2C_{10}H_6]\}_4$. *J. Am. Chem. Soc.* **2001**, *123*, 11162–11167.

Beck, B.; Picard, A.; Herdtweck, E.; Domling, A. Highly Substituted Pyrrolidinones and Pyridones by 4-CR/2-CR Sequence. *Org. Lett.* **2004**, *6*, 39–42.

Besser, J. M.; Brumbaugh, W. G.; Kemble, N. E.; May, T. W.; Ingersoll, C. G. Effects of Sediment Characteristics on the Toxicity of Chromium(III) and Chromium(VI) to the Amphipod *Hyalella azteca*. *Environ. Sci. Technol.* **2004**, *38*, 6210–6216.

Boesten, W. H. J.; Seerden, J.-P. G.; de Lange, B.; Dielemans, H. J. A.; Elsenberg, H. L. M.; Kaptein, B.; Moody, H. M.; Kellogg, R. M.; Broxterman, Q. B. Asymmetric Strecker Synthesis of α-Amino Acids via a Crystallization-Induced Asymmetric Transformation using (R)-Phenylglycine Amide as Chiral Auxiliary. *Org. Lett.* **2001**, *3*, 1121–1124.

Bogialli, S.; Curini, R.; Di Corcia, A.; Nazzari, M.; Tamburro, D. A Simple and Rapid Assay for Analyzing Residues of Carbamate Insecticides in Vegetables and Fruits: Hot Water Extraction Followed by Liquid Chromatography-Mass Spectrometry. *J. Agric. Food Chem.* **2004**, *52*, 665–671.

Bol, A. A.; Meijerink, A. Luminescence Quantum Efficiency of Nanocrystalline $ZnS:Mn^{2+}$. 2. Enhancement by UV Irradiation. *J. Phys. Chem. B* **2001**, *105*, 10203–10209.

Borovik, A. S.; Bott, S. G.; Barron, A. R. Arene-Mercury Complexes Stabilized Aluminum and Gallium Chloride: Synthesis and Structural Characterization. *J. Am. Chem. Soc.* **2001**, *123*, 11219–11228.

Boyd, K. Getting an Inside Look at Cells' Chemistry. *Science* **1998**, *279*, 1856.

Brown, E. C.; Borden, W. T. What Accounts for the Remarkable Difference between Silabenzene and Phosphabenzene in Stability toward Dimerization? *Organometallics* **2000**, *19*, 2208–2214.

Buerge, I. J.; Buser, H.-R.; Muller, M. D.; Poiger, T. Behavior of the Polycyclic Musks HHCB and AHTN in Lakes, Two Potential Anthropogenic Markers for Domestic Wastewater in Surface Waters. *Environ. Sci. Technol.* **2003**, *37*, 5636–5644.

Bushey, J. T.; Nallana, A. G.; Driscoll, C. T.; Choi, H.-D.; Holsen, T. Enhancement of Atmospheric Mercury Deposition by Plants within a Northern Forest Landscape, USA. *Proceedings of the 233rd American Chemical Society National Meeting,* Chicago, IL, March 25–29, 2007.

Cabras, P.; Caboni, P.; Cabras, M. Analysis by HPLC of Ryanodine and Dehydroryanodine Residues on Fruits and in Ryania Powdery Wood. *J. Agric. Food Chem.* **2001**, *49*, 3161–3163.

Cameron, K. S.; Fielding, L. NMR Diffusion Spectroscopy as a Measure of Host-Guest Complex Association Constants and as a Probe of Complex Size. *J. Org. Chem.* **2001**, *66*, 6891–6895.

Cantrill, S. J.; Youn, G. J.; Stoddart, J. F.; Williams, D. J. Supramolecular Daisy Chains. *J. Org. Chem.* **2001**, *66,* 6857–6872.

Carroll, J.; Jonsson, E. N.; Ebel, R.; Hartman, M. S.; Holman, T. R.; Crews, P. Probing Sponge-Derived Terpenoids for Human 15-Lipoxygenase Inhibitors. *J. Org. Chem.* **2001**, *66,* 6847–6851.

Castlehouse, H.; Smith, C.; Raab, A.; Deacon, C.; Meharg, A. A.; Feldmann, J. Biotransformation and Accumulation of Arsenic in Soil Amended with Seaweed. *Environ. Sci. Technol.* **2003**, *37,* 951–957.

Celis, R.; Hermosín, M. C.; Cornejo, J. Heavy Metal Adsorption by Functionalized Clays. *Environ. Sci. Technol.* **2000**, *34,* 4593–4599.

Chen, X.-B.; Fang, W.-H.; Phillips, D. L. Theoretical Studies of the Photochemical Dynamics of Acetylacetone: Isomerization, Dissociation, and Dehydration Reactions. *J. Phys. Chem. A* **2006**, *110,* 4434–4441.

Cheng, A.-C.; Huang, T.-C.; Lai, C.-S.; Kuo, J.-M.; Huang, Y.-T.; Lo, C.-Y.; Ho, C.-T.; Pan, M.-H. Pyrrolidine Dithiocarbamate Inhibition of Luteolin-Induced Apoptosis through Up-Regulated Phosphorylation of Akt and Caspase-9 in Human Leukemia HL-60 Cells. *J. Agric. Food Chem.* **2006**, *54,* 4215–4221.

Cheng, H.-H.; Lai, M.-H.; Hou, W.-C.; Huang, C.-L. Antioxidant Effects of Chromium Supplementation with Type 2 Diabetes Mellitus and Euglycemic Subjects. *J. Agric. Food Chem.* **2004**, *52,* 1385–1389.

Chouchane, S.; Snow, E. T. In Vitro Effect of Arsenical Compounds on Glutathione-Related Enzymes. *Chem. Res. Toxicol.* **2001**, *14,* 517–522.

Clancy, C. M. R.; Krogmeier, J. R.; Pawlak, A.; Rozanowska, M.; Sarna, T.; Dunn, R. C.; Simon, J. D. Atomic Force Microscopy and Near-Field Scanning Optical Microscopy Measurements of Single Human Retinal Lipofuscin Granules. *J. Phys. Chem. B* **2000**, *104,* 12098–12101.

Clarke, W.; Chowdhuri, A. R.; Hage, D. S. Analysis of Free Drug Fractions by Ultrafast Immunoaffinity Chromatography. *Anal. Chem.* **2001**, *73,* 2157–2164.

Cooper, D. J.; Husband, F. A.; Mills, E. N. C.; Wilde, P. J. Role of Beer Lipid-Binding Proteins in Preventing Lipid Destabilization of Foam. *J. Agric. Food Chem.* **2002**, *50,* 7645–7650.

Cortes, J. M.; Sanchez, R.; Diaz-Plaza, E. M.; Villen, J.; Vazquez, A. Large Volume GC Injection for the Analysis of Organophosphorus Pesticides in Vegetables using the Through Oven Transfer Adsorption Desorption (TOTAD) Interface. *J. Agric. Food Chem.* **2006**, *54,* 1997–2002.

Creaser, C. S.; Lamarca, D. G.; Brum, J.; Werner, C.; New, A. P.; dos Santos, L. M. F. Reversed-Phase Membrane Inlet Mass Spectrometry Applied to the Real-Time Monitoring of Low Molecular Weight Alcohols in Chloroform. *Anal. Chem.* **2002**, *74,* 300–304.

Crespo, A.; Marti, M. A.; Roitberg, A. E.; Amzel, L. M.; Estrin, D. A. The Catalytic Mechanism of Peptidylglycine α-Hydroxylating Monooxygenase Investigated by Computer Simulation. *J. Am. Chem. Soc.* **2006**, *128,* 12817–12828.

Curtis, R. A.; Blanch, H. W.; Prausnitz, J. M. Calculation of Phase Diagrams for Aqueous Protein Solutions. *J. Phys. Chem. B* **2001**, *105,* 2445–2452.

Daglia, M.; Tarsi, R.; Papetti, A.; Grisoli, P.; Dacarro, C.; Pruzzo, C.; Gazzani, G. Antiadhesive Effect of Green and Roasted Coffee on *Streptococcus mutans*' Adhesive Properties on Saliva-Coated Hydroxyapatite Beads. *J. Agric. Food Chem.* **2002**, *50,* 1225–1229.

Daiguebonne, C.; Kerbellec, N.; Bernot, K.; Gerault, Y.; Deluzet, A.; Guillou, O. Synthesis, Crystal Structure, and Porosity Estimation of Hydrated Erbium Terephthalate Coordination Polymers. *Inorg. Chem.* **2006**, *45,* 5399–5406.

Dane, A. J.; Havey, C. D.; Voorhees, K. J. The Detection of Nitro Pesticides in Mainstream and Sidestream Cigarette Smoke using Electron Monochromator-Mass Spectrometry. *Anal. Chem.* **2006**, *78,* 3227–3233.

Danil de Namor, A. F.; Al Rawi, N.; Piro, O. E.; Castellano, E. E.; Gil, E. New Lower Rim Calix(4)arene Derivatives with Mixed Pendent Arms and Their Complexation Properties for Alkali-Metal Cations. Structural, Electrochemical, and Thermodynamic Characterization. *J. Phys. Chem. B* **2002**, *106,* 779–787.

Dávalos, J. Z.; Baer, T. Thermochemistry and Dissociative Photoionization of $Si(CH_3)_4$, $BrSi(CH_3)_3$, $ISi(CH_3)_3$, and $Si_2(CH_3)_6$ Studied by Threshold Photoelectron-Photoion Coincidence Spectroscopy. *J. Phys. Chem. A* **2006**, *110,* 8572–8579.

Davis, M. I.; Orville, A. M.; Neese, F.; Zaleski, J. M.; Lipscomb, J. D.; Solomon, E. I. Spectroscopic and Electronic Structure Studies of Protocatechuate 3,4-Dioxygenase: Nature of Tyrosinate-Fe(III) Bonds and Their Contribution to Reactivity. *J. Am. Chem. Soc.* **2002**, *124,* 602–614.

Dedeloudis, C.; Fransaer, J.; Celis, J.-P. Surface Force Measurements at a Copper Electrode/Electrolyte Interface. *J. Phys. Chem. B* **2000**, *104,* 2060–2066.

Dellinger, B.; Pryor, W. A.; Cueto, R.; Squadrito, G. L.; Hegde, V.; Deutsch, W. A. Role of Free Radicals in the Toxicity of Airborne Fine Particulate Matter. *Chem. Res. Toxicol.* **2001**, *14,* 1371–1377.

Demko, Z. P.; Sharpless, K. B. Preparation of 5-Substituted 1H-Tetrazoles from Nitriles in Water. *J. Org. Chem.* **2001**, *66,* 7945–7950.

Dhanaraj, J.; Jagannathan, R.; Kutty, T. R. N.; Lu, C.-H. Photoluminescence Characteristics of $Y_2O_3:Eu^{3+}$ Nanophosphors Prepared using Sol-Gel Thermolysis. *J. Phys. Chem. B* **2001**, *105,* 11098–11105.

D'hooghe, M.; Van Brabandt, W.; De Kimpe, N. New Synthesis of Propargylic Amines from 2-(Bromomethyl)aziridines. Intermiediacy of 3-Bromoazetidinium Salts. *J. Org. Chem.* **2004**, *69,* 2703–2710.

Dick, L. W., Jr.; McGown, L. B. Aptamer-Enhanced Laser Desorption/Ionization for Affinity Mass Spectrometry. *Anal. Chem.* **2004**, *76,* 3037–3041.

Dóka, O.; Bicanic, D. D.; Dicko, M. H.; Slingerland, M. A. Photoacoustic Approach to Direct Determination of the Total Phenolic Content in Red Sorghum Flours. *J. Agric. Food Chem.* **2004**, *52,* 2133–2136.

Dong, Y.; Steffenson, B. J.; Mirocha, C. J. Analysis of Ergosterol in Single Kernel and Ground Grain by Gas Chromatography-Mass Spectrometry. *J. Agric. Food Chem.* **2006**, *54,* 4121–4125.

Dooley, G. P.; Prenni, J. E.; Prentiss, P. L.; Cranmer, B. K.; Andersen, M. E.; Tessari, J. D. Identification of a Novel Hemoglobin Adduct in Sprague Dawley Rats Exposed to Atrazine. *Chem. Res. Toxicol.* **2006**, *19,* 692–700.

Dreuw, A. Influence of Geometry Relaxation on the Energies of the S_1 and S_2 States of Violaxanthin, Zeaxanthin, and Lutein. *J. Phys. Chem. A* **2006**, *110,* 4592–4599.

Dunbar, R. C.; Moore, D. T.; Oomens, J. IR-Spectroscopic Characterization of Acetophenone Complexes with Fe^+, Co^+, and Ni^+ using Free-Electron-Laser IRMPD. *J. Phys. Chem. A* **2006**, *110,* 8316–8326.

Duncan, L. K.; Vikesland, P. J. Characterization of the Size, Shape, Crystallinity, and Surface Charge of C60 Aggregates Formed in Aqueous Systems. *Proceedings of the 233rd American Chemical Society National Meeting,* Chicago, IL, March 25–29, 2007.

Dyer, D. CAREER: Career Plan for Research in Polar Organic Materials with an Integrated Emphasis on Industrial Problem Solving, 2001 (NSF CHE 0094195).

Erel, E.; Aubriet, F.; Finqueneisel, G.; Muller, J.-F. Capabilities of Laser Ablation Mass Spectrometry in the Differentiation of Natural and Artificial Opal Gemstones. *Anal. Chem.* **2003**, *75,* 6422–6429.

Fabjan, N.; Rode, J.; Kosir, I. J.; Wang, Z.; Zhang, Z.; Kreft, I. Tartary Buckwheat (*Fagopyrum tataricum* Gaertn.) as a Source of Dietary Rutin and Quercitrin. *J. Agric. Food Chem.* **2003,** *51,* 6452–6455.

Fairbrother, H. CAREER: Exploring the Mechanisms of Organic Surface Modification Processes, 2000 (NSF CHE 9985372).

Fajgar, R.; Roithova, J.; Pola, J. Trimethylsilyl Group Migrations in Cryogenic Ozonolysis of Trimethylsilylethene: Evidence for Nonconcerted Primary Ozonide Decomposition Pathway. *J. Org. Chem.* **2001,** *66,* 6977–6981.

Farinha, J. P. S.; Picarra, S.; Miesel, K.; Martinho, J. M. G. Fluorescence Study of the Coil-Globule Transition of a PEO Chain in Toluene. *J. Phys. Chem. B* **2001,** *105,* 10536–10545.

Fattorusso, E.; Santelia, F. U.; Appendino, G.; Ballero, M.; Taglialatela-Scafati, O. Polyoxygenated Eudesmanes and *trans*-Chrysanthemanes from the Aerial Parts of *Santolina insularis*. *J. Nat. Prod.* **2004,** *67,* 37–41.

Ferguson, P. L.; DeMarco, A. Aggregation and Sorptive Properties of Single-Walled Carbon Nanotubes in the Estuarine Environment. *Proceedings of the 233rd American Chemical Society National Meeting,* Chicago, IL, March 25–29, 2007.

Filippova, M.; Duerksen-Hughes, P. J. Inorganic and Dimethylated Arsenic Species Induce Cellular p53. *Chem. Res. Toxicol.* **2003,** *16,* 423–431.

Finley, J. W.; Ip, C.; Lisk, D. J.; Davis, C. D.; Hintze, K. J.; Whanger, P. D. Cancer-Protective Properties of High-Selenium Broccoli. *J. Agric. Food Chem.* **2001,** *49,* 2679–2683.

Finney, N. CAREER: New Approaches to the Development of Fluorescent Chemosensors for Heavy Metal Ions, 1999 (NSF CHE 9876333).

Fraile, P.; Garrido, J.; Ancin, C. Influence of a *Saccharomyces cerevisiae* Selected Strain in the Volatile Composition of Rosé Wines. Evolution during Fermentation. *J. Agric. Food Chem.* **2000,** *48,* 1789–1798.

Fredriksson, S.-Å.; Hulst, A. G.; Artursson, E.; de Jong, A. L.; Nilsson, C.; van Baar, B. L. M. Forensic Identification of Neat Ricin and of Ricin from Crude Castor Bean Extracts by Mass Spectrometry. *Anal. Chem.* **2005,** *77,* 1545–1555.

Fujiwara, M.; Chie, K.; Sawai, J.; Shimizu, D.; Tanimoto, Y. On the Movement of Paramagnetic Ions in an Inhomogeneous Magnetic Field. *J. Phys. Chem. B* **2004,** *108,* 3531–3534.

Galbraith, J. M.; Gaspar, P. P.; Borden, W. T. What Accounts for the Difference between Singlet Phenylphosphinidene and Singlet Phenylnitrene in Reactivity toward Ring Expansion? *J. Am. Chem. Soc.* **2002,** *124,* 11669–11674.

Gambaro, A.; Manodori, L.; Zangrando, R.; Cincinelli, A.; Capodaglio, G.; Cescon, P. Atmospheric PCB Concentrations at Terra Nova Bay, Antarctica. *Environ. Sci. Technol.* **2005,** *39,* 9406–9411.

George, S. K.; Schwientek, T.; Holm, B.; Reis, C. A.; Clausen, H.; Kihlberg, J. Chemoenzymatic Synthesis of Sialylated Glycopeptides Derived from Mucins and T-Cell Stimulating Peptides. *J. Am. Chem. Soc.* **2001,** *123,* 11117–11125.

Goldberg, J.; Jin, Q.; Ambroise, Y.; Satoh, S.; Desharnais, J.; Capps, K.; Boger, D. L. Erythropoietin Mimetics Derived from Solution Phase Combinatorial Libraries. *J. Am. Chem. Soc.* **2002,** *124,* 544–555.

Goldstein, I. J. Lectin Structure-Activity: The Story Is Never Over. *J. Agric. Food Chem.* **2002,** *50,* 6583–6585.

Gorman, J. Molecular Chemistry Takes a New Twist. *Sci. News* **2001,** *159,* 340.

Gridelet, E.; Lorquet, A. J.; Locht, R.; Lorquet, J. C.; Leyh, B. Hydrogen Atom Loss from the Benzene Cation. Why Is the Kinetic Energy Release So Large? *J. Phys. Chem. A* **2006,** *110,* 8519–8527.

Grundl, T. J.; Aldstadt, J. H., III; Harb, J. G.; St. Germain, R. W.; Schweitzer, R. C. Demonstration of a Method for the Direct Determination of Polycyclic Aromatic Hydrocarbons in Submerged Sediments. *Environ. Sci. Technol.* **2003,** *37,* 1189–1197.

Gudmundsdottir, A. CAREER: Photolysis of Alkylazides in Solution and in Crystals, 2001 (NSF CHE 0093622).

Gunes, G.; Liu, R. H.; Watkins, C. B. Controlled-Atmosphere Effects on Postharvest Quality and Antioxidant Activity of Cranberry Fruits. *J. Agric. Food Chem.* **2002,** *50,* 5932–5938.

Guo, Z.; Zhang, Q.; Zou, H.; Guo, B.; Ni, J. A Method for the Analysis of Low-Mass Molecules by MALDI-TOF Mass Spectrometry. *Anal. Chem.* **2002,** *74,* 1637–1641.

Gurrieri, S.; Miceli, L.; Lanza, C. M.; Tomaselli, F.; Bonomo, R. P.; Rizzarelli, E. Chemical Characterization of Sicilian Prickly Pear (*Opuntia ficus indica*) and Perspectives for the Storage of Its Juice. *J. Agric. Food Chem.* **2000,** *48,* 5424–5431.

Haas, C.; Drent, J. The Interface between a Protein Crystal and an Aqueous Solution and Its Effects on Nucleation and Crystal Growth. *J. Phys. Chem. B* **2000,** *104,* 368–377.

Haes, A. J. ACS Division of Analytical Chemistry Graduate Fellowship: The Characterization and Development of the Localized Surface Plasmon Resonance Nanosensor, 2003.

Hafez, A. M.; Taggi, A. E.; Dudding, T.; Lectka, T. Asymmetric Catalysis on Sequentially-Linked Columns. *J. Am. Chem. Soc.* **2001,** *123,* 10853–10859.

Hageman, K. J.; Istok, J. D.; Field, J. A.; Buscheck, T. E.; Semprini, L. In Situ Anaerobic Transformation of Trichlorofluoroethene in Trichloroethene-Contaminated Groundwater. *Environ. Sci. Technol.* **2001,** *35,* 1729–1735.

Hageman, K. J.; Simonich, S. L.; Campbell, D. H.; Wilson, G. R.; Landers, D. H. Atmospheric Deposition of Current-Use and Historic-Use Pesticides in Snow at National Parks in the Western United States. *Environ. Sci. Technol.* **2006,** *40,* 3174–3180.

Han, S. M.; Ashurst, W. R.; Carraro, C.; Maboudian, R. Formation of Alkanethiol Monolayer on Ge(111). *J. Am. Chem. Soc.* **2001,** *123,* 2422–2425.

Hanmura, T.; Ichihashi, M.; Kondow, T. Reaction of Benzene Molecule on Size-Selected Nickel Cluster Ions. *J. Phys. Chem. A* **2002,** *106,* 4525–4528.

Hansen, N.; Klippenstein, S. J.; Miller, J. A.; Wang, J.; Cool, T. A.; Law, M. E.; Westmoreland, P. R.; Kasper, T.; Kohse-Hoinghaus, K. Identification of C_5H_x Isomers in Fuel-Rich Flames by Photoionization Mass Spectrometry and Electronic Structure Calculations. *Phys. Chem. A* **2006,** *110,* 4376–4388.

Harpp, K. CAREER: Teaching through Research: Plume-Ridge Interactions in the Galápagos as Focus of an Integrated Interdisciplinary Chemistry Curricular Initiative, 1998 (NSF CHE 9733597).

Hedin-Dahlström, J.; Rosengren-Holmberg, J. P.; Legrand, S.; Wikman, S.; Nicholls, I. A. A Class II Aldolase Mimic. *J. Org. Chem.* **2006,** *71,* 4845–4853.

Heimann, A. C.; Jakobsen, R. Experimental Evidence for a Lack of Thermodynamic Control on Hydrogen Concentrations during Anaerobic Degradation of Chlorinated Ethenes. *Environ. Sci. Technol.* **2006,** *40,* 3501–3507.

Hergenrother, P. CAREER: Combinatorial Chemistry in the Classroom and Laboratory: Identification of Novel Small Molecule Ligands for Apoptotic Proteins, 2002 (NSF CHE 0134779).

Hoffmann, R. *The Metamict State;* University of Central Florida Press: Orlando, 1987.

Hotze, E. M.; Badireddy A. R.; Chellam, S.; Wiesner, M. R. Singlet Oxygen and Superoxide Production by Three Types of Aqueous Fullerene Suspensions. *Proceedings of the 233rd American Chemical Society National Meeting,* Chicago, IL, March 25–29, 2007.

Houser, R. P. CAREER: Models for Metal Active Sites in Proteins, 2001 (NSF CHE 0094079).

Hoye, T. R.; Zhao, H. Method for Easily Determining Coupling Constant Values: An Addendum to "A Practical Guide to First-Order Multiplet Analysis in ^{1}H NMR Spectroscopy". *J. Org. Chem.* **2002,** *67,* 4014–4016.

Huang, X.-T.; Chen, Q.-Y. Ethyl α-Fluoro Silyl Enol Ether: Stereoselective Synthesis and Its Aldol Reaction with Aldehydes and Ketones. *J. Org. Chem.* **2002**, *67*, 3231–3234.

Huange, G.; Ouyang, J.; Delanghe, J. R.; Baeyens, W. R. G.; Dai, Z. Chemiluminescent Image Detection of Haptoglobin Phenotyping after Polyacrylamide Gel Electrophoresis. *Anal. Chem.* **2004**, *76*, 2997–3004.

Huggett, R. J.; Stegeman, J. J.; Page, D. S.; Parker, K. R.; Woodin, B.; Brown, J. S. Biomarkers in Fish from Prince William Sound and the Gulf of Alaska: 1999–2000. *Environ. Sci. Technol.* **2003**, *37*, 4043–4051.

Isas, J. M.; Langen, R.; Haigler, H. T.; Hubbell, W. L. Structure and Dynamics of a Helical Hairpin and Loop Region in Annexin 12: A Site-Directed Spin Labeling Study. *Biochemistry* **2002**, *41*, 1464–1473.

Ito, M.; Matsuumi, M.; Murugesh, M. G.; Kobayashi, Y. Scope and Limitation of Organocuprates, and Copper or Nickel Catalyst-Modified Grignard Reagents for Installation of an Alkyl Group onto *cis*-4-Cyclopentene-1,3-diol Monoacetate. *J. Org. Chem.* **2001**, *66*, 5881–5889.

Jackson, E. A.; Scott, L. T. A Synthetic Organic Approach to Uniform Carbon Nanotubes: Significant Progress toward Two Armchair Targets. *Proceedings of the 233rd American Chemical Society National Meeting*, Chicago, IL, March 25–29, 2007.

Jarvis, R. M.; Goodacre, R. Discrimination of Bacteria using Surface-Enhanced Raman Spectroscopy. *Anal. Chem.* **2004**, *76*, 40–47.

Jiao, L.; Smith, K. M. A New Synthetic Route to 2,2'-Bipyrroles. *Proceedings of the 233rd American Chemical Society National Meeting*, Chicago, IL, March 25–29, 2007.

Jöbstl, E.; Howse, J. R.; Fairclough, J. P. A.; Williamson, M. P. Noncovalent Cross-Linking of Casein by Epigallocatechin Gallate Characterized by Single Molecule Force Microscopy. *J. Agric. Food Chem.* **2006**, *54*, 4077–4081.

Johnsen, A. R.; de Lipthay, J. R.; Reichenberg, F.; Sorensen, S. J.; Andersen, O.; Christensen, P.; Binderup, M.-L.; Jacobsen, C. S. Biodegradation, Bioaccessibility, and Genotoxicity of Diffuse Polycyclic Aromatic Hydrocarbon (PAH) Pollution at a Motorway Site. *Environ. Sci. Technol.* **2006**, *40*, 3293–3298.

Johnson, J. S. CAREER: Discovery of New Enantioselective Dipolar Cycloadditions from Small Ring Heterocycles, 2003 (NSF CHE 0239363).

Joubert, J.; Fleurat-Lessard, P.; Delbecq, F.; Sautet, P. Simulating Temperature Programmed Desorption of Water on Hydrated γ-Alumina from First-Principles Calculations. *J. Phys. Chem. B* **2006**, *110*, 7392–7395.

Jozefaciuk, G.; Muranyi, A.; Fenyvesi, E. Effect of Cyclodextrins on Surface and Pore Properties of Soil Clay Minerals. *Environ. Sci. Technol.* **2001**, *35*, 4947–4952.

Jozefaciuk, G.; Muranyi, A.; Fenyvesi, E. Effect of Randomly Methylated β-Cyclodextrin on Physical Properties of Soils. *Environ. Sci. Technol.* **2003**, *37*, 3012–3017.

Katritzky, A. R.; Button, M. A. C. Efficient Syntheses of Thiochromans via Cationic Cycloadditions. *J. Org. Chem.* **2001**, *66*, 5595–5600.

Kinsel, G. CAREER: Analytical Applications of MALDI Mass Spectrometry to the Study of Surface-Protein Interactions, 1999 (NSF CHE 9876249).

Kizil, R.; Irudavaraj, J.; Seetharaman, K. Characterization of Irradiated Starches by using FT-Raman and FTIR Spectroscopy. *J. Agric. Food Chem.* **2002**, *50*, 3912–3918.

Kloepper, K. D.; Onuta, T.-D.; Amarie, D.; Dragnea, B. Field-Induced Interfacial Properties of Gold Nanoparticles in AC Microelectrophoretic Experiments. *J. Phys. Chem. B* **2004**, *108*, 2547–2553.

Kohen, A. CAREER: Protein Dynamics and Hydrogen Tunneling in Enzymatic Catalysis, 2002 (NSF CHE 0133117).

Kothe, U.; Rodnina, M. V. Delayed Release of Inorganic Phosphate from Elongation Factor Tu Following GTP Hydrolysis on the Ribosome. *Biochemistry* **2006**, *45*, 12767–12774.

Kouassi, G. K.; Irudayaraj, J. Magnetic and Gold-Coated Magnetic Nanoparticles as a DNA Sensor. *Anal. Chem.* **2006,** *78,* 3234–3241.

Koundouras, S.; Marinos, V.; Gkoulioti, A.; Kotseridis, Y.; van Leeuwen, C. Influence of Vineyard Location and Vine Water Status on Fruit Maturation of Nonirrigated Cv. Agiorgitiko (*Vitis vinifera* L.). Effects on Wine Phenolic and Aroma Components. *J. Agric. Food Chem.* **2006,** *54,* 5077–5086.

Kozukue, N.; Han, J.-S.; Lee, K.-R.; Friedman, M. Dehydrotomatine and α-Tomatine Content in Tomato Fruits and Vegetative Plant Tissues. *J. Agric. Food Chem.* **2004,** *52,* 2079–2083.

Kristovich, R.; Knight, D. A.; Long, J. F.; Williams, M. V.; Dutta, P. K.; Waldman, W. J. Macrophage-Mediated Endothelial Inflammatory Responses to Airborne Particulates: Impact of Particulate Physicochemical Properties. *Chem. Res. Toxicol.* **2004,** *17,* 1303–1312.

Kubicki, J. D. Molecular Simulations of Benzene and PAH Interactions with Soot. *Environ. Sci. Technol.* **2006,** *40,* 2298–2303.

Kubo, I.; Fujita, K.; Lee, S. H. Antifungal Mechanism of Polygodial. *J. Agric. Food Chem.* **2001,** *49,* 1607–1611.

Kuijt, J.; van Teylingen, R.; Nijbacker, T.; Ariese, F.; Brinkman, U. A. T.; Gooijer, C. Detection of Nonderivatized Peptides in Capillary Electrophoresis using Quenched Phosphorescence. *Anal. Chem.* **2001,** *73,* 5026–5029.

Kujawinski, E. B.; Hatcher, P. G.; Freitas, M. A. High-Resolution Fourier Transform Ion Cyclotron Resonance Mass Spectrometry of Humic and Fulvic Acids: Improvements and Comparisons. *Anal. Chem.* **2002,** *74,* 413–419.

Kunert, M.; Friese, K.; Weckert, V.; Markert, B. Lead Isotope Systematics in *Polytrichum formosum*: An Example from a Biomonitoring Field Study with Mosses. *Environ. Sci. Technol.* **1999,** *33,* 3502–3505.

Kuwata, K. T.; Valin, L. C.; Converse, A. D. Quantum Chemical and Master Equation Studies of the Methyl Vinyl Carbonyl Oxides Formed in Isoprene Ozonolysis. *J. Phys. Chem. A* **2005,** *109,* 10710–10725.

Langof, L.; Ehrenfreund, E.; Lifshitz, E. Continuous-Wave and Time-Resolved Optically Detected Magnetic Resonance Studies of Nonetched/Etched InP Nanocrystals. *J. Phys. Chem. B* **2002,** *106,* 1606–1612.

Laureles, L. R.; Rodriguez, F. M.; Reano, C. E.; Santos, G. A.; Laurena, A. C.; Mendoza, E. M. T. Variability in Fatty Acid and Triacylglycerol Composition of the Oil of Coconut (*Cocos nucifera* L.) Hybrids and Their Parentals. *J. Agric. Food Chem.* **2002,** *50,* 1581–1586.

Laurencin, D.; Villanneau, R.; Gérard, H.; Proust, A. Experimental and Theoretical Study of the Regiospecific Coordination of Ru^{II} and Os^{II} Fragments on the Lacunary Polyoxometalate $[\alpha\text{-}PW_{11}O_{39}]^{7-}$. *J. Phys. Chem. A* **2006,** *110,* 6345–6355.

Lebow, S; Williams, R.; Lebow, P. Effect of Simulated Rainfall and Weathering on Release of Preservative Elements from CCA Treated Wood. *Environ. Sci. Technol.* **2003,** *37,* 4077–4082.

Lecchi, S.; Allen, K. E.; Pardo, J. P.; Mason, A. B.; Slayman, C. W. Conformational Changes of Yeast Plasma Membrane H^+-ATPase during Activation by Glucose: Role of Threonine-912 in the Carboxy-Terminal Tail. *Biochemistry* **2005,** *44,* 16624–16632.

Lee, A.; Gavrin, L. K.; Provencher, B. A.; McKew, J. C. Synthesis of Pyrazolo[1,5-α]pyrimidine Regioisomers. *Proceedings of the 233rd American Chemical Society National Meeting,* Chicago, IL, March 25–29, 2007.

Lee, J. CAREER: Mechanistic Studies of Nucleotide Reactivity, 2001 (NSF CHE 0092215).

Lee, J.-H.; Landrum, P. F. Application of Multi-component Damage Assessment Model (MDAM) for the Toxicity of Metabolized PAH in *Hyalella azteca*. *Environ. Sci. Technol.* **2006,** *40,* 1350–1357.

Lee, S.-J.; Noble, A. C. Characterization of Odor-Active Compounds in Californian Chardonnay Wines using GC-Olfactometry and GC-Mass Spectrometry. *J. Agric. Food Chem.* **2003,** *51,* 8036–8044.

LeMagueres, P.; Im, H.; Dvorak, A.; Strych, U.; Benedik, M.; Krause, K. L. Crystal Structure at 1.45 Å Resolution of Alanine Racemase from a Pathogenic Bacterium, *Pseudomonas aeruginosa*, Contains Both Internal and External Aldimine Forms. *Biochemistry* **2003,** *42,* 14752–14761.

Li, C.-J.; Meng, Y. Grignard-Type Carbonyl Phenylation in Water and Under an Air Atmosphere. *J. Am. Chem. Soc.* **2000,** *122,* 9538–9539.

Liberty, A. M.; Hart, P. E.; Neto, C. C. Ursolic Acid and Proanthocyanidins from Cranberry (*Vaccinium macrocarpon*) Inhibit Colony Formation and Proliferation in HCT-116 and HT-29 Colon and MCF-7 Breast Tumor Cells. *Proceedings of the 233rd American Chemical Society National Meeting,* Chicago, IL, March 25–29, 2007.

Lifshitz, A.; Tamburu, C.; Suslensky, A.; Dubnikova, F. Decomposition of Anthranil. Single Pulse Shock-Tube Experiments, Potential Energy Surfaces and Multiwell Transition-State Calculations. The Role of Intersystem Crossing. *Phys. Chem. A* **2006,** *110,* 8248–8258.

Lin, D.; Zhu, L.; Luo, L. Factors Affecting Transfer of Polycyclic Aromatic Hydrocarbons from Made Tea to Tea Infusion. *J. Agric. Food Chem.* **2006,** *54,* 4350–4354.

Lissens, G.; Thomsen, A. B.; De Baere, L.; Verstraete, W.; Ahring, B. K. Thermal Wet Oxidation Improves Anaerobic Biodegradability of Raw and Digested Biowaste. *Environ. Sci. Technol.* **2004,** *38,* 3418–3424.

Liu, J.; Alvarez, J.; Ong, W.; Roman, E.; Kaifer, A. E. Phase Transfer of Hydrophilic, Cyclodextrin-Modified Gold Nanoparticles to Chloroform Solutions. *J. Am. Chem. Soc.* **2001,** *123,* 11148–11154.

Liu, Q.; Qu, Y.; Van Antwerpen, R.; Farrell, N. Interaction of Polynuclear Platinum Anticancer Agents. Implications for Cellular Uptake. *Biochemistry* **2006,** *45,* 4248–4256.

Llompart, M.; Pazos, M.; Landín, P.; Cela, R. Determination of Polychlorinated Biphenyls in Milk Samples by Saponification—Solid-Phase Microextraction. *Anal. Chem.* **2001,** *73,* 5858–5865.

Löfgren, C.; Guillotin, S.; Evenbratt, H.; Schols, H.; Hermansson, A.-M. Effects of Calcium, pH, and Blockiness on Kinetic Rheological Behavior and Microstructure of HM Pectin Gels. *Biomacromolecules* **2005,** *6,* 646–652.

Lombardi-Boccia, G.; Lucarini, M.; Lanzi, S.; Aguzzi, A.; Cappelloni, M. Nutrients and Antioxidant Molecules in Yellow Plums (*Prunus domestica* L.) from Conventional and Organic Productions: A Comparative Study. *J. Agric. Food Chem.* **2004,** *52,* 90–94.

Lorigan, G. A. CAREER: Investigating Membrane Proteins with Magnetic Resonance Spectroscopy, 2002 (NSF CHE 0133433).

Luthe, G.; Leonards, P. E. G.; Reijerink, G. S.; Liu, H.; Johansen, J. E.; Robertson, L. W. Monofluorinated Analogues of Polybrominated Diphenyl Ethers as Analytical Standards: Synthesis, NMR, and GC-MS Characterization and Molecular Orbital Studies. *Environ. Sci. Technol.* **2006,** *40,* 3023–3029.

Ly, T. N.; Shimoyamada, M.; Yamauchi, R. Isolation and Characterization of Rosmarinic Acid Oligomers in *Celastrus hindsii* Benth Leaves and Their Antioxidative Activity. *J. Agric. Food Chem.* **2006,** *54,* 3786–3793.

Lyon, L. A. CAREER: Bioresponsive Hydrogel Nanoparticles, 2000 (NSF CHE 9984012).

Mahmoud, A. A.; Natarajan, S. S.; Bennett, J. O.; Mawhinney, T. P.; Wiebold, W. J.; Krishnan, H. B. Effect of Six Decades of Selective Breeding on Soybean Protein Composition and Quality: A Biochemical and Molecular Analysis. *J. Agric. Food Chem.* **2006,** *54,* 3916–3922.

Mansfeldt, C. B.; Bott, C. B.; Holbrook, R. D. Behavior and Removal of Multiwalled Carbon Nanotubes during Simulated Drinking Water Treatment Processes. *Proceedings of the 233rd American Chemical Society National Meeting,* Chicago, IL, March 25–29, 2007.

Martos, P. A.; Pawliszyn, J. Sampling and Determination of Formaldehyde using Solid-Phase Microextraction with On-Fiber Derivatization. *Anal. Chem.* **1998,** *70,* 2311–2320.

Mas-Torrent, M.; Rodriguez-Mias, R. A.; Sola, M.; Molins, M. A.; Pons, M.; Vidal-Gancedo, J.; Veciana, J.; Rovira, C. Isolation and Characterization of Four Isomers of a C60 Bisadduct with a TTF Derivative. Study of Their Radical Ions. *J. Org. Chem.* **2002,** *67,* 566–575.

Mathiassen, S. K.; Kudsk, P.; Mogensen, B. B. Herbicidal Effects of Soil-Incorporated Wheat. *J. Agric. Food Chem.* **2006,** *54,* 1058–1063.

Matsumoto, H.; Ichiyanagi, T.; Iida, H.; Ito, K.; Tsuda, T.; Hirayama, M.; Konishi, T. Ingested Delphinidin-3-rutinoside Is Primarily Excreted to Urine as the Intact Form and to Bile as the Methylated Form in Rats. *J. Agric. Food Chem.* **2006,** *54,* 578–582.

McClellan, J. E.; Murphy, J. P., III; Mulholland, J. J.; Yost, R. A. Effects of Fragile Ions on Mass Resolution and on Isolation for Tandem Mass Spectrometry in the Quadrupole Ion Trap Mass Spectrometer. *Anal. Chem.* **2002,** *74,* 402–412.

McCue, K. Beer Foam Stabilization Proteins, 2002. American Chemical Society Web site. http://www.chemistry.org/portal/a/c/s/1/feature_pro.html?id=54cf9b78110b11d7f48b6ed9fe800100 (accessed July 2007).

McLean, K. J.; Warman, A. J.; Seward, H. E.; Marshall, K. R.; Girvan, H. M.; Cheesman, M. R.; Waterman, M. R.; Munro, A. W. Biophysical Characterization of the Sterol Demethylase P450 from *Mycobacterium tuberculosis*, Its Cognate Ferredoxin, and Their Interactions. *Biochemistry* **2006,** *45,* 8427–8443.

McMurry, J. *Organic Chemistry,* 6th ed.; Brooks/Cole-Thomson Learning: Belmont, CA, 2004.

Meadows, E. S.; De Wall, S. L.; Barbour, L. J.; Gokel, G. W. Alkali Metal Cation-Interactions Observed by using a Lariat Ether Model System. *J. Am. Chem. Soc.* **2001,** *123,* 3092–3107.

Meharg, A. A.; Rahman, M. M. Arsenic Contamination of Bangladesh Paddy Field Soils: Implications for Rice Contribution to Arsenic Consumption. *Environ. Sci. Technol.* **2003,** *37,* 229–234.

Mehinagic, E.; Royer, G.; Symoneaux, R.; Jourjon, F.; Prost, C. Characterization of Odor-Active Volatiles in Apples: Influence of Cultivars and Maturity Stage. *J. Agric. Food Chem.* **2006,** *54,* 2678–2687.

Meijer, S. N.; Halsall, C. J.; Harner, T.; Peters, A. J.; Ockenden, W. A.; Johnston, A. E.; Jones, K. Organochlorine Pesticide Residues in Archived UK Soil. *Environ. Sci. Technol.* **2001,** *35,* 1989–1995.

Meininger, D. P.; Rance, M.; Starovasnik, M. A.; Fairbrother, W. J.; Skelton, N. J. Characterization of the Binding Interface between the E-Domain of Staphylococcal Protein A and an Antibody Fv-Fragment. *Biochemistry* **2000,** *39,* 26–36.

Mileni, M.; Haas, A. H.; Mäntele, W.; Simon, J.; Lancaster, C. R. D. Probing Heme Propionate Involvement in Transmembrane Proton Transfer Coupled to Electron Transfer in Dihemic Quinol:Fumarate Reductase by ^{13}C-Labeling and FTIR Difference Spectroscopy. *Biochemistry* **2005,** *44,* 16718–16728.

Miles, J. E. C.; Ramsewak, R. S.; Nair, M. G. Antifeedant and Mosquitocidal Compounds from Delphinium × cultorum Cv. Magic Fountains Flowers. *J. Agric. Food Chem.* **2000,** *48,* 503–506.

Mitchell, C. A.; Yu, L.; Ward, M. D. Selective Nucleation and Discovery of Organic Polymorphs through Epitaxy with Single Crystal Substrates. *J. Am. Chem. Soc.* **2001,** *123,* 10830–10839.

Mitsumi, M.; Murase, T.; Kishida, H.; Yoshinari, T.; Ozawa, Y.; Toriumi, K.; Sonoyama, T.; Kitagawa, H.; Mitani, T. Metallic Behavior and Periodical Valence Ordering in a MMX Chain Compound, $Pt_2(EtCS_2)_4I$. *J. Am. Chem. Soc.* **2001,** *123,* 11179–11192.

Monteiro, C.; Hervé du Penhoat, C. Translational Diffusion of Dilute Aqueous Solutions of Sugars as Probed by NMR and Hydrodynamic Theory. *J. Phys. Chem. A* **2001**, *105*, 9827–9833.

Montembault, A.; Viton, C.; Domard, A. Rheometric Study of the Gelation of Chitosan in Aqueous Solution without Cross-Linking Agent. *Biomacromolecules* **2005**, *6*, 653–662.

Murahashi, T.; Nakashima, H.; Nagai, T.; Mino, Y.; Okuno, T.; Jalil, M. A.; Kurosawa, H. Stereoretentive Elimination and Trans-olefination of the Dicationic Dipalladium Moiety $[Pd_2L_n]^{2+}$ Bound on 1,3,5-Trienes. *J. Am. Chem. Soc.* **2006**, *128*, 4377–4388.

Muzikar, J.; van de Goor, T.; Kenndler, E. The Principle Cause for Lower Plate Numbers in Capillary Zone Electrophoresis with Most Organic Solvents. *Anal. Chem.* **2002**, *74*, 434–439.

Napolitano, A.; Di Donato, P.; Prota, G. Zinc-Catalyzed Oxidation of 5-S-Cysteinyldopa to 2,2'-Bi(2H-1,4-benzothiazine): Tracking the Biosynthetic Pathway of Trichochromes, the Characteristic Pigments of Red Hair. *J. Org. Chem.* **2001**, *66*, 6958–6966.

Neitola, R.; Pakkanen, T. A. Ab Initio Studies on the Atomic-Scale Origin of Friction between Diamond (111) Surfaces. *J. Phys. Chem. B* **2001**, *105*, 1338–1343.

Nenes, A.; Asa-Awuku, A.; Padro, L. T. Characterizing the Interactions of Water Vapor with Carbonaceous Aerosol. *Proceedings of the 233rd American Chemical Society National Meeting*, Chicago, IL, March 25–29, 2007.

Nico, P. S.; Fendorf, S. E.; Lowney, Y. W.; Holm, S. E.; Ruby, M. V. Chemical Structure of Arsenic and Chromium in CCA-Treated Wood: Implications of Environmental Weathering. *Environ. Sci. Technol.* **2004**, *38*, 5253–5260.

Nico. P. S.; Werner, M.; Anastasio, C.; Marcus, M. A. Chemical Speciation of Chromium in Ambient Aerosol Particles. *Proceedings of the 233rd American Chemical Society National Meeting*, Chicago, IL, March 25–29, 2007.

Nicolaides, A.; Enyo, T.; Miura, D.; Tomioka, H. *p*-Phenylenecarbenonitrene and Its Halogen Derivatives: How Does Resonance Interaction between a Nitrene and a Carbene Center Affect the Overall Electronic Configuration? *J. Am. Chem. Soc.* **2001**, *123*, 2628–2636.

Nie, X.; Wang, G. Synthesis and Self-Assembling Properties of Diacetylene-Containing Glycolipids. *J. Org. Chem.* **2006**, *71*, 4734–4741.

Ninio, R.; Lewinsohn, E.; Mizrahi, Y.; Sitrit, Y. Changes in Sugars, Acids, and Volatiles during Ripening of Koubo [*Cereus peruvianus* (L.) Miller] Fruits. *J. Agric. Food Chem.* **2003**, *51*, 797–801.

Offenberg, J. H.; Eisenreich, S. J.; Chen, L. C.; Cohen, M. D.; Chee, G.; Prophete, C.; Weisel, C.; Lioy, P. J. Persistent Organic Pollutants in the Dusts That Settled across Lower Manhattan after September 11, 2001. *Environ. Sci. Technol.* **2003**, *37*, 502–508.

Ou, Y. H.; Chung, P. C.; Chang, Y.-C.; Ngo, F. Q. H.; Hsu, K.-Y.; Chen, F.-D. Butachlor, a Suspected Carcinogen, Alters Growth and Transformation Characteristics of Mouse Liver Cells. *Chem. Res. Toxicol.* **2000**, *13*, 1321–1325.

Ozen, B. F.; Mauer, L. J. Detection of Hazelnut Oil Adulteration using FT-IR Spectroscopy. *J. Agric. Food Chem.* **2002**, *50*, 3898–3901.

Pal, S. K.; Itkis, M. E.; Reed, R. W.; Oakley, R. T.; Cordes, A. W.; Tham, F. S.; Siegrist, T.; Haddon, R. C. Synthesis, Structure and Physical Properties of the First One-Dimensional Phenalenyl-Based Neutral Radical Molecular Conductor. *J. Am. Chem. Soc.* **2004**, *126*, 1478–1484.

Park, J. W.; Song, H. E.; Lee, S. Y. Face Selectivity of Inclusion Complexation of Viologens with β-Cyclodextrin and 6-O-(2-Sulfonato-6-naphthyl)-β-cyclodextrin. *J. Phys. Chem. B* **2002**, *106*, 7186–7192.

Parra, R. D.; Yoo, B.; Wemhoff, M. Conformational Stability of a Model Macrocycle Tetraamide: An Ab Initio Study. *J. Phys. Chem. A* **2006**, *110*, 4487–4494.

Patrick, D. CAREER: Liquid Crystal Imprinting, 2000 (NSF CHE 9985428).

Pavia, D. L.; Lampman, G. M.; Driz, G. S.; Engel, R. G. *Introduction to Organic Laboratory Techniques: Small Scale Approach;* Saunders College Publishing: Orlando, FL, 1998; p 72.

Pazos, M.; Lois, S.; Torres, J. L.; Medina, I. Inhibition of Hemoglobin- and Iron-Promoted Oxidation in Fish Microsomes by Natural Phenolics. *J. Agric. Food Chem.* **2006**, *54*, 4417–4423.

Peck, A. M.; Hornbuckle, K. C. Gas-Phase Concentrations of Current-Use Pesticides in Iowa. *Environ. Sci. Technol.* **2005**, *39*, 2952–2959.

Peippo, P.; Hagren, V.; Lovgren, T.; Tuomola, M. Rapid Time-Resolved Fluoroimmunoassay for the Screening of Narasin and Salinomycin Residues in Poultry and Eggs. *J. Agric. Food Chem.* **2004**, *52*, 1824–1828.

Pelander, A.; Ojanperä, I.; Laks, S.; Rasanen, I.; Vuori, E. Toxicological Screening with Formula-Based Metabolite Identification by Liquid Chromatography/Time-of-Flight Mass Spectrometry. *Anal. Chem.* **2003**, *75*, 5710–5718.

Phares, D. J. Collection of Ultrafine Aerosols by Electrostatic Classification for Size-Resolved Chemical Analysis. *Proceedings of the 233rd American Chemical Society National Meeting,* Chicago, IL, March 25–29, 2007.

Philp, D. Dynamic Covalent Chemistry of Boron-Containing Heteroaromatic Systems. *Proceedings of the 233rd American Chemical Society National Meeting,* Chicago, IL, March 25–29, 2007.

Plaper, A.; Jenko-Brinovec, S.; Premzl, A.; Kos, J.; Raspor, P. Genotoxicity of Trivalent Chromium in Bacterial Cells. Possible Effects on DNA Topology. *Chem. Res. Toxicol.* **2002**, *15*, 943–949.

Pophristic, V.; Goodman, L. Hyperconjugation Not Steric Repulsion Leads to the Staggered Structure of Ethane. *Nature* **2001**, *411*, 565–568.

Pou-Amerigo, R.; Orti, E.; Merchan, M.; Rubio, M.; Viruela, P. M. Electronic Transitions in Tetrathiafulvalene and Its Radical Cation: A Theoretical Contribution. *J. Phys. Chem. A* **2002**, *106*, 631–640.

Pray, T. R.; Reiling, K. K.; Demirjian, B. G.; Craik, C. S. Conformational Change Coupling the Dimerization and Activation of KSHV Protease. *Biochemistry* **2002**, *41*, 1474–1482.

Prevedouros, K.; Jones, K. C.; Sweetman, A. J. Estimation of the Production, Consumption, and Atmospheric Emissions of Pentabrominated Diphenyl Ether in Europe between 1970 and 2000. *Environ. Sci. Technol.* **2004**, *38*, 3224–3231.

Quickenden, T. I.; Green, T. A.; Lennon, D. Luminescence from UV-Irradiated Amorphous H_2O Ice. *J. Phys. Chem.* **1996**, *100*, 16801–16807.

Raczyńska, E. D.; Darowska, M. Experimental and Theoretical Evidence of Basic Site Preference in Polyfunctional Superbasic Amidinazine: N^1,N^1-Dimethyl-N^2-β-(2-pyridylethyl)formamidine. *J. Org. Chem.* **2004**, *69*, 4023–4030.

Ranatunge, R. R.; Augustyniak, M.; Bandarage, U. K.; Earl, R. A.; Ellis, J. L.; Garvey, D. S.; Janero, D. R.; Letts, L. G.; Martino, A. M.; Murty, M. G.; Richardson, S. K.; Schroeder, J. D.; Shumway, M. J.; Tam, S. W.; Trocha, A. M.; Young, D. V. Synthesis and Selective Cyclooxygenase-2 Inhibitory Activity of a Series of Novel, Nitric Oxide Donor-Containing Pyrazoles. *J. Med. Chem.* **2004**, *47*, 2180–2193.

Rawashdeh-Omary, M. A.; Omary, M. A.; Patterson, H. H.; Fackler, J. P. Excited-State Interactions for $[Au(CN)_2^-]_n$ and $[Ag(CN)_2^-]_n$ Oligomers in Solution. Formation of Luminescent Gold–Gold Bonded Excimers and Exciplexes. *J. Am. Chem. Soc.* **2001**, *123*, 11237–11247.

Richmond, S. A.; Lindstrom, J. E.; Braddock, J. F. Assessment of Natural Attenuation of Chlorinated Aliphatics and BTEX in Subarctic Groundwater. *Environ. Sci. Technol.* **2001**, *35*, 4038–4045.

Riggleman, S.; DeShong, P. Application of Silicon-Based Cross-Coupling Technology to Triflates. *J. Org. Chem.* **2003**, *68*, 8106–8109.

Rodrigues, J. A. R.; Abramovitch, R. A.; de Sousa, J. D. F.; Leiva, G. C. Diastereoselective Synthesis of Cularine Alkaloids via Enium Ions and an Easy Entry to Isoquinolines by Aza–Wittig Electrocyclic Ring Closure. *J. Org. Chem.* **2004**, *69*, 2920–2928.

Rohrer, C.; Majoni, S. Effects of Shelf-Life on Phytonutrients in Beer Beverages. *Proceedings of the 233rd American Chemical Society National Meeting*, Chicago, IL, March 25–29, 2007.

Rose-Petruck, C. CAREER: Ultrafast X-ray Imaging of Molecular Dynamics in Solution: A Research Program That Enhances Students' Learning, 2000 (NSF CHE 9984890).

Ruffle, S. V.; Michalarias, I.; Li, J.-C.; Ford, R. C. Inelastic Incoherent Neutron Scattering Studies of Water Interacting with Biological Macromolecules. *J. Am. Chem. Soc.* **2002**, *124*, 565–569.

Russell, L.; Bahadur, R. Predicting Nanoparticle Interfaces with Molecular Dynamics. *Proceedings of the 233rd American Chemical Society National Meeting*, Chicago, IL, March 25–29, 2007.

Saito, S.; Kobayashi, S. Highly Anti-selective Catalytic Aldol Reactions of Amides with Aldehydes. *J. Am. Chem. Soc.* **2006**, *128*, 8704–8705.

Saito, T.; Kobayashi, T. Conformational Change in Azobenzene in Photoisomerization Process Studied with Chirp-Controlled sub-10-fs Pulses. *J. Phys. Chem. A* **2002**, *106*, 9436–9441.

Sakai, N.; Gerard, D.; Matile, S. Electrostatics of Cell Membrane Recognition: Structure and Activity of Neutral and Cationic Rigid Push-Pull Rods in Isoelectric, Anionic, and Polarized Lipid Bilayer Membranes. *J. Am. Chem. Soc.* **2001**, *123*, 2517–2524.

Sang, S.; Lao, A.; Wang, Y.; Chin, C.-K.; Rosen, R. T.; Ho, C.-T. Antifungal Constituents from the Seeds of *Allium fistulosum* L. *J. Agric. Food Chem.* **2002**, *50*, 6318–6321.

Sanov, A. CAREER: Structure and Dynamics of Negative Ions via Photoelectron Imaging Spectroscopy, 2002 (NSF CHE 0134631).

Satoh, A. Y.; Trosko, J. E.; Masten, S. J. Epigenetic Toxicity of Hydroxylated Biphenyls and Hydroxylated Polychlorinated Biphenyls on Normal Rat Liver Epithelial Cells. *Environ. Sci. Technol.* **2003**, *37*, 2727–2733.

Schauer, J. J.; Fraser, M. P.; Cass, G. R.; Simoneit, B. R. T. Source Reconciliation of Atmospheric Gas-Phase and Particle-Phase Pollutants during a Severe Photochemical Smog Episode. *Environ. Sci. Technol.* **2002**, *36*, 3806–3814.

Schön, E.; Zhang, X.; Zhou, Z.; Chisholm, M. H.; Chen, P. Gas-Phase and Solution-Phase Polymerization of Epoxides by Cr(salen) Complexes: Evidence for a Dinuclear Cationic Mechanism. *Inorg. Chem.* **2004**, *43*, 7278–7280.

Schwahn, D.; Willner, L. Phase Behavior and Flory-Huggins Interaction Parameter of Binary Polybutadiene Copolymer Mixtures with Different Vinyl Content and Molar Volume. *Macromolecules* **2002**, *35*, 239–247.

Schwikowski, M.; Barbante, C.; Doering, T.; Gaeggeler, H. W.; Boutron, C.; Schotterer, U.; Tobler, L.; Van de Velde, K.; Ferrari, C.; Cozzi, G.; Rosman, K.; Cescon, P. Post-17th-Century Changes of European Lead Emissions Recorded in High-Altitude Alpine Snow and Ice. *Environ. Sci. Technol.* **2004**, *38*, 957–964.

Seeram, N. P.; Adams, L. S.; Hardy, M. L.; Heber, D. Total Cranberry Extract versus Its Phytochemical Constituents: Antiproliferative and Synergistic Effects against Human Tumor Cell Lines. *J. Agric. Food Chem.* **2004**, *52*, 2512–2517.

Senda, T.; Ogasawara, M.; Hayashi, T. Rhodium-Catalyzed Asymmetric 1,4-Addition of Organoboron Reagents to 5,6-Dihydro-2(1H)-pyridinones. Asymmetric Synthesis of 4-Aryl-2-piperidinones. *J. Org. Chem.* **2001**, *66*, 6852–6856.

Service, R.F. Drug Industry Looks to the Lab Instead of Rainforest and Reef. *Science* **1999**, *285*, 186.

Shadpour, H.; Soper, S. A. Two-Dimensional Electrophoretic Separation of Proteins using Poly(methyl methacrylate) Microchips. *Anal. Chem.* **2006**, *78,* 3519–3527.

Shi, H.; Liu, S.; Miyake, M.; Liu, K. J. Ebselen Induced C6 Glioma Cell Death in Oxygen and Glucose Deprivation. *Chem. Res. Toxicol.* **2006**, *19,* 655–660.

Shie, J.-L.; Lin, J.-P.; Chang, C.-Y.; Wu, C.-H.; Lee, D.-J.; Chang, C.-F.; Chen, Y.-H. Oxidative Thermal Treatment of Oil Sludge at Low Heating Rates. *Energy Fuels* **2004**, *18,* 1272–1281.

Short, J. W.; Maselko, J. M.; Lindeberg, M. R.; Harris, P. M.; Rice, S. D. Vertical Distribution and Probability of Encountering Intertidal *Exxon Valdez* Oil on Shorelines of Three Embayments within Prince William Sound, Alaska. *Environ. Sci. Technol.* **2006**, *40,* 3723–3729.

Silva, F. A. M.; Borges, F.; Guimaraes, C.; Lima, J. L. F. C.; Matos, C.; Reis, S. Phenolic Acids and Derivatives: Studies on the Relationship among Structure, Radical Scavenging Activity, and Physicochemical Parameters. *J. Agric. Food Chem.* **2000**, *48,* 2122–2126.

Singh, S.; Wegmann, J.; Albert, K.; Muller, K. Variable Temperature FT-IR Studies of n-Alkyl Modified Silica Gels. *J. Phys. Chem. B* **2002**, *106,* 878–888.

Song, Y.; Singh, R.; Guo, Z. X. A First-Principles Study of the Electronic Structure and Stability of a Lithium Aluminum Hydride for Hydrogen Storage. *J. Phys. Chem. B* **2006**, *110,* 6906–6910.

Song, Y.-A.; Hsu, S.; Stevens, A. L.; Han, J. Continuous-Flow pI-Based Sorting of Proteins and Peptides in a Microfluidic Chip using Diffusion Potential. *Anal. Chem.* **2006**, *78,* 3528–3536.

Spain, E. CAREER: Atomic Force Microscopy Studies of Transition Metal Chalcogenide Deposition using Translationally Hot Atoms, 1997 (NSF CHE 9703345).

Spanget-Larsen, J.; Gil, M.; Gorski, A.; Blake, D. M.; Waluk, J.; Radziszewski, J. G. Vibrations of the Phenoxyl Radical. *J. Am. Chem. Soc.* **2001**, *123,* 11253–11261.

Spengler, J.; Anderle, F.; Bosch, E.; Grasselli, R. K.; Pillep, B.; Behrens, P.; Lapina, O. B.; Shubin, A. A.; Eberle, H. J.; Knozinger, H. Antimony Oxide-Modified Vanadia-Based Catalysts—Physical Characterization and Catalytic Properties. *J. Phys. Chem. B* **2001**, *105,* 10772–10783.

Stockton, J. D.; Merkert, M. C.; Kellaris, K. V. A Complex of Chaperones and Disulfide Isomerases Occludes the Cytosolic Face of the Translocation Protein Sec61p and Affects Translocation of the Prion Protein. *Biochemistry* **2003**, *42,* 12821–12834.

Subramanian, R.; Boparai, P.; Bond, T. Charring of Organic Compounds during Thermal-Optical Analysis: What Can We Learn about the Carbonaceous Aerosol? *Proceedings of the 233rd American Chemical Society National Meeting,* Chicago, IL, March 25–29, 2007.

Swenson, R. E.; Sowin, T. J.; Zhang, H. Q. Synthesis of Substituted Quinolines using the Dianion Addition of N-Boc-Anilines and α-Tolylsulfonyl-α,β-Unsaturated Ketones. *J. Org. Chem.* **2002**, *67,* 9182–9185.

Taatjes, C. A. Uncovering the Fundamental Chemistry of Alkyl + O_2 Reactions via Measurements of Product Formation. *J. Phys. Chem. A* **2006**, *110,* 4299–4312.

Takats, Z.; Nanita, S. C.; Cooks, R. G.; Schlosser, G.; Vekey, K. Amino Acid Clusters Formed by Sonic Spray Ionization. *Anal. Chem.* **2003a**, *75,* 1514–1523.

Takats, Z.; Nanita, S. C.; Schlosser, G.; Vekey, K.; Cooks, R. G. Atmospheric Pressure Gas-Phase H/D Exchange of Serine Octamers. *Anal. Chem.* **2003b**, *75,* 6147–6154.

Tam, J. P.; Yu, Q.; Yang, J.-L. Tandem Ligation of Unprotected Peptides through Thiaprolyl and Cysteinyl Bonds in Water. *J. Am. Chem. Soc.* **2001**, *123,* 2487–2494.

Tateo, F.; Bononi, M. Fast Determination of Sudan I by HPLC/APCI-MS in Hot Chilli, Spices, and Oven-Baked Foods. *J. Agric. Food Chem.* **2004**, *52,* 655–658.

Taylor, W. G.; Zulyniak, H. J.; Richards, K. W.; Acharya, S. N.; Bittman, S.; Elder, J. L. Variation in Diosgenin Levels among 10 Accessions of Fenugreek Seeds Produced in Western Canada. *J. Agric. Food Chem.* **2002**, *50,* 5994–5997.

Tikk, M.; Tikk, K.; Tørngren, M. A.; Meinert, L.; Aaslyng, M. D.; Karlsson, A. H.; Andersen, H. J. Development of Inosine Monophosphate and Its Degradation Products during Aging of Pork of Different Qualities in Relation to Basic Taste and Retronasal Flavor Perception of the Meat. *J. Agric. Food Chem.* **2006**, *54*, 7769–7777.

Tongraar, A.; Sagarik, K.; Rode, B. M. Effects of Many-Body Interactions on the Preferential Solvation of Mg^{2+} in Aqueous Ammonia Solution: A Born-Oppenheimer Ab Initio QM/MM Dynamics Study. *J. Phys. Chem. B* **2001**, *105*, 10559–10564.

Tsipis, C. A.; Karipidis, P. A. Mechanism of a Chemical Classic: Quantum Chemical Investigation of the Autocatalyzed Reaction of the Serendipitous Wöhler Synthesis of Urea. *J. Am. Chem. Soc.* **2003**, *125*, 2307–2318.

Tuckerman, M. CAREER: Theoretical Investigations of Chemical Processes in Bulk Crystals and on Surfaces, 1999 (NSF CHE 9875824).

Tuulmets, A.; Nguyen, B. T.; Panov, D.; Sassian, M.; Jarv, J. Kinetics of the Grignard Reaction with Silanes in Diethyl Ether and Ether-Toluene Mixtures. *J. Org. Chem.* **2003**, *68*, 9933–9937.

Usugi, S.; Yorimitsu, H.; Shinokubo, H.; Oshima, K. Disulfidation of Alkynes and Alkenes with Gallium Trichloride. *Org. Lett.* **2004**, *6*, 601–603.

Vaisman, E.; Cook, R. L.; Langford, C. H. Characterization of a Composite Photocatalyst. *J. Phys. Chem. B* **2000**, *104*, 8679–8684.

Valliant, J. F.; Schaffer, P.; Stephenson, K. A.; Britten, J. F. Synthesis of Boroxifen, a Nido-carborane Analogue of Tamoxifen. *J. Org. Chem.* **2002**, *67*, 383–387.

Van Berkel, G. J.; Sanchez, A. D.; Quirke, J. M. E. Thin-Layer Chromatography and Electrospray Mass Spectrometry Coupled using a Surface Sampling Probe. *Anal. Chem.* **2002**, *74*, 6216–6223.

Varlet, V.; Knockaert, C.; Prost, C.; Serot, T. Comparison of Odor-Active Volatile Compounds of Fresh and Smoked Salmon. *J. Agric. Food Chem.* **2006**, *54*, 3391–3401.

Veldhuyzen, W. F.; Shallop, A. J.; Jones, R. A.; Rokita, S. E. Thermodynamic versus Kinetic Products of DNA Alkylation as Modeled by Reaction of Deoxyadenosine. *J. Am. Chem. Soc.* **2001**, *123*, 11126–11132.

Venkateswara Rao, P.; Holm, R. H. Synthetic Analogues of the Active Sites of Iron–Sulfur Proteins. *Chem. Rev.* **2004**, *104*, 527–560.

Vesely, P.; Lusk, L.; Basarova, G.; Seabrooks, J.; Ryder, D. Analysis of Aldehydes in Beer using Solid-Phase Microextraction with On-Fiber Derivatization and Gas Chromatography/Mass Spectrometry. *J. Agric. Food Chem.* **2003**, *51*, 6941–6944.

Vieth, M.; Siegel, M. G.; Higgs, R. E.; Watson, I. A.; Robertson, D. H.; Savin, K. A.; Durst, G. L.; Hipskind, P. A. Characteristic Physical Properties and Structural Fragments of Marketed Oral Drugs. *J. Med. Chem.* **2004**, *47*, 224–232.

Vitòria, L.; Otero, N.; Soler, A.; Canals, À. Fertilizer Characterization: Isotopic Data (N, S, O, C, and Sr). *Environ. Sci. Technol.* **2004**, *38*, 3254–3262.

Voets, J.; Bervoets, L.; Blust, R. Cadmium Bioavailability and Accumulation in the Presence of Humic Acid to the Zebra Mussel, *Dreissena polymorpha*. *Environ. Sci. Technol.* **2004**, *38*, 1003–1008.

Vyvyan, J. CAREER: Synthesis of Allelopathic Agents as Leads to New Agrochemicals, 2001 (NSF CHE 0094378).

Walker, R. CAREER: Surface-Mediated Solvation at Hydrophobic and Hydrophilic Interfaces, 2001 (NSF CHE 0094246).

Walter, R. I. Absence of Detectable Freely Diffusing Radicals during the Formation of an Aromatic Grignard Reagent. *J. Org. Chem.* **2000**, *65*, 5014–5015.

Walz, J. Y.; Ruckenstein, E. Comparison of the van der Waals and Undulation Interactions between Uncharged Lipid Bilayers. *J. Phys. Chem. B* **1999**, *103*, 7461–7468.

Wang, X. H.; Li, J.-G.; Kamiyama, H.; Moriyoshi, Y.; Ishigaki, T. Wavelength-Sensitive Photocatalytic Degradation of Methyl Orange in Aqueous Suspension over Iron(III)-Doped TiO_2 Nanopowders under UV and Visible Light Irradiation. *J. Phys. Chem. B* **2006,** *110,* 6804–6809.

Wang, Y.-S.; Beyer, B. M.; Senior, M. M.; Wyss, D. F. Characterization of Autocatalytic Conversion of Precursor BACE1 by Heteronuclear NMR Spectroscopy. *Biochemistry* **2005,** *44,* 16594–16601.

Warren, T. CAREER: Metal-Ligand Multiple Bonding in Later, First Row Complexes, 2002 (NSF CHE 0135057).

Webber, J. S.; Jackson, K. W.; Parekh, P. P.; Bopp, R. F. Reconstruction of a Century of Airborne Asbestos Concentrations. *Environ. Sci. Technol.* **2004,** *38,* 707–714.

Wei, A.; Mura, K.; Shibamoto, T. Antioxidative Activity of Volatile Chemicals Extracted from Beer. *J. Agric. Food Chem.* **2001,** *49,* 4097–4101.

Weinberger, D. A.; Higgins, T. B.; Mirkin, C. A.; Stern, C. L.; Liable-Sands, L. M.; Rheingold, A. L. Terthienyl and Poly-terthienyl Ligands as Redox-Switchable Hemilabile Ligands for Oxidation-State-Dependent Molecular Uptake and Release. *J. Am. Chem. Soc.* **2001,** *123,* 2503–2516.

Wertz, D. L.; Smith, E. R. On the Molecular-Level Interactions between Pittsburgh No. 8 Coal and Several Organic Liquids. *Energy Fuels* **2003,** *17,* 1423–1428.

Weston, D. P.; You, J.; Lydy, M. J. Distribution and Toxicity of Sediment-Associated Pesticides in Agriculture-Dominated Water Bodies of California's Central Valley. *Environ. Sci. Technol.* **2004,** *38,* 2752–2759.

Wi, S.; Frydman, L. Heteronuclear Recoupling in Solid-State Magic-Angle-Spinning NMR via Overtone Irradiation. *J. Am. Chem. Soc.* **2001,** *123,* 10354–10361.

Wong, B. M.; Thom, R. L.; Field, R. W. Accurate Inertias for Large-Amplitude Motions: Improvements on Prevailing Approximations. *J. Phys. Chem. A.* **2006,** *110,* 7406–7413.

Wong, E. L. S.; Goding, J. J. Charge Transfer through DNA: A Selective Electrochemical DNA Biosensor. *Anal. Chem.* **2006,** *78,* 2138–2144.

Wright, V. A.; Patrick, B. O.; Schneider, C.; Gates, D. P. Phosphorus Copies of PPV: π-Conjugated Polymers and Molecules Composed of Alternating Phenylene and Phosphaalkene Moieties. *J. Am. Chem. Soc.* **2006,** *128,* 8836–8844.

Wu, X.; Beecher, G. R.; Holden, J. M.; Haytowitz, D. B.; Gebhardt, S. E.; Prior, R. L. Lipophilic and Hydrophilic Antioxidant Capacities of Common Foods in the United States. *J. Agric. Food Chem.* **2004,** *52,* 4026–4037.

Xu, G.; Aksay, I. A.; Groves, J. T. Continuous Crystalline Carbonate Apatite Thin Films. A Biomimetic Approach. *J. Am. Chem. Soc.* **2001,** *123,* 2196–2203.

Xue, Y.; Kim, C. K. Effects of Substituents and Solvents on the Reactions of Iminophosphorane with Formaldehyde: Ab Initio MO Calculation and Monte Carlo Simulation. *J. Phys. Chem. A* **2003,** *107,* 7945–7951.

Yamashita, N.; Kannan, K.; Taniyasu, S.; Horii, Y.; Okazawa, T.; Petrick, G.; Gamo, T. Analysis of Perfluorinated Acids at Parts-per-Quadrillion Levels in Seawater using Liquid Chromatography-Tandem Mass Spectrometry. *Environ. Sci. Technol.* **2004,** *38,* 5522–5528.

Yan, B.; Zhao, J.; Leopold, K.; Zhang, B.; Jiang, G. Structure-Dependent Response of a Chemiluminescence Nitrogen Detector for Organic Compounds with Adjacent Nitrogen Atoms Connected by a Single Bond. *Anal. Chem.* **2007,** *79,* 718–726.

Yang, T. C.; Wolfe, M. D.; Neibergall, M. B.; Mekmouche, Y.; Lipscomb, J. D.; Hoffman, B. M. Substrate Binding to NO-Ferro-Naphthalene 1,2-Dioxygenase Studied by High-Resolution Q-Band Pulsed ^{2}H-ENDOR Spectroscopy. *J. Am. Chem. Soc.* **2003,** *125,* 7056–7066.

Yang, Z.; Kollman, J. M.; Pandi, L.; Doolittle, R. F. Crystal Structure of Native Chicken Fibrinogen at 2.7 Å Resolution. *Biochemistry* **2001,** *40,* 12515–12523.

Yasuda, Y.; Mizusawa, H.; Kamimura, T. Frequency Response Method for Investigation of Kinetic Details of a Heterogeneous Catalyzed Reaction of Gases. *J. Phys. Chem. B* **2002,** *106,* 6706–6712.

Ye, L.; Landen, W. O.; Eitenmiller, R. R. Liquid Chromatographic Analysis of All-trans-Retinyl Palmitate, β-Carotene, and Vitamin E in Fortified Foods and the Extraction of Encapsulated and Nonencapsulated Retinyl Palmitate. *J. Agric. Food Chem.* **2000,** *48,* 4003–4008.

Yu, C. J.; Wan, Y.; Yowanto, H.; Li, J.; Tao, C.; James, M. D.; Tan, C. L.; Blackburn, G. F.; Meade, T. J. Electronic Detection of Single-Base Mismatches in DNA with Ferrocene-Modified Probes. *J. Am. Chem. Soc.* **2001,** *123,* 11155–11161.

Zhang, G.; Hamaker, B. R. A Three Component Interaction among Starch, Protein, and Free Fatty Acids Revealed by Pasting Profiles. *J. Agric. Food Chem.* **2003,** *51,* 2797–2800.

Zhou, S.; Barnes, I.; Zhu, T.; Bejan, I.; Benter, T. Kinetic Study of the Gas-Phase Reactions of OH and NO_3 Radicals and O_3 with Selected Vinyl Ethers. *J. Phys. Chem. A.* **2006,** *110,* 7386–7392.

Zimmerman, H. E.; Alabugin, I. V. Energy Distribution and Redistribution and Chemical Reactivity. The Generalized Delta Overlap-Density Method for Ground State and Electron Transfer Reactions: A New Quantitative Counterpart of Electron-Pushing. *J. Am. Chem. Soc.* **2001,** *123,* 2265–2270.

Zuo, Y.; Chen, H.; Deng, Y. Isolation and Identification of Flavonol Glycosides in American Cranberry Fruit using HPLC and GC-MS. *Proceedings of the 233rd American Chemical Society National Meeting,* Chicago, IL, March 25–29, 2007.

Zysman-Colman, E.; Harpp, D. N. Optimization of the Synthesis of Symmetric Aromatic Tri- and Tetrasulfides. *J. Org. Chem.* **2003,** *68,* 2487–2489.

References

Alley, M. *The Craft of Scientific Presentations: Critical Steps To Succeed and Critical Errors To Avoid;* Springer: New York, 2003.

Alley, M. *The Craft of Scientific Writing,* 3rd ed.; Springer: New York, 1996.

Anholt, R. R. H. *Dazzle'em with Style: The Art of Oral Scientific Presentation;* W. H. Freeman: Oxford, U.K., 1994.

Beal, H.; Trimbur, J. *A Short Guide to Writing about Chemistry,* 2nd ed.; Pearson Education: New York, 2001.

Bhatia, V. *Analyzing Genre: Language Use in Professional Settings;* Longman: London, 1993.

Bhatia, V. *Worlds of Written Discourse: A Genre-Based View;* Continuum: London, 2004.

Biber, D.; Conrad, S.; Reppen, R. Corpus-Based Approaches to Issues in Applied Linguistics. *Appl. Linguist.* **1994,** *15,* 168–189.

Biber, D.; Conrad, S.; Reppen, R. *Corpus Linguistics: Investigating Structure and Use;* Cambridge University Press: Cambridge, U.K., 1998.

Bowker, L.; Pearson, J. *Working with Specialized Language: A Practical Guide to using Corpora;* Routledge: New York, 2002.

Bowman, J. P.; Branchaw, B. P. *How To Write Proposals That Produce;* Oryx Press: Phoenix, AZ, 1992.

Coghill, A. M.; Garson, L. R., Eds. *The ACS Style Guide: Effective Communication of Scientific Information,* 3rd ed.; American Chemical Society: Washington, DC, 2006.

Connor, U.; Mauranen, A. Linguistic Analysis of Grant Proposals: European Union Research Grants. *Engl. Specif. Purposes* **1999**, *18,* 47–62.

Crismore, A.; Farnsworth, R. Metadiscourse in Popular and Professional Science Discourse. In *The Writing Scholar: Studies in Academic Discourse;* Nash, W., Ed.; Sage: Newbury Park, CA, 1990; pp 118–136.

Dodd, J. S., Ed. *The ACS Style Guide: A Manual for Authors and Editors,* 2nd ed.; American Chemical Society: Washington, DC, 1997.

Hill, S. S.; Soppelsa, B. F.; West, G. K. Teaching ESL Students To Read and Write Experimental-Research Papers. *TESOL Quarterly* **1982**, *16,* 333–347.

Houp, K. W.; Pearsall, T. E.; Tebeaux, E.; Dragga, S. *Reporting Technical Information,* 11th ed.; Oxford University Press: New York, 2006.

Huckin, T. N. Surprise Value in Scientific Discourse. ERIC Document Reproduction Service No. ED284291, 1987.

Hyland, K. Hedging in Academic Writing and EAP Textbooks. *Engl. Specif. Purposes* **1994,** *13,* 239–256.

Hyland, K. Talking to the Academy: Forms of Hedging in Science Research Articles. *Written Commun.* **1996,** *13,* 251–281.

Hyland, K. *Hedging in Scientific Research Articles;* John Benjamins: Amsterdam, 1998.

Hyland, K. Genre: Language, Context, and Literacy. *Annu. Rev. Appl. Linguist.* **2002,** *22,* 113–135.

Hyland, K. *Disciplinary Discourses: Social Interactions in Academic Writing;* University of Michigan Press: Ann Arbor, 2004a.

Hyland, K. *Genre and Second Language Writing;* University of Michigan Press: Ann Arbor, 2004b.

Hyland, K. *English for Academic Purposes: An Advanced Resource Book;* Routledge: London, 2006.

Johns, A. M. *Genre in the Classroom: Multiple Perspectives;* Lawrence Erlbaum: Mahwah, NJ, 2002.

Paltridge, B. *Genre, Frames, and Writing in Research Settings;* John Benjamins: Amsterdam, 1997.

Paradis, J. G.; Zimmerman, M. L. *The MIT Guide to Science and Engineering Communication;* MIT Press: Cambridge, MA, 1997.

Swales, J. M. *Genre Analysis: English in Academic and Research Settings;* Cambridge University Press: Cambridge, U.K., 1990.

Swales, J. M. *Research Genres: Exploration and Applications;* Cambridge University Press: Cambridge, U.K., 2004.

Truss, L. *Eats, Shoots & Leaves: The Zero Tolerance Approach to Punctuation;* Gotham Books: New York, 2003.

Turk, C. Do You Write Impressively? *Bull. Br. Ecol. Soc.* **1978,** *9,* 5–10.

Index

In page references, *f* indicates figures, *s* indicates schemes, and *t* indicates tables. In some instances, excerpts are included in page spans.

Abbreviations
- vs. acronyms, 20, 73–74
- to avoid, 74–75
- capitalization of, 72, 76*t*–77*t*, 625
- *CASSI,* 558, 558*t*
- of chemical structures, 77*t*
- in conference abstracts, 287
- of funding agencies, 368
- of instrumental techniques, 76*t*
- in journal article abstracts, 250–251, 256–257
- for journal names, 558, 558*t*
- language tip, 601–604
- list of common, 76*t*–77*t*
- not needing definition, 74, 602*t*
- parentheses, use with, 73–74
- periods, use with, 74, 396, 602*t*
- plural forms of, 88, 603
- in posters, 300, 305, 306, 319
- in project summaries, 509
- of statistical symbols, 77*t*
- as a subcomponent of Writing Conventions, 7*t*, 19–20
- in titles, 250–251, 264, 517
- of units of measure, 76*t*–77*t*
- *See also* Compound labels, Units of measure

"Ab initio", without italics, 96, 495
Abstract. *See* Conference abstracts; Journal articles, abstract; Poster, abstract; Proposals, Project Summary
Acknowledgments section
- examples of, 29
- in journal articles, 45
- in posters, 297, 331

Acronyms. *See* Abbreviations
Active voice. *See* Voice
"Affect" vs. "effect", 634–636
Affiliations. *See* Author affiliations
Alignment, in tables, 534*t*
"Although" vs. "while", 650–652
"Among" vs. "between", 636–638
Analytical data, reporting in Methods sections of journal articles, 80
"And", use with punctuation, 627–628, 629–630
Apparatus
- in journal articles, describing, 90–95, 251, 257
- plural form of, 91, 156*t*
- in posters, describing, 302

Audience
- definition of, 8
- different types of, 7–11, 10*f*
- *See also* Audience and Purpose

Audience and Purpose
- as a component of genre analysis, 7*t*
- for conference abstracts, 276–277
- for journal articles, overview, 35, 45
- overview, 7–11, 10*f*
- for posters, overview, 294, 296
- for proposals, overview, 359, 375–376
- *See also* Conciseness; Detail, level of; Formality, level of; *Journal articles, individual sections*; *Posters, individual sections*; Proposal, Project Summary; Word Choice

Author affiliations
- in conference abstracts, 287
- in posters, examples of, 295*f*, 350*f*, 351*f*, 352*f*
- in posters, font size, 346*t*, 353*t*

Author-date citations. *See* Citations
Author guidelines. *See* Guidelines for authors, Information for Authors

Author lists
in conference abstracts, 287
in posters, examples of, 295*f*, 350*f*, 351*f*, 352*f*
in posters, font size, 346*t*, 353*t*
in references, 559*t*, 560, 561, 562, 561*t*–562*t*, 564, 565*t*, 567
See also "et al."
Axes, in graphs, 527*t*, 528*t*

Bar graphs. *See* Graphs
"Because" vs. "since", 650–652
"Between" vs. "among", 636–638
Boldface type
in compound labels, 75, 137–138, 284–285, 540*t*
in posters, 346
in reactions and schemes, 78, 540*t*
in references, 560
Broader impacts, 372–373
See also Proposals, Outcomes and Impacts section; Proposals, Project Summary
Bulleted lists. *See* Lists
"But", use with punctuation, 627–628, 629–630

Capitalization
of abbreviations, 72, 76*t*–77*t*, 625
in captions, 527*t*
of chemical compounds, 72, 74*t*
examples of, 74*t*, 76*t*–77*t*, 624–625
of "Figure", "Table", "Scheme", 74*t*, 524, 527*t*, 533, 536*t*, 539, 540*t*
of genus and species names, 70, 74*t*
of molecular formulas, 72
rules, 71–73, 74*t*, 624–626
of seasons, 70
of table titles, 536*t*
in tables, column headings, 535*t*
of titles, 262
of units from surnames, 73, 74*t*
See also Graphs, References, Tables
Captions. *See* Graphs
CASSI, 558, 558*t*
Celsius, 73, 76*t*, 609, 625
Chemical Abstracts Service Source Index. *See CASSI*
Chemicals
in journal articles, describing, 63, 66–69
in posters, describing, 302
cis, italicization and capitalization, 72
Citations
author–date format, 554–555
authors' names, referring to, 549–551
"et al.", 550–551
formats for, overview, 549
italic numbers in parentheses format, 552–554
purpose, 545
quantity of, 546, 549
superscript number format, 551–552
See also Citing the literature, References
Citing the literature
in conference abstracts, 288
direct quotes, avoiding, 216–217, 545
in posters, overview, 325, 327*f*, 329–330, 332
primary literature vs. sources of general knowledge, 34, 405, 546–548, 549
in proposals, overview, 377, 392*f*, 405, 437*f*, 508
what and what not to cite, 216–217, 546–548, 549
See also Citations; Journal articles, abstract; Journal articles, Discussion section; Journal articles, Introduction section; Paraphrasing; Posters, Discussion section; Posters, Introduction section; References
Colon
for conciseness, 230–231
in headings, avoiding, 393
language tip, 626–629
for lists, introducing, 397
in titles, 263–264
Color
in abstract graphics, 254
in graphics, overview, 527*t*, 528*t*, 534*t*
in journal articles, 528*t*
in posters, 319, 348–349, 528*t*
in proposals, 405, 406, 444, 528*t*
Columns. *See* Tables, formatting
Comma
with "and" or "but", 627–628, 629–630
with "et al.", 550, 617–618
with introductory words and phrases, 233–234, 235*t*
language tip, 629–632
in references, 560
with three or more items in a list, 395, 627–628, 630
Commonly confused words. *See individual words*
"Compose" vs. "comprise", 638–641
Compound labels
as compound abbreviations, 75, 137–138, 423, 603
examples of, 140–141, 145–146, 284, 285, 304
in journal article abstracts, 255
numbering, 75
in posters, 304, 318
in proposals, 423
in schemes, 540, 540*s*, 540*t*
"Comprise" vs. "compose", 638–641
Conciseness
achieving, 36–43, 39*t*, 227–231
language tip, 584–586

parentheses, to achieve, 40, 585
"respectively", to achieve, 151–152, 230
as a subcomponent of Audience and Purpose, 7t
in titles, 262–263
See also Nominalizations, Revising and editing
Concise writing. *See* Conciseness
Conclusions. *See* Journal articles, Discussion section, concluding; Posters, Discussion section, concluding; Proposals, Outcomes and Impacts section, concluding
Conference abstracts
abbreviations in, 287
acceptance criteria, 276–277
audience and purpose, 276–277
author list and affiliations, 287
citations in, 278, 288
compound labels in, 284–285
definition of, 273, 274
graphics in, 283, 284, 285
highlighting different moves in, 278–279
highlighting gaps in, 282–283
highlighting methods in, 280–281, 282
highlighting results in, 279–280, 285
highlighting a synthesis in, 283, 284, 285
Instructions for Authors, example, 275–276
vs. journal article abstracts, 275, 277, 278, 287–288
keywords, 288
numerical values in, 278, 279–280
organization, moves and submoves, 278–279, 278*f*
title, 246*t*–247*t*, 286–287
verb tense, 288
voice, 288
"we", 288
word limit, 274
Content, science, as a component of genre analysis, 7*t*, 22–23
Contractions, avoiding in scientific writing, 38
"Current work", 204–205, 223, 223*t*

Dangling graphics, 122, 319, 320
Dangling headings, 580
"Data" as singular or plural noun, 154–155, 155*t*, 616
Degree symbol
with Celsius, 73, 76*t*, 88, 609
not with kelvin, 73, 76*t*, 609
"Deionized" instead of "DI", 69, 75
Detail, level of
in journal articles vs. lab reports, 61, 66–67, 68, 79, 81–82, 82*t*, 84–85, 90
in posters, 298–299, 302
in proposals, 464–465
as a subcomponent of Audience and Purpose, 7*t*
Detection limits, reporting, 132–133
Direct quotes, avoiding, 216–217, 545
See also Paraphrasing
Discussion section. *See* Journal articles, Discussion section; Posters, Discussion section

Editing. *See* Revising and editing
"Effect" vs. "affect", 634–636
Error terms, reporting, 252, 257
"et al.", 43, 77*t*, 550–551, 617–618
Ethical guidelines. *See* Plagiarism
"Experiment" vs. "work", 61, 398, 399*t*
Experimental section. *See* Journal articles, Methods section; Posters, Methods section
Expert audience. *See* Audience, different types of

"Farther" vs. "further", 641–643
"Fewer" vs. "less", 643–644
"Figure"
calling out in text, 74*t*, 524
in captions, 527*t*
Figures, 524–526, 527*t*–528*t*
See also Graphs
Fill-the-gap statements
in journal articles, 205–206, 223–226, 223*t*, 224*t*
in proposals, 420, 421
First person. *See* "We"
Fluid writing, 233–234, 235*t*, 586–588
Font and font size
in graphs, 528*t*
in posters, 344*t*, 345*f*, 346*t*, 353*t*
in proposals, captions, 450
in tables, 535*t*
See also Posters, designing
Formal vocabulary, 588–590
See also Words to avoid
Formality, level of
in journal articles, 61, 66–67
as a subcomponent of Audience and Purpose, 7*t*
See also Detail, level of; Formal vocabulary
Formatting
as a subcomponent of Writing Conventions, 7*t*, 19
See also Citations, Graphs, Headings, References, Schemes, Tables
FTIR
abbreviation, 76*t*
as an adjective, 93–94
data, reporting, 91–94
See also IR
Funding agencies, 367–368

"Further" vs. "farther", 641–643
Future tense. *See* Verb tense

Gap statements
examples of, 205*t*
in journal articles, 205, 221–222, 249–250
in posters, 327
in proposals, 371, 420–421
GC/MS, describing in journal article methods sections, 94–95
Gender-neutral language, 492
General audience. *See* Audience, different types of
General reaction conditions, describing in journal article Methods sections, 71
Genre analysis
of audience and purpose, 7–11
components of, 7*t*
definition of, 7
of grammar and mechanics, 20–21
of organization, 12–18, 14*t*, 15*f*, 16*t*, 17*f*
of science content, 22–23
of writing conventions, 18–20
Genres
addressed in this book, 6
definition of, 6–7
examples of, 6, 8–9, 10*f*
See also Genre analysis
Genus and species names, 70, 74*t*
Grammar and Mechanics, as a component of genre analysis, 7*t*, 20–22
See also Capitalization, Colon, Comma, Hyphenated two-word modifiers, *individual troublesome words,* Parallelism, Period, Plural words, Semicolon, Split infinitives, Subject-verb agreement
Grant proposal. *See* Proposal
Graphics
calling out in text, 122–123, 524
in conference abstracts, 283, 284
dangling, 122, 319, 320
in journal article abstracts, 254, 255
in posters, 302, 309, 314, 319
in proposals, 375, 444, 445–446, 447, 450
as a subcomponent of Science Content, 7*t*, 23
See also Color; Graphs; Posters, Results section; Schemes; Tables
Graphs
axes, 527*t*, 528*t*
calling out in text, 122–123, 524
captions, 310, 319, 527*t*
examples of, "before" and "after", 529*f*, 530*f*
formatting, 527, 527*t*–528*t*
legends, 528*t*
misleading, 526
numbering, 524
overcrowding, 526
purpose and use, 524–526
when to use, 314, 525
See also Posters, Results section
Guidelines for authors
example of, for posters, 336–337
examples of, for proposals, 385, 504–505
See also Information for Authors

Headings
IMRD format, 44–45
in journal articles, 44–45, 167, 243
in posters, 297, 325, 338, 339*f*, 341*f*, 346*t*, 353*t*
in proposals, 376–378, 377*t*, 378*t*
for References section, 557, 567
for table columns, 534*t*–535*t*
See also Parallelism; Subheadings; *Proposals, individual sections*
Hedging
in journal articles, 118, 164, 190
language tip, 590–594
in posters, 325
in proposals, 484
Hourglass structure, 46–47, 46*f*
Hyphenated two-word modifiers, 181, 263, 632–634

"I", 288, 421, 425, 514–515
Illustrations. *See* Graphics; Graphs; Proposals, Experimental Approach section; Proposals, Goals and Importance section
IMRD format. 44–45
See also Headings; Hourglass structure
Infinitives, split, 619–620
Information for Authors, 45–46, 524
Instructions for Authors. *See* Information for Authors
Instrumentation. *See* Apparatus
Intellectual merit, 371–372
"Interestingly", 130
Internal standards, reporting in journal articles, 68–69
Internet sources, citing, 563
Introduction. *See* Journal articles, Introduction section; Posters, Introduction section; Proposals, Goals and Importance section
in vitro, 134
in vivo, 134
IR
data, reporting, 80
not needing definition, 73, 74
See also FTIR
"Irradiation" vs. "radiation", 456
Italic and non-italic type
ab initio, 96

cis and *trans*, 72
in citations, 552–554
"et al.", 77*t*, 550
m/z, 77*t*
ortho, meta, para (o, m, p), 72
R vs. (*R*), 72, 142
in references, 560, 562
in vitro, 134
in vivo, 134
"Its" vs. "it's", 644–645

Journal article
audience and purpose, 35
vs. lab report, 61, 66–67, 79–83, 82*t*, 84–85, 90
move structures, 62*f*, 120*f*, 166*f*, 204*f*, 245*f*
organization, 44–48, 46*f*
peer-reviewed, 33–34
See also Journal articles, individual sections
Journal articles, abstract
abbreviations in, 250–251, 256–257
audience and purpose, 242, 244
citing the literature in, 249, 257
compound labels in, 255
concluding, 246, 252, 253
double-spacing, 257
error terms in, 252, 257
examples of, 243, 254–256, 259–260
gaps, identifying, 249
graphics in, 254, 255
heading, 243
instrumentation, describing, 251, 257
keywords, 244, 257
methods, describing, 250–251
in organic journals, 254–256
organization, 245–246, 245*f*
present perfect, 258
principal findings, describing, 252–256
purpose and accomplishments, describing, 248–249
research area, describing, 249–250
verb tense, 257–258
voice, 258
"we", 258
word limit, 242
Journal articles, conclusions, 167–168, 172, 174
See also Journal articles, Discussion section, concluding
Journal articles, Discussion section
audience and purpose, 45, 164, 165–166
citing the literature in, 166*f*, 167, 175–176, 177
combined Results and Discussion sections, 112–114, 166, 170–171, 172–173
concluding, 167–168, 174, 177, 178, 182, 185
hand-waving arguments, 171, 176, 178
hedging, 164, 190
mechanisms, proposing, 178, 181–182, 183, 184–185
organization, 166–168, 166*f*
overinterpretation, 164, 171
results and discussion, aligning, 167
results, interpreting, 170–171, 175–177
Results section, transitioning from, 166, 175, 176–177
schemes, 182, 183, 183*s*, 184–185, 186, 186*s*
separate Discussion sections, 166, 175–177, 179–181
subheadings, 167
synthesis (organic), discussing, 181–183, 184–185, 186
take-home message, 167, 168, 172, 177
verb tense, 187, 189*t*
voice, 187, 189*t*
"we", 188–189
Journal articles, Introduction section
audience and purpose, 45, 200, 202–203
background information, providing, 204*f*, 216–219
citing the literature in, 204–205, 204*f*, 209, 210–215, 216–219, 228–229
conciseness, 227–231
concluding, 226
"current work", 204–205, 223, 223*t*
direct quotes, avoiding, 216–217
first sentence, 207–208, 209
gap, filling, 205–206, 223–226, 223*t*, 224*t*
gap, identifying, 205, 205*t*, 221–222
importance, establishing, 210–212
linking words and phrases, 223, 223*t*, 233–234, 235*t*
in organic journals, 206, 224, 226
organization, moves and submoves, 204–206, 204*f*
paraphrasing, 202
present perfect, 214–215, 214*t*
research topic, identifying, 207–208
"researcher", avoiding, 228–229
results, previewing, 204*f*, 206, 226
verb tense, 209, 213–215, 223, 224*t*, 225*t*, 226
voice, 214*t*, 225*t*
"we", 205–206, 223, 224*t*
words to avoid, 209, 224, 228–229
See also Citations, References
Journal articles, Methods section
analytical data, reporting, 80–81
apparatus, describing, 90–95
audience and purpose, 45, 61
chemicals and reagents, describing, 63, 66–69
detail, level of, 61, 66–67, 68, 79, 81–82, 82*t*, 84, 90
"experiment" vs. "work", 61

Journal articles, Methods section (*continued*)
formality, level of, 61, 66–67
general reaction conditions, describing, 71
vs. lab reports, 61, 66–67, 68, 79, 81–82, 82*t*, 84–85, 90
materials, describing, 63, 66–71
numerical methods, describing, 64, 95–97
operational parameters, reporting, 91–95
organization, moves and submoves, 62–65, 62*f*
other names for, 58
procedures, describing, 63–64, 79–80, 81–83, 82*t*, 84, 89–90
QA/QC, reporting, 63, 89–90
samples, describing, 69–71
standard solutions, reporting, 68–69
subheadings, 63, 64
synthesis (organic), describing, 79–80, 81–83, 82*t*
vendors, reporting, 66–67, 68, 69
verb tense, 97–98, 100*t*
voice, 99–100, 99*f*, 100*t*
Journal articles, Results section
audience and purpose, 45, 112, 118–119
broad-to-narrow approach, 131–132, 133
chemical (organic) synthesis, describing, 137–138, 140–141, 145–146
combined vs. separate Results and Discussion sections, 112–114
description vs. interpretation, 118–119
detection limits, reporting, 132–133
graphics, calling out, 122–123
hedging, 118
logical vs. chronological order, 127, 128
Methods section, transitioning from, 122–123
multiple sets of results, 120–121, 124
organization, moves and submoves, 120, 120*f*
purpose, 112
quantitative vs. qualitative language, 153
representative data, 125, 136
story of scientific discovery, telling, 124, 126–127, 136
subheadings, 122
verb tense, 147–148, 148*t*
voice, 148, 148*t*
"we", 149–151, 149*f*
See also Graphs, Tables
Journal articles, title
abbreviations in, 250–251, 264
capitalization in, 262
colons in, 263–264
conciseness, 262–263
examples of, 246*t*–247*t*, 263–264
hyphenated two-word modifiers in, 263
keywords in, 257, 264
purpose, 244
"the", "a", "an", avoiding, 262–263
word limit, 241–242
X of Y by Z pattern, 246, 246*t*–247*t*, 262–263
Justification. *See* Posters, designing

Keywords
in conference abstracts, 288
in journal article abstracts, 244, 257
in titles, 257, 264
use of, 44

Lab reports vs. journal articles, 61, 66–67, 68, 79, 81–82, 82*t*, 84–85, 90
"Lead" vs. "Led", 411
Leading zero
alignment, in tables, 534*t*
with numeric decimals, 88, 305
"Led" vs. "Lead", 411
Legends, 528*t*
"Less" vs. "fewer", 643–644
Linking words and phrases, 233–234, 235*t*
Lists
in journal articles, avoiding, 66–67
in posters, Discussion section, 322, 323, 325
in posters, Introduction section, 328, 329
in posters, Methods section, 302, 303, 304, 305
in posters, Results section, 317–318, 319
in proposals, Goals and Importance section, 394–396
in proposals, Outcomes and Impacts section, 488–490, 492
See also Colon, Parallelism, Series
Literature Cited. *See* References
Literature values, 457
Logos, in posters, 319, 349

Margins, 580
Materials, describing in Methods sections, 63, 66–71, 302
Materials and Methods section. *See* Journal articles, Methods section; Posters, Methods section; Proposals, Experimental Approach section
Mechanics. *See* Grammar and mechanics
Methods section. *See* Journal articles, Methods section; Posters, Methods section; Proposals, Experimental Approach section
Move structures
in conference abstracts, 278*f*
in journal articles, 62*f*, 120*f*, 166*f*, 204*f*, 245*f*
as organizational flow charts, 15–18, 15*f*, 17*f*
in posters, 298*f*, 309*f*, 322*f*, 327*f*
in proposals, 392*f*, 437*f*, 482*f*, 507*f*
See also Moves and submoves
Moves and submoves
in conference abstracts, 278–279

definition of, 13
in journal articles, 62–65, 120, 166–168, 169, 204–206, 245–246
overview, 12–18, 14*t*, 16*t*
in posters, 297, 298–299, 308–311, 322, 327
in proposals, 378, 378*t*, 391–392, 437, 482–483, 507–508
required vs. optional, 17, 17*f*
See also Move structures, Organization

NMR
abbreviating, 74
data, reporting, 80–81
Nominalizations
for conciseness, 41–43, 231
examples of, 41*t*, 484
language tip, 594–597
Numbered lists. *See* Lists
Numbering
of citations, 218
of compounds, 75, 78
of figures, 524
of references, 557, 559
of schemes, 539
a series of items, 231, 613
of tables, 533
in tables, the entry column, 535*t*
Numbers
at beginning of sentence, 86, 87, 305, 608
as citations, 551–554
as compound labels, 75, 603
language tip, 608–611
numerical vs. word form, 86–88, 608–609
in a series, 86, 88, 609
and symbols, spacing, 88
in tables, 534*t*, 535*t*, 536*t*
and units, spacing, 86–87, 88, 305, 608–609
See also Citations, Leading zero, References
Numerical methods, describing in Methods sections, 64, 95–97, 302
Numerical references. *See* References

Operational parameters, reporting
examples of, 92–95
in journal articles, 91–92
in posters, 302
in proposals, 465
Ordinal language
in journal articles, avoiding, 84–85
in proposals, 397–398, 455
Organization
as a component of genre analysis, 7*t*
journal articles, 44–48, 46*f*
overview, 12–18
posters, 297
proposals, 376–378, 378*t*
See also Headings, Move structures, Moves and submoves
ortho, meta, para (o, m, p), italicization and capitalization, 72
"Our", 364, 401, 452, 497, 514–515

Paper. *See* Journal article
Parallelism
in headings and subheadings, 306, 454
language tip, 612–616
in lists, 305, 306, 307, 490
in series, 231
as a subcomponent of Grammar and Mechanics, 7*t*
Paraphrasing, 202
Parentheses
abbreviations and acronyms, introducing, 73–75
in citations, 552–554
with compound labels, 603
conciseness, achieving, 40
figures and tables, calling out, 122–123
vendors, reporting, 66–69
Passive voice. *See* Voice
Past tense. *See* Verb tense
Peer review
practice, 107–109, 161–162, 197–198, 238–239, 269
process, 33–34
Peer-reviewed journals, 33–34
Percent sign, spacing, 87, 88
Period
with abbreviations or acronyms, 74
with "et al.", 43, 550
in references, 560, 562, 565
with units of measure, 608
Personal pronouns. *See* "I", "Our", "We"
"Phenomena" vs. "phenomenon", 441
Plagiarism, 543
Plural words
abbreviations, 603
"data", 154–155, 155*t*, 616
tricky plurals, 154–156, 155*t*, 156*t*, 616–619
units of measure, 88
Poster
abbreviations in, 300, 305, 306, 319
abstract, 297
acknowledgments, 331
audience and purpose, 294, 296
examples, 295*f*, 350*f*, 351*f*, 352*f*
headings, 338
move structures, 298*f*, 309*f*, 322*f*, 327*f*
organization, 297
references, 332

Poster (*continued*)
sections of , 297
title, 286, 287, 353*t*
See also Conference abstracts; Posters, designing; *Posters, individual sections*
Poster presentation. *See* Poster
Posters, designing
artwork, 349
backgrounds, 348–349, 353*t*
color, 319, 348–349, 353*t*
dimensions, 337–338
examples, 295*f*, 350*f*, 351*f*, 352f, 353*t*
font size, 346, 346*t*, 353*t*
fonts, 344, 344*t*, 345*f*, 353*t*
fonts, special effects, 346–347
guidelines, example of, 336–337, 353*t*
headings and subheadings, 338, 339*t*, 341*t*, 353*t*
justification, text, 347, 347*f*, 353*t*
layout, 337–342, 339*f*, 340*f*, 341*f*, 342*f*, 353*t*
logos, 319, 349
text boxes, 348, 353*t*
See also Graphs, Tables
Posters, Discussion section
audience and purpose, 321, 322
citing the literature in, 322, 325
combined vs. separate Results and Discussion sections, 322
concluding, 322, 323–324
examples, in full posters, 295*f*, 350*f*, 351*f*, 352*f*
headings, 322, 325
hedging, 325
list vs. paragraph, 322, 323, 324, 325
organization, 322, 322*f*
results, highlighting, 322
take-home message, 322
verb tense, 325
voice, 326
"we", 326
See also Citations, References
Posters, Introduction section
accomplishments, previewing, 327
audience (viewer), 327
background information, sharing, 327
citations in, 327, 327*f*, 329–330, 332
examples, in full posters, 295*f*, 350*f*, 351*f*, 352*f*
gap, identifying, 327
importance, establishing, 327
lists, avoiding, 328, 329
organization, 327, 327*f*
present perfect, 330
references in, 332
research area, introducing, 327
subheadings, 328
verb tense, 330
voice, 330
"we", 330
See also Citations, References
Posters, Methods section
abbreviations in, 300, 305, 306
examples, in full posters, 295*f*, 350*f*, 351*f*, 352*f*
instruments, describing, 302
vs. journal article Methods section, 302, 303–304
lists in, 302, 303, 304, 305
materials, describing, 298, 302
numbers and units, 305
numerical methods, describing, 302
organization, 298–299, 298*f*
procedures, describing, 302
subheadings, 302, 306
synthesis (organic), describing, 303–304
verb tense, 306
voice, 306
"we", 306
See also Compound labels
Posters, Results section
abbreviations, in limited space, 319
audience and purpose, 308–309
examples, in full posters, 295*f*, 350*f*, 351*f*, 352*f*
graphics, examples of, 312–313, 315–316, 317, 318
graphics, overview, 309, 310, 311, 314, 319, 320
lists in, 317–318, 319
organization, 308–311, 309*f*
raw data, 309
results, sharing, 308–311
verb tense, 319
voice, 319
"we", 319
See also Compound labels, Graphs, Lists, Tables
"Precede" vs. "proceed", 645–648
Present perfect
in journal articles, 214–215, 214*t*, 258
in posters, 330
in proposals, 418, 419*t*, 425, 425*t*, 442, 443*t*, 452, 452*t*
See also Verb tense
Present tense. *See* Verb tense
Primary literature, 34, 546–547
"Principal" vs. "principle", 648–650
"Principle" vs. "principal", 648–650
"Proceed" vs. "precede", 645–648
Project summary. *See* Proposals, Project Summary
"Promising", 448
Proofreading. *See* Revising and editing
Proposal
audience and purpose, 359, 375–376
broader impacts, 372–373
color, 405, 406, 444
content, overview, 363

funding agencies, 367–368
headings and subheadings, 376–378, 377*t*, 378*t*
intellectual merit, establishing, 371–372
lists in, 394–396
move structures, 392*f*, 437*f*, 482*f*, 507*f*
need, establishing, 370
organization, 376–378, 377*t*, 378*t*
references, 378
rejections, reasons for, 373–374, 374*t*
See also Proposals, individual sections; RFP
Proposals, Experimental Approach section
detail, level of, 464–465
headings, 377*t*, 378*t*, 433–434, 438, 438*t*, 444, 444*t*, 454, 454*t*
illustrations, examples of, 445, 446, 447, 450, 486
as a main section of the Project Description, 378, 378*t*
obstacles, anticipating, 469–471
organization, 437, 437*f*
preliminary results, sharing, 440, 443–451
prior accomplishments, sharing, 438–440
procedures and instrumentation, describing, 457–460, 462–469
"promising", 448
promising results, reminding readers of, 455–456
proposed methodology, describing, 454–456, 457–460, 462–471
"recently", 448–449
verb tense, 442, 443*t*, 452, 452*t*, 474
voice, 442, 443*t*, 452, 452*t*, 474
"we", 452
Proposals, Goals and Importance section
background information, providing, 407–409, 411–413, 414–416
citing textbooks, 405
fundamental concepts, explaining, 404–406
gap statements, 205*t*, 420–424
gaps, identifying and filling, 205*t*, 420–424
goals and objectives, stating, 392–399, 401
goals vs. objectives, 393
headings, 377*t*, 378*t*, 387, 393, 393*t*, 403, 403*t*, 420
illustrations, 405, 406
importance, establishing, 402–416
as a main section of the Project Description, 378, 378*t*
organization, 391–392, 392*f*
proposed work, introducing, 420–424
research area, identifying, 403–404
research story, developing, 404–416
SAM test, 394
verb tense, 401, 401*t*, 418, 419*t*, 425, 425*t*
voice, 401, 401*t*, 418, 419*t*, 425, 425*t*
"we" , 401, 418, 421, 425
words to avoid, 398–399, 399*t*
See also Citations, References
Proposals, Outcomes and Impacts section
broader impacts, describing, 492–496
concluding, 479, 480, 492–496
expected outcomes, listing, 488–491
headings, 377*t*, 378*t*, 479, 483, 483*t*, 488–489, 489*t*, 492, 493*t*
hedging, 484, 590–594
lists, 488–490, 492
as a main section of the Project Description, 378, 378*t*
organization, 482–483, 482*f*
timeline, presenting, 483–486
verb tense, 488, 488*t*, 492, 497, 497*t*
voice, 488, 488*t*, 492, 497, 497*t*
"we", 497
Proposals, Project Description
main sections of, 378, 378*t*
as a major division of the proposal, 377
overview, 378
Proposals, Project Summary
vs. abstracts, 502, 503
audience and purpose, 501, 502–503, 508
author guidelines, examples, 504–505
broader impacts, summarizing, 506, 507*f*, 508
citations, absence of, 508
heading, 501
as a major division of the proposal, 378
methods, summarizing, 508–512
organization, 507–508, 507*f*
significance, emphasizing, 508–512
topic and goals, introducing, 508–512
verb tense, 515, 515*t*–516*t*
voice, 515, 515*t*–516*t*
"we", 514–515
Proposals, References Cited, as a major division of the proposal, 378
See also References
Proposals, title, 517–518
See also Journal articles, title
"Prove", 118, 164, 399*t*
See also Hedging
Punctuation
as subcomponent of Grammar and Mechanics, 7*t*, 20–21
See also Capitalization, Colon, Comma, Hyphenated Two-Word Modifiers, Period, Semicolon
Purpose. *See* Audience and Purpose

QA/QC, reporting in journal articles, 63, 89–90
Quality Assurance/Quality Control. *See* QA/QC
Quotations. *See* Direct quotes

(*R*) as prefix in chemical names, 72, 142
R for radical or residue, 137–138, 140, 142
"Radiation" vs. "irradiation", 456
Reagents
in journal articles, describing, 63, 66–69
in posters, describing, 302
"Recently"
in conference abstracts, 283
in proposals, 448–449
Redundancy, eliminating, 573–575
See also Conciseness, Wordiness
Refereed journals, 33–34
Reference standards, reporting in journal articles, 68–69
References
alphabetical format, books, 564–565, 565*t*, 566–567
alphabetical format, journal articles, 564–565, 565*t*, 566–567
CASSI, 558, 558*t*
Chemical Abstracts Service Source Index, 558, 558*t*
for electronic sources, 563
formatting, overview, 557–558, 567
heading, 557, 567
numerical format, books, 559, 561–562, 561*t*–562*t*
numerical format, journal articles, 559–560, 559*t*
parallel to citations, 557
purpose, 557
section, in journal articles, 557, 567
section, in posters, 557, 567
section, in proposals, 557, 567
Request for application. *See* RFP
Request for proposal. *See* RFP
Research article. *See* Journal article
Research proposal. *See* Proposal
"Researcher", avoiding, 228–229
"Respectively", 151–152, 230, 597–599
Results and Discussion sections, combined vs. separate, 112–114, 322
Results section. *See* Journal articles, Results section; Posters, Results section; Proposals, Experimental Approach section
Revising and editing
adherence to external guidelines, 579
appearance, 580–581
awkward sentences, 586–588
common errors, 577–578
conciseness, 575–576
follow-through, 573
formatting, 580–581
organization, 571–572
redundancy, 573–575
science content, 578–579
tense and voice, 576
wordiness, 575–576
RFA (Request for application). *See* RFP
RFP (Request for proposal)
ACS Division of Analytical Chemistry Graduate Fellowship, 361
definition, 369
generic, 379–380
National Cancer Institute (NCI) Quick Guide for Grant Applications, 385
NSF Collaborative Research in Chemistry (CRC), 383–384
NSF Faculty Early Career Development Award in Chemistry (CAREER), 380–381

(*S*) as prefix in chemical names, 72
Samples, describing in Methods section, 69–71
Schemes
calling out in text, 74*t*, 539
compound labels in, 540, 540*s*, 540*t*
examples, 56*s*, 143*s*, 183*s*, 186*s*, 324*s*, 540*s*
formatting, 539–540, 540*t*
in journal articles, Discussion section, 182, 183*s*, 184–185, 186, 186*s*
numbering, 539
purpose, 539
titles, 540*t*
Science content, as a component of genre analysis, 7*t*, 22–23
Scientific audience. *See* Audience, different types of
Scientific paper. *See* Journal article
Scientific poster. *See* Poster
Scientific words, plural and singular, 154–156, 155*t*, 156*t*, 616–619
Seasons, capitalization of, 70
Semicolon, 626–629
Series
numbering of items in, 231
with numbers 10 or greater, 86, 88, 609
parallelism, 88, 231, 612–616
punctuation, 627–628, 630
with "respectively", 597–599
units of measure as final item, 609
See also Lists
"Since" vs. "because", 650–652
Singular vs. plural words, 154–156, 155*t*, 156*t*, 616–619
Spacing
after initials, in references, 560
double vs. single, 567, 580
of numbers and symbols, 87, 88
of numbers and units, 86–87, 88, 305, 608–609
with percent sign, 87, 88

of variables and symbols, 88
"Spectra" vs. "spectrum", 616–617
"Spectrum" vs. "spectra", 616–617
Spelling out
abbreviations, 601–603
authors' first names, 287, 560
journal names, in References, 332, 560, 558, 558*t*
numbers, 608–609
units of measure, 86–88, 608–609
Split infinitives, 619–620
Standard solutions, reporting in journal articles, 68–69
Statistical methods, reporting in journal articles, Methods section, 64, 95–97
Stock solutions, reporting in journal articles, 68–69
Student audience. *See* Audience, different types of
Subheadings
in journal articles, 63, 64, 122, 167
in posters, 302, 306, 328, 338
in proposals, 376–378, 377*t*, 378*t*
See also Headings
Subject-verb agreement
language tip, 620–624
as a subcomponent of Grammar and Mechanics, 7*t*
with units of measure, 81, 87
Submoves. *See* Moves and submoves
Superscript numbers, in citations, 551–552

Tables
alignment, 534*t*
calling out in text, 74*t*, 533
columns, 534*t*–535*t*
equations in, 139, 146
examples of, "before" and "after", 537*f*
font, 535*t*
footnotes, 535*t*
formatting, 534, 534*t*–536*t*
gridlines, 535*t*
numbering, 533
purpose, 533
repeated values, 536*t*
titles, 536*t*
when to use, 314, 533
See also Posters, Results section
Take-home message, 167, 168, 172, 177, 322
Tense. *See* Verb tense
"Than" vs. "then", 653–654
"That" vs. "which", 654–657
"Then" vs. than", 653–654
Time, units of. *See* Units of time
Timeline. *See* Proposals, Outcomes and Impacts section
Titles
for conference abstracts, 286–287
for posters, 286–287, 353*t*
for proposals, 517–518
for schemes, 540*t*
for tables, 536*t*
See also Journal articles, title; X of Y by Z pattern
trans, italicization and capitalization, 72
"Truth", 118, 164
See also Hedging
Two-word modifiers, hyphenation, 181, 263, 632–634

Unambiguous writing, 599–601
Units of measure
abbreviating vs. spelling out, 87, 88, 602
abbreviations, 76*t*–77*t*
in axes labels, 527*t*
definition of, 87
language tip, 608–611
plural forms, 88
rules, 86–88
spacing, between numbers and units, 86–87, 88, 305, 608–609
surnames as, 73
in tables, column headings, 534*t*
verb agreement, 81, 87
Units of time
abbreviating vs. spelling out, 71, 76*t*–77*t*, 602, 608–609
in axes labels, 527*t*
plural forms, 608–609
See also Units of measure
UV–vis, 74*t*, 494

Vendors
in journal articles, 66–69, 257
in posters, 302
Verb tense
in conference abstracts, 288
in journal articles, 100*t*, 148*t*, 189*t*, 225*t*, 225*t*
in posters, 306, 319, 325, 330
in proposals, 401*t*, 419*t*, 425*t*, 443*t*, 452*t*, 488*t*, 497*t*, 515*t*–516*t*
as a subcomponent of Writing Conventions, 7*t*, 20
See also Journal articles, individual sections; Present perfect; *Proposals, individual sections*
"Very", 153–154
Vocabulary, formal, 588–590
Voice
active vs. passive, 99–100, 99*f*
in conference abstracts, 288
in journal articles, 99*f*, 100*t*, 148*t*, 189*t*, 214*t*, 225*t*
language tip, 604–608

Voice (*continued*)
in posters, 306, 319, 326, 330
present perfect, 214–215, 214*t*
in proposals, 401*t*, 419*t*, 425*t*, 443*t*, 452*t*, 488*t*, 497*t*, 515*t*–516*t*
as a subcomponent of Writing Conventions, 7*t*, 20
See also Journal articles, individual sections; *Proposals, individual sections*

"We". *See* Conference abstracts; "I"; *Journal articles, individual sections*; "Our"; *Posters, individual sections*; *Proposals, individual sections*
"Which" vs. "that", 654–657
"While" vs. "although", 650–652
Word choice
formal vocabulary, 399*t*, 588–590
linking words and phrases, 223, 223*t*, 233–234, 235*t*
quantitative vs. qualitative terms, 153
as a subcomponent of Audience and Purpose, 7*t*, 8–9
vs. word usage, 21
See also Hedging, Nominalizations, Words to avoid
Word usage
as a subcomponent of Grammar and Mechanics, 7*t*, 21
See also individual troublesome words, Plural words, Word choice
Wordiness
in titles, 262–263
ways to eliminate, 36–43, 39*t*, 41*t*, 227–231
See also Conciseness, Nominalizations, Revising and editing
Words to avoid, 150, 209, 224, 262, 398–399, 399*t*
See also "Experiment" vs. work", Formal vocabulary, Hedging, Ordinal language, "Prove", "Researcher", "Truth", "Very"
"Work" vs. "experiment", 61, 398, 399*t*
Writing conventions, as a component of genre analysis, 7*t*, 18–20
See also Abbreviations, Verb tense, Voice

X of Y by Z pattern, titles, 246, 246*t*–247*t*, 262–263
X-ray, 494

Zero
before decimal points, 88, 305
concentration, reporting, 132–133

CPSIA information can be obtained at www.ICGtesting.com
Printed in the USA
BVOW06s2336250813

329387BV00003B/14/P

9 780195 305074

2425 33

CEDAR CREST COLLEGE LIBRARY
3 1543 50220 1520